Frugivory and seed dispersal

Advances in vegetation science 15

The titles published in this series are listed at the end of this volume.

Frugivory and seed dispersal: ecological and evolutionary aspects

Edited by

T.H. FLEMING AND A. ESTRADA

Reprinted from Vegetatio, volume 107/108

Kluwer Academic Publishers

Dordrecht / Boston / London

Library of Congress Cataloging-in-Publication Data

Frugivory and seed dispersal : ecological and evolutionary aspects / edited by T.H. Fleming and A. Estrada.
p. cm. -- (Advances in vegetation science ; 15)
"Reprinted from Vegetatio."
Includes index.
ISBN 0-7923-2141-3 (HB : alk. paper)
1. Seeds--Dispersal--Congresses. 2. Frugivores--Ecology--Congresses. 3. Frugivores--Evolution--Congresses. 4. Mutualism (Biology)--Congresses. 5. Coevolution--Congresses. 6. Animal-plant relationships--Congresses. I. Fleming, Theodore H. II. Estrada, Alejandro. III. Series: Advances in vegetation science ; v. 15.
QK929.F79 1993
582.13'0452482--dc20 92-46138

Published by Kluwer Academic Publishers,
P.O. Box 17, 3300 AA Dordrecht, The Netherlands.

Kluwer Academic Publisher incorporates
the publishing programmes of
D. Reidel, Martinus Nijhoff, Dr W. Junk and MTP Press.

Sold and distributed in the U.S.A. and Canada
by Kluwer Academic Publishers Group,
101 Philip Drive, Norwell, MA 02061, U.S.A.

In all other countries, sold and distributed
by Kluwer Academic Publishers,
P.O. Box 322, 3300 AH Dordrecht, The Netherlands.

Printed on acid-free paper

Printed in Belgium

Contents

Dedication

This book is dedicated to the current and future generations of students of the frugivore-seed dispersal mutualism. Their research and education efforts will bear important fruits in the global struggle to preserve biodiversity.

Acknowledgements

This book is the product of the Second International Symposium-Workshop on Frugivores and Seed Dispersal held at Catemaco, Veracruz, Mexico, in June 1991. The conference was supported by funds from the U.S. National Science Foundation, the Instituto de Biologia of the Universidad Nacional Autonoma de Mexico, and the Consejo Nacional de Ciencia y Tecnologia de Mexico. We are grateful to the staff of the Los Tuxtlas Biological Station for their enthusiastic logistical support before, during, and after the conference. We especially thank Rosamond Coates-Estrada for her assistance in organizing and conducting the conference.

Vegetatio **107/108**: xi–xii, 1993.

General introduction

Several years have passed since the First International Symposium-Workshop on Frugivores and Seed Dispersal was held at the Los Tuxtlas Biological Station in Veracruz, Mexico. That symposium was convened to bring together for the first time many of the major researchers studying an ecological interaction of considerable theoretical and applied importance, the mutualism between fleshy-fruited plants and their seed dispersers. Modern study of this mutualism dates from the early 1970s, and by 1985 it was time to assess the status of the field. What did we know about this mutualism in theory and practice and in what directions should this field be moving?

Major points to emerge from that synthesis (Estrada & Fleming 1986) include the following:

1. Plants do not appear to closely track changing abiotic and biotic environments via gradual changes in phenology and fruit and seed structure; phylogenetic inertia appears to be strong in fleshy-fruited plants.
2. In addition to its specific fruit characteristics, the fruiting environment (density and crop sizes of fruiting plants) around a plant can influence its rate of fruit removal and its seed dispersal pattern by influencing vertebrate food choice.
3. Secondary dispersal, often by insects, is important for many vertebrate-dispersed plants.
4. Though the importance of seed dispersal for plant fitness can be demonstrated theoretically, it is often hard to document in practice.
5. Seed predation and pollination mechanisms can override the importance of seed dispersal as factors influencing the genetic structure of plant populations.
6. Gap formation is often crucial for the successful establishment of seedlings, but the timing of dispersal does not appear to be tightly linked with the timing of gap formation.
7. Keystone plant resources may control the species richness of communities of frugivores.

Researchers and students left the Los Tuxtlas conference with renewed enthusiasm for learning more about ecological mechanisms and evolutionary consequences of the fruit-frugivore mutualism. Partially as a result of that conference, the publication rate of articles dealing with fruits, frugivores, and seed dispersal continued to increase, as reflected by the number of papers on these topics appearing in one major journal, *Ecology* (Fig. 1). By the early 1990s, however, publication rate in this field seemed to be on the decrease, suggesting that this field will never be as 'hot' as, for example, the study of plant-herbivore (sensu stricto) interactions. Nonetheless, because of its clear importance for the regeneration of tropical forests, the study of frugivory and seed dispersal seems to have found a modest niche in the pantheon of modern ecology. This discipline appears to have a relatively stable core of senior researchers and, to judge from the encouraging world-wide response of graduate students to the second conference, a growing legion of young researchers.

Like its predecessor, major themes of this book are four-fold: historical and theoretical aspects of frugivory and seed dispersal, fruiting and dispersal strategies of fleshy-fruited plants, foraging and physiological strategies of frugivores, and the demographic and community consequences of seed dispersal. Unlike its predecessor, this book includes a broader geographic coverage (but still lacks information from the Asian tropics owing to Mark Leighton's failure to contribute a written version of his conference paper), is richer in theoretical-evolutionary papers, and contains a greater number of progress

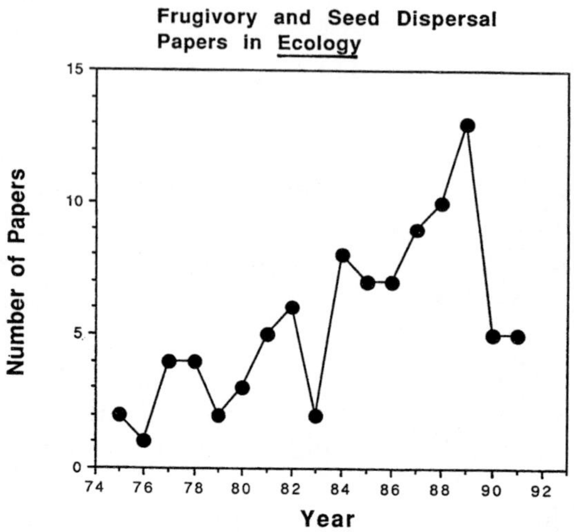

Fig. 1. The number of papers dealing with seed dispersal, fruits, and frugivores published in *Ecology* each year between 1975 and 1991.

reports from important long-term studies. To judge from the diversity of approaches researchers are using to study the frugivore-seed dispersal mutualism, this field is very much alive and well despite the relatively low number of papers being published each year. Significant progress is being made towards understanding the complex links between frugivore foraging decisions and their consequences for the reproductive success of fleshy-fruited plants. It will be exciting to review the status of this field in another six years.

References

Estrada, A. & Fleming, T. H. (eds.) 1986. Frugivores and Seed Dispersal. Dr W. Junk Publishers, Dordrecht, The Netherlands.

Part 1

Historical and theoretical aspects of frugivory and seed dispersal

Introduction

Any scientific discipline needs a theoretical framework to guide its development and to sharpen the questions its researchers pursue. In biology, evolution is the grand theoretical framework, and an historical perspective is necessary to understand present-day biological conditions. In its formative years, the modern study of the fruit-frugivore mutualism was guided by the 'specialist-generalist' paradigm developed by D. Snow, D. McKey, and H. Howe. Howe reviews the current status of this evolutionary paradigm and points out that it has been dismissed by many workers before being adequately tested. This is because ecologists working with the tropical plants and frugivorous birds for which the paradigm was originally developed rarely measure the seed dispersal effectiveness of different disperser species. He indicates that this paradigm still has heuristic value and suggests that several additional ecological paradigms, including the concept of keystone species of plants and frugivores and the role that frugivores play in density-dependent mortality in tropical trees, are worth studying.

The concept of seed dispersal quality has been central to discussions of fruit-frugivore coevolution. Schupp thoroughly reviews data bearing on this concept, constructs a hierarchical framework for viewing disperser effectiveness, and points out that disperser effectiveness depends on both the quantity and quality of seed dispersal. Effectiveness, in turn, affects both evolutionary and ecological relationships between dispersers and their food plants.

Historically, three hypotheses (escape, colonization, and directed dispersal) have been put forth to explain the adaptive significance of seed dispersal. Venable and Brown discuss these hypotheses and develop three models to describe how dispersal functions to reduce predation risk and to escape the negative consequences of crowding and high concentrations of relatives. They point out that the different functions of dispersal are sensitive to spatial, temporal, and genetic variation in plant populations and discuss methods for field-testing their models.

Most researchers currently view the fruit-frugivore mutualism as being the product of diffuse coevolution. F. G. Stiles and Roselli analyze ecological and evolutionary aspects of the interaction between neotropical melastome (Melastomataceae) fruits and their avian dispersers, mainly tanagers and manakins. They propose that tanagers, which are fruit 'mashers' and whose patterns of species diversity coicide with those of melastomes, have selected for small seeds in melastome fruits. Coevolution between these plants and birds appears to be 'diffuse' but involves considerable parallelism and convergence between groups of species in different geographic areas.

Charles-Dominique examines the relationship between the degree of dietary specialization in frugivorous birds and mammals in French Guiana and the species richness of their major food plants. Plant genera eaten by specialized vertebrates are inordinately rich in species, an association that Charles-Dominique ascribes to stepwise coevolution featuring bursts of evolution followed by long periods of evolutionary stability.

Jordano develops a protocol for examining plant-frugivore interactions in a biogeographic context and applies it to the frugivorous thrushes and junipers of Europe. He notes that species with strong interactions of mutual dependence show low values of biogeographic congruence, which limits the possibilities for pairwise, specific coevolution. Mutual evolutionary effects among groups of plant and frugivore species are possible, however, whenever species are ecologically interchangeable in their selective effects and other constraints on coevolution are not operating.

Vegetatio **107/108**: 3–13, 1993.
T. H. Fleming and A. Estrada (eds). Frugivory and Seed Dispersal: Ecological and Evolutionary Aspects.

Specialized and generalized dispersal systems: where does 'the paradigm' stand?

Henry F. Howe
Biological Sciences (M/C 066), University of Illinois at Chicago, Box 4348, Chicago, IL 60680, USA

Keywords: Ecology, Frugivory, Seed dispersal, Tropics

Abstract

I explore the 'specialization' versus 'generalization' paradigm in frugivory and seed dispersal. This view predicts that some tropical trees produce nutritious fruits adapted for use by a small coterie of specialized frugivores that provide reliable seed dissemination. Other tree species are expected to offer superabundant fruits of lower nutritional reward, relying instead on common opportunistic frugivores that are individually less reliable, but collectively disperse seeds effectively. Though widely referenced, many aspects of 'the paradigm' are untested with tropical trees and avian frugivores, primarily because plant ecologists rarely determine whether 'specialist' or 'generalist' foragers are responsible for different patterns of seed distribution, while students of foraging behavior rarely determine the effects seed dispersal by different animals for survival of seeds or seedlings of 'specialist' or 'generalist' trees.

Ecological paradigms provide alternative hypotheses, without evolutionary arguments. 'Keystone' species have ecological effects disproportionate to their abundance; it is important for management considerations to know whether fruiting trees or frugivores serve as keystone mutualists in tropical forests. Alternatively, the extent to which vertebrate seed dispersers influence density-dependent seed, seedling, sapling, or adult mortality may have important consequences for spatial dispersion and population dynamics of tree species in tropical forests.

Introduction

More than 20 years have passed since Snow (1971), Janzen (1970), and Connell (1971) published seminal insights for the study of frugivory and seed dispersal. Snow developed the thesis that plant species with large seeds invest more in the nutritional value of fruits to promote dispersal by frugivores than do plant species with small seeds. He further suggested that fruit-eating birds evolved to accomodate different levels of plant investment in fruit, perhaps exemplifying 'mutual evolution' between plants and avian dispersal agents. Janzen and Connell argued that seeds carried away from high seed density near parent trees escaped disproportionate mortality, thereby providing a selective advantage to local dispersal. Snow identified characteristics of birds and plants that might coevolve, while Janzen and Connell offered a plausible rationale why reliable seed dispersal should matter.

Snow in ornithology and Janzen and Connell in plant ecology catalyzed the attempt to integrate traditionally separate fields. McKey (1975) further discussed physiological mechanisms by which specialist frugivores might evolve gentle means of handling seeds, abetted by tree species that maximized passage rate through the gut to avoid digestion. In emphasizing both the probabilistic natures of both frugivory and seedling establishment, Howe and Estabrook (1977) argued that specialists might be expected to deliver seeds from disproportionate mortality in large aggregations of seeds or seedlings around parent trees, much as Janzen (1970) and Connell (1971) predicted. Opportunists might be expected to provide effective dispersal for plant species which, by virtue of extended seed dormancy, resistence of seeds or seedlings to density-dependent mortality, or a strategy of 'gap colonization' rather than 'escape from parents,' had less demanding requirements that could be met by a wide variety of dispersal agents.

Together these papers defined 'The Paradigm' (Table 1), posing a contrast between trees that produce scarce but especially nutritious fruits that entrain specialized and reliable dispersal agents, and other tree species that produce common but less nutritious fruits that appeal to individually less reliable but collectively more common species of opportunistic dispersal agents. This contrast of characters and proposed 'strategies' posed alternative hypotheses about fruit use and seed success in a corner of natural history that, until 1970–1971, was devoid of hypotheses. Such extremes along a continuum of fruit-eating and fruit-producing 'strategies' were expected to occur in species-rich communities in which dispersal agents 'had a choice' of a wide variety of fruit resources, and in which fruiting plants of roughly similar life form experiencing comparable physiological constraints could draw upon any of dozens of potential dispersal agents.

What works?

Discussion of extreme 'strategies' could either imply coevolution of species or taxa, as proposed by Snow (1971) and McKey (1975), or important ecological interdependence of species with little history of evolutionary interaction (see Janzen

Table 1. 'The Paradigm' of suggested suites of characters associated with specialization and generalization among avian frugivores and tropical trees, as evolved from Janzen (1970), Snow (1971), McKey (1975), Howe & Estabrook (1977), and Howe & Smallwood (1982).

Specialized dispersal systems	
Trees	**Birds**
Often large seeds (e.g. > 1 g), not mechanically tough	Often large (e.g. > 250 g), reduced gizzard, intestine
Fruits nutritious, and energy-rich, with high lipid or protein content	Depend on fruits for critical portion of diet, reduced use of animal food
Low annual fecundity (e.g. median crop 100–5,000)	Small populations of a few (e.g. 5–10) species/community
Consistently high fruit removal	Efficient daily use of crops
Extended fruiting seasons make use of a limited disperser assemblage	Specialists entrained to consistently use few fruit species per day
Seed dispersal away from parents critical for recruitment	A few reliable specialists most effective for tree recruitment
Generalized dispersal systems	
Trees	**Birds**
Often small seeds (e.g. < 0.5 g), mechanically tough	Often small (e.g. < 50 g), generalized gut
Fruits less energy-rich, high carbohydrate and water	Fruits complement primarily non-fruit diet
High annual fecundity (e.g. median crop 20,000–1,000,000 +)	Large populations or many (e.g. 20–100) species/community
Low and variable seed removal	Some species may fail to visit days or weeks at a time
Sharply peaked fruiting season, drawing numerous opportunists to a superabundant display	May use many fruiting species, switch opportunistically, or complement mostly insect diet
Seed dispersal may or may not be critical; many seeds dormant	Many frugivores play roughly similar roles

1980). The first empirical demand of the dialectic is demonstration of the degree to which birds and trees actually do depend on each other. The second demand is to interpret from comparative evidence whether dependence is due to transient ecological conditions, diffuse coevolution of broad taxa of birds and trees, or even coevolution of genera or species of birds with genera or species of trees.

Isolated aspects of 'the paradigm' have been confirmed, but generally suffer from the fact that the marriage of zoology and plant ecology has not yet been consummated. Zoologists still show little interest in seed dispersal and seedling recruitment, as contrasted with diet selection and coincidental seed removal. Plant ecologists still show little interest in the animals that determine patterns of seedling recruitment. Lack of integration ensures that most evidence for or against predictions of the paradigm (Table 1) is fragmentary, and therefore that most tests are indirect and incomplete.

Best established is the advantage to local seed dispersal hypothesized by Janzen (1970) and Connell (1971), and elaborated by Fleming and Heithaus (1981) and Howe and Smallwood (1982). Virtually wherever a quantitative test has been attempted, seeds or seedlings near large fruiting adults fare less well than members of a cohort a few to a few tens of meters away. This 'escape advantage' may be due to disproportionate seed or seedling predation by insects or mammals, to pathogen infestation, or to competition within the cohort under or near parent trees (reviews by Clark & Clark 1984; Howe 1986a; Augspurger 1988). To date, most plant ecologists working on the problem have been content to demonstrate the presence or absence of an effect of dispersal distance on seed or seedling survival. The importance of evident variations in the effect within tree species is generally unknown (Howe 1992). Because few studies use comparable methods, the importance of differences in mortality between species is speculative (Janzen 1970; Hubbell 1979; Howe 1989).

Similarly, ornithological studies suggest differing degrees of 'specialization' on fruits (reviewed by Howe 1986a, 1989; also Fleming 1986). In some cases one to three bird species in species-rich communities take far more fruits from given tree species than other frugivores. Instances of apparent local specialization from the plant perspective, in which trees attract a small coterie of fruit specialists with a limited offering of extremely fatty (60% lipid) or protein-rich fruit pulp, appear to apply to *Virola* and other members of the Myristicaceae, large-seeded species of *Casearia* (Flacourteaceae), some species of Lauraceae, and probably large-seeded species in other tropical families. The pattern may be contrasted with systems in which several to many species of birds and often mammals take small portions of superabundant crops of fruits that have individually low nutrition. Examples include *Tetragastris* and *Protium* (Burseraceae), *Ficus* and *Cecropia* (Moraceae), and probably a large number of genera that produce small berrylike fruits (e.g. Melastomaceae, Rubiaceae).

It must be emphasized, however, that evidence for or against this scheme is weak. Clear exceptions to the scheme exist on even casual inspection; dozens of opportunistic birds take tiny seeds of *Stemmadenia donnell-smithii* from an extraordinarily lipid-rich pulp (McDiarmid *et al.* 1977). According to 'the paradigm,' fatty pulp should accompany large seeds, and be found and used most effectively by large specialized frugivores. In virtually all cases, it is not known which dispersal agents actually disseminate seeds to the most favorable sites, from the plant perspective, because plant ecologists do not study the dispersal process or know seed or seedling patterns traceable to different vertebrate dispersal agents (Schupp 1992). Ornithologists rarely perform even basic nutritional analysis of fruit pulp, and almost never estimate fruit crops over time with repeatable sampling designs.

Differences of interpretation also crop up. Wheelwright (1983) rejected the idea that quetzals (*Pharomachrus mocinno*) in highland Costa Rica were 'specialized' on species of the Lauraceae because the birds use fruits of several virtually indistinguishable species at any one time. Such evidence might otherwise suggest strong specialization on the taxon, if not one species within it. Quetzals do not breed without Lauraceae, evidence for at least local dependence on this plant family. The influence of quetzals on plant recruitment is unknown, but future studies might easily show that some tree species in Costa Rican Lauraceae depend on quetzals for seed dispersal and ultimately seedling recruitment, while other relatives depend on other birds. Half of the story is untold. Quetzal use of several Lauraceae is consistent with either the view that quetzals adjust to and come to depend on members of the Lauraceae in ecological time with little evolutionary interaction, or that coevolution involves mutually beneficial adaptations among taxa of higher than the species (Janzen 1980; Snow 1981; Schemske 1983; Howe 1984a; Herrera 1985; Howe & Westley 1988). Competition among almost indistinguishable plant species for the same limited disperser resource might even represent convergence of fruit characters to use an especially effective class of dispersal agents, as Schemske argued (1981) for the convergence of floral characters of *Costus* (Zingiberaceae) species that use one pollinator. In cases of possible convergence and coadaptation of taxa higher than species, usual meanings of 'specialist' versus 'generalist' terminology, rooted in the notion of interactions between species pairs as contrasted with interactions among more than two species, do not serve well. Regarding quetzals and lauraceous trees, non-evolutionary and evolutionary alternatives cannot be distinguished until more comparative evidence of fruit use and its consequences is available from this and other sites.

Few studies of tropical tree dispersal actually attempt to ascertain the relative importance of different dispersal agents for seedling recruitment (Schupp 1992). The *Virola nobilis* (once *V. surinamensis*) system in Panama has received the most attention (Howe 1992 [this volume] and references therein). As predicted by 'the paradigm,' this large-seeded species with a rich fatty aril is dispersed by only 8 of 80 frugivores on Barro Colorado Island. Of these, only three (two toucans and a guan) consistently remove substantial numbers of seeds beyond a zone of nearly complete seed and seedling mortality $\leq$ 15 m from the trees. Smaller motmots and trogons regurgitate seeds in viable condition, but leave almost all in 'the zone of death' under the parental tree crowns.

It is important to recognize that the enormous variety of phenomena included under the term 'frugivory' and 'seed dispersal' are not equally relevant to possible contrasts of strategies in species-rich communities. For instance, exceptional studies concern frugivory at temperate trees in species-poor communities (e.g. Herrera & Jordano 1981; Herrera 1984a; Jordano 1987), of temperate or tropical mistletoes of quite different life histories than trees (Reed 1986, 1989, Larson 1991), or of trees with fruits eaten by large mammals that impose different physiological challenges on seeds than much smaller birds and affect seed ecology differently (e.g. rhinos: Dinerstein & Wemmer 1988; see Howe 1985, 1989). These broaden the universe of known fruit and frugivore interactions in many important ways, but do not address 'the paradigm' directly. None were undertaken in communities even as rich as that of Barro Colorado Island, with only 80 common vertebrate frugivores and $\pm$ 250 tree species bearing fleshy fruits (see Croat 1978). Only the occasional quantitative study explores the effects, or absence of distinctive effects, of different birds on seed dispersal and seedling recruitment of *trees* in *species-rich* tropical communities.

The general message is that a case can be made for 'the paradigm' from fragmentary evidence, but key elements are rarely tested. Few workers try to relate disperser constancy to its influence on seed and seedling demography. Comprehensive studies that do are often not readily applicable to 'the paradigm' because they are drawn from communities in which neither bird nor plant has much choice of fruit or dispersal agents, life histories or life forms of the plants do not suggest comparable allocation patterns (e.g. expecting patterns in understory shrubs like those of canopy trees), or dispersal agents such as ants or large ungulates impose dramatically different conditions on seed dispersal than would birds or mammals of moderate size visualized in 'the paradigm.'

What is missing?

Despite citation in studies of avian foraging, diet choice, seed dispersal in temperate regions, and plant ecology, 'the paradigm' is rarely tested with birds and tropical trees for which it was intended. This is unfortunate. As a device prescribing testable alternative hypotheses, 'the paradigm' has driven a great deal of productive research (see Howe 1989; Schupp 1992), and is as yet hardly tested. What are the first prioirities for extending and refining the specialization versus generalization framework?

Model systems
The process of testing a dialectic will probably confirm the existence of extreme examples, such as the *Virola nobilis* system alluded to above as contrasted with figs and their relatives (see Charles-Dominique 1986; Estrada and Coates-Estrada 1986). More importantly it will define the limits of dispersal reliability in a wide variety of 'systems' (a tree population and animals that enhance or hinder recruitment) that fall between the extremes (Table 1). Comprehensive studies of frugivory, seed dispersal, and seedling recruitment for 50–100 tropical dispersal systems, most usefully undertaken by biologists within developing countries, would immeasurably increase understanding of the ecological and perhaps adaptive limits of phenomena that apply to tens of thousands of tree species in the Old and New World Tropics.

Wide variation in both disperser activity and seedling recruitment (Howe 1980, 1990a, 1992) require that such studies include replicated sampling of at least several (e.g. 6–12) individual trees in a population. As in other branches of ecology, a truly quantitative study of patterns of fruit production, fruit depletion by animals, seed deposition by animals, and seed and seedling survival with a sample of 25–50 individual trees would undoubtedly yield a degree of certainty and rigor not apparent in the baseline studies attempted to date.

Comparative studies
Clearest insights often emerge from comparative studies of similar species in the same or different places (see Charles-Dominique 1986). Much as unreplicated experimental studies preclude distinctions between treatment and incubator or field effects, even a large number of unreplicated studies of taxonomically and geographically disjunct 'model systems' cannot clearly distinguish inherent characteristics of 'a system' from its immediate ecological circumstances. Comparative studies of the same system in different localities, of taxonomically related systems in the same place, and of taxonomically related systems in different geographical contexts would quickly sharpen the debate over such perennial issues in the study of frugivory and seed dispersal as the extent of co-evolution of species.

First, the same species should be studied in different places. For instance, the overwhelming effect of one toucan (*Ramphastos swainsonii*) as compared with another (*R. sulfuratus*) and a guan (*Penelope purpurascens*) in the *Virola nobilis* system on Barro Colorado Island, Panama depended entirely on the number of seeds carried, which in turn was a function of the number of visits made (*Ramphastos* vs. *Penelope*) and the number of seeds carried away per visit (larger *R. swainsonii* vs. smaller *R. sulfuratus*). These three birds were indistinguishable in the 'quality' of seed dissemination *per seed carried*, though each was far better per seed carried than other dispersal agents (Howe 1992; Schupp 1992). Because relative abundance of bird populations varies, either the smaller toucan *R. sulfuratus* or the guan *Penelope purpurascens* or some other large bird (here 300 g) with equivalent habits could easily drive the system in another locality.

Moreover, plant taxa vary, and indeed taxonomic status is often in dispute. *Virola* trees in Panama recently considered conspecific with *V. surinamensis* in South America (Croat 1978) have much larger fruits than in the Amazon; the binomial *V. surinamensis* is once again reserved for the South American trees, with a return to *V. nobilis* for the Central American tree with large fruits (Duke 1962). In any event, one might expect smaller dispersal agents to predominate in Brazil, where smaller seed size does not present a barrier to dispersal agents ≥ 300 g.

Finally, seedling demographies vary because seed dispersal patterns and sources of mortality vary from place to place. Without curculionid weevils (*Conotrachelus*) now common on Barro Colorado Island, differences in dispersal quality by motmots (*Baryphthengus martii*), trogons (*Trogon massena*) and toucans might be negligible from the plant vantage because seeds near the parent could survive as well as those 30–40 m away. Without currently high densities of mammalian seed and seedling predators, quite different effects might be expected (Howe 1992).

Similar questions could be asked of dispersal systems centered on other groups of taxonomically similar species. The most rapid progress will be with taxa that have already been studied by students of frugivory or seed dispersal, or both. For instance, *Casearia corymbosa* (Flacourtiaceae) trees in wet (Howe 1977) and dry (Howe & Vande Kerckhove 1979) Costa Rican forests show stark differences in frugivore activity, but seed dispersal and seedling survival were not evaluated carefully. A great deal is known about the use of figs (Moraceae) and pipers (Piperaceae) by bats (Fleming *et al.* 1977; Morrison 1978; Fleming 1986, 1988), and of figs by monkeys (Milton 1980; Estrada & Coates-Estrada 1986), making comparative studies of use and its consequences for fig seedling establishment and survival most timely. Such questions should ultimately be asked of all model 'dispersal systems,' but there is special value in exploring similarities and differences in widespread systems that have already received extensive study. Second, related dispersal systems could profitably be evaluated in the same place. For instance, Wheelwright (1983, 1986) has provided a classic comparison of fruiting phenologies and fruit use in members of the Lauraceae in highland Costa Rica; an analysis of roles of frugivores on seedling demography would be most useful. Do other dispersal agents eclipse quetzals in importance for some trees in this group? Are quetzals more beneficial for some lauraceous trees than others in total quantity of seeds carried? Per seed carried? Both? Similar questions could be asked of *Virola* species on Barro Colorado Island, which appear to share similar dispersal agents (e.g. Howe 1981, 1983). For instance, masked tityras (*Tityra semifasciata*; 85 g) easily eat fruits of the small-seeded (*V. sebifera*, but they cannot swallow the larger seeds (2–3 g) of *V. nobilis*. This bird is a common and apparently effective and reliable dispersal agent of *V. sebifera*, but is a 'fruit thief' of *V. nobilis* because it strips off the arils without dispersing the seeds. Such evaluations in a number of genera or families would determine how finely partitioned the dispersal niche is for related taxa within a given locality from the perspectives of both dispersal agents and plants.

Third, related tree systems should be studied with comparable methods over broad geographical regions. Many tree taxa are pantropical, including some genera (e.g. *Casearia* in the Flacourteaceae or *Beilschmedia* in the Lauraceae; *Virola* and *Myristica* in the Myristicaceae are almost indistinguishable). Comprehensive geographical comparisons of the dispersal systems of closely related tree taxa would be invaluable for determining the extent of ecological interdependence, and perhaps coevolution, between dispersal agents and tropical trees. One cannot be sure that differences ascribed to dispersal systems centered on quite different tree taxa reflect differences *between* taxa, unless the extent of variation *within* taxa is known.

Phenologies

Howe and Estabrook (1977) predicted that tropical trees competing for a limited assemblage of connoisseurs should have extended fruiting seasons, while those competing for a wide variety of opportunists should have sharply peaked fruiting seasons. The idea is promising because casual natural history suggests that some taxa (e.g. 'specialized' *Virola* with 10–15 week fruiting seasons vs 'generalized' *Ficus* with one week seasons) fit the prediction. Initial tests were not promising, both because most fruiting phenologies appeared to be 'fairly' peaked (there is no quantitative baseline for comparison), and because such other factors as seasonal changes in weather or exigencies of pollination might have at least as much influence on fruiting phenology as the disperser assemblage (Howe & Smallwood 1982). However, no test with comparable species, adjusted for taxonomic and life-form bias, has yet been attempted. Do sympatric members of a large genus or family with very different disperser assemblages differ in phenology? No one knows. Do different species of similar life form (shrubs, treelets, canopy trees) selected from many taxa with dramatically different disperser assemblages, differ as predicted in fruiting phenologies? No one knows.

Plant investment and diet specialization

The tenet that disproportionate investment by plants in fruits is associated with disproportionate use by specialist frugivores (Snow 1971; McKey 1975) is open to dispute. One reason for an unfortunate research void is the difficulty of determining allocation in trees. Simple distinctions between 'lipid-rich' or 'sugar-rich' fruits may or may not reflect allocation continua of tree species facing quite different ecological situations and carrying with them quite different phylogenetic biases. For instance, the 'reproductive cost' of allocation to a fatty fruit pulp in an emergent canopy tree could be similar to or even much less than that of apparently less nutritious pulp in an understory tree species with 15% of the leaf area, mostly in the shade. Canopy trees with ample light and understory trees in the shade undoubtedly have quite different physiological costs of allocation to fruits of a given energetic content, if only because photosynthetic resources differ so much. A related issue, diet specialization, has been explored in far greater theoretical detail (e.g. Martin 1985), but not in enough practical detail to understand the physiology of diet choice and preference. Long-lived non-symbiotic animal mutualists essentially never utilize just one species of plant (Snow 1981; Schemske 1983; Howe 1984a; Herrera 1985; Fleming 1986; Howe & Westley 1988), but it is not yet clear what the appropriate range of specialization might be, much less what proximate factors dictate fruit choice in the field. Advances in teasing apart factors influencing fruit choice in aviary environments (e.g. Moermond and Denslow 1983; Martinez del Rio *et al.* 1988, 1989) are intriguing, and must be considered along with other variables that might influence foraging in a forest full of fruits of widely different qualities and quantities (see Martin 1985). In the *V. nobilis* example, for instance, toucans apparently favor individual trees with high aril to seed ratios when fruits are most scarce (Howe 1983; see Howe 1986b, Tables 8 and 11). This appears contrary to foraging theory, unless one considers the transit costs of finding widely separated trees in a highly competitive foraging environment.

Much could be learned through study and thoughtful interpretation of basic foraging patterns. Do some avian frugivores feed on species of only three or four fruits at a time, while some feed on 20 or 30? This is the operational definition of 'specialization' implied in the frugivory literature. Or is preference the feature to be measured? Are the handful of bird species that strip dry forest *Casearia corymbosa* (Howe & Vande Kerckhove 1979), or wet forest *Virola sebifera* (Howe 1981) and *Virola nobilis* (Howe & Vande Kerckhove 1981) early in the morning 'specialists,' because they later go on to forage for other 'cheaper' foods that are rarely depleted? From the plant perspective a bird is a specialist species if it consistently prefers and depletes its fruit over that of a competitor tree species. But the plant perspective might not be of interest to an ornithologist not interested in plant seedling recruitment, if the bird depletes fruits at one tree species and then goes on to forage at several (or several dozen) more.

What is needed here is enough insight from a large enough set of dispersal systems to define 'specialization' and 'opportunism' in ecologically relevant terms. Classic studies with radio-monitored bats have determined actual diet use in the field (Fleming *et al.* 1977; Fleming & Heithaus 1981; Fleming 1988), but the digestive physiology and the actual foraging routines of tropical fruit-eating birds are rarely subjects of serious study (e.g. Murray 1988).

Is 'the paradigm' worth pursuing?

One measure of effectiveness is whether a conceptual framework in science, a scientific paradigm of Kuhn (1970), offers hypotheses worth testing. Another is whether the framework itself contributes insights to related phenomena. The last measure of worth is whether a better alternative paradigm exists.

The 'specialization' vs 'generalization' paradigm has much to offer because it frames testable hypotheses that require the integrated study of animal use of fruits and consequences of that use for plant reproduction. This value exists whether specialization and generalization refer to short-term but locally important ecological relationships, coevolution of higher taxa of plants and animals, or even symmetrical or asymmetrical coevolution of bird species or genera with plant species or genera. Study of untested questions discussed here would advance understanding of the evolutionary ecology of frugivory and seed dispersal.

Pursuit of 'the paradigm' has yielded useful insight, whether confirmatory or not. Its criticism (Wheelwright & Orians 1982; Moermond & Denslow 1985) greatly clarified the biological issues involved, and its extension (Howe 1985, 1989; Schupp 1992) elaborated new suites of testable hypotheses. Even attempts to test its predictions in ecological communities that are not strictly comparable have been extremely productive. Herrera provided the key insight that tropical fruits are not really 'richer' than temperate fruits (Herrera 1981; cf. Snow 1971; McKey 1975), and further discovered very efficient dispersal of temperate fruits that are not especially nutritious by birds that are not obviously specialized for frugivory (Herrera & Jordano 1981; Herrera 1984a, b). If 'the paradigm' has any validity in species-rich tropical communities with a wide range of frugivores and animal-dispersed trees, dispersal phenomena in species-poor Spanish scrub follow different rules.

Alternative frameworks

Ecological rationales offer promising alternative paradigms that can and should be pursued, whatever the fate of the adaptive arguments inherent in the 'specialization' and 'generalization' paradigm. Study of 'pivotal' or 'keystone' mutualists and tests of the presence or absence of density-dependence as a tool in understanding forest dynamics both require integration of frugivore and plant ecology.

Of general significance is the potential importance of 'pivotal' or 'keystone' mutualists in community stability and conservation biology. A 'keystone species' is one with an ecological impact in a community that is clearly disproportionate to its abundance (Paine 1966). A 'keystone mutualist' provides resources for a wide variety of species, which without it would face drastically reduced numbers or even local extinction (Howe 1977, 1984b; Gilbert 1980; Terborgh 1986). 'Specialized' dispersal agents might determine whether a tropical tree species survives or becomes extinct in a given locality. Alternatively, uncommon or rare trees will support a number of frugivores, including many unimportant for the recruitment of the 'pivotal' or 'keystone' tree in question, through annual periods of seasonal fruit scarcity (Howe 1977, 1984b). Tropical conservation biologists might simply ask which species of frugivores and plants are necessary to maintain stable tropical forest associations. Representative questions to be answered include: To what degree do different tree populations maintain frugivores that are critical dispersal agents for other tree populations at the site? What role do different vertebrate frugivores play in maintaining different tree populations? Important work need not assume coevolution to point out the presence or absence of a central role that plant/frugivore interactions could play in determining the abundance of tree and frugivore species, and that consequently may define the management protocols for conservation of tropical forest tracts.

Of equal importance, but not yet developed, are the implications of density-dependent seed or seedling mortality for forest community dynamics. Current tests of density-dependence, used as probes for the presence or absence of equilibrial tree assemblages, find little evidence that adult trees depress survival or growth of conspecific saplings, and consequently little evidence of density-dependent mortality that might suggest equilibrium tree communities (Hubbell *et al.* 1990). Given the axiom that virtually all plant mortality, density-dependent or not, occurs in seed and seedling stages (Harper 1977), the potential contributions of integrated analyses of frugivory and seed dispersal are obvious. Adult tree and sapling interactions are not likely to offer much insight because most density-dependent mortality (i.e. the 'escape advantage' discussed here) occurs years before a cohort reaches the sapling stage. That is, rampant density-dependent mortality may occur in early stages and thereby determine possible patterns of adult dispersion, but never be detected by tests of effects of adults on conspecific saplings. Adult and sapling relationships *will* be important where low mortality of earlier stages leaves many larger plants of a species in direct competition with each other.

In short, ecological studies of frugivores and seed dispersal have much to offer, independent of adaptive or evolutionary considerations. Keystone mutualisms are plausible but have not yet been described in depth, and tests of density-dependence using seed and seedling demographies are all but absent from the relevant literature (but see Howe 1992). Integrated and comprehensive studies of frugivory and its influence on seedling recruitment should be central to each of these alternative paradigms, but have not been widely applied to either.

Challenge

Where is the study of frugivory and seed dispersal going? The field is now an uncomfortable juxtaposition of a special case of animal foraging theory with particularly messy plant demography. Vigorous tests of alternative hypotheses, derived from a general framework, will be necessary to catalyze further research. Either the evolutionary 'specialization' paradigm or the ecological 'keystone' or 'forest dynamics' paradigms, or other frameworks not yet recognized, must be the basis of future work. Any of the three frameworks now available will require workers to overcome sectarian botanical and zoological biases to a degree not yet common. Otherwise, the ubiquitous and important phenomena of fruit-eating and seed dissemination by animals in tropical forests will drift to the anecdotal pre-1970 periphery of mainstream ecology. A branch of ecological science cannot mature without vigorous tests of alternative hypotheses, derived from paradigms relevant to general issues of scientific concern.

Acknowledgements

Field work of my research cited here was supported by the Smithsonian Tropical Research Institute and the National Science Foundation. Writing was supported by the University of Illinois. T. H. Fleming and E. W. Schupp provided numerous constructive criticisms of the manuscript.

References

Augspurger, C. K. 1988. Impact of pathogens on natural plant populations. In: Plant Population Ecology, Davy, A. J., Hutchings, M. J., and Watkinson, A. R. Blackwell (eds.), Scientific, Oxford. pp. 413–433.

Charles-Dominique, P. 1986. Inter-relations between frugivorous vertebrates and pioneer plants: *Cecropia*, birds and bats in French Guiana. In: Frugivory and seed dispersal, A. Estrada and T. H. Fleming (eds.), Dr. W. Junk Publ., Boston, pp. 119–135.

Clark, D. B. & Clark, D. A. 1984. Spacing dynamics of a tropical rain forest tree, evaluation of the Janzen-Connell model, Am. Nat. 124: 769–788.

Connell, J. H. 1971. On the role of natural enemies in preventing competitive exclusion in some marine animals and in rain forest trees. In: Dynamics of populations, P. J. den Boer and P. R. Gradwell (eds.), PUDOC, Wageningen. pp. 298–312.

Croat, T. B. 1978. The Flora of Barro Colorado Island. Stanford University Press, Stanford.

Dinerstein, E. & Wemmer, C. M. 1988. Fruits *Rhinoceros* eat: Megafaunal seed dispersal on a South Asian flood plain. Ecology 69: 1768–1775.

Duke, J. 1962. Myristicaceae. A. Missouri Bot. Gard. 49: 214–225.

Estrada, A. & Coates-Estrada, R. 1986. Frugivory in howling monkeys (*Alouatta palliata*) at Los Tuxtlas, Mexico: Dispersal and the fate of seeds. In: Frugivory and seed dispersal, A. Estrada and T. H. Fleming (eds.), Dr. W. Junk Publ. pp. 93–105.

Fleming, T. H. 1986. Opportunism vs. specialization: The evolution of feeding strategies in frugivorous bats. In: Frugivory and seed dispersal, A. Estrada and T. H. Fleming (eds.), Dr. W. Junk Publ. pp. 105–118.

Fleming, T. H. 1988. The Short-tailed Fruit Bat: A study in plant-animal interactions. University of Chicago Press, Chicago.

Fleming, T. H., Heithaus, E. R., & Sawyer, W. B. 1977. An experimental analysis of thc food location behavior of frugivorous bats. Ecology 58: 619–627.

Fleming, T. H. & Heithaus, E. R. 1981. Frugivorous bats, seed shadows, and the structure of tropical forests. Biotropica (Suppl.) 13: 45–53.

Harper, J. L. 1977. The population biology of plants. Academic, London.

Herrera, C. M. 1981. Are tropical fruits more rewarding to dispersers than temperate ones? American Naturalist 118: 896–907.

Herrera, C. M. 1984a. A study of avian frugivores, bird-dispersed plants, and their interaction in Mediterranean scrublands. Ecological Monographs 54: 1–23.

Herrera, C. M. 1984b. Adaptation to frugivory of Mediterranean avian seed dispersers. Ecology 65: 609–617.

Herrera, C. M. 1985. Determinants of plant-animal coevolution: the case of mutualistic dispersal of seeds by vertebrates. Oikos 44: 132–141.

Herrera, C. M. & Jordano, P. 1981. *Prunus mahaleb* and birds: The high efficiency seed dispersal system of a temperate fruiting tree. Ecological Monographs 51: 203–218.

Howe, H. F. 1977. Bird activity and seed dispersal of a tropical wet forest tree. Ecology 58: 539–550.

Howe, H. F. 1980. Monkey dispersal and waste of a neotropical fruit. Ecology 61:944–959.

Howe, H. F. 1981. Dispersal of a neotropical nutmeg (*Virola sebifera)* by birds. Auk 98: 88–98.

Howe, H. F. 1983. Annual variation in a neotropical seed dispersal system. In: Tropical rain forest: ecology and management. Sutton, S. L., Whitmore, T. C., and Chadwick, A. C. Blackwell (eds.) Scientific, Oxford. pp 211–277.

Howe, H. F. 1984a. Constraints on the evolution of mutualisms. American Naturalist 123: 764–777.

Howe, H. F. 1984b. Implications of seed dispersal by animals for management of tropical reserves. Biological Conservation 30: 261–281.

Howe, H. F. 1985. Gomphothere fruits: A critique. American Naturalist 125: 853–865.

Howe, H. F. 1986a. Seed dispersal by fruit-eating birds and mammals. In: Seed Dispersal. Murray, D. R. (ed.) Academic Press, Sydney, pp. 123–189.

Howe, H. F. 1986b. Consequences of seed dispersal by birds: A case study from Central America. Journal of the Bombay Natural History Society (supplement) 83: 19–42.

Howe, H. F. 1989. Scatter- and clump-dispersal and seedling demography: hypothesis and implications. Oecologia 79: 417–426.

Howe, H. F. 1990a. Seed dispersal by birds and mammals: implications for seedling demography. In: Reproductive Ecology of Tropical Forest Plants, K. S. Bawa and M. Hadley (eds.), UNESCO & Parthenon Publ. Group, Paris. pp. 191–218.

Howe, H. F. 1990b. Survival and growth of juvenile *Virola surinamensis* in Panama: effects of herbivory and canopy closure. J. Trop. Ecology 6: 259–280.

Howe, H. F. 1993. Aspects of variation in a neotropical seed dispersal system. Vegetatio 107/108: 149–162.

Howe, H. F. & Estabrook, G. F. 1977. On intraspecific competition for dispersal agents in tropical trees. American Naturalist 111: 817–832.

Howe, H. F. & Richter, W . 1982. Effect of seed size on seedling size in a neotropical nutmeg (*Virola surinamensis*): A within and between crop analysis. Oecologia 53: 347–351.

Howe, H. F., Schupp, E. W., & Westley, L. C. 1985. Early consequences of seed dispersal for a neotropical tree (*Virola surinamensis).* Ecology 66: 781–791.

Howe, H. F. & Smallwood, J. 1981. Ecology of seed dispersal. Ann. Rev. Ecol. Syst. 13: 201–228.

Howe, H. F. & Vande Kerckhove, G. A. 1979. Fecundity and seed dispersal by birds of a tropical tree. Ecology 60: 180–189.

Howe, H. F. & Vande Kerckhove, G. A. 1981. Removal of nutmeg (*Virola surinamensis*) crops by birds. Ecology 62: 1093–1106.

Howe, H. F. & Westley, L. C. 1988. Ecological relationships of plants and animals. Oxford, New York.

Hubbell, S. P. 1979. Tree dispersion, abundance, and diversity in a tropical dry forest. Science 203: 1299–1309.

Hubbell, S. P., Condit, R. & Foster, R. B. 1990. Presence and absence of density dependence in a neotropical tree community. Proc. Royal Soc. (London). B: 330: 269–281.

Janzen, D. H. 1970. Herbivores and the number of tree species in tropical forests. Am. Nat. 104: 501–528.

Janzen, D. H. 1980. When is it coevolution? Evolution 34: 611–612.

Jordano, P. 1987. Avian fruit removal: effects of fruit variation, crop size, and insect damage. Ecology 68: 1711–1723.

Kuhn, T. 1970. The structure of scientific revolutions. Second Ed. Univ. Chicago Press, Chicago.

Larson, D. 1991. The phainopepla/mistletoe seed dispersal mutualism. Ph.D. Dissertation, U. of Illinois at Chicago.

Manasse, R. S. & Howe, H. F. 1983. Competition for dispersal agents among tropical trees: influences of neighbors. Oecologia 59: 185–190.

Martin, T. E. 1985. Resource selection by tropical frugivorous birds: Integrating multiple interactions. Oecologia 66: 563–573.

Martinez del Rio, B. R. Stevens, D. E. Daneke & Andreadis, P. T. 1988. Physiological correlates of preference and aversion for sugars in three species of birds. Physiological Zoology 61: 222–229.

Martinez del Rio, C., W. H. Karasov, & Levey, D. J. 1989. Physiological basis and ecolo;gical consequences of sugar preferences in cedar waxwings. Auk 106: 64–71.

McKey, D. 1975. The ecology of coevolved seed dispersal systems. In: Coevolution of animals and plants. L. E. Gilbert and P. Raven (eds.), U. Texas Press, Austin. pp. 159–191.

Milton, K. 1980. The foraging strategy of howler monkeys. Columbia University Press, New York.

Moermond, T. & Denslow, J. 1985. Neotropical avian frugivores: Patterns of behavior, morphology, and nutrition with consequences for fruit selection. In: Neotropical Ornithology. P.A. Buckley, M.S. Foster, E.S. Morton, R.S. Ridgely, and N. G. Smith (eds.), A.O.U. Monographs, Allen Press. pp. 865–897.

Morrison, D. W. 1978. Foraging ecology and energetics of the frugivorous bat *Artibeus jamaicensis*. Ecology 59: 716–723.

Murray, K. G. 1988. Avian seed dispersal of three neotropical gap-dependent plants. Ecological Monographs 58: 271–298.

Paine, R. T. 1966. Food web complexity and species diversity. American Naturalist 100: 65–75.

Reid, N. 1986. Pollination and seed dispersal of mistletoes (Loranthaceae) by birds in southern Australia. In: The dynamics partnership: Birds and plants in southern Australia. H. A. Ford and D. C. Paton (eds.). Government Printer, South Australia. pp. 64–173.

Reid, N. 1989. Dispersal of a mistletoe by a honeyeater and a flowerpecker: Components of seed dispersal quality. Ecology 70: 137–145.

Schemske, D. W. 1981. Floral convergence and pollinator sharing in two bee-pollinated tropical herbs. Ecology 62: 946–954.

Schemske, D. W. 1983. Limits to specialization and coevolution in plant-animal mutualisms. In: Coevolution, M. H. Nitecki (ed.). Univ. of Chicago Press, Chicago. pp. 67–111.

Schupp, E. W. 1993. Quantity, quality, and the effectiveness of seed dispersal. Vegetatio 107/108:15–29.

Snow, D. W. 1971. Evolutionary aspects of fruit-eating in birds. Ibis 113:194–202.

Snow, D. W. 1981. Tropical frugivorous birds and their food plants: A world survey. Biotropica 13: 1–14.

Terborgh, J. 1986. Community aspects of frugivory in tropical forests. In: Frugivory and seed dispersal. A. Estrada and T. H. Fleming (eds.), Dr. W. Junk, Boston. pp. 371–384.

Wheelwright, N. T. 1983. Fruits and the ecology of resplendent quetzals. Auk 100: 286–301.

Wheelwright, N. T. 1986. A seven-year study of individual variation in fruit production in tropical bird-dispersed tree species in the family Lauraceae. In: Frugivory and seed dispersal. A. Estrada and T. H. Fleming (eds.). Dr. W. Junk Publ., Boston. pp. 21–36.

Wheelwright, N. T. & Orians, G. 1982. Seed dispersal by animals: contrasts with pollen dispersal, problems of terminology, and constraints on coevolution. Am. Nat. 119: 402–413.

Vegetatio **107/108**: 15–29, 1993.
T. H. Fleming and A. Estrada (eds). Frugivory and Seed Dispersal: Ecological and Evolutionary Aspects.

Quantity, quality and the effectiveness of seed dispersal by animals

Eugene W. Schupp
Savannah River Ecology Laboratory, Drawer E, Aiken, SC 29802, USA; Present address: Department of Range Science and The Ecology Center, Utah State University, Logan, UT 84322-5230, USA

Keywords: Consequences of seed dispersal, Plant fitness, Seed dispersal effectiveness, Seed dispersal quality, Seed dispersal quantity, Seed disperser

Abstract

Disperser effectiveness is the contribution a disperser makes to the future reproduction of a plant. Although it is a key notion in studies of seed dispersal by animals, we know little about what determines the effectiveness of a disperser. The role of the present paper is to review the available information and construct a hierarchical framework for viewing the components of disperser effectiveness.

Effectiveness has both quantitative and qualitative components. The quantity of seed dispersal depends on (A) the number of visits made to the plant by a disperser and (B) the number of seeds dispersed per visit. The quality of seed dispersal depends on (A) the quality of treatment given a seed in the mouth and in the gut and (B) the quality of seed deposition as determined by the probability that a deposited seed will survive and become an adult. In this paper I review the ways disperser behavior, morphology and physiology can influence these major components of disperser effectiveness, and when data permit present preliminary analyses of relationships among components.

Introduction

Central to studies of the ecological and evolutionary consequences of seed dispersal by animals is the notion that species of dispersers differ in their effectiveness (McKey 1975, Howe & Estabrook 1977, Snow 1981, Wheelwright & Orians 1982, Levey 1987). Effectiveness, the contribution a disperser makes to plant fitness, depends on the quantity of seed dispersed and the quality of dispersal provided each seed (Herrera & Jordano 1981). Progress is being made towards understanding the attributes of effective dispersers and the extent to which effectiveness varies among species (e.g., Davidar 1987; Reid 1989; Izhaki *et al.* 1991), but many questions remain unanswered. For example, are there highly effective dispersers that maximize both the quantity and the quality of dispersal, or are all deficient in some way? Is quantity or quality more important in determining the effectiveness of dispersal? Are species of dispersers inherently of high or low quality, or does quality differ with plant species dispersed, habitat or year? These and other critical questions have only begun to be addressed. The primary role of this paper is to consolidate the available information and to provide a framework for addressing disperser effectiveness. Note that the same approach is useful for other mutualisms as well (Herrera 1987, 1989a).

Disperser effectiveness

Many important terms in the literature of seed dispersal have been inconsistently applied. 'Efficiency,' for example, has referred to the number of seeds dispersed (Coates-Estrada & Estrada 1988), the quality of dispersal (Reid 1989) and the product of the two (Howe & Estabrook 1977). Just as frequently, several terms have been used for the same concept; 'quality' (McKey 1975; Levey 1987), 'reliability' (Howe & Estabrook 1977), 'goodness' (Janzen 1983b) and 'efficiency' (Howe & Estabrook 1977; Fialho 1990) have all been used more or less as synonyms of effectiveness as defined here. A useful framework for studying the effectiveness of seed dispersal must be built on a logical and consistent terminology. Table 1 presents a hierarchical outline of the major factors likely to influence effectiveness, suggesting the most appropriate terms.

Disperser effectiveness is ideally defined in terms of the contribution a disperser makes to the future reproduction of a plant. As such, the best measure is the number of new adult plants produced by the dispersal activities of a disperser, relative to the number produced by others. Because reliable estimates of the production of new adults can be very difficult, this is to a large extent only a heuristic definition at present. An empirical measure based on seedlings (Reid 1989) is far more tractable and may be nearly equivalent in many systems. In others, however, an emphasis on seedlings may prove misleading. The consequences of a particular pattern of seed dispersal can last far beyond the seedling stage: nearness to conspecifics can affect tropical tree growth even in the 8–16 cm DBH size class (Hubbell & Foster 1990) and differences in *Prunus mahaleb* growth and survival across a range of microhabitats are evident at least up to the size of first reproduction (Schupp, E. W., unpubl. data). Studies of effectiveness will be more difficult, but more enlightening, if viewed as presented here. The key lies in incorporating thorough analyses of plant demography into seed dispersal studies.

Empirically, we will consider:

$$\text{Effectiveness} = (\text{Quantity}) \times (\text{Quality}),$$

where quantity is the number of seeds dispersed and quality is the probability that a dispersed seed will produce a new reproductive adult. The quantity of dispersal is a function of (A) the number of visits made by a disperser and (B) the number of seeds dispersed per visit (Table 1). The quality of dispersal is a function of (A) the quality of treatment in the mouth and gut and (B) the quality of seed deposition (Table 1). Each of these components is in turn influenced by aspects of the biologies of the animal dispersers and the plants being dispersed.

Note that for both dispersal agents and plants, the present definition allows effectiveness to be considered at the scale of individuals, populations, or species. All levels will be addressed, but the following sections will primarily consider disperser populations interacting with plant populations.

Quantity of seed dispersal

Discussion of patterns of variation in the quantity of seed dispersal focuses on 16 data sets covering 15 species of trees, shrubs and vines from the eastern United States, Mexico, Central America and Spain (Fig. 1). Studies are limited to those

Table 1. A hierarchical outline of the major components of disperser effectiveness.

- I. Quantity of seed dispersal
  - A. Number of visits
    1. abundance of disperser
    2. diet
    3. reliability of visitation
  - B. Number of seeds dispersed per visit
    1. number of seeds handled per visit
    2. probability of dispersing a handled seed
- II. Quality of seed dispersal
  - A. Quality of treatment
    1. destroy or pass seeds intact
    2. alter percent or rate of germination
  - B. Quality of deposition
    1. movement patterns
       - a. habitat and microsite selection
       - b. rate and directionality of movement
    2. deposition patterns
       - a. rate and pattern of deposition
       - b. seed (diet) mixing

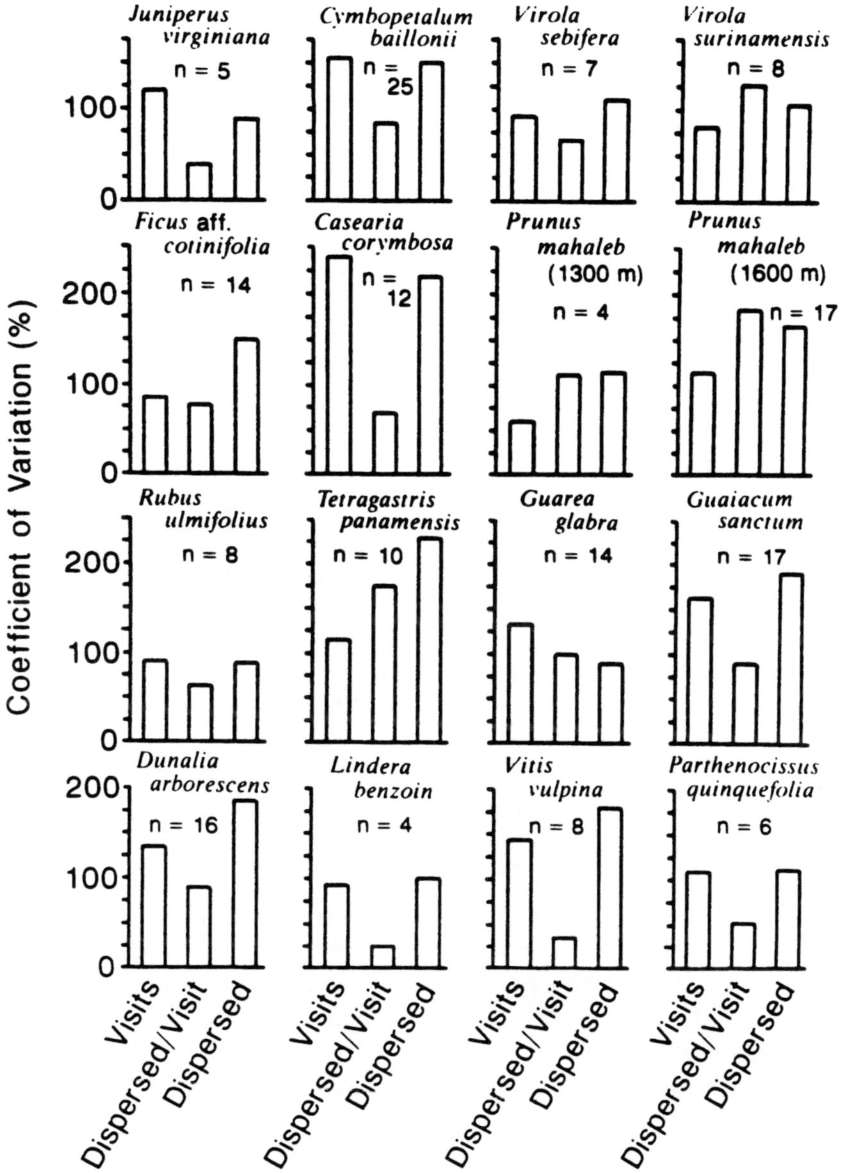

Fig. 1. Among-species variability in the quantitative traits of dispersal exhibited by the disperser assemblages visiting 16 species of plants. Shown are coefficients of variation for the number of visits, the number of seeds dispersed per visit and the total number of seeds dispersed. See text for details on the data sets. n = the number of species of dispersers used. Sources are: *Juniperus virginiana* (Holthuijzen & Sharik 1985), *Cymbopetalum baillonii* (Coates-Estrada & Estrada 1988), *Virola sebifera* (Howe 1981), *Virola surinamensis* (Howe & Vande Kerckhove 1981), *Ficus* aff. *cotinifolia* (Coates-Estrada & Estrada 1986), *Casearia corymbosa* (Howe & Vande Kerckhove 1979), *Prunus mahaleb* at 1300 m elevation (Herrera & Jordano 1981), *P. mahaleb* at 1600 m elevation (Jordano, P. & Schupp, E. W., unpubl. data), *Rubus ulmifolius* (Jordano 1982), *Tetragastris panamensis* (Howe 1980), *Guarea glabra* (Howe & De Steven 1979), *Guaiacum sanctum* (Wendelken & Martin 1987), *Dunalia arborescens* (Cruz 1981), and *Lindera benzoin*, *Vitis vulpina* and *Parthenocissus quinquefolia* (Malmborg & Willson 1988).

with reasonable estimates of the number of visits, the number of seeds 'dispersed' per visit and the total number of seeds 'dispersed' by individual species of dispersers. The accuracies of estimates vary, as do definitions of 'visits' and 'dispersal,' so generalities must be considered tentative.

Number of visits

Differences among disperser species in the number of visits made to a plant can be due to differences in abundance, diet, and reliability of visitation (Table 1). Species of dispersers differ extensively in local abundances (Fleming *et al.* 1977; Greenberg 1981; Herrera 1984, 1988; Thomas 1991); in one study, net captures of common avian dispersers of *Rubus ulmifolius* varied from four for *Sylvia melanocephala* to 184 for *S. atricapilla* (Jordano 1982). Such large differences in abundances contribute to differences in visitation.

Disperser species also differ in the importance of fruit in the diet and in the choice of individual fruit species. Species within local assemblages (Jordano 1982; Herrera 1984), families (e.g., Phyllostomatidae; Fleming 1988) and even genera (e.g., *Sylvia*; Jordano 1987) show continuous variation from near total frugivory to the addition of occasional fruits to a diet of insects, vertebrates, pollen or nectar. Frugivorous species also differ in which species of fruit they consume. In Panama (Howe & DeSteven 1979; Howe 1980, 1981, 1986b; Greenberg 1981), an Illinois woodlot (Malmborg & Willson 1988), west Africa (Gautier-Hion *et al.* 1985) and southern Spain (Herrera 1989b) frugivorous species differed in the use of available fruits. Disperser size (Jordano 1987) and digestive physiology (Martinez del Rio & Restrepo, this volume), fruit presentation (Moermond & Denslow 1985), concentrations of nutrients (Fleming 1988) and chemical deterrents (Howe & Vande Kerckhove 1981), and the packaging (Murray *et al.*, this volume) and passage rates (Sorensen 1984) of seeds have all been implicated in fruit choice. Regardless of the causes, choice will play a role in determining visitation.

Disperser reliability spans a range of temporal and spatial scales. Temporally, a reliable disperser dependably visits plants throughout the day, the season and the years. Variation among species of dispersers has been observed at all scales. Kantak (1981) noted that some Mexican birds fed on fruit throughout the day, but others fed only part of the day. At La Selva, Costa Rica *Tityra semifasciata* was the only disperser that reliably visited *Casearia corymbosa* throughout the season, and consequently was the most abundant visitor overall (Howe 1977). Reliability on the scale of years can take several forms. A disperser may be abundant one year and absent the next, as in the periodic southern invasions of Waxwings (*Bombycilla garrulus*) beyond their normal wintering range (Snow & Snow 1988). Species may be reliably present each year, but unreliably abundant. In southern Spain, autumn abundances of the birds *Erithacus rubecula* and *Sylvia atricapilla* varied more than four-fold over nine years (Herrera 1988). Similarly, the disperser assemblage of *Virola surinamensis* in Panama was consistent over four years, but relative abundances of the species differed significantly (Howe 1986b). Even if a disperser is reliably present and abundant each year, it may be an unreliable visitor due to year-to-year shifts in fruit choice (Malmborg & Willson 1988).

Spatially, a reliable disperser will dependably visit all individuals of all populations throughout the range of the plant species. As with temporal reliability, variation among seed dispersers exists at all scales. Observing 14 *Virola sebifera* in Panama, Howe (1981) recorded three disperser species visiting all trees and four visiting only three or fewer trees. On a larger scale, disperser assemblages differed for *Casearia corymbosa* in wet and dry forests in Costa Rica (Howe 1977; Howe & Vande Kerckhove 1979), and for *Prunus mahaleb* at 1300 m and 1600 m elevation in the Sierra de Cazorla of Spain (Herrera & Jordano 1981; Jordano, P. & Schupp, E. W., unpubl. data). At the largest scale, the geographical range of a disperser may or may not (Wheelwright 1988; Jordano, this volume) encompass the range of a plant species it disperses.

Patterns of variation in disperser abundances, diets and reliabilities of visitation combine to yield overall patterns of visitation to fruiting plants. In all systems studied, species of seed dispersers differed in the number of feeding visits made (Fig. 1); coefficients of variation for the 16 data sets range from 59–239% with a median of 114%.

Though extensive, these estimates of variation are minimal. Restricting the data sets to dispersal agents for which feeding (dispersal) rate information existed biased the data against rare visitors, decreasing the CV's. Additionally, while published lists of dispersers reliably record abundant visitors, they lack some to many infrequent visitors. Rarefaction may be useful for estimating the degree of completeness of sampling (P. Jordano, pers. comm.), but increased effort should add at least a few species to most lists (see Wheelwright *et al.* 1984), increasing estimates of variation. The rate at which increased sampling adds new species will depend on whether the added effort is at the same trees, new trees or new populations.

Fig. 2 shows the proportion of the total visits made by individual disperser species to representative tree species with small (*Prunus mahaleb*, 1300 m elevation), medium (*Virola sebifera*) and large (*Dunalia arborescens*) disperser assemblages. In large assemblages (> 10 disperser species), most species make few visits, while several make over half of all visits. With smaller assemblages there are proportionally fewer species of infrequent visitors, but most visits are still made by a few species.

Number of seeds dispersed per visit

The number of seeds dispersed per visit is the product of the number of seeds handled and the probability a handled seed is dispersed (Table 1). A key characteristic affecting both is handling method. Classifications of handling methods have been initiated for bats (Bonaccorso & Gush 1987)

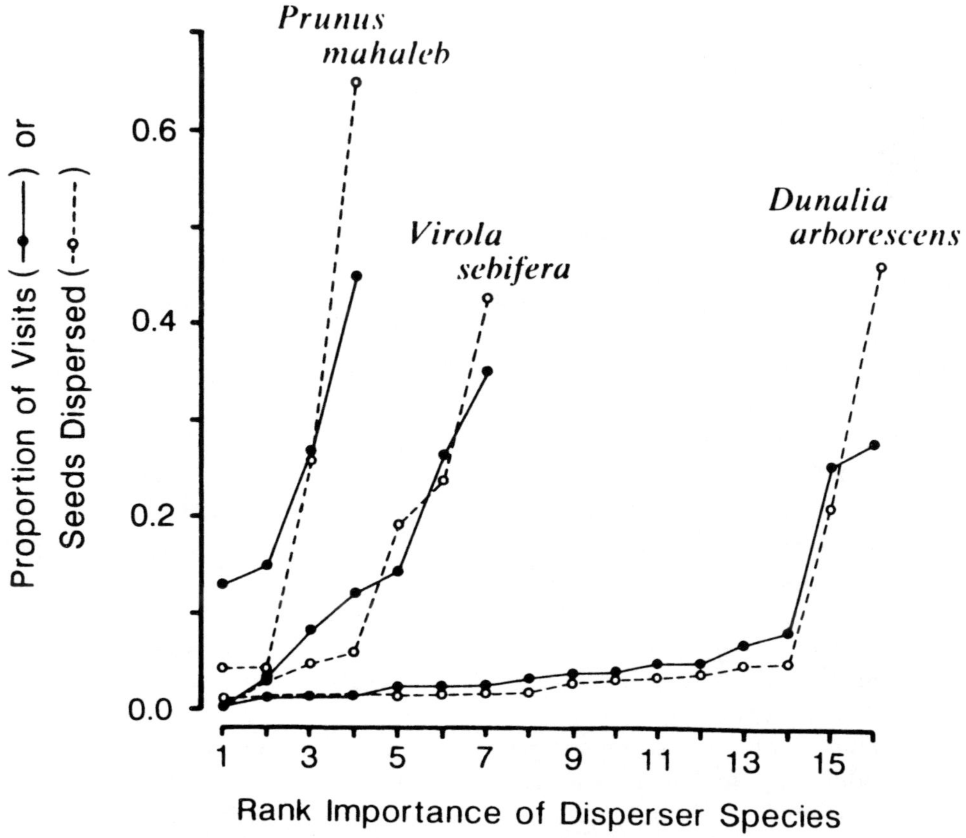

Fig. 2. The proportion of visits made by a species and the proportion of total seeds dispersed by a species, with dispersal agents ranked from smallest to largest proportions (rank importance of disperser species). Representative small, medium and large disperser assemblages are shown. See Fig. 1 legend for sources of data.

and primates (Corlett & Lucas 1990), but I will focus on birds, for which more information is available. Frugivorous birds may be categorized as 'swallowers,' 'mashers' or 'biters' (Trainer & Will 1984; Moermond & Denslow 1985; Foster 1987; Levey 1987).

Swallowers may reject plucked fruits, but generally swallow fruits and included seeds whole so they have a relatively high probability of dispersing handled seeds. Mashers manipulate the fruit in the mouth and ingest juice and pulp along with a variable number of seeds; the proportion swallowed decreases from 1.0 for very small-seeded species to 0.0 for large-seeded species (Levey 1987). Consequently, mashers should have a high probability of dispersing small seeds, but a low probability of dispersing large seeds. Biters remove bits of pulp by biting or pecking a fruit that is either still attached or plucked and held against a branch. Seeds are not swallowed, and are only rarely dispersed when a biter carries the fruit to another tree for feeding. Larger species are more likely to be swallowers, while smaller species tend to be mashers or biters (Trainer & Will 1984; Foster 1987; Snow & Snow 1988).

At least for swallowers, the number of fruits handled per visit roughly increases with disperser size (Howe & De Steven 1979; Jordano 1982; Murray 1988). For dispersers of equivalent size, however, handling method should be a major determinant of the maximum number of fruits that can be handled in a feeding bout. Since they reject seeds and much of the attached pulp, biters will fill the gut less rapidly and potentially handle more seeds per visit than a swallower. This expected pattern may, however, be counteracted by the longer handling times per fruit for biters (Levey 1987). Mashers should be intermediate, but more like swallowers when feeding on small-seeded species and more like biters when feeding on large-seeded fruits (Levey 1987).

The potential number of seeds dispersed per visit is further modified by the length of a visit. Staying in a tree longer will result in more seeds being handled, but also more being regurgitated or defecated beneath the parent (Howe 1981; Pratt & Stiles 1983). Most dispersers leave a feeding tree quickly, however, because of aggressive displacement (Herrera & Jordano 1981; Howe 1981), fear of predators (Howe 1979; Snow & Snow 1988; Fleming 1988) or as part of an intentional diet-mixing strategy (Jordano 1987, 1988; Loiselle 1990). Few dispersers appear to stay in a tree long enough to regurgitate or defecate seeds from that feeding bout (Wheelwright 1991), but for those with longer visit lengths, gut processing becomes an important trait. A swallowed seed may be regurgitated or defecated; regurgitation is most likely for large seeds and small dispersers (Johnson *et al.* 1985). Gut passage time for defecation varies considerably (Herrera 1984; Jordano 1987) but is almost always longer than for regurgitation (Johnson *et al.* 1985; Levey 1986, 1987). Given the same visit length, then, a disperser that regurgitates may be less likely to disperse seeds than one that defecates.

Although it is not now possible to dissect out the relative contributions of handling method, visit length and seed processing, disperser species do vary in the number of seeds dispersed per visit (Fig. 1). Coefficients of variation range from 26–183% with a median of 80%. The variation is less than in the number of visits made, but important.

Quantity of seed dispersal: reprise

Species vary greatly in the total number of seeds dispersed (Fig. 1); the median CV is 130.5% and the range from 88–229%. As with the number of visits, the distributions of total numbers of seeds dispersed tend to be skewed; a few species disperse many seeds but most disperse a small to moderate number (Fig. 2).

Two relationships are of further interest. First, do the quantitative traits covary such that species that make many visits tend to also disperse many seeds per visit? The answer appears to be no. Of the 16 data sets, the number of visits was significantly correlated with the number of seeds dispersed per visit only in *Guaiacum sanctum* ($r_s = .690$, $n = 17$, $p < 0.005$). The typical disperser is deficient in one way or the other.

Second, what is the major determinant of the

quantitative importance of a disperser to a plant? The number of visits a disperser makes appears to be a slightly better predictor of the total quantity of seed dispersed than is the number of seeds dispersed per visit (Table 2). Spearman rank correlations involving the number of visits were generally larger and more likely to be significant than correlations involving the number of seeds dispersed per visit. This suggestive trend does not mean that the number of seeds dispersed per visit is a trivial component of dispersal, however.

Quality of seed dispersal

The quality of seed dispersal is a function of (A) the quality of seed treatment in the mouth and gut and (B) the quality of seed deposition, or probability a deposited seed survives and produces a new adult (Table 1). Wheelwright and Orians (1982) suggested that differences among species in the quality of dispersal may be relatively small, but this need not be true. Dispersers vary in seed treatment, and in the size, shape and heterogeneity of seed shadows produced (Willson 1986), and there are many opportunities for some attribute of a disperser to affect the probability of a seed safely completing the long, perilous passage to adulthood.

Quality of treatment

Seed treatment can affect the quality of dispersal by destroying seeds or by altering patterns of germination (Table 1). Although in reality one grades into the other, they will be discussed as distinct effects. In some dispersal systems, seed processing in the bill or mouth is detrimental; seed-eating birds (Coates-Estrada & Estrada 1986) and mammalian carnivores (Herrera 1989b) feeding on fruit crush variable numbers of seeds while swallowing the remaining undamaged. Effects of treatment in the gut, however, are probably more pervasive.

Species of frugivorous birds that are usually considered 'legitimate' seed dispersers seldom damage seeds during gut passage (Howe 1986a). In contrast, seed-eating birds destroy vast quan-

Table 2. Spearman rank correlation coefficients for correlations between estimates of the total number of seeds dispersed by a species and either the number of seeds dispersed per visit or the total number of visits made for the 16 species of plants in Fig. 1. n = number of disperser species used in analyses. * = $p<0.05$, ** = $p<0.01$, *** = $p<0.001$.

Species	n	Correlation between total # seeds dispersed and	
		# dispersed/visit (r_s)	# visits (r_s)
Juniperus virginiana	5	−0.700	0.900
Cymbopetalum bailonii	25	0.470*	0.729***
Virola sebifera	7	0.714	0.714
Virola surinamensis	8	0.833*	0.695
Ficus aff. *cotinifolia*	14	0.975***	0.642*
Casearia corymbosa	12	−0.149	0.965***
Prunus mahaleb	4	0.800	0.800
Prunus mahaleb	17	0.320	0.867***
Rubus ulmifolius	8	0.714	0.905**
Tetragastris panamensis	10	0.406	0.828**
Guarea glabra	14	0.026	0.882***
Guaiacum sanctum	17	0.905***	0.913***
Dunalia arborescens	16	0.493	0.532*
Lindera benzoin	4	0.949	0.800
Vitis vulpinia	8	0.814*	0.898**
Parthenocissus quinquefolia	6	0.406	0.928*

tities in the grinding gizzard, but pass some in viable condition (Lambert 1989); the proportion destroyed varies with both bird and seed species (Krefting & Roe 1949). Less data are available for mammals, but it appears that at least some monkeys pass most to all seeds in a viable condition while ungulates often destroy substantial quantities (Howe 1986a). *Tapirus bairdii* and feral horses and cows destroyed, respectively, 78%, 44–83% and 14–21% of the *Enterolobium cyclocarpum* seeds ingested (Janzen 1981, 1982). The differences are at least partly a function of the length of time seeds are retained in the gut.

Studies of germination have focused on comparisons of germination with and without ingestion by dispersers. Of interest here, however, are the few studies that address the effects of different disperser species on germination of a given plant species. In 12 of 49 'trials,' species of dispersers differed significantly in their effects on percent germination (Holthuijzen & Sharik 1985; Reid 1989; Barnea *et al.* 1990, 1991; Izhaki & Safriel 1990). Lieberman & Lieberman (1986) presented a complex study of 16 fruit species and six animal species and reported no overall effect of animal species, but several figures in their paper are suggestive of a difference for some plant species. As with the destruction of seeds, differences among species in effects on percent germination may result from differences in gut retention (Barnea *et al.* 1991).

Dispersers may also alter germination rate (Izhaki & Safriel 1990), and some detected differences in percent germination may simply reflect differences in germination rate on the scale of years. The consequences of altering the speed of germination are likely complex. Potential effects include changes in the length of exposure to post-dispersal seed predators and in the probability of encountering ephemeral seedling microsites.

Quality of deposition

This topic is usually discussed in terms of the probability a seed will be deposited in a suitable site (McKey 1975; Howe & Estabrook 1977; Herrera & Jordano 1981; Wheelwright & Orians 1982). Sites are not simply suitable or unsuitable, however; they vary continuously with respect to seed survival, germination, and seedling growth and survival (Janzen 1983a). Further, the suitability of a site may differ for seeds and seedlings (Schupp & Frost 1989). Seed dispersal should not be viewed in terms of the probability of hitting the right spot, but rather in terms of the creation of a continuous array of survival probabilities, each associated with a dispersed seed.

Wheelwright & Orians (1982) suggested that species of dispersers differ relatively little in quality because suitable sites for dispersal are unpredictable in space and time, and no seed has much of a chance of surviving to produce a new adult. It is true that locations of 'suitable' sites are unpredictable (Schupp 1988a; Whelan *et al.* 1991), but some sites are predictably associated with higher probabilities of survival than are others (Schupp *et al.* 1989). It is also true that the probability of any seed producing a new adult is exceedingly small, but the range of probabilities represented by the different seeds in a population may span orders of magnitude. There are ample opportunities for dispersers to vary substantially in the seed shadows they produce, and consequently in the distributions of survival probabilities they create.

Of interest in this section are attributes of seed dispersers that determine 'where' seeds fall in the broad sense: the biotic and physical conditions into which seeds are deposited. I will consider movement patterns and deposition patterns (Table 1). Though not a clear dichotomy, this distinction will facilitate the discussion.

The major movement pattern affecting where seeds fall is habitat selection. Although they are habitat generalists in a relative sense, dispersers select among habitats on a variety of scales (Herrera 1985). Selection can occur between treefall gaps and forest understory (Charles-Dominique 1986; Hoppes 1988; Levey 1988; Malmborg & Willson 1988; Thomas 1991), between mature and second growth patches (McDiarmid *et al.* 1977; Thomas 1991), or along moisture and vegetation gradients (Karr & Freemark 1983). On a more local scale, species of frugivorous birds differ in post-feeding microhabitat use (Sorensen 1981; Jordano 1982; Izhaki *et al.* 1991; Schupp, E. W., unpubl. data, see Fig. 3). At an even smaller scale, species of ants can differ in the types of substrates to which they disperse seeds (Horvitz 1981, Horvitz & Schemske 1986). The habitats and microsites in which seeds are deposited vary in suitability to plants. The characteristics of the physical environment that influence habitat selection by dispersers also influence the probability of a seed surviving, germinating and growing to adulthood (Augspurger 1983; Sork 1985, 1987; Murray 1988; Schupp 1988a, b; Willson 1988; Schupp & Frost 1989; Alvarez-Buylla & Martinez-Ramos 1990; Howe 1990; Vazquez-Yanes & Orozco-Segovia 1990; Willson & Whelan 1990; Fisher *et al.* 1991; Whelan *et al.* 1991).

The size, shape and density of a seed shadow

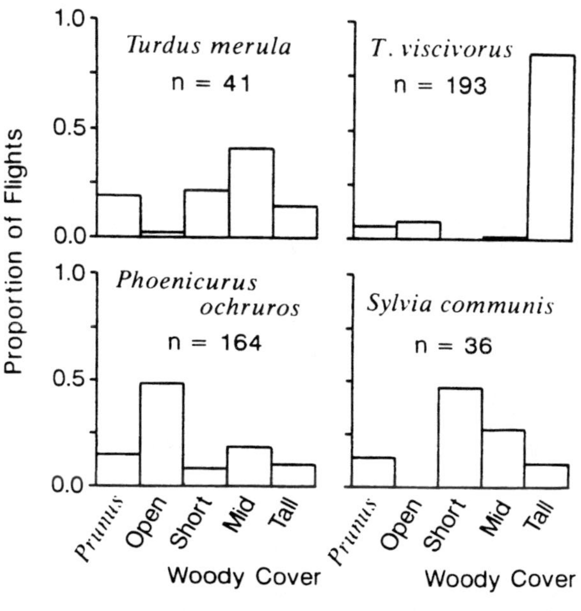

Fig. 3. Differences in distributions of post-foraging microhabitat use for four representative seed dispersers of *Prunus mahaleb* at 1600 m in the Sierra de Cazorla, Spain. Distributions are based on the first destination of a bird after feeding on fruit and leaving the tree (Schupp, E. W., unpubl. data). n = number of flights recorded.

is further influenced by the directionality and rate of movement away from the fruiting tree, and by the rate and pattern of seed deposition. Some disperser species move directly to other fruiting trees, others to isolated perches (Howe & Primack 1975; Coates-Estrada & Estrada 1988; Fleming 1988). Some move continuously, others intersperse quiescent periods with short periods of rapid movement (Murray 1988). Some fly short distances to a perch for seed processing, others fly longer distances (Cruz 1981; Howe 1986b; Coates-Estrada & Estrada 1988; Schupp, E. W., unpubl. data, see Fig. 4). Overlying these movement patterns are variations in the rate and pattern of seed deposition. Regurgitated seeds are deposited rapidly (Levey 1986), while defecated seeds may be passed in minutes by small birds (Levey 1986), hours to days by monkeys (Milton 1981) or weeks to months by ungulates (Janzen 1982). Seeds may further be deposited singly or in clumps. *Trogon rufus* defecates *Tetragastris panamensis* seeds individually while *Alouatta palliata* defecates them in piles of up to 60 (Howe 1980). Some species deposit the entire gut load in a single defecation, others spread it over several defecations (Izhaki *et al.* 1991). Given that growth and survival depend in some manner on distance from a conspecific adult (Clark & Clark 1984; Howe *et al.* 1985; Coates-Estrada & Estrada 1988; Hubbell & Foster 1990) and on density of conspecific seeds and seedlings (Howe 1980; Webb & Willson 1985; Andersen 1987; Lewis 1987), movement and deposition patterns can have major impacts on the quality of seed deposition.

Diet-mixing by dispersers adds complexity to the local environment facing a dispersed seed (Fleming & Heithaus 1981; Loiselle & Blake 1990). Both the probability of being deposited in a fecal clump with another species and the identity of neighbors in that clump vary with disperser species. Seed species richness is greater in the guts of large than of small sylviid warblers (Jordano 1987), and greater in defecations of Costa Rican mashers than of swallowers (Loiselle 1990). Certain combinations of seeds are found more often than expected by chance, and dispersers

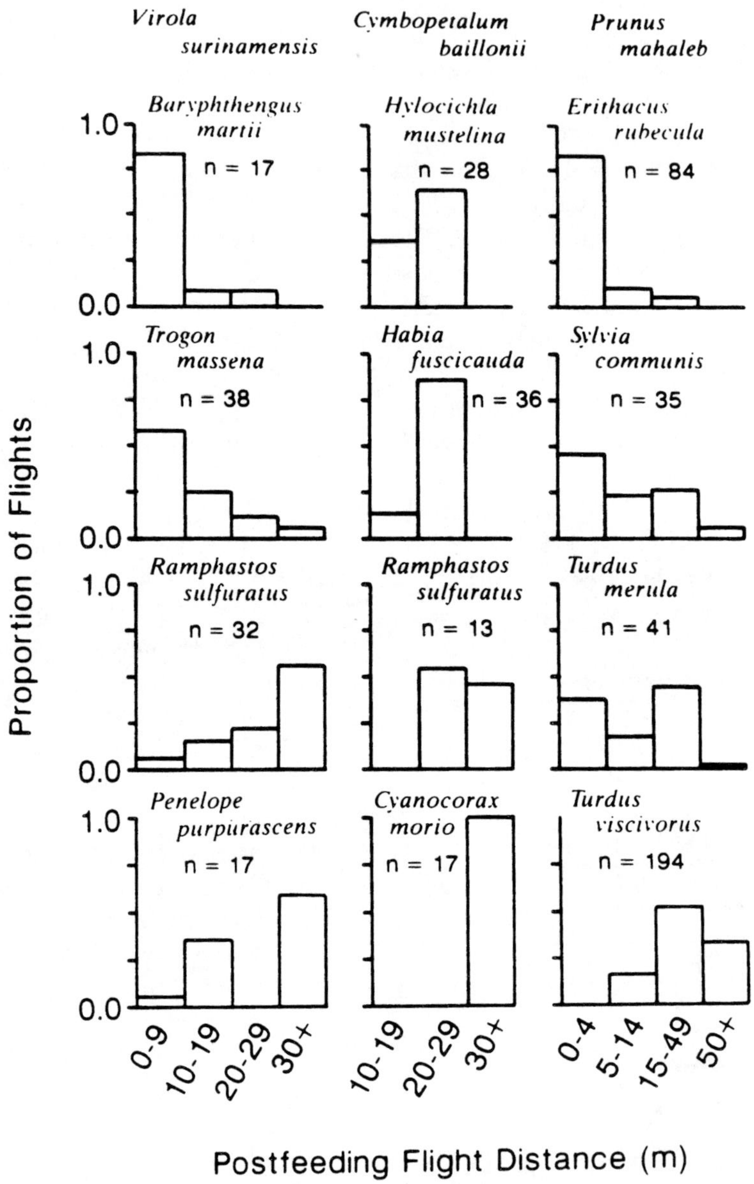

Fig. 4. Distributions of distances flown in first flight from feeding tree for representative bird species dispersing *Virola surinamensis* (Howe 1986b), *Cymbopetalum baillonii* (Coates-Estrada & Estrada 1988) and *Prunus mahaleb* (Schupp, E. W., unpubl. data). n = the number of flights observed for a disperser species.

differ in the combinations generated (Jordano 1988; Loiselle 1990). Interspecific competition among seedlings in these mixed species clumps may be an important factor affecting the quality of deposition. If some species are consistently superior competitors, the vigor of a seedling will partly depend on the identity of its neighbors, a function of the species that dispersed that seed (Loiselle 1990).

Which disperser attributes have the greatest impact on the quality of deposition depend on the reason dispersal is advantageous to the plant. Howe & Smallwood (1982) proposed three advantages of dispersal: (1) escape, (2) colonization and (3) directed dispersal. If escape from the vicinity of the parent is a critical advantage of dispersal, traits promoting movement out of the parent and away from conspecifics will be most important. If colonization of unpredictable disturbances is crucial, characteristics promoting widespread dispersal will have a dominant impact. If specific but predictable conditions are needed for establishment and survival, dispersal directed towards the necessary habitat or microsite will be the crucial characteristic. Adequately assessing which disperser traits are most important requires more detailed knowledge about the consequences of various patterns of seed dispersal than is currently available. Recent studies have, however, supported the notion that the characteristics associated with high-quality deposition are not fixed, but rather depend on the ecology of the plant.

Howe *et al.* (1985) presented evidence that the quality of deposition of *Virola surinamensis* seeds is largely a function of how far birds fly before perching to process seeds. Murray (1988) suggested that for gap-dependent plants, however, the quality of deposition primarily depends on rates of movement and gut passage; species moving rapidly and depositing seeds over an extended period are most likely to disperse seeds to present and future gaps. Finally, in southeastern Brazil the quality of deposition of *Erythroxylum ovalifolium* seeds is determined by plant microsite requirements. Seeds are dispersed to the moist interiors of terrestrial bromeliads by the frog *Hyla truncata* and to open, sandy soil by the lizard *Tropidurus torquatus* (Fialho 1990); experimental 'dispersal' yielded higher germination and establishment for 'frog-dispersed' than for 'lizard-dispersed' seeds. Empiricists have only begun to address the quality of deposition, but these and other studies demonstrate that innovative approaches can yield exciting insights.

Disperser effectiveness: reprise

Both quantity and quality are critical components of effectiveness, but a paucity of detailed studies hinders an assessment of the relative importance of each at present. Recent studies have begun to yield preliminary answers, however. Reid's (1989) study of dispersal of the mistletoe *Amyema quandang* by the birds *Dicaeum hirundinaceum* and *Acanthagenys rufogularis* is one of the most thorough. *Acanthagenys rufogularis* was less reliable seasonally but more abundant, and as a result was quantitatively more important than *D. hirundinaceum*. Compared to seeds dispersed by *D. hirundinaceum*, those dispersed by *A. rufogularis* were less likely to germinate and less likely to land on a perch, but perches on which they did land were more likely to be suitable. The outcome was no detectable difference between the species in the quality of dispersal. As a consequence, *A. rufogularis* was the more effective disperser because it provided the greater quantity of dispersal.

Quantity does not appear to be the key component of disperser effectiveness in all systems, however. *Phainopepla nitens* was overwhelmingly the most effective disperser of the mistletoe *Phoradendron californicum* because it was both quantitatively and qualitatively superior to either of the other two dispersers at the site (Larson 1991).

Few data exist from systems involving more species of dispersers and less discrete qualities of dispersal sites, but useful preliminary calculations can be made from Howe's work with *Virola surinamensis*. Using data on the quantity of dispersal from Howe & Vande Kerckhove (1981), flight distances from Howe (1986b), and distance-dependent survival from Howe *et al.* (1985), I estimated both quantitative and qualitative components of disperser effectiveness. This analysis assumed the quality of dispersal was determined by the distance a seed was dispersed, and effectiveness could be represented by the estimated number of 12-wk seedlings produced. Spearman rank correlations suggest effectiveness was related more strongly to the quality ($r_s = 0.90$, $p = 0.10$, $n = 5$) than to the quantity ($r_s = 0.10$, $p > 0.50$, $n = 5$) of dispersal. If the definition of dispersal was relaxed to include all handled seeds to account for the small but real probability of survival beneath the parent, effectiveness was still more strongly correlated with quality ($r_s = 0.93$, $p < 0.05$, $n = 6$) than quantity ($r_s = 0.52$, $p > 0.20$, n = 6). Despite the crudeness of the analyses, there is evidence that in some systems quality is the dominant component of effectiveness.

Empirical studies of seed dispersal have emphasized the quantity of seeds dispersed by different species, often with the assumption that quantity is a strong correlate of effectiveness. The available evidence suggests, however, that this assumption may frequently be invalid. Investigations of the quantity of dispersal are important, but are not surrogates for studies of effectiveness; more emphasis should be placed on quantifying the consequences of dispersal by different disperser species. Effectiveness can only be evaluated with detailed studies of the natural histories of the interacting organisms; studies that interweave the behavior, morphology and physiology of dispersers with the demography of plants. The present review has revealed many gaps in our knowledge of effectiveness, but has also shown that the answers are not beyond our reach.

Disperser effectiveness: epilogue

This paper has covered characteristics of dispersers that more or less directly determine disperser effectiveness, but more complex relationships exist. The interactions of secondary dispersal agents such as dung beetles (Estrada & Coates-Estrada 1986) and ants (Byrne & Levey, this volume) with the seed shadows produced by vertebrates may greatly alter the effectiveness of primary dispersers. Defense of a fruit source (Snow & Snow 1984) can curtail the number of disperser species contributing to the seed shadow and delay the timing of dispersal. As a consequence of fruit selection based on external traits, disperser species may sometimes disperse different-sized seeds (Jordano 1984), which can in turn influence the success of dispersal. These are but a fraction of the intriguing complexities involved in disperser effectiveness.

Acknowledgements

My perspective has been influenced by my work with H. F. Howe, C. M. Herrera and P. Jordano. I thank them for the interactions and experiences they have shared. Equally important are my experiences on Barro Colorado Island, Panama and in the Sierra de Cazorla, Spain. This paper has benefited from support by the Smithsonian Tropical Research Institute and the National Science Foundation, and was written while the author held a Department of Energy Alexander Hollaender Distinguished Postdoctoral Fellowship administered by Oak Ridge Associated Universities. T. Fleming, H. Howe, D. Imm, P. Jordano, K. Kirkman, B. Loiselle, K. Moorhead, A. Schnabel, R. Sharitz, and M. Vaitkus helped with clarity and brevity. Manuscript preparation was assisted by contract DE-AC09-76SROO-819 between the U. S. Department of Energy and the University of Georgia's Savannah River Ecology Laboratory. This paper is dedicated to the memory of Margarita Cobo.

References

Alvarez-Buylla, E. R. & Martinez-Ramos, M. 1990. Seed bank versus seed rain in the regeneration of a tropical tree. Oecologia 84: 314-325.

Andersen, A. N. 1987. Effects of seed predation by ants on seedling densities at a woodland site in SE Australia. Oikos 48: 171–174.

Augspurger, C. K. 1983. Seed dispersal of the tropical tree, *Platypodium elegans*, and the escape of its seedlings from fungal pathogens. J. Ecol. 71: 759–771.

Barnea, A., Yom-Tov, Y. & Friedman, J. 1990. Differential germination of two closely related species of *Solanum* in response to bird ingestion. Oikos 57: 222–228.

Barnea, A., Yom-Tov, Y. & Friedman, J. 1991. Does ingestion by birds affect seed germination? Functional Ecology 5: 394–402.

Bonaccorso, F. J. & Gush, T. J. 1987. Feeding behaviour and foraging strategies of captive phyllostomid fruit bats: an experimental study. J. Animal Ecol. 56: 907–920.

Charles-Dominique, P. 1986. Inter-relations between frugivorous vertebrates and pioneer plants: *Cecropia*, birds and bats in French Guyana. In: Estrada, A. & Fleming, T. H. (eds.), Frugivores and seed dispersal, pp. 119–135. Dr W. Junk Publ, Dordrecht.

Clark, D. A. & Clark, D. B. 1984. Spacing dynamics of a tropical rain forest tree: evaluation of the Janzen-Connell model. Amer. Natur. 124: 769–788.

Coates-Estrada, R. & Estrada, A. 1986. Fruiting and frugivores at a strangler fig in the tropical rain forest of Los Tuxtlas, Mexico. J. Trop. Ecol. 2: 349–357.

Coates-Estrada, R. & Estrada, A. 1988. Frugivory and seed dispersal in *Cymbopetalum baillonii* (Annonaceae) at Los Tuxtlas, Mexico. J. Trop. Ecol. 4: 157–172.

Corlett, R. T. & Lucas, P. W. 1990. Alternative seed-handling strategies in primates: seed-spitting by long-tailed macaques (*Macaca fascicularis*). Oecologia 82: 166–171.

Cruz, A. 1981. Bird activity and seed dispersal of a montane forest tree (*Dunalia arborescens*) in Jamaica. Biotropica 13 (suppl.): 34–44.

Davidar, P. 1987. Fruit structure in two neotropical mistletoes and its consequences. Biotropica 19: 137–139.

Estrada, A. & Coates-Estrada, R. 1986. Frugivory in howling monkeys (*Allouatta palliata*) at Los Tuxtlas, Mexico: dispersal and fate of seeds. In: Estrada, A. & Fleming, T. H. (eds.), Frugivores and seed dispersal, pp. 93–104. Dr W. Junk, Publ., Dordrecht.

Fialho, R. F. 1990. Seed dispersal by a lizard and a treefrog – effect of dispersal site on seed survivorship. Biotropica 22: 423–424.

Fisher, B. L., Howe, H. F. & Wright, S. J. 1991. Survival and growth of *Virola surinamensis* yearlings: water augmentation in gap and understory. Oecologia 86: 292–297.

Fleming, T. H. 1988. The short-tailed fruit bat. A study in plant-animal interactions. Univ. Chicago Press, Chicago.

Fleming, T. H., Heithaus, E. R. & Sawyer, W. B. 1977. An experimental analysis of the food location behavior of frugivorous bats. Ecology 58: 619–627.

Fleming, T. H. & Heithaus, E. R. 1981. Frugivorous bats, seed shadows, and the structure of tropical forests. Biotropica 13 (suppl.): 45–53.

Foster, M. S. 1987. Feeding methods and efficiencies of selected frugivorous birds. Condor 89: 566–580.

Gautier-Hion, A., Duplantier, J. M., Quris, R., Feer, F., Sourd, C., Decoux, J.-P., Dubost, G., Emmons, L. H., Erard, C., Hecketsweiler, P., Moungazi, A., Roussilhon, C. & Thiollay, J.-M. 1985. Fruit characters as a basis of fruit choice and seed dispersal in a tropical forest vertebrate community. Oecologia 65: 324–337.

Greenberg, R. 1981. Frugivory in some migrant tropical forest wood warblers. Biotropica 13: 215–223.

Herrera, C. M. 1984. A study of avian frugivores, bird-dispersed plants, and their interaction in Mediterranean scrublands. Ecol. Monog. 54: 1–23.

Herrera, C. M. 1985. Habitat – consumer interactions in frugivorous birds. In: Cody, M. (ed.), Habitat selection in birds, pp. 341–365. Academic Press, Orlando.

Herrera, C. M. 1987. Components of pollinator 'quality': comparative analysis of a diverse insect assemblage. Oikos 50: 79–90.

Herrera, C. M. 1988. Variaciones anuales en las poblaciones de pajaros frugivoros y su relacion con la abundancia de frutos. Ardeola 35: 135–142.

Herrera, C. M. 1989. Pollinator abundance, morphology, and flower visitation rate: analysis of the 'quantity' component in a plant-pollinator system. Oecologia 80: 241–248.

Herrera, C. M. 1989. Frugivory and seed dispersal by carnivorous mammals, and associated fruit characteristics, in undisturbed Mediterranean habitats. Oikos 55: 250–262.

Herrera, C. M. & Jordano, P. 1981. *Prunus mahaleb* and birds: the high-efficiency seed dispersal system of a temperate fruiting tree. Ecol. Monog. 51: 203–218.

Holthuijzen, A. M. A. & Sharik, T. L. 1985. The avian seed dispersal system of eastern red cedar (*Juniperus virginiana*). Can. J. Bot. 63: 1508–1515.

Hoppes, W. G. 1988. Seedfall pattern of several species of bird-dispersed plants in an Illinois woodland. Ecology 69: 320–329.

Horvitz, C. C. 1981. Analysis of how ant behaviors affect germination in a tropical myrmecochore *Calathea microcephala* (P.&E.) Koernicke (Marantaceae): microsite selection and aril removal by neotropical ants, *Odontomachus*, *Pachycondyla*, and *Solenopsis* (Formicidae). Oecol. 51: 47–52.

Horvitz, C. C. & Schemske, D. W. 1986. Seed dispersal of a neotropical myrmecochore: variation in removal rates and dispersal distance. Biotropica 18: 319–323.

Howe, H. F. 1977. Bird activity and seed dispersal of a tropical wet forest tree. Ecology 58: 539–550.

Howe, H. F. 1979. Fear and frugivory. Amer. Natur. 114: 925–931.

Howe, H. F. 1980. Monkey dispersal and waste of a neotropical fruit. Ecology 61: 944–959.

Howe, H. F. 1981. Dispersal of a neotropical nutmeg (*Virola sebifera*) by birds. Auk 98: 88–98.

Howe, H. F. 1986a. Seed dispersal by fruit-eating birds and mammals. In: Murray, D. (ed.), Seed dispersal, pp. 123–189. Academic Press, New York.

Howe, H. F. 1986b. Consequences of seed dispersal by birds: a case study from Central America. J. Bombay Nat. Hist. Soc. 83 (suppl.): 19–42.

Howe, H. F. 1990. Survival and growth of juvenile *Virola surinamensis* in Panama: effects of herbivory and canopy closure. J. Trop. Ecol. 6: 259–280.

Howe, H. F. & De Steven, D. 1979. Fruit production, migrant bird visitation, and seed dispersal of *Guarea glabra* in Panama. Oecologia 39: 185–196.

Howe, H. F. & Estabrook, G. F. 1977. On intraspecific competition for avian dispersers in tropical trees. Amer. Natur. 111: 817–832.

Howe, H. F. & Primack, R. B. 1975. Differential seed dispersal by birds of the tree *Casearia nitida* (Flacourtiaceae). Biotropica 7: 278–283.

Howe, H. F., Schupp, E. W. & Westley, L. C. 1985. Early consequences of seed dispersal for a neotropical tree (*Virola surinamensis*). Ecology 66: 781–791.

Howe, H. F. & Smallwood, J. 1982. Ecology of seed dispersal. Ann. Rev. Ecol. Syst. 13: 201–228.

Howe, H. F. & Vande Kerckhove, G. A. 1979. Fecundity and seed dispersal of a tropical tree. Ecology 60: 180–189.

Howe, H. F. & Vande Kerckhove, G. A. 1981. Removal of wild nutmeg (*Virola surinamensis*) crops by birds. Ecology 62: 1093–1106.

Hubbell, S. P. & Foster, R. B. 1990. Structure, dynamics, and equilibrium status of old-growth forest on Barro Colorado Island. In: Gentry, A. H. (ed.), Four neotropical rainforests, pp. 522–541. Yale University Press, New Haven.

Izhaki, I. & Safriel, U. N. 1990. The effect of some mediterranean scrubland frugivores upon germination patterns. J. Ecol. 78: 56–65.

Izhaki, I., Walton, P. B. & Safriel, U. N. 1991. Seed shadows generated by frugivorous birds in an eastern mediterranean scrub. J. Ecol. 79: 575–590.

Janzen, D. H. 1981. Digestive seed predation by a Costa Rican Baird's tapir. Biotropica 13 (Suppl.): 59–63.

Janzen, D. H. 1982. Differential seed survival and passage rates in cows and horses, surrogate Pleistocene dispersal agents. Oikos 38: 150–156.

Janzen, D. H. 1983a. Seed and pollen dispersal by animals: convergence in the ecology of contamination and sloppy harvest. Biol. J. Linn. Soc. 20: 103–113.

Janzen, D. H. 1983b. Dispersal of seeds by vertebrate guts. In: Futuyma, D. J. & Slatkin, M. (eds.), Coevolution, pp. 232–262. Sinauer Ass., Inc., Sunderland.

Johnson, R. A., Willson, M. F., Thompson, J. N. & Bertin, R. I. 1985. Nutritional values of wild fruits and consumption by migrant frugivorous birds. Ecology 66: 819–827.

Jordano, P. 1982. Migrant birds are the main seed dispersers of blackberries in southern Spain. Oikos 38: 183–193.

Jordano, P. 1984. Seed weight variation and differential avian dispersal in blackberries *Rubus ulmifolius*. Oikos 43: 149–153.

Jordano, P. 1987. Frugivory, external morphology and digestive system in mediterranean sylviid warblers *Sylvia* spp. Ibis 129: 175–189.

Jordano, P. 1988. Diet, fruit choice and variation in body condition of frugivorous warblers in mediterranean scrubland. Ardea 76: 193–209.

Kantak, G. E. 1981. Temporal feeding patterns of some tropical frugivores. Condor 83: 185–187.

Karr, J. R. & Freemark, K. E. 1983. Habitat selection and environmental gradients: dynamics in the stable tropics. Ecology 64: 1481–1494.

Krefting, L. W. & Roe, E. I. 1949. The role of some birds and mammals in seed germination. Ecol. Monog. 19: 269–286.

Lambert, F. R. 1989. Pigeons as seed predators and dispersers of figs in a Malaysian lowland forest. Ibis 131: 521–527.

Larson, D. 1991. Ecology of desert mistletoe seed dispersal. Dissertation, Univ. Illinois at Chicago, Chicago, IL.

Levey, D. J. 1986. Methods of seed processing by birds and seed deposition patterns. In: Estrada, A. & Fleming, T. H. (eds.), Frugivores and seed dispersal, pp. 147–158. Dr W. Junk Publ., Dordrecht.

Levey, D. J. 1987. Seed size and fruit-handling techniques of avian frugivores. Amer. Natur. 129: 471–485.

Levey, D. J. 1988. Treefall gaps in a tropical wet forest and the distribution of understory birds and plants. Ecology 69: 1076–1089.

Lewis, D. M. 1987. Fruiting patterns, seed germination, and distribution of *Sclerocarya caffra* in an elephant-inhabited woodland. Biotropica 19: 50–56.

Lieberman, M. & Lieberman, D. 1986. An experimental study of seed ingestion and germination in a plant – animal assemblage in Ghana. J. Trop. Ecol. 2: 113–126.

Loiselle, B. A. 1990. Seeds in droppings of tropical fruit-eating birds: importance of considering seed composition. Oecologia 82: 494–500.

Loiselle, B. A. & Blake, J. G. 1990. Diets of understory fruit-eating birds in Costa Rica: seasonality and resource abundance. Studies in Avian Biology 13: 91–103.

Malmborg, P. K. & Willson, M. F. 1988. Foraging ecology of avian frugivores and some consequences for seed dispersal in an Illinois woodlot. Condor 90: 173–186.

McDiarmid, R. W., Ricklefs, R. E. & Foster, M. S. 1977. Dispersal of *Stemmadenia donnell-smithii* (Apocynaceae) by birds. Biotropica 9: 9–25.

McKey, D. 1975. The ecology of coevolved seed dispersal systems. In: Gilbert, L. E. & Raven, P. H. (eds.), Coevolution of animals and plants, pp. 159–191. Univ. Texas Press, Austin.

Milton, K. 1981. Food choice and digestive strategies of two sympatric primate species. Amer. Natur. 117: 496–505.

Moermond, T. C. & Denslow, J. S. 1985. Neotropical avian frugivores: patterns of behavior, morphology, and nutrition, with consequences for fruit selection. In: Buckley, P. A., Foster, M. S., Morton, E. S., Ridgley, R. S. & Buckley, F. G. (eds.), Neotropical ornithology. Ornith. Monog. No. 36, pp. 865–897.

Murray, K. G. 1988. Avian seed dispersal of three neotropical gap-dependent plants. Ecol. Monog. 58: 271–298.

Pratt, T. K. & Stiles, E. W. 1983. How long fruit-eating birds stay in the plants where they feed: implications for seed dispersal. Amer. Natur. 122: 797–805.

Reid, N. 1989. Dispersal of mistletoes by honeyeaters and flowerpeckers: components of seed dispersal quality. Ecology 70: 137–145.

Schupp, E. W. 1988a. Factors affecting post-dispersal seed survival in a tropical forest. Oecol. 76: 525–530.

Schupp, E. W. 1988b. Seed and early seedling predation in the forest understory and in treefall gaps. Oikos 51: 71–78.

Schupp, E. W. & Frost, E. J. 1989. Differential predation of *Welfia georgii* seeds in treefall gaps and the forest understory. Biotropica 21: 200–203.

Schupp, E. W., Howe, H. F., Augspurger, C. K. & Levey, D. J. 1989. Arrival and survival in tropical treefall gaps. Ecology 70: 562–564.

Snow, B. K. & Snow, D. W. 1984. Long-term defence of fruit by Mistle Thrushes *Turdus viscivorus*. Ibis 126: 39–49.

Snow, B. K. & Snow, D. W. 1988. Birds and berries. A study of an ecological interaction. T & A D Poyser Ltd., Calton.

Snow, D. W. 1981. Tropical frugivorous birds and their food plants: a world survey. Biotropica 13: 1–14.

Sorensen, A. E. 1981. Interactions between birds and fruit in a temperate woodland. Oecologia 50: 242–249.

Sorensen, A. E. 1984. Nutrition, energy and passage time: experiments with fruit preference in European blackbirds (*Turdus merula*). J. Anim. Ecol. 53: 545–557.

Sork, V. L. 1985. Germination response in a large-seeded neotropical tree species, *Gustavia superba* (Lecythidaceae). Biotropica 17: 130–136.

Sork, V. L. 1987. Effects of predation and light on seedling establishment in *Gustavia superba*. Ecology 68: 1341–1350.

Thomas, S. C. 1991. Population densities and patterns of habitat use among anthropoid primates of the Ituri forest, Zaire. Biotropica 23: 68–83.

Trainer, J. M. & Will, T. C. 1984. Avian methods of feeding on *Bursera simaruba* (Burseraceae) fruits in Panama. Auk 101: 193–194.

Vazquez-Yanes, C. & Orozco-Segovia, A. 1990. Ecological significance of light controlled seed germination in two contrasting tropical habitats. Oecologia 83: 171–175.

Webb, S. L. & Willson, M. F. 1985. Spatial heterogeneity in post-dispersal predation on *Prunus* and *Uvularia* seeds. Oecologia 67: 150–153.

Wendelken, P. W. & Martin, R. F. 1987. Avian consumption of *Guaiacum sanctum* fruit in the arid interior of Guatemala. Biotropica 19: 116–121.

Wheelwright, N. T. 1988. Four constraints on coevolution between fruit-eating birds and fruiting plants: a tropical case history. In: Ouellet, H. (ed.), Acta XIX Congressus Internationalis Ornithologici, pp. 827–845.

Wheelwright, N. T. 1991. How long do fruit-eating birds stay in plants where they feed? Biotropica 23: 29–40.

Wheelwright, N. T. & Orians, G. H. 1982. Seed dispersal by animals: contrasts with pollen dispersal, problems of terminology, and constraints on coevolution. Amer. Natur. 119: 402–413.

Wheelwright, N. T., Haber, W. A., Murray, K. G. & Guindon, C. 1984. Tropical fruit-eating birds and their food plants: a survey of a Costa Rican lower montane forest. Biotropica 16: 173–192.

Whelan, C. J., Willson, M. F., Tuma, C. A. & Souza-Pinta, I. 1991. Spatial and temporal patterns of postdispersal seed predation. Can. J. Bot. 69: 428–436.

Willson, M. F. 1986. Avian frugivory and seed dispersal in eastern North America. Curr. Ornith. 3: 223–279.

Willson, M. F. 1988. Spatial heterogeneity of post-dispersal survivorship of Queensland rainforest seeds. Aust. J. Ecol. 13: 137–146.

Willson, M. F. & Whelan, C. J. 1990. Variation in postdispersal survival of vertebrate-dispersed seeds: effects of density, habitat, location, season, and species. Oikos 57: 191–198.

Vegetatio **107/108**: 31–55, 1993.
T. H. Fleming and A. Estrada (eds). Frugivory and Seed Dispersal: Ecological and Evolutionary Aspects.

The population-dynamic functions of seed dispersal

D. L. Venable[1] & J. S. Brown[2]
[1] *Department of Ecology and Evolutionary Biology, University of Arizona, Tuscon, AZ 85721, USA;*
[2] *Department of Biological Sciences, University of Illinois, Box 4348, Chicago, IL 60680, USA*

Keywords: Allocation costs, Bet-hedging, Breeding systems, Density dependence, Dispersal, Metapopulations, Patch structure, Sib competition

Abstract

We summarize some of the population-dynamic consequences of the mosaic structure of plant populations for the evolution of seed dispersal. A fairly elaborated set of theoretical ideas exist regarding the evolution of dispersal and we have synthesized some of them in an attempt to make them more accessible to field ecologists. We consider the relationship of these general theoretical ideas to our understanding of fruit and seed dispersal.

We develop three related models to describe the similarities and differences in how dispersal functions for risk reduction (bet hedging), escaping the negative consequences of crowding, and escaping high concentrations of relatives. We also briefly discuss directed dispersal as a fourth population-dynamic aspect of dispersal. Dispersal can have a risk-reducing function only when there is global (metapopulation) temporal variance in success. Dispersal to escape the negative consequences of crowding requires only spatial and local temporal environmental variation. Dispersal for escaping high concentrations of relatives requires no environmental variation, but does require genetic population structure. Directed dispersal, defined as non-random into particular patch types contingent on the expectation of local success, is always valuable when possible and represents an advantage independent the others which can occur with random dispersal.

In an effort to accommodate for the differences between simple mathematical models and the behavior of complex natural fruit and seed dispersal systems we have discussed the following issues: actual patterns of patch structure and dispersal distance; the implications of plant cosexuality, perenniality, and allocation costs of dispersal structures; and the impact of the detailed nature of density dependence, breeding systems, and genetic structure. We briefly compare the population-dynamic functions of dispersal presented here with the widely cited functions of colonization, escape, and directed dispersal. Finally, we suggest how the theoretical models can be used with field data to estimate the fitness consequences of dispersal.

Introduction

As pointed out by John Harper (1977), an advantage of studying the population biology of plants is that they sit in one place and 'wait to be counted'. Yet this same fixity of position is a double-edged sword. Rather than experiencing the average habitat quality of some potentially communal home range, plants experience the abiotic conditions of the local safe site or patch where they germinated. Likewise, rather than interacting with a variety of individuals and experiencing the average density of their population, plants interact with specific individuals within a canopy or root crown diameter (Silander & Pacala 1990). Even different spatial distributions with the same overall density can have dramatically different consequences for population dynamics (Harper 1961; Bergelson 1990). Thus, the 'other edge' of fixity of position is that an understanding of plant population dynamics requires a understanding of the patchy spatial distributions of plants. Central to this is the phenomenon of seed dispersal.

To appreciate the status of research on the population-dynamic functions of seed dispersal, it is instructive to compare progress in this field to that in other branches of plant reproductive ecology. Plant reproductive ecology encompasses the areas of breeding systems, pollination, seed dispersal, dormancy and establishment, and possibly vegetative reproduction (though many prefer to consider the latter as simply a form of growth). Breeding system and pollination ecology have grown rapidly in the last 15 years due to increased interest in experimental plant population biology and an application of insights from ecological and evolutionary theory to the topic of plant reproduction (Lovett Doust & Lovett Doust 1988). The theoretical developments include a careful use of 'selection thinking', often phrased in terms of the ESS concept, and of ideas derived from the cost-of-sex concept, sex-ratio theory, and Bateman's principle (Charnov 1982; Lloyd 1983). The evolutionary ecology of seed and fruit dispersal has remained mainly at the natural history level with considerable interest generated by plant-animal interactions and patterns seemingly explained by coadaptation (Murray 1986; Estrada & Fleming 1986). The integration of theoretical and experimental approaches aimed at understanding fitness implications of evolutionary strategies has been less prominent in the seed dispersal literature than in the pollination and breeding system literature (but see Alvarez-Buylla & García-Barrios 1991; Olivieri & Gouyon 1985; and Horvitz & Schemske 1986 for some work in this direction). The reasons for the differences in approach and rates of progress in these two branches of plant evolutionary ecology are no doubt varied. The fitness consequences of seed dispersal is an inherently complicated issue involving spatio-temporal patterns of microhabitat suitability (Cohen & Levin 1991) with no simple 'target' of the dispersal event (c.f. the 'target' of pollination – to land on a compatible stigma [Wheelwright & Orians 1982]). Also there are fewer simple rules of the kind that have been so useful in breeding system theory, such as the equalization of male and female fitness at the population level (Lloyd 1983).

We have at present a fairly well-elaborated, if somewhat scattered and diffuse, theoretical literature on how patterns of spatial and temporal heterogeneity favor the evolution of greater or lesser dispersibility (Johnson & Gaines 1990). Here we tie some of these diverse threads together into a more unified story and explore how this theory can be used to understand and investigate seed and fruit dispersal. A closer interaction of theoretical and experimental population-dynamic approaches may spur a period of more rapid advance in our understanding of seed dispersal similar to that seen for pollination and breeding systems ecology.

A few years ago when writing a review of the evolutionary ecology of seed banks, Venable (1989) was able to summarize the results of research on dormancy strategies by referring in a general way to a few intuitive formulae. We intended to take a similar approach in the present chapter, but immediately ran into trouble. First of all, the population-dynamic consequences of dispersal are more complicated than those of dormancy. While dormancy can be explored in models of spatially homogeneous environments experiencing temporal variation, the very nature of dispersal requires a consideration of population subdivision and the details of spatio-temporal variation. Thus as recently as 15 years ago John Harper (1977) was able to accurately state that, 'Attempts to determine the fittest dispersal strategies for specified environmental regimes have so far proved too complex to be handled other than through numerical experiments on digital computers.' Today this is no longer true and many insights can be derived from a few simple analytical formulae. However, because of the inherent complications of dispersal models, the origin and logic behind them are less intuitive at first glance than for the models of dormancy. Finally, the three basic models to which we refer are probably less accessible elsewhere to the empirical ecologist interested in fruit and seed dispersal. As a result we have decided to provide a little more substance to the theoretical synthesis presented here.

We will begin by presenting three related models which explore three basic population-dynamic functions of dispersal for organisms in spatially structured populations, and then discussing a fourth function. We will use simple analytical models in an attempt to shed some light on the general nature and function of dispersal. For the sake of simplicity and hopefully clarity we will model each function in isolation to clearly distinguish its effects from those of other functions. Later in the chapter we will return and ask what are the consequences of relaxing some of the more important assumptions of the models and hopefully, in this manner, begin to get a feel for the importance of these processes in fruit dispersal systems. The four functions of dispersal that we will explore or discuss are risk reduction or bet hedging, escaping the negative consequences of crowding, escaping the negative consequences of high concentrations of relatives, and directed dispersal. These functions contrast somewhat with the widely cited functions of dispersal: colonization, escape from the vicinity of the parent, and directed dispersal (Howe & Smallwood 1982). We will briefly discuss these contrasts later in the chapter.

Models

In each model we concern ourselves with the aggregate behavior of a 'metapopulation' (*sensu* Levins 1970) structured into a set of separate 'cells' or 'patches'.

A risk-reduction model

In this model we only consider density-independent processes, thus we exclude the possibility of selective effects due to escape from crowding or interactions with relatives. We consider n patches, each of equal area. Conditions in each patch vary over time and we will assume, at least initially, that the variability in each patch is independent of variation in the others and can be described by the same random variable. Most of these assumptions can be easily relaxed but they are useful for examining the main points.

A recursive formula gives the number of individuals in a single patch, j, at time $t+1$ as a function of the number of seeds in patch j and in all of the other patches at time t,

$$N_{j(t+1)} = (1 - D)N_{j(t)}S_{ij} + \frac{Da}{n} \sum_{k=1}^{n} N_{k(t)} S_{ik} . \quad (1)$$

In (1), D is the 'dispersal fraction' i.e. the proportion of the seeds produced by an adult plant that disperse out of the home patch and into a 'dispersal pool'. For simplicity, we assume an annual life history so that S_{ij} describes the 'success' of seeds at reproducing (i.e. the product of their survival and fecundity) in patch j experiencing conditions i. Thus $i = 1, \ldots, m$ indexes a random variable responsible for the independent variation in each of the patches. The seeds in a patch come from 2 sources: *in situ* reproduction, and dispersal into the patch. Seeds leaving the home patch enter a dispersal pool from which seeds fall evenly among all patches (i.e. an equal proportion of dispersing seeds ($1/n$) goes to each patch). It is assumed that dispersing seeds experience some cost of dispersal indexed by $a \in [0, 1]$ (i.e. dispersal is cost-free if $a = 1$). This formulation provides a caricature of the dispersal process that captures its essence though in a somewhat exaggerated fashion (i.e. the idea that dispersal involves sampling and averaging over a variety of sites).

We will briefly sketch 2 simple cases that illustrate most of the important points: no dispersal and 100% dispersal. With no dispersal, (1) reduces to

$$N_{j(t+1)} = N_{j(t)} S_{ij} \,. \tag{2}$$

The per capita growth rate in patch j will equal

$$\lambda_{ij} = \frac{N_{j(t+1)}}{N_{j(t)}} = S_{ij} \,. \tag{3}$$

The per capita growth rate for the metapopulation will equal the sum over all patches of seeds at time $t+1$ divided by the sum over all patches of seeds produced at time t,

$$\lambda_{\text{meta}} = \frac{\sum_{j=1}^{n} N_{j(t)} S_{ij}}{\sum_{j=1}^{n} N_{j(t)}} = \sum_{j=1}^{n} \left(\frac{N_{j(t)} S_{ij}}{\sum_{j=1}^{n} N_{j(t)}} \right)$$

$$= \sum_{j=1}^{n} \rho_j S_{ij} = \sum_{j=1}^{n} \rho_j \lambda_{ij} \tag{4}$$

where ρ_j equals $N_{j(t)}/\Sigma_j N_{j(t)}$, the proportion of the global seed population that is found in patch j.

Equation (4) is the formula for the per capita growth rate of the whole metapopulation for a single year. But since the conditions in each patch vary from year to year, this growth rate will vary from year to year, and thus we must consider its average over time. Since population growth is multiplicative over time, we must use a multiplicative average, i.e. the geometric mean. The geometric mean (also population growth generally) is variance sensitive, as is illustrated in the simple example in Table 1. In fact, the geometric mean can be approximated as a variance-discounted arithmetic mean (Real 1980).

$$GEO(\lambda) \approx E(\lambda) - k\sigma^2(\lambda) \tag{5}$$

where $GEO(\cdot)$ stands for 'the geometric mean of' and $E(\cdot)$stands for the 'the expected value of' or 'arithmetic average of' and k is a constant. This expression states that the geometric mean (temporal average) of the population growth rate is smaller that the arithmetic mean growth rate by an amount proportional to the magnitude of the year-to-year variance in growth rate. 'Bet hedging' refers to a sacrifice of $E(\lambda)$ to increase fitness by reducing $\sigma^2(\lambda)$. Since this term has been used quite loosely in the literature to refer to a number of situations which do not involve a tradeoff between mean and variance (Seger & Brockmann 1987), we will use the term 'risk reduction' to refer to bet hedging in this more technical sense.

Table 1. An example demonstrating that fitness is sensitive to temporal variance.

	Demographic success	
	Year 1 2 3	per capita growth after 3 years
Phenotype A*	2 × 5 × 8	= 80
Phenotype B	4 × 4.5 × 5	= 90

* In this example, the phenotype with higher mean and variance of success (A) actually had lower growth than the one with lower mean and variance (B).

We can understand the risk-reducing consequences of dispersal by understanding the effect of dispersal on the tradeoff between mean and variance of λ_{meta}. In what follows we rely on some of the simple rules of probability for expected values and variances which the reader may need to review to follow this section in detail (see e.g. Ostle & Mensing 1975). The expected value of λ_{meta} is

$$E(\lambda_{\text{meta}}) = E\left(\sum_{j=1}^{n} \rho_j S_{ij}\right) = \sum_{j=1}^{n} \rho_j \mathrm{E}(\mathrm{S}_{ij})$$

$$= E(S_i) \sum_{j=1}^{n} \rho_j = E(S_i) \,. \tag{6}$$

Note that the expected value of the per capita growth of the metapopulation is the same as the expected value of the per capita growth rate of a particular patch

$$E(\lambda_{ij}) = E(S_{ij}) = E(S_i) \,. \tag{7}$$

While the variance of a single patch is $\sigma^2(S_i)$, the variance of the metapopulation is

$$\sigma^2(\lambda_{\text{meta}}) = \sigma^2\left(\sum_{j=1}^{n} \rho_j S_{ij}\right)$$

$$= \sum_{j=1}^{n} \sum_{k=1}^{n} \rho_j \rho_k \sigma(S_{ij} S_{ik})\,. \qquad (8)$$

where $\sigma(x, y)$ refers to the covariance of x and y. Since patches are assumed to be independent and S is identically distributed across patches, this reduces to

$$\sigma^2(\lambda_{\text{meta}}) = \sum_{j=1}^{n} \rho_j^2 \sigma^2(S_{ij}) = \sigma^2(S_i) \sum_{j=1}^{n} \rho_j^2\,. \qquad (9)$$

The term $\Sigma\rho_j^2$ is smaller the more patches there are and the more similar the ρ_j's are (remember ρ_j's are the proportion of the global seed population found in each patch). The intuition behind this can be understood by recognizing that $\Sigma\rho_j^2$ is an information statistic, biologically analogous (and mathematically equal to) the reciprocal of Simpson's diversity index. Remember that a diversity index is higher the more species there are and the more evenly distributed their abundances are. Similarly, $\Sigma\rho_j^2$ is lower the more patches there are and the more evenly distributed individuals are among patches.

Thus the conclusion for the density-independent risk-reduction model with zero dispersal are: 1) the expected value of λ is the same for one patch or many, but; 2) the variance of λ_{meta} (the global temporal variance) is a declining function of patch number and of the evenness of the spread of individuals among patches; 3) since the geometric-mean population growth can be expressed as a variance-discounted arithmetic mean, and the arithmetic mean is the same for 1 patch or many, the long-term-average population growth will be higher for metapopulations with more patches and with seeds more evenly spread among patches.

Now consider what happens with 100% dispersal. Since the model assumes uniform dispersal and equal patch size, an equal number of dispersing seeds lands in each patch. With 100% dispersal, all patches will have the same growth rate in a given year (which comes entirely from the global dispersal pool). Thus the metapopulation growth rate equals the local patch growth rate.

$$\lambda_{ij} = \frac{a}{n} \sum_{j=1}^{n} S_{ij};\ \lambda_{\text{meta}} = \frac{1}{n} \sum_{j=1}^{n} \left(\frac{a}{n} \sum_{j=1}^{n} S_{ij}\right)$$

$$= \left(\frac{a}{n} \sum_{j=1}^{n} S_{ij}\right) \sum_{j=1}^{n} \frac{1}{n}$$

$$= \frac{a}{n} \sum_{j=1}^{n} S_{ij} = \lambda_{ij}\,. \qquad (10)$$

The mean and the variance of the growth rate are

$$E(\lambda_{\text{meta}}) = E(\lambda_j) = E\left(\frac{a}{n} \sum_{j=1}^{n} S_{ij}\right)$$

$$= a \sum_{j=1}^{n} \frac{E(S_i)}{n} = aE(S_i) \sum_{j=1}^{n} \frac{1}{n}$$

$$= aE(S_i) \qquad (11)$$

$$\sigma^2(\lambda_{\text{meta}}) = \sigma^2(\lambda_j) = \sigma^2\left(\sum_{j=1}^{n} \frac{a}{n} S_{ij}\right)$$

$$= \sum_{j=1}^{n} \left(\frac{a}{n}\right)^2 \sigma^2(S_{ij}) = \frac{a^2 \sigma^2(S_i)}{n}\,. \qquad (12)$$

Compared to the case with no dispersal, with 100% dispersal the expected value of λ is lower by an amount determined only by the cost of dispersal (if dispersal is cost-free, the mean is the same for 0 and 100% dispersal).

Thus in this density-independent model, the only way dispersal can be of benefit is via reduction in the global temporal variance and the positive effect this has on geometric-mean fitness. The variance terms for 0 and 100% dispersal are similar and, aside from the cost of dispersal term, differ only in the term describing the distribution of seeds among patches. With 100% dispersal, seeds are distributed uniformly such that $\Sigma_j\rho_j^2$

takes on its lowest possible value for a given number of patches, $1/n$ (analogously, a Simpson's index takes on its highest possible value for a given number of species [maximum 'information'] when all species have equal abundance).

In summary, in this risk-reduction model, dispersal decreases arithmetic mean fitness. Thus, dispersal can only evolve for risk reduction if there is global (metapopulation) temporal variance and if the reduction in global temporal variance brought about by dispersal more than compensates for the decrease in arithmetic-mean fitness. Global temporal variance declines with the number of patches and with a more even distribution of seeds among patches. The risk-reducing function of dispersal is to reduce global temporal variance at the expense of lower arithmetic-mean fitness by spreading seeds more evenly among patches.

Escaping the negative consequences of crowding

Escaping the effects of crowding is a separate factor favoring dispersal that is functionally independent of risk reduction, and arises from different causes. To illustrate this we will construct a model with crowding effects but no risk reduction and use it to explore the evolution of dispersal. We will make a number of simplifying assumptions in order to get a general analytical model which will reveal some general properties of dispersal as a mechanism to escape from crowding. Let there be n patches, each of equal area. Let density dependence be described by a 'constant-' or 'reciprocal-yield law'. Density dependence acts locally within patches and 'total yield' can vary from site to site and from year to year. Thus this model takes into account the local nature of density interactions championed by Harper (1977) and Silander & Pacala (1990). The model is similar to one given by Levin *et al.* (1984).

In density-dependent models, individual fitness depends not only on an individual's own dispersibility, but also on the dispersibility and the resulting population dynamics of the plants with which it interacts. To deal with this context- and frequency-dependent selection, we must either use explicitly genetic models or an ESS (evolutionarily stable strategy) approach. Since, as ecologists, we seldom know the genetic details underlying the phenomena we study, and since population-genetic models are usually more complicated (Taper and Case, 1992), we will use the ESS approach. This permits the incorporation of greater ecological sophistication, maintains mathematical tractability, and in most instances captures the essence of individual-level natural selection (Vincent & Brown 1988).

One simple way to construct a model to explore the influence of dispersal on avoiding crowding is to consider competition for a fixed number of 'safe sites'. We can define a function (known as a fitness-generating function, *sensu* Vincent & Brown 1988) which gives the number of post-competition safe sites occupied by plants with an arbitrary dispersal type, D, (sometimes called the 'mutant' strategy) within a metapopulation dominated by individuals using strategy $\hat{D}$ (the 'resident' strategy). For example, we might assume that seeds compete for safe sites at germination. Alternatively, safe sites could represent canopy sites for adults or neighborhood areas for reproduction. For this model it is critical that the density-dependent effect be local (within patches) but it is not critical which life stage is represented by N (though this does become critical in the next model).

To avoid the evolution of dispersal as a mechanism for risk-reduction, in this model we assume that the number of patches, n, is very large; i.e. $n \rightarrow \infty$. As the number of patches gets large, the variance over time of the per capita rate of increase of the metapopulation approaches zero and the geometric-mean fitness approaches the arithmetic mean. That is, the long-term-average per capita growth rate of the metapopulation is the same as the spatial average across patches in any one year (cf. flip a fair coin once for each of an infinite number of patches and you will get 50% heads each year and a temporal average of 50%).

To avoid the evolution of dispersal as a mechanism for escaping sib interactions, assume that the number of safe sites per patch, N, is sufficiently large that the contribution of any one plant to the seed production of the patch is negligible (an analogous assumption is often made in sex allocation problems to eliminate local mate competition, a related phenomenon, cf. Lloyd 1984). Relaxation of this assumption forms the topic of the next section. Under the assumptions of many patches and many safe sites within a patch, we can now formulate the fitness-generating function.

The expected per capita growth rate of an individual using strategy D can be described by the number of safe sites occupied in the following year for each safe site occupied during the present year. To simplify our analysis and without loss of generality, we assume that all patches have a same number of safe sites. The per capita increase in the number of safe sites can be separated into a contribution from *in situ* reproduction and a contribution from seeds dispersing out of the patch. We will assume 'fair-lottery density dependence' (i.e. that safe sites are won in proportion to propagule availability). Thus the probability of an individual 'winning' a particular safe site within its natal patch is given by the proportion of all seed landing in the patch that are produced by the individual. This is given by dividing the number of the individual's non-dispersing seeds, $(1-D)S_i$, by the sum of all non-dispersing seeds produced within the patch, $N(1-\hat{D})S_i$, and all seeds that disperse into the patch from elsewhere, $anN\hat{D}\bar{S}/n$. Note that the seeds dispersing into the patch come from sampling the spatial average of the environment and so the average seed production of a dispersing seed's parent is $\bar{S}$. The total number of safe sites won by the mutant individual within its natal patch, Y_{home}, is the product of this 'proportion won' and the number of safe sites:

$$Y_{\text{home}} = N \cdot \frac{(1-D)\,S_i}{N(1-\hat{D})\,S_i + aN\hat{D}\bar{S}} = \frac{(1-D)\,S_i}{(1-\hat{D})\,S_i + a\hat{D}\bar{S}}. \tag{13}$$

Assuming fair-lottery density dependence, the probability of an individual 'winning' a particular safe site within a non-natal patch is given by the ratio of the number of its dispersing seeds landing in the patch divided by the sum of all non-dispersing seeds produced within the patch, $N(1-\hat{D})S_j$, and all seeds that disperse into the patch from elsewhere, $anN\hat{D}\bar{S}/n$. Assuming uniform dispersal, the total number of dispersing seeds landing in all of the patches experiencing conditions j will be $p_j aDS_i$. Since there are N safe sites to be 'won' in each patch, the dispersing seeds will capture a total of $Np_j aDS_i / (N(1-\hat{D})S_j + aN\hat{D}\bar{S})$ safe sites summed over all the patches experiencing conditions j. Summed over all patches types, the total number of safe sites won by the individual in non-natal patches, Y_{away}, is

$$Y_{\text{away}} = \sum_{j=1}^{m} p_j \cdot \frac{NaDS_i}{N(1-\hat{D})S_j + aN\hat{D}\bar{S}} = \sum_{j=1}^{m} \frac{p_j aDS_i}{(1-\hat{D})S_j + a\hat{D}\bar{S}}. \tag{14}$$

The increase of an individual using strategy D in a particular year will be given by the sum of Y_{home} and Y_{away} , which will vary depending on the conditions, i, experienced in the home patch. Since an arbitrary individual using strategy D might occur in any patch type, $i = 1, \ldots, m$, its *expected* per capita increase in a metapopulation composed of individuals using $\hat{D}$ is given by $\Sigma p_i(Y_{\text{home}} + Y_{\text{away}})$. Thus the fitness-generating function for this model is

$$G(D,\hat{D}) = \sum_{i=1}^{m} p_i \left[\frac{(1-D)S_i}{(1-\hat{D})S_i + a\hat{D}\bar{S}} + \sum_{j=1}^{m} \frac{p_j aDS_i}{(1-\hat{D})S_j + a\hat{D}\bar{S}} \right]$$

$$= \sum_{i=1}^{m} \frac{p_i(1-D)S_i}{(1-\hat{D})S_i + a\hat{D}\bar{S}} + \sum_{j=1}^{m} \sum_{i=1}^{m} \frac{p_j p_i aDS_i}{(1-\hat{D})S_j + a\hat{D}\bar{S}}$$

$$= \sum_{i=1}^{m} \frac{p_i(1-D)S_i}{(1-\hat{D})S_i + a\hat{D}\bar{S}}$$

$$+ \sum_{i=1}^{m} \frac{p_j aD\bar{S}}{(1-\hat{D})S_j + a\hat{D}\bar{S}}$$

$$= \sum_{i=1}^{m} p_i \left(\frac{(1-D)S_i + aD\bar{S}}{(1-\hat{D})S_i + a\hat{D}\bar{S}}\right). \qquad (15)$$

This fitness-generating function is simple and useful in its final form, though it may not be very intuitive at first glance. It is the formula for the expected per capita rate of increase in the number of safe sites occupied by an individual using an arbitrary dispersal strategy D in a metapopulation with strategy $\hat{D}$ where all plants experience fair-lottery density dependence in local patches. The derivation tells us that this proportional increase in safe sites can be calculated simply as an average of the ratio of the reproductive success that a mutant and resident can expect to have in each patch type. The numerator of this ratio can be thought of as the non-dispersing seeds produced locally by the mutant strategy plus the dispersing seeds of the mutant sampled from all patch types. The denominator is the same for the resident strategy. Thus the fairly complicated density-dependent rate of increase we desire reduces to a fairly intuitive average of the ratio of per capita mutant and resident reproductive success. This suggests a simple way of calculating expected fitness for different dispersal strategies from easily-collected field data, which we will discuss in a subsequent section of this paper.

When the average of this ratio over all patch types is greater than 1, the dispersal type D will increase in frequency. In the bet-hedging model of the previous section, dispersal reduced variance at the expense of the cost of dispersal. In the present model, spatial and local temporal variances are being reduced rather than global temporal variance (which we have explicitly assumed away by making a very large number of patches). The selective force here is to escape from the negative effects of density which are created by spatial variation in success. With no dispersal, there will be much variance in density between patches prior to the action of density dependence. Even the uniform (non-directed) dispersal used in these models will tend to move seeds from areas of high density to areas of low density. This is because in any given year, most mutant seeds are produced in areas that will have high seed density in the following year since resident plants do well at the same times and places as mutants. Thus, if a mutant disperses more seeds than the resident, it can have high per capita increases when it arrives in sites with low *in situ* success. Dispersal is a way of averaging across variable patches, which allows plants to escape the negative effects of spatial variation, but plants must pay the cost of dispersal. Increased dispersal evolves when the benefits of landing in patches where residents did poorly (because of spatial and local temporal environmental variation) outweigh the cost of dispersal. Some simple numerical examples to illustrate this are given in Table 2.

The examples in Table 2 also illustrate that the fitness of a given dispersal type, D, depends the dispersal strategy of others, $\hat{D}$ (compare cases 2 and 3). In such situations we can use the ESS concept to find the fitness-maximizing dispersibility. An ESS is a strategy which, when common, cannot be invaded by rare alternative strategies (Maynard-Smith & Price 1973). The ESS problem can be conceived of informally as follows (Fig. 1). The dispersal strategy D represents the strategy of an individual in a metapopulation of individuals using $\hat{D}$. Natural selection should favor the strategy D that maximizes fitness in this context. As long as there is genetic variance for dispersibility, such selection will, in turn, change the dispersal strategy of the population, $\hat{D}$. This process of shifting $\hat{D}$ towards the dispersal strategy that maximizes individual fitness (which may also be changing as $\hat{D}$ changes) should continue until the two coincide (Fig. 2). At this point (the ESS) there will be no further selection for change because the dispersal strategy that maximizes the fitness of the individual is the same as the population's dispersal strategy. Any other rare alternative dispersibility will have lower fitness (at the

Table 2. Some examples with the model for escaping the negative effects of crowding showing how dispersal can increase fitness by reducing spatial variance[a].

	Patch types				
Case	1	2	3	Mean[b]	
1	8/8	5/5	3/3	1	Mutant and resident are identical
2	5/8	5/5	5/3	1.1	Greater dispersal of the mutant provides cost-free variance reduction (dispersal invades).
3	5/9	5/5	5/1	2.1	Greater variance in the success of the less dispersive resident (dispersal invades).
4	4/8	4/5	4/3	0.88	Cost of dispersal is too high for the variance reduction achieved.
5	4.56/8	4.56/5	4.56/3	1	Break-even cost of dispersal.

[a] The formula for this model calculates the average of the ratio for each patch type of mutant over resident demographic success $\Sigma p_i(M/R)$ where mutant success is given by $M = (1 - D)S_i + aD\bar{S}$ and resident success is given by $R = (1 - \hat{D})S_i + a\hat{D}\bar{S}$. (see equation (15) and its explanation in the text). For these examples assume three equiprobable patch types, $p_1 = p_2 = p_3 = 1/3$ and that the resident strategy is non-dispersal, i.e. $\hat{D} = 0$. For case 1 the mutant strategy is non-dispersal ($D = 0$) but for the other cases the mutant individual disperses all of its seeds ($D = 1$). Note that complete dispersal of the seeds of the mutant individual eliminates variance in the numerator. In case 2 and 3, this is done at no cost, but in cases 4 and 5, the cost of dispersal reduces the mean success of the mutant individual.

[b] The mean is the mutant fitness defined as the *per capita* increase in number of 'safe sites' occupied by the mutant.

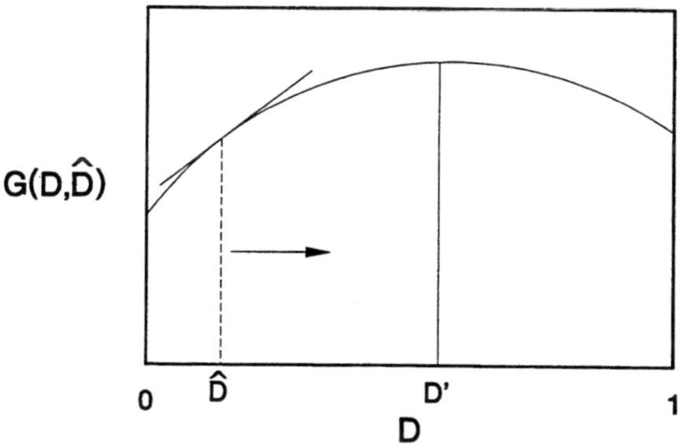

Fig. 1. $G(D, \hat{D})$, the fitness of a 'mutant' individual with dispersal D is plotted against D for a given dispersal fraction $\hat{D}$ in the rest of the metapopulation. In the example illustrated, individuals with dispersibility slightly higher than $\hat{D}$ have higher fitness than the rest of the metapopulation (as indicated by the positive fitness slope at $\hat{D}$ and the position of the fitness maximizing mutant dispersibility, D' to the right of $\hat{D}$, The arrow indicates the direction of selection on $\hat{D}$. The shape of the mutant fitness curve may change as $\hat{D}$ evolves.

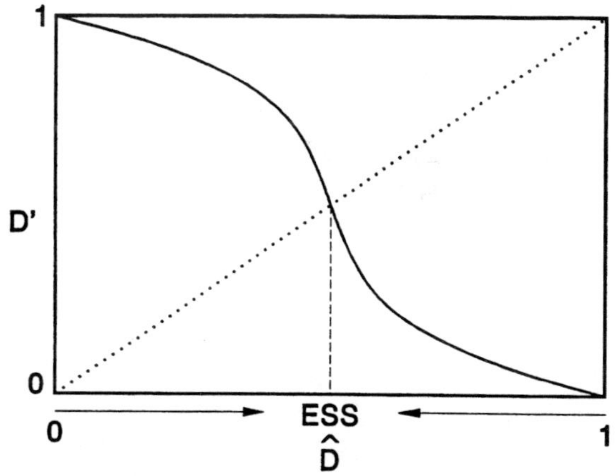

Fig. 2. A hypothetical plot of dispersibility that maximizes individual fitness (D') for the full range of possible dispersibilities in the metapopulation ($\hat{D}$). For any dispersibility to the left of the ESS, selection will favor a shift to the right (indicated by arrow) because the dispersibility that maximizes individual fitness is greater than the resident dispersibility. Likewise, for any resident dispersibility to the right of the ESS, selection will favor a shift to the left because the individual-fitness-maximizing dispersibility is lower. The ESS occurs where the 2 coincide.

ESS there is no selection for a unilateral change in strategy). Computationally, the ESS can be found by applying an ESS maximum principle (Vincent & Brown 1988) and setting the derivative of the fitness-generating function with respect to the individual's dispersibility, D, equal to zero (to determine the dispersibility that maximizes an individual's fitness). The solution to the resulting expression must then be found for the situation in which $D = \hat{D} = D^*$, the ESS dispersibility.

Before solving for the ESS we will ask a related question: what is required for dispersal to be of any value at all in this escape-from-density system, i.e., under what circumstances will dispersal be favored in a metapopulation without dispersal? This is the same as asking, when is $\partial G/\partial D\,|_{D=\hat{D}=0} > 0$ (c.f. Fig. 1)? Calculating the derivative gives:

$$\frac{\partial G}{\partial D} = \sum_{i=1}^{m} p_i \left(\frac{-S_i + a\bar{S}}{S_i(1-D_r) + aD_r\bar{S}} \right)$$

$$\left.\frac{\partial G}{\partial D}\right|_{D=\hat{D}=0} = a \sum_{i=1}^{m} p_i \frac{\bar{S}}{S_i} - 1\,. \quad (16)$$

Thus $\partial G/\partial D\,|_{D=\hat{D}=0} > 0$ requires that

$$a \sum_{i=1}^{m} p_i \frac{\bar{S}}{S_i} > 1 \Rightarrow a\bar{S} \sum_{i=1}^{m} p_i \frac{1}{S_i} > 1 \Rightarrow \frac{a\bar{S}}{H(S_i)} > 1 \quad (17)$$

where $H(S_i)$ is the harmonic mean of S (the harmonic mean is the reciprocal of the average of reciprocals, i.e. $1/E(1/S_i)$). Thus dispersal will be favored in a metapopulation without dispersal whenever the harmonic mean of success at reproduction is less than the dispersal-cost-discounted average success. As long as there is any variation, the harmonic mean is lower than the geometric mean which is lower than the arithmetic mean. These differences among means disappear in the absence of variance and increase with increasing variance. Thus in the absence of variation in S_i (spatial and local temporal variance) dispersal cannot evolve in this model. The large the variance, the easier it is for (17) to be true and consequently the easier it is for a dispersing morph to invade. Also, the greater the cost of dispersal (lower a), the harder it is for dispersal to be favored. As long as there is any variance in success, cost-free dispersal would be favored in a resident population with no dispersal. The greater the variance, the greater the cost of dispersal that can be tolerated to obtain improved dispersibility. Thus (17) proves the intuition given above and in Table 2.

The other 'boundary' question is, when, if ever, can dispersal go to fixation, i.e. when will $D^* = 1$ be the ESS? Solving for $\partial G/\partial D\,|_{D=\hat{D}=1} \geq 0$ gives $a \geq 1$. Thus D^* can only be 1 if dispersal is cost-free (i.e. if $a = 1$).

If, for a particular system, we determine that the ESS dispersibility is greater than zero and less than 1, we can calculate it by setting $\partial G|\partial D = 0$ and then substituting $D = \hat{D} = D^*$. This gives the ESS dispersibility implicitly as,

$$\sum_{i=1}^{m} p_i \frac{a\bar{S}}{S_i(1-D^*) + aD^*\bar{S}} = \sum_{i=1}^{m} p_i \frac{S_i}{S_i(1-D^*) + aD^*\bar{S}}\,. \quad (18)$$

The formula can be used to calculate D^* for different parameter values and to explore how changing parameter values changes D^*. A simpler expression for the ESS can be obtained by assuming 2 patch types, $i = 1, 2$:

$$D^* = \frac{p_i}{1 - \frac{a\bar{S}}{S_2}} + \frac{p_2}{1 - \frac{a\bar{S}}{S_1}}\,. \quad (19)$$

Since $\partial D^*/\partial a$ is always positive for this expression, the smaller the cost of dispersal (i.e. the larger the values of a), the higher the ESS dispersibility, all else being equal. The relationship of D^* to the magnitude of environmental variation (i.e. the difference between S_1 and S_2) is less straight forward, though if variance is sufficiently large, D^* will increase with variance. This is not necessarily true when variance is small (cf. Levin *et al.* 1984). The driving force for the evolution of dispersal in this model is escaping spatio-temporal variance. Thus it makes sense that for parameter values at which variance is not very important, other negative factors associated with dispersal may outweigh the variance escaping benefits of dispersal. Such factors include a high cost of dispersal and a high probability that dispersing seed will land in an unfavorable patch.

If there is complete reproductive failure in one of the patch types so that $S_1 = 0$, then (19) reduces to

$$D^* = \frac{p_1}{1 - ap_2}\,. \quad (20)$$

From this equation we see that the greater the probability of complete reproductive failure in patches, the greater the ESS dispersibility (as long as dispersal is not completely cost-free). Also, the greater the cost of dispersal the lower the ESS dispersibility.

We can obtain an even simpler expression by adding the assumption that $p_1 = p_2 = 0.5$: $D^* = 1/(2 - a)$. This means that if the probability of encountering a patch with complete reproductive failure is 0.5, the ESS dispersal is to disperse at least half of the seeds, even if it is almost suicidal to do so (i.e. even if a is very low). The same ESS formula will show up again in the next model, but the two results should not be confused. Here the driving force for the evolution of high dispersal rates is the high probability of total failure for non-dispersing seeds and there are no sib interactions. In the next model, the high dispersal rates evolve to reduce sib interactions in a constant environment.

In summary, because the problem of crowding is inherently frequence-dependent, it should be thought of in ESS terms (i.e. by analyzing the fitness of an individual given a particular average or resident strategy) or else in explicitly genetic terms. In a density-dependent model where crowding is a problem, dispersal can evolve in the absence of the bet-hedging effect which we showed in the previous model to be due to global temporal variance. While, global temporal variance is no longer required, there must be spatial and local temporal environmental variation to make it possible for dispersal to alleviate crowding. Dispersal will evolve when the benefits to a mutant of landing in the uncrowded patches where residents did poorly outweigh the costs of dispersal.

Escaping the negative consequences of high concentrations of siblings

In the previous model bet hedging was eliminated by having no global temporal variance in fitness and sib-interaction effects were eliminated by allowing so many individuals per patch that an individual's seed production represented a negligible part of the patch total. Now we will relax this latter assumption to explore the effects of sib-density interactions (c.f. Hamilton & May 1977; Schoen & Lloyd 1983).

We will start with most assumptions of the previous model such as constant-yield density effects and patches of equal area. To avoid bet hedging we assume many patches. However, we will ensure that the selective factors involved in the previous model do not operate by assuming temporal constancy, i.e. $i = 1$, $S_i = S$.

The per capita growth rate of an individual using strategy D can be described in a fashion similar to that in the previous model focused on crowding. We consider the growth rate of an arbitrary individual with dispersibility, D in a metapopulation in which others have dispersibility, $\hat{D}$. Per capita growth rate in the sum of sites gained in the natal patch, Y_{home}, and sites gained in non-notal patches through dispersal, Y_{away}. In determining Y_{home}, we now include the effect of the individual plant with dispersibility D on the sum of all non-dispersing seeds produced within the patch. The sum of all non-dispersing seeds is now given by: $(1 - D)S + (N - 1)(1 - \hat{D})S$. Assuming fair-lottery density dependence, the probability of an individual 'winning' a particular safe site within its natal patch is given by the ratio of its non-dispersing seed, $(1 - D)S$, divided by the sum of all non-dispersing seeds produced within the patch and all seeds that disperse into the patch form elsewhere, $anN\hat{D}S/n$. The total number of safe sites won by an individual within its natal patch, Y_{home}, is the product of this ratio and the number of safe sites:

$$Y_{\text{home}} = N \cdot \frac{(1 - D)S}{(1 - D)S + (N - 1)(1 - \hat{D})S + aN\hat{D}S} = \frac{N(1 - D)}{(1 - D) + (N - 1)(1 - \hat{D}) + aN\hat{D}}. \tag{21}$$

Assuming fair-lottery density dependence, the probability of an individual winning a particular safe site within a non-natal patch is given by the ratio of the number of its dispersing seeds landing in the patch divided by the sum of all non-dispersing seeds produced within the patch, $N(1-\hat{D})S$, and all seeds that disperse into the patch from elsewhere $anN\hat{D}S/n$. Even though the total number of dispersing seeds produced by one mutant individual, aDS, may be large, we assume that the number landing in any particular non-natal patch is small. Thus it need not be considered in the denominator of the proportion of safe sites captured by the mutant. The total number of safe sites won by this individual within non-natal patches, Y_{away}, is the product of this proportion and the number of safe sites, summed over all patches:

$$Y_{away} = \frac{aDNS}{N(1-\hat{D})S + aN\hat{D}S} = \frac{aD}{(1-\hat{D}) + a\hat{D}}. \quad (22)$$

The per capita growth rate of mutant individual using an arbitrary strategy D in a population composed of individuals using $\hat{D}$ is given by the expected value of the sum of Y_{home} and Y_{away}. This is the fitness-generating function for the sib interaction model:

$$G(D,\hat{D}) = \frac{(1-D)N}{(N-1)(1-\hat{D}) + Na\hat{D} + (1-D)} + \frac{aD}{1-\hat{D}+a\hat{D}}. \quad (23)$$

Remember that, in the ESS is zero dispersal for the environment specified in this model (no local or global temporal or spatial environmental variation). Let us now ask, in this model with sib interactions, under what conditions (if any) will a non-dispersing strategy be invadable by a dispersal strategy with D greater than zero? These conditions are found by solving for $\partial G/\partial D\,|_{D=\hat{D}=0} > 0$:

$$\frac{\partial G}{\partial D} = \frac{-N[(N-1)(1-\hat{D}) + Na\hat{D} + 1 - D] + (1-D)N}{[(N-1)(1-\hat{D}) + Na\hat{D} + 1 - D]^2} + \frac{a}{1-\hat{D}+a\hat{D}}$$

$$\left.\frac{\partial G}{\partial D}\right|_{D=\hat{D}=0} = \frac{-N(N-D) + (1-D)N}{(N-D)^2} + a$$

$$\left.\frac{\partial G}{\partial D}\right|_{D=\hat{D}=0} > 0 \Rightarrow \frac{N - N^2 + a(N-D)^2}{(N-D)^2} > 0$$

which implies,

$$aN^2 > N^2 - N \;\Rightarrow\; N(1-a) < 1. \quad (24)$$

Thus, a non-zero dispersal type can invade a resident non-dispersal type whenever $N(1-a)<1$; otherwise, $D^*=0$ is the ESS. This means that, for dispersal to evolve via sib-density effects, N, the number of safe sites per patch, must be small or the cost of dispersal must be small.

Before solving for the ESS we will explore what happens at the other boundary, $D=1$, by determining under what conditions (if any) dispersal will go to fixation (i.e. when will $D^*=1$ be the ESS?). This is the same as solving $\partial G/\partial D\,|_{D=\hat{D}=1} \geq 0$:

$$\left.\frac{\partial G}{\partial D}\right|_{D=\hat{D}=1} = \frac{-N(Na+1-D) + (1-D)N}{(Na+1-D)^2} + 1 \geq 0. \quad (25)$$

With some algebraic manipulation this reduces to $a \geq 1$ which implies that $D=1$ can only be an ESS (and a neutrally stable one at that) if there is no cost to dispersal. Otherwise, the ESS dispersal will be less than 1 (this is the same condition for $D^*=1$ as for the previous model).

The ESS dispersal fraction can be calculated using the ESS maximum principle. Set $\partial G/\partial D = 0$ for $D=\hat{D}=D^*$ and solve for D^*:

$$\frac{(N-1-ND^*+D^*+NaD^*)}{N(1-D^*+aD^*)^2} = \frac{a}{1-D^*+aD^*}$$

$$D^* = \frac{1-N(1-a)}{1-N(1-a)^2}. \quad (26)$$

which can take on any value between 0 and 1 depending on parameter values (assuming that $N(1-a)<1$; otherwise, the ESS is $D^*=0$).

Thus, under sib competition, dispersal can be favored even in the absence of spatial and local temporal environmental variance. What favors dispersal is variance in how many relatives occur in a particular patch. With small N, the non-dispersing offspring of a single individual may form a large fraction of the individuals interacting within the natal patch. There are diminishing returns that can be achieved in the natal patch by an individual having lower and lower dispersal since the individual would only be gaining a higher and higher probability of winning a limited 'prize' (much as in the case of local mate competition in models of the evolution of sex allocation, see Lloyd 1984). As modelled, there are no diminishing returns to an individual from dispersal. Dispersing seeds land in low densities in patches where the mutant individual did not reproduce. While avoiding diminishing returns in the natal patch, these seeds must pay the cost of dispersal which is an inherent constant disadvantage of dispersal. The ESS dispersal fraction occurs where the returns to an individual from non-dispersal diminish to the point where they equal the constant cost-discounted returns on dispersal. If N is large, returns on non-dispersal diminished less rapidly and complete non-dispersal may be favored unless dispersal is relatively cost-free. If the cost of dispersal is very small (a close to 1) the ESS dispersal fraction declines almost linearly with increasing N (within the permitted range of N given by $N(1-a)<1$).

If $N=1$, $D^*=1/(2-a)$ which is the result obtained by Hamilton and May (1977). In this case the ESS dispersal fraction varies from 1/2 to 1 even if the cost of dispersal is very high (i.e. even if a is very low; this same ESS condition was found in the previous model, but under a very different set of assumptions). However, if N is the slightest bit greater than 1, there will be a maximum cost (minimum a) beyond which $D^*=0$. Also, low dispersal fractions may be the ESS if N is larger than 1. Because of the critical importance of the magnitude of N, the life stage at which density limitation occurs becomes critical in these sib models (since there are likely to be many more seedlings per patch than adults).

In summary, in this sib interaction model, dispersal can evolve in the absence of spatial and local temporal variation, but only if the number of individuals per patch is small or the cost of dispersal is small. If $N=1$, a high dispersal fraction is favored even in the face of virtually suicidal costs involved in dispersal. Yet as long as $N>1$ by even a tiny amount, there is a maximum cost beyond which dispersal will not be favored, and low dispersal fractions may be ESS's. The life stage at which density limitation occurs is critical in these models.

Directed dispersal

We will not say much about directed dispersal other than to relate it to the general modelling scheme presented so far. In the previous models, dispersal was considered to be random with respect to the occurrence of favorable conditions in patches. Directed dispersal refers to the fact that, by attracting the appropriate vector, or having the appropriate morphology, a plant may be able to vary its dispersal into a patch or habitat depending on the likelihood of its success in that patch. There are two aspects to the definition of directed dispersal. First, seeds must land disproportionately in a subset of patch types. Secondly, in order to have 'adaptive' directed dispersal, these patch types must be a favorable subset. The possibility has been discussed extensively in the empirical seed dispersal literature (Howe & Smallwood 1982; Hanzawa *et al.* 1988) and this kind of behavior is regularly considered in models of animal movement (Swingland & Greenwood 1983).

This situation is readily modelled in terms of conditional probabilities. If some cue is available upon which a plant can condition its dispersal into a patch, the plant should use it to the extent that it is a successful predictor of growth conditions. For example, ants should be used as vectors to the extent that ant middens are better-than-average sites for growth (Hanzawa *et al.* 1988), and 'gulpers' (*sensu* Moermond & Denslow 1985) should be used as vectors to the extent that associated fecal matter functions as fertilizer. In the risk-reduction and escape-from-crowding models, directed dispersal implies that D can vary contingent upon some cue (e.g. presence of a gap or ant nest) which is predictive of either poor reproduction in the recent past (and thus indicative of low density in the present) or high reproduction in the near future (indicating high potential success). Theoretically, dispersal into a patch could be contingent on the density of relatives, thus extending the directed dispersal concept to sib-interaction models. Yet it is difficult to conceive of a realistic mechanism to achieve this. If dispersing seeds arrive nonrandomly at a favorable subset of patches, selection for dispersibility will be stronger than otherwise predicted in the previous models. Predictive dispersal is directly analogous to the phenomenon of predictive dormancy (Venable & Lawlor 1980; Venable 1989).

Real seed dispersal systems

In the previous section we presented three heuristic models in an attempt to explain four general population-dynamic functions of dispersal. With these caricatures of the dispersal process we tried to capture the essential population-dynamic properties of dispersal in models that are analytically tractable and simple enough that we can readily grasp the intuition behind the results. Yet an understanding of the population-dynamic consequences of seed and fruit dispersal requires an accommodation for the differences between the assumptions of the relatively simple mathematical models and the behavior of complex natural seed dispersal systems. In the rest of the paper we would like to discuss this accommodation.

Patch structure, dispersal distance, and implications for bet hedging

To understand the fitness consequences of seed dispersal, we must know the actual patterns of spatial and temporal variance in the environments experienced by plants as well as know how dispersal distance corresponds to the scale of spatial and temporal variance. These are big questions for which we only have partial answers at present. Review articles have been written on each of these issues (White 1979; Pickett & White 1985; Willson, this volume; Levin & Kerster 1974).

Ecological patch structure is a species-specific property consisting of biotic and abiotic components. Thus it is most readily explored on a species-by-species basis, focussing on plant responses rather than measuring environmental parameters (the phytometer approach; Antonovics & Primack 1982; Clements & Goldsmith 1924).

Real patch structures are likely to be more or less continuous with spatial and temporal autocorrelations. Such correlations are expected to have strong effects on the fitness consequences of dispersal (c.f. Venable & Brown 1988 for an analysis of their effects on a risk-reduction model of dispersal). Real patch structures are likely to be hierarchical, with different processes occurring at different spatial and temporal scales. For example, solitary treefalls typically occur at small spatial and temporal scales, while hurricanes or epidemics may result in a superimposed patch structure occurring at a much larger spatial and temporal scale.

Mathematical tools are available for succinctly describing how the texture of environmental variation changes at different spatial and temporal scales (e.g. applied fractal geometry; Sugihara & May 1990). There are probably some systems, for example those involving dominant plant species in low diversity communities, for which a good idea of patch structure could be obtained from fractal geometric analyses of aerial photographs (c.f. Krummel *et al.* 1987). Such analyses can be performed on time series data as well, which might be useful for determining how much variation is occurring at different temporal scales in local patches (Sugihara & May 1990). However, since the property undergoing variation is a plant species-specific perception of environmental variation, the immediate problem for most plant demographic systems is the initial collection of the proper data, not its succinct mathematical description.

Real seed dispersal patterns tend to be leptokurtic, often described successfully by a negative exponential distribution, with most seeds travelling on a scale of meters (Willson, this volume). Limited seed dispersal creates an interesting dilemma with respect to the risk-reducing function of dispersal. Remember that dispersal only evolves for risk-reduction in systems in which global temporal variance is important and that global temporal variance declines with patch number. It would seem that risk reduction may not be an important function of seed dispersal for most widespread plant species because of one of the following factors. First, a widespread plant species occurs in so many patches that global temporal variance may be low. How many patches are 'many'? Recall from the risk-reduction model above that, assuming independent patches and uniform plant density across patches, global temporal variance equals the local temporal variance divided by patch number. Under these assumptions, global temporal variance is dramatically reduced with 20–50 patches. Accounting for spatial and temporal autocorrelation of conditions and the uneven distribution of plants among patches, the number of patches necessary to dramatically reduce global temporal variance will increase by some generally unknown amount (though calculable in specific cases).

If global temporal variance is high in a widely distributed species, it is probably generated by processes occurring at a spatial scale not readily travelled by dispersing seeds. For example, weather phenomena, such as El Niño events, are likely to create global temporal variance for many plant species, but the spatial scale of variation is so great that few seeds can be expected to travel the necessary distances (c.f. Ellner & Shmida 1981). Under these circumstances, risk-reducing adaptations are likely to evolve, but the plant traits affected are more likely to be such things as seed dormancy or fire resistance rather than seed dispersibility (see Venable & Brown 1988 for a discussion of the fitness interactions of dispersal, dormancy, and other traits, such as seed size with respect to risk reduction).

These comments refer only to widespread plant species. Seed dispersal might have a very important risk-reducing function in narrowly distributed species, and the role of dispersal in risk reduction should not be ignored when considering conservation measures for such species (c.f. Simberloff 1989).

This discussion leads logically to another possible risk-reducing role of dispersal, i.e. increasing the number of patches by expanding the species range. While a logical possibility, we did not mention this previously because we are not aware of any evidence supporting it. This would be an interesting empirical question to pursue further: do related taxa which differ in dispersibility also differ systematically in the sizes of their range of distribution?

The implications of plant cosexuality for seed dispersal

To measure dispersal distances, most plant ecologists find some technique for measuring the distances each seed travels from its mother (Willson, this volume). The resulting distribution is considered to give the seed dispersal curve for that species. It is important to recognize that this approach only describes dispersal for approximately 1/2 of the offspring of an outcrossing cosexual plant, i.e. those produced through female function. The other 1/2 typically travel much further than we usually think. Seeds resulting from a plant's paternal function fall in the seed shadows of all of the mates of the plant in question and, by definition travel a greater distance and cover a much larger area (Fig. 3; c.f. Schoen & Lloyd 1983). These seeds carry just as many parental genes as the ones actually filled on the parent, and they should be considered when estimating seed dispersal distance as part of any attempt to understand the fitness consequences of seed dispersal.

Since most plant species are cosexual, most plants will produce (and disperse) seeds via both male and female function. This observation is a direct result of the recognition by breeding system ecologists that we have been ignoring male function in estimates of plant fitness (Bertin 1988; Stanton & Galloway 1991). For example, uniparental plants (selfers and asexuals) will only dis-

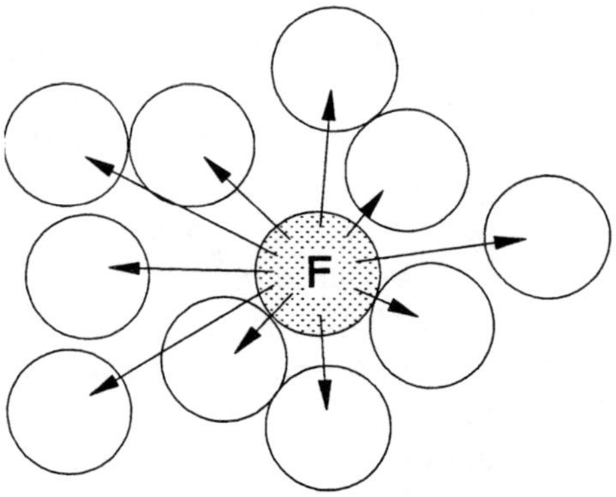

Fig. 3. Dispersal shadows for a plant in question (stippled) and its potential pollen recipients. Seeds produced through female function will land in the stippled seed shadow. Yet seed produced through male function will land in the open shadows and travel farther from the plant in question.

perse seeds to the stippled seed shadow in Fig. 3, whereas males of dioecious populations will only disperse seeds to the outer unstippled shadows. Monomorphic outbreeding cosexuals will disperse roughly 1/2 of their offspring to the stippled circle and 1/2 to the outer circles. Thus, counter to normal assumptions, seed dispersal distances for cosexual plants depend intimately on the nature of pollination and breeding systems.

The population-dynamic consequences of dispersing seeds through male function are different from those for seeds produced via female function in that these 'male' seeds cannot numerically swamp patches in which local seed production is low. This is because the seeds actually fill on the local plants which are experiencing low seed production. The swamping effect of dispersal occurs via paternity, i.e. while few seeds are produced locally, most of them will be fathered by plants from nearby patches undergoing conditions more favorable for reproduction.

The models presented above only considered female function. Technically, they are accurate for plants which have the same dispersal fractions for seeds produced via male and female functions. In practice this is likely to mean agamospermous or selfing species. ESS models could be created that explicitly treat the dispersal of seeds produced through male as well as female function.

The impact of perenniality

The models presented above were framed for simplicity in terms of an annual life cycle. Since most plant species are perennials, we might ask what the likely impacts of perenniality are for the fitness consequences of seed dispersal. We will take a step towards answering this problem by analyzing the simplest case, that of Type II survivorship or constant mortality (cf. Comins 1982). As long as adult survival is less variable than reproductive success, perenniality will buffer variance (cf. Chesson & Huntley 1988). This means that the risk-reducing function of dispersal is likely to be less important for perennial plants than for annuals. The interaction of perenniality and dispersal as adaptations for reducing risk has been analyzed in depth for one form of perenniality, the production of long-lived seed banks (Cohen & Levin 1985; Venable & Brown 1988).

For the models of general density effects and sib-interactions it can be shown that perenniality does not necessarily affect the ESS conditions nor the conditions for invasion. For the general density effects model, let q stand for the mortality of adults, so that $1 - q$ is adult survival. Note that the proportion of safe sites available for colonization or establishment will equal q. The equation (15) becomes

$$
\begin{aligned}
G(D, \hat{D}) &\\
&= \sum_{i=1}^{m} p_i \left((1 - q) + q \, \frac{S_i(1 - D) + aD\bar{S}}{S_i(1 - \hat{D}) + a\hat{D}\bar{S}} \right) \\
&= 1 - q + q \sum_{i=1}^{m} p_i \left(\frac{S_i(1 - D) + aD\bar{S}}{S_i(1 - \hat{D}) + a\hat{D}\bar{S}} \right) \qquad (27)
\end{aligned}
$$

and (16) becomes

$$
\frac{\partial G}{\partial D} = q \sum_{i=1}^{m} p_i \left(\frac{- S_i + a\bar{S}}{S_i(1 - \hat{D}) + a\hat{D}\bar{S}} \right). \qquad (28)
$$

Thus the slope of the adaptive landscape with respect to dispersal will be shallower, the greater adult survivorship is. Yet the conditions for invasion and the ESS conditions will not change since q will cancel out of these conditions (i.e. the peaks and valleys of the adaptive landscape will be in the same place but the topography will be flatter).

The intuition behind the lower slopes is this: greater adult survival increases the importance of survival adaptations (competitive ability, etc.) and reduces the importance of reproduction and dispersal. ESS conditions do not change because sites occupied by adults will be unavailable for establishment, but within the fewer available safe sites for establishment, the same dynamics will be going on (e.g. the same proportions of mutant and resident seeds) and the need for dispersal will be the same.

Perenniality can be introduced into the sib-interaction model as follows:

$$G(D, \hat{D}) = 1 - q + \frac{q\,N(1 - D)}{(N - 1)(1 - \hat{D}) + Na\hat{D} + (1 - D)} + \frac{q\,aD}{1 - \hat{D} + a\hat{D}}. \quad (29)$$

If N is large, this model can be conceived of as indicating that in each patch, a proportion $1 - q$ of the adults survive to the next year. If N is small (e.g. if $N = 1$), the same model could indicate that the adults survive in a proportion $1 - q$ of the patches and die in a proportion q of the patches.

In this model, all terms containing D, the dispersibility of a mutant individual, are multiplied by q. Thus, once again, all terms in the expression for $\partial G/\partial D$ will contain q thus the slope of mutant fitness with respect to a change in mutant dispersibility will be shallower for a perennial than for an annual by an amount given by q. Once gain, q cancels out of the ESS and invasion conditions.

If adults survive from one year to the next there will be fewer sites in a patch that are available for establishment. Since sib-density effects are more important when the number of sites per patch is small, it might at first seem counter-intuitive that perenniality does not favor increased dispersal in a sib-interaction model. Yet, in sib models, a low number of safe sites has its impact on the proportion that seeds of the mutant individual constitute of all seeds competing for safe sites in the patch. This proportion will be determined by the total number of sites in which seeds are produced in a patch and not by the number of safe sites available for establishment.

We can conclude this analyse by stating that perenniality, by reducing variance, will tend to reduce the importance of the risk-reducing function of dispersal. Yet perenniality *per se* does not necessarily change the ESS dispersibility for escaping crowding or sib interactions. Perenniality does reduce the fitness impact of a change in seed dispersibility, simply because seed reproduction becomes a less important vehicle for persistence from one year to the next.

Several caveats should be mentioned. This analysis of the effects of perenniality was done on an 'all-else-being-equal' basis. Other complications can occur. For example, it might be reasonable in some cases to assume that perennials are larger than annuals, but that patch size is the same. This would result in fewer safe sites per patch (lower N) and thus stronger sib-density effects favoring greater dispersal.

Other complications are likely to arise with Type I or Type II survivorship curves which must be analyzed with age- or stage-specific matrix models. For example, the ESS dispersibility with a Type I survivorship curve would probably be to disperse most seeds produced at early ages but to manufacture a declining dispersal fraction as the probability of dying increases. Similarly, the ESS might be to increase dispersibility with age for a Type III curve. It would be interesting to examine empirical examples suggesting age-specific shifts in dispersal patterns in light of these ideas. There clearly are age-specific changes in dispersibility in some species due to changes in plant height (e.g. wind, adhesive, and ballast systems) and fruit crop size (e.g. frugivore systems). Yet the most obvious such changes are more likely to be proximate correlates of growth rather than adaptive shifts in dispersibility as a function of changing probabilities of mortality. Nevertheless, the possibility of adaptive shifts should not be discarded out of hand.

Another potential complication is that, by creating habitat texture (i.e. vacant and occupied sites), perenniality creates a substrate for directed or contingent dispersal. If mechanisms are available to direct dispersing seeds to unoccupied microsites, then greater dispersibility could be favored for perennials. But this would be due to greater predictive dispersal in the perennial system not to perenniality *per se*. The above analysis assumes that dispersal is random with respect to site favorability.

Two kinds of dispersal cost

Most models of dispersal in the theoretical literature represent the cost of dispersal in a form that is not fully applicable to the seed and fruit dispersal problem. As usually modelled, only dispersing individuals pay the cost of dispersal (but see Cohen & Motro 1989). This is a reasonable representation of the mortality cost of dispersal (e.g. the production of seeds destroyed during passage through a vertebrate gut, or eaten by a scatterhoarder). But, the allocation costs of dispersal, the proportion of fruit pulp, adhesive barbs, wind-borne parachutes, etc. are typically paid by dispersing and non-dispersing fruits alike (an exception can be made for those dispersal polymorphisms in which only dispersing morphs have dispersal structures, e.g. Venable & Levin 1985 and Venable *et al.* 1987). It is easier to conceive of seed and fruit dispersal problems if both kinds of costs are included.

The two types of costs operate differently and have different effects on the evolution of dispersal. The mortality cost of dispersal is adequately represented by a which is a proportional reduction in the number of successfully dispersing seeds. Allocation costs ultimately involve fitness tradeoffs, so we will consider that, for a given amount of resources, a plant can either produce a larger seed crop with lower allocation to dispersal structure such as high quality pulp, or a smaller seed crop with higher allocation to auxiliary dispersal structures (cf. Schupp, this volume). If seed set is S_i in the absence of dispersal allocation, then with dispersal allocation, seed set will be $S_i \cdot A_s/(A_s + A_d)$ where A_s stands for seed allocation, A_d stands for dispersal allocation and $A_s/(A_s + A_d)$ is the cost of dispersal calculated as the proportion of the reproductive allocation still going to seeds (i.e. the proportion of seed set retained with a given dispersal allocation). Assume that dispersal increases with allocation to dispersal according to some function $D = h(A_d)$. The inverse function, $A_d = g(D)$, gives the allocation required for a given level of dispersal. Using this formulation, the allocation cost of dispersal can be expressed as a function of the dispersibility, $f(D) = A_s/(A_s + g(D))$. By choosing the appropriate units, the allocation cost of dispersal can be reexpressed for mathematical convenience as $f(D) = 1/(1 + g(D))$.

The allocation cost of dispersal is paid by dispersing and non-dispersing fruits alike. Thus the escape-from-crowding model can be reformulated to include both kinds of costs as follows:

$$G(D, \hat{D}) = \sum_{i=1}^{m} p_i \left(\frac{S_i f(D)(1 - D) + aD\bar{S}f(D)}{S_i f(\hat{D})(1 - \hat{D}) + a\hat{D}\bar{S}f(\hat{D})} \right). \tag{30}$$

For comparison to the previous version of the model which only considers the mortality cost of dispersal we will recalculate the conditions under which a dispersing morph can invade a non-dispersing resident population:

$$\frac{\partial G}{\partial D} = (-f(D) + f'(D)(1 - D)) \times \sum_{i=1}^{m} \frac{p_i S_i}{S_i f(\hat{D})(1 - \hat{D}) + a\hat{D} f(\hat{D})\bar{S}} + a\bar{S}(f(D) + Df'(D)) \times \sum_{i=1}^{m} \frac{p_i}{S_i f(\hat{D})(1 - \hat{D}) + a\hat{D}\bar{S}f(\hat{D})}. \tag{31}$$

where $f'(D)$ stands for $\partial f(D)/\partial D$. Thus,

$$\left.\frac{\partial G}{\partial D}\right|_{D = \hat{D} = 0} > 0 \Rightarrow \frac{a\bar{S}}{H(S)} > 1 - f'(D). \tag{32}$$

where $H(S)$ stands for the harmonic mean of S_i. Thus the invasion conditions are the same as those for the previous case with no mortality costs but with an additional term representing the allocation costs of dispersal to the right of the 1. Since $f(D)$ is a decreasing function of $D, f'(D)$ is negative and the righthand site of (32) is greater than 1. Thus, the conditions for invasion of a dispersing morph are more restrictive than when only mortality costs are included.

It can also be shown that $D^* = 1$ will never be the ESS:

$$\left.\frac{\partial G}{\partial D}\right|_{D=\hat{D}=1} \geq 0 \Rightarrow -\sum_{i=1}^{m} \frac{p_i S_i f}{a\bar{S}f}$$

$$+\sum_{i=1}^{m} \frac{p_i a\bar{S}f'}{a\bar{S}f} + \sum_{i=1}^{m} \frac{p_i a\bar{S}f}{a\bar{S}f} \geq 0$$

$$\Rightarrow \frac{1}{a} - \frac{f'}{f} \leq 1 . \tag{33}$$

Since $a \leq 1$ and $f' < 0$, this expression can never be satisfied and $D^* = 1$ will never be the ESS. Cohen & Motro (1989) reached similar conclusions in an analysis of a sib-competition model with no spatial or temporal variance.

The nature of density dependence

We have frequently used some fairly awkward and vague expressions such as 'sib-density effects' or 'escaping the effects of density' instead of simpler expression such as 'sib competition'. This is because the models actually treat general density effects which could just as readily be due to density-dependent predation as to competition.

For analytical simplicity the previous models assumed constant yield or exact density compensation. This corresponds quite well to the empirically observed phenomenon of constant-yield competition in plants (Kira *et al.* 1953). Yet there are many ways in which density effects could operate, and the details will affect the quantitative predictions of the models. For the escape-from-crowding model, the value of dispersal is in arriving at patches in which residents did poorly in the previous year, but which will experience favorable conditions in the present year. If one or a few plants cannot realize all of the potential yield of a patch, selection favoring dispersal will be weaker than in the simple analytical constant-yield models. Many models of constant-yield competition include the possibility that yield is not fully realized at the lowest densities (reviewed in Watkinson 1980). The degree to which this low-density effect is important will depend on the relationship between plant size and patch size (i.e. single individuals of species with small maximum plant size will not able to fully realize the potential yield of a large patch).

Another possibility is that density effects may occur with undercompensation such that total yield is higher at higher densities. With undercompensation, there will be less reason to escape crowded patches and selection favoring dispersal will be weaker than in the constant-yield model. If density effects are overcompensating such that total yield is lower at higher densities, the selection for dispersal to escape crowding or sib interactions will be stronger. For example, overcompensating density-dependent predation is the mechanism in the 'escape-from-density' version of Janzen's escape hypothesis. These or other specific density relationships could be inserted into computer-simulation versions of the models presented above.

In order to evaluate the importance of sib-density effects in natural systems, the life stage at which density-dependent limitation is occurring must be known. The reasoning behind this statement is as follows. The mechanism favoring dispersal in this model is that there are diminishing returns from retaining more and more non-dispersing seeds in the parental patch. The diminishing returns are due to the fact the mutant's seeds represent a large fraction of seeds in the parental patch. N is the number of individuals per patch at the limiting life-stage and escape from sib-density effects is only an important function of dispersal when N is small. Since seedling densities may be orders of magnitude higher than adult densities, N will be much larger if safe sites for seedling establishment are limiting rather than canopy sites for adults. Thus, the life stage at which density-dependent limitation occurs is critical for determining the influence of sib interactions.

The consequences of the breeding system and genetic structure

While the ideas presented here have been predominantly ecological, there are some significant junctures with population genetics. First of all, as mentioned above, on average 1/2 of the offspring of a cosexual plant are produced via male function. These seeds are produced on the mates of the parent in question and consequently the dispersal distance for 1/2 of the offspring depend intimately on the breeding system. This particular juncture between the breeding system and the dispersal biology of plants is critical to an understanding of the population-dynamic consequences of dispersal and it has not been explicitly recognized in much of the seed dispersal literature.

Secondly, the function of dispersal for escape from high concentrations of relatives depends on the genetic structure of the population. Several studies have demonstrated how the strength of the sib-interaction effect depends on the genetic relationship between the individual with the mutant dispersal type and the seeds being left behind (Frank 1986; Taylor 1988). Frank (1986) used Price's (1972) method for analyzing selection in subdivided populations. He showed that $D^* = (\rho - (1-a))/(\rho - (1-a)^2)$ where ρ is Hamilton's (1972) regression coefficient of relatedness of the genotype controlling dispersal (the mother plant in our case) onto a member of the cohort of offspring chosen randomly before dispersal. ρ is a measure of the relatedness of individuals within a patch. It declines with increasing number of individuals per patch and increases with the level of inbreeding. Thus variation in the breeding system will affect the ESS value of dispersal in a sib-interaction model. Under our assumption that the proportion of dispersing offspring is determined by the mother and with the added assumption that there is no genetic correlation among the mothers in a site or the pollen they have received, Frank's result is equal to ours (26), i.e. $\rho = 1/N$. Since spatial genetic structure is expected to accumulate in metapopulations with fractional dispersal, genetic correlations are likely to exist among plants in a site. This favors greater dispersibility and (26) becomes a lower bound on the ESS dispersal rate. Taylor (1988) formulates the problem in inclusive fitness terms, emphasizing the 'altruistic' nature of dispersal in a sib-interaction model. Dispersing seeds have the same expected reproductive success as non-dispersing ones, yet they must pay the mortality cost of dispersal. In exchange, they improve the prospects of success for those seeds left behind, some of which are their relatives. This phrasing of the problem permits the examination of parent/offspring conflict over offspring dispersal (the mother wants more), with all of the potential intricacies created by the different genetic constitutions of the different seed and fruit tissues (cf. Queller 1983).

Colonization, escape from parents, and directed dispersal revisited

In this paper we have discussed four fundamental population-dynamic functions of dispersal which differ from the three functions of dispersal which are cited in the majority of papers on seed dispersal: colonization, escape, and directed dispersal (Howe & Smallwood 1982). The four population-dynamic functions presented here are based on theoretical results. They are conceptually distinct and have some fundamental population-dynamic reality. For example, risk reduction requires global temporal variance, escape from crowding requires spatial and local temporal variance, while escape from sib-density effects requires only spatial structure and varying concentration of relatives. Colonization, escape, and directed dispersal refer to a mix of ideas that have had important impact on empirical research in dispersal ecology (the 'case making' approach of Howe, this volume). Thus the escape hypothesis is built around Janzen (1970) and Connell's (1971) hypothesis regarding seed predation in relation to distance from tropical trees. The colonization hypothesis has its roots in early ideas on plant life histories in successional environments (Salisbury 1942; Baker & Stebbins 1965). The directed dispersal hypothesis emerged directly from studies of the natural history of scatterhoarders, mistletoes, and ant-dispersed plants (Vander Wall & Balda 1977; Docters van Leeuwen 1954; Handel 1978).

The colonization hypothesis relates most directly to the escape-from-crowding models. Only limited success is possible in densely occupied patches and occupied patches will eventually disappear. Windfall reproductive profits can be obtained from successful dispersal to empty or sparsely occupied patches.

The escape hypothesis is often phrased in terms of either *distance*-dependent success or *density*-dependent success. Distance dependence refers to negative effects of a long-lived parent (usually a tree) either in terms of competitive suppression or attraction of species-specific predators. A strict distance-from-parent effect is perhaps best considered in terms of the escape from crowding model with perenniality, i.e. occupied safe sites or canopy sites are unavailable for regeneration until the occupant dies. Curiously, perenniality does not necessarily favor increased dispersibility in the models presented here. This is because, in our models, all habitable patches are potentially occupied by adults and the probability of adult mortality creating an open site is the same everywhere.

Density-dependent versions of the escape hypothesis are often phrased in terms of what we would call an escape-from-crowding or sib-interaction model with over-compensating density dependence. At the highest densities, mortality is so intense that absolute regeneration is lower than at intermediate densities.

Directed dispersal represents a distinct mechanism (i.e. 'predictive dispersal') from the other three which all can operate with random dispersal. If a plant can vary its dispersal into a patch or habitat depending on the likelihood of its success in that patch (perhaps by utilizing an appropriate vector or morphology) then it is clearly fitness enhancing to do so.

Interactions between theoretical and empirical studies

Gaining an empirical understanding of the fitness consequences of fruit dispersal is a daunting task. Where do seeds go? What are their fates throughout the life of the plants? How do different density/frequency/distance dependent factors affect the fitness of dispersing seeds? What are the costs involved in dispersal, both in terms of allocation to dispersal structures and in terms of dispersal related mortality? How does all of this vary in space and time? Understanding all of these factors and processes for any single species seems all but hopeless.

One approach to understanding these difficult to measure phenomena is through conceptual exploration of the phenomena under consideration, and that is the approach taken in this chapter so far. But ultimately, the ideas must be 'unpacked' and interfaced with empirical results for specific systems. Some interesting work has been done in this area with modelling frameworks other than those advocated here. Murray (1988) has investigated three neotropical gap-dependent plants using data on dormancy, dispersal, and forest dynamics in a simulation model aimed at determining the fitness consequences of bird dispersal. Alvarez-Buylla & García-Barrios (1991) has integrated field data on patch structure and seed behavior into models of the metapopulation dynamics of the pioneer rainforest tree, *Cecropia obtusifolia*. Olivieri & Gouyon (1985) used a metapopulation approach to model the evolution of dispersibility of the temperate weeds, *Carduus pycnocephalus* and *C. tenuiflorus*. They successfully predicted trends with population age in the ratios of dispersing and non-dispersing seed morphs. Horvitz & Schemske (1986) utilize a metapopulation model to predict the fitness consequences of dispersal and dormancy for *Calathea ovandensis* a tropical herb.

The models presented here can provide a powerful guide to empiricists trying to determine the fitness consequences of dispersal. The models are simple to use and calculations with field data are easily carried out. The data needed to fit to the models is among the easiest to collect and the models point the way to answering difficult questions about seed dispersal.

In the section on escape from crowding we showed that the expected rate of increase in site occupancy of an individual with a mutant dispersal type is given simply by the sum over all patch types of the ratio for each patch type of the reproductive success of mutant and resident seeds (15). The reproductive success in each patch type of mutant and resident seeds can be estimated from field data on seed rain in different patches or quadrats and inserted in the following equation:

$$G(D, \hat{D}) = \sum_{\substack{\text{quadrats} \\ \text{or patches}}} \left(\frac{\text{mutant seed density}}{\text{resident seed density}} \right). \quad (34)$$

As before, $G(D, \hat{D})$ gives the expected fitness of an individual adopting the dispersibility represented by the numerator in a population with the dispersibility represented by the denominator under the assumptions of the escape-from-crowding model.

Various combinations of actual and hypothetical seed distributions can be inserted in the numerator and denominator of (34). For example, the spatial distribution of seeds that would occur if all seeds fell under the parent plant could be used, or the distribution assuming uniform dispersal. One could then ask what the fitness advantage would be of a mutant individual with uniform dispersal in a resident population with the actual dispersibility. Alternatively, one could ask what the fitness advantage of a mutant with the actual dispersibility might be in a population without dispersal. Many other questions could be generated in a similar fashion. For example, what would be the fitness consequence of a change in the slope of the dispersal curve (c.f. Mary Willson chapter).

The actual data required include: 1) the spatial distribution of reproductive plants (and preferably an estimate of their fecundity), 2) the spatial distribution of seed fall (without necessarily knowing where each seed comes from) or the spatial distribution of seeds on the soil shortly after dispersal, and 3) an estimate of the dispersal curve. Actually, the calculations could be made with either 2) or 3) but if both are available, some cross-validation of the data are possible (e.g. it should be possible to generate 2) from 1) and 3) since applying the dispersal curves to the adult distributions should yield something close to the actual spatial distribution of seeds).

One practical consideration that comes up in calculating (34) is what to do with 0's in the denominator. This represents the situation in which a mutant dispersing seeds would have complete access to the resources of a patch in which there are no residents. The field biologist would have to determine, for the particular patch size assumed, what, N, the maximum site occupancy in that patch type would be. N represents an upper limit on the gain in site occupancy in a particular patch. The calculation would be to sum over patch types either N or the ratio of mutant divided by resident seed density, which ever is smaller.

Elements of sophistication can be progressively added to such a model. The fitness gain accrued by changing morphology could be calculated if measurements were available on the impact of dispersal morphology on seed spatial distributions or on the shape of the dispersal curve (c.f. Augspurger, this volume). Allocation costs of changing morphology could be estimated (Augspurger, this volume) as could mortality costs. At each step of model building one can test the impact of modifying assumptions about unknown aspects of the plant biology (such as the detailed nature of density dependence) or one could set out to measure the unknown aspect.

Exercises of this nature provide first-order empirical estimates of the fitness consequences of dispersal and also focus attention on the biological issues in need of further study to increase realism and sophistication of understanding. Another useful aspect of this approach is the focus it places on studying all aspects of the dispersal problem in the same experimental system. Currently bits and pieces of information are available for a variety of plant species, but for any single species, large gaps remain.

In this section we have emphasized the model for escape from crowding. The other functions could be explored as well. To investigate the role of risk reduction, measurements for a large number of patches would have to be made in a variety of year types. Sib-interaction models require a knowledge of the partition of seeds in a patch into those produced *in situ* and those coming from elsewhere (c.f. (23)). Evidence on the genetic structure of populations could be incorporated as well. The role of directed dispersal can be estimated by quantifying the non-randomness of deposition patterns and the increased demographic success in the target patch types relative to random patches.

Conclusion

In summary, the field is in an interesting position where a fairly sophisticated set of theoretical ideas exists that can be applied to the question of the fitness consequences of seed and fruit dispersal in spatially and temporally varying plant metapopulations. If this theory is carefully 'unpacked' and applied to frugivory problems it can provide a framework for thinking about the processes involved in the evolution of seed dispersal. We have tried to flesh out some of the implications for real dispersal systems, yet many interesting issues remain untouched. For example, different dispersal syndromes have systematic differences in allocation and mortality costs and in the evenness and directedness of dispersal. Thus dispersal vectors should differ in their relative contributions to risk reduction, escape from crowding, escape from sib interactions, and directed dispersal. Some dispersal mechanisms involve the concentrated clumping of seeds and the details determine the degree of risk reduction, escape from crowding, and escape from sib interactions achieved. We will have to leave these and many other implications of these ideas to the reader's imagination.

The models presented here should also be very useful for guiding and simplifying empirical research efforts aimed at estimating the fitness consequences of fruit and seed dispersal. Considerable insights can be achieved with relatively easy-to-collect data and the models point the way to add progressive layers of realism by probing more deeply in to an array of critical biological processes.

Acknowledgements

The following people contributed with either helpful discussions or comments on drafts: W. A. Mitchell, G. A. Fox, M. Martinez-Ramos, J. X. Becerra. The research was supported in part by NSF grants BSR-8516971 and BSR-9107324.

References

Alvarez-Buylla, E. R. & García-Barrios, R. 1991. Seed and forest dynamics: a theoretical framework and an example from the Neotropics. Am. Nat. 137: 133–154.

Antonovic, J. & Primack, R. 1982. Experimental ecological genetics in *Plantago:* VI. The demography of seedling transplants of *P. lanceolata*. J. Ecol; 70: 55–75.

Baker, H. G. & Stebbins, G. L. 1965. Genetics of colonizing species. Academic Press, New York.

Bertin, R. I. 1988. Paternity in plants. In: J. Lovett Doust & L. Lovett Doust, (eds.) Plant reproductive ecology: patterns and strategies, pp. 30–59. Oxford University Press, Oxford.

Bergelson, J. 1990. Life after death: site pre-emption by the remains of *Poa annua*. Ecology 71: 2157–2165.

Charnov, E. L. 1982. The theory of sex allocation. Monographs in population biology 18, Princeton University Press, Princeton.

Chesson, P. L. & Huntly, N. 1988. Community consequences of life-history traits in a variable environment. Ann. Zool. Fennici 25: 5–16.

Clements, F. E. & Goldsmith, G. W. 1924. The phytometer method in Ecology. Carnegie Institute of Washington Publication No. 356.

Cohen, D. & Levin, S. A. 1985. The interaction between dispersal and dormancy strategies in varying and heterogeneous environments. In: E. Teromato & M. Yomaguti, (eds.) Mathematical topics in population biology, morphogenesis, and neuroscience. pp. 110–122. Proceed. Internat. Symp. Kyoto, 1985.

Cohen, D. & Levin, S. A. 1991. Dispersal in patchy environments: the effects of temporal and spatial structure. Theor. Pop. Biol. 39: 63–99.

Cohen, D. & Motro, U. 1989. More on optimal rates of dispersal: taking into account the cost of the dispersal mechanism. Am. Nat. 134: 659–663.

Comins, H. N. 1982. Evolutionarily stable strategies for localized dispersal in two dimensions. J. Theor. Biol. 94: 579–606.

Connell, J. H. 1971. On the role of natural enemies in preventing competitive exclusion in some marine animals and in rain forest trees. In: P. J. Den Boer, G. Gradwell, (eds.) Dynamics of Populations. pp. 298–312. PUDOC, Wageningen.

Docters van Leeuwen, W. M. 1954. On the biology of some Loranthaceae and the role birds play in their life-history. Beaufortia 4: 105–208.

Ellner, S. & Shmida, A. 1981. Why are adaptations for long-range seed dispersal rare in desert plants? Oecologia (Berl.) 51: 133–144.

Estrada, A. & Fleming, T. H. 1986. Frugivores and seed dispersal. Dr W. Junk Publishers, Dordrecht.

Frank, S. A. 1986. Dispersal polymorphisms in subdivided populations. J. Theor. Biol. 122: 303–309.

Hamilton, W. D. 1972. Altruism and related phenomena, mainly in social insects. Ann. Rev. Ecol. Syst. 3: 193–232.

Hamilton, W. D. & May, R. M. 1977. Dispersal in stable habitats. Nature 269: 578–581.

Handel, S. N. 1978. The competitive relationship of three woodland sedges and its bearing on the evolution of ant-dispersal of *Carex pedunculata*. Evolution 32: 151–163.

Hanzawa, F. M., Beattie, A. J. & Culver, D. C. 1988. Directed dispersal: demographic analysis of an ant-seed mutualism. Am. Nat. 131: 1–13.

Harper, J. L. 1961. Approaches to the study of plant competition. In: F. L. Milthorpe, (ed.), Mechanisms in biological competition. Symp. Soc. exp. Biol. 15: 1–39.

Harper, J. L. 1972. The population biology of plants. Academic Press, London.

Howe, H. F. & Smallwood, J. 1982. Ecology of seed dispersal. Ann. Rev. Ecol. Sys. 13: 201–228.

Horvitz, C. C. & Schemske, D. W. 1986. Seed dispersal and environmental heterogeneity in a neotropical herb: a model of population and patch dynamics. In: A. Estrada & T. H. Fleming, (eds.) Frugivores and seed dispersal. Dr W. Junk Publishers, Dordrecht.

Janzen, D. H. 1970. Herbivores and the number of tree species in tropical forests. Am. Nat. 104: 501–528.

Johnson, M. & Gaines, M. S. 1990. Evolution of dispersal: theoretical models and empirical tests using birds and mammals. Ann. Rev. Ecol. Syst. 21: 449–480.

Kira, T., Ogawa, H. & Shinozaki, K. 1953. Intraspecific competition among higher plants. 1. Competition-density-yield inter-relationships in regularly dispersed populations. J. Inst. Polytech. Osaka Cy. Univ. D. 4: 1–16.

Krummel, J. R., Gardner, R. H., Sugihara, G. & O'Neill, R. V. 1987. Landscape patterns, in a disturbed environment. Oikos 48: 321–324.

Levin, D. A. & Kerster, H. W. 1974. Gene flow in seed plants. Evol. Biol. 7: 139–220.

Levin, S. A., Cohen, D. & Hastings, A. 1984. Dispersal strategies in patchy environments. Theor. Popul. Biol. 26: 165–191.

Levins, R. 1970. Extinction. Some mathematical questions in biology. In: M. Gerstenhaber, (ed.) Lectures on mathematics in the life sciences. pp. 77–107. American Mathematical Society, Providence, RI.

Lloyd, D. G. 1983. Evolutionarily stable sex ratios and sex allocations. J. Theor. Biol. 105: 525–539.

Lloyd, D. G. 1984. Gender allocations in outcrossing cosexual plants. In: R. Dirzo & J. Sarukhan, (eds.), Perspectives on plants population ecology. pp. 166–187. Sinauer, Sunderland, MS.

Lovett-Doust, J. & Lovett-Doust, L. 1988. Plant reproductive ecology: patterns and strategies. Oxford U. Press, New York.

Maynard-Smith, J. & Price, G. R. 1973. The logic of animal conflict. Nature, Lond. 246: 15–18.

Moermond, T. C. & Denslow, J. S. 1985. Neotropical avian frugivores: patterns of behavior, morphology, and nutrition, with consequences for fruit selection. In: P. A. Buckley, M. S. Foster, E. S. Morton, R. S. Ridgely & N. G. Buckley (eds.) Neotropical Ornithology. Orithological Monographs 36: 865–807.

Murray, D. R. 1986. Seed dispersal. Academic Press, Sydney.

Murray, K. G. 1988. Avian seed dispersal of three neotropical gap-dependent plants. Ecol. Monogr. 58: 271–298.

Olivieri, I. & Gouyon, P. H. 1985. Seed dimorphism for dispersal: theory and implications. In: J. Haeck & J. W. Woldendorp, (eds.) Structure and functioning of plant populations/2. Phenotypic and genotypic variation in plant populations. North-Holland Publishing Co., Amsterdam.

Ostle, B. & Mensing, R. W. 1975. Statistics in research. (3rd ed.) Iowa State University Press, Ames.

Pickett, S. T. A. & White, P. S. 1985. The ecology of natural disturbance and patch dynamics. Academic Press, Orlando FL.

Price, G. R. 1972. Extension of covariance selection mathematics. Ann. hum. Genet. 35: 485–490.

Queller, D. 1983. Kin selection and conflict in seed maturation. J. Theor. Biol. 100: 153–172.

Real, L. A. 1980. Fitness, uncertainty, and the role diversification in evolution and behavior. Am. Nat. 115: 623–635.

Salisbury, E. J. 1942. The reproductive capacity of plants. Bell, London.

Schoen, D. J. & Lloyd, D. G. 1983. The selection of cleistogamy and heteromorphic diaspores. Biol. J. Linn. Soc. 23: 303–323.

Seger, J. & Brockmann, H. J. 1987. What is bet-hedging? Oxford Surveys in Evol. Biol. 4: 183–211.

Silander, J. A. & Pacala, S. W. 1990. The application of plant population dynamic models to understanding plant competition. In: J. B. Grace & D. Tilman, (eds.), Perspectives on plant competition. pp. 67–92. Academic Press, San Diego.

Simberloff, D. 1989. The contribution of population and community biology to conservation science. Ann. Rev. Ecol. Syst. 19: 473–512.

Stanton, M. & Galloway, L. F. 1991. Natural selection and allocation to reproduction in flowering plants. In: M. Mangel (ed.) Sex Allocation and sex change: experiments and models. Providence: American Math. Soc., pp. 1–50.

Sugihara, G. & May, R. M. 1990. Applications of fractals in ecology. TREE 5: 79–86.

Swingland, I. R. & Greenwood, P. J. (eds.) 1983. The ecology of animal movement. Clarendon, Oxford.

Taper, M. L. & Case, T. J. 1992. Models of character displacement and the theoretical robustness of taxon cycles. Evolution 46: 317–333.

Taylor, P. D. 1988. An inclusive fitness model for dispersal of offspring. J. Theor. Biol. 130: 363–378.

Vander Wall, S. B. & Balda, R. P. 1977. Coadaptations of the Clark's Nutcracker and the piñon pine for efficient seed harvest and dispersal. Ecol. Monogr. 47: 89–111.

Venable, D. L. 1989. Modelling the evolutionary ecology of seed banks. In: M. A. Leck, V. T. Parker & R. L. Simpson (eds.). The ecology of soil seed banks, pp. 67–87. Academic Press, San Diego.

Venable, D. L. & Brown, J. S. 1988. The selective interactions of dispersal, dormancy, and seed size as adaptations for reducing risk in variable environments. Am. Nat. 131: 360–384.

Venable, D. L., Búrquez, A., Corral, G., Morales, E. & Espinoza, F. 1987. Ecology of seed heteromophism in *Heterosperma pinnatum* in Central Mexico. Ecology 68: 65–76.

Venable, D. L. & Lawlor, L. R. 1980. Delayed germination in desert annuals: escape in space and time. Oecologia (Berl.) 46: 272–282.

Venable, D. L. & Levin, D. A. 1985. Ecology of seed polymorphism in *Heterotheca latifolia:* Achene structure, germination, dispersal. J. Ecol. 73: 133–145.

Vincent, T. L. & Brown, J. S. 1988. The evolution of ESS theory. Ann. Rev. Ecol. Syst. 19: 423–444.

Watkinson, A. R. 1980. Density-dependence in single-species populations of plants. J. Theor. Biol. 83: 345–357.

Wheelwright, N. T. & Orians, G. H. 1982. Seed dispersal by animals: contrasts with pollen dispersal, problems of terminology, and constraints on coevolution. Am. Nat. 119: 402–413.

White, P. S. 1979. Pattern, process, and natural disturbance in vegetation. Bot. Rev. 45: 229–299.

Vegetatio **107/108**: 57–73, 1993.
T. H. Fleming and A. Estrada (eds). Frugivory and Seed Dispersal: Ecological and Evolutionary Aspects.

Consumption of fruits of the Melastomataceae by birds: how diffuse is coevolution?

F. Gary Stiles [1] & Loreta Rosselli [2]
[1] *Instituto de Ciencias Naturales, Universidad Nacional de Colombia, A.A. 7495, Bogotá, Colombia;*
[2] *Empresa de Acueducto y Alcantarillado de Bogotá, A.A. 80186, Bogotá, Colombia*

Keywords: Coevolution, Frugivory, Manakins, Melastomataceae, Neotropics, Tanagers

Abstract

The family Melastomataceae shows two major modes of dispersal for its small seeds: wind for capsular fruits, and birds for berry-type fruits. Distribution patterns of these two dispersal types differ. We focus on evolutionary diversification of bird-dispersed melastomes in the New World, in relation to that of their avian dispersers. We first examine in detail patterns of melastome fruit availability and consumption by understory birds in a Costa Rican rain forest. Although a relatively small portion of total fruit available in most months, melastomes comprised a major component of the diet of all bird groups studied, most notably manakins and most tanagers, which in turn were the major consumers of melastome fruits. Examination of data from several other Neotropical areas confirms the importance of the manakins and tanagers as the avian frugivores most closely associated with the melastomes. These two groups differ in their manner of fruit consumption, being 'gulpers' and 'mashers', respectively. The patterns of species richness of small mashers in several Neotropical areas conform very closely to the distribution patterns of bird-dispersed melastomes; both are most speciose in wet forests of middle elevations. Gulpers, especially manakins, are most diverse at low elevations. Thus, the small-seeded melastome berry appears to have evolved primarily in association with masher-type frugivores such as tanagers; with the later addition of manakins to their disperser spectrum, the melastomes may have increased greatly in local abundance, but not (yet?) in diversity, in the wet lowland neotropics.

Introduction

Most interactions between fruit and consumer/disperser species in a community tend to be weak and asymmetrical: either a given fruit contributes only a small proportion of the frugivore's diet, or the latter eats and disperses only a small part of the total fruit crop, or both (review in Jordano 1987). The relatively few strong interactions often signify a considerable degree of ecological interdependence between plant and frugivore (e.g. Howe 1977; Foster 1977), but the extent to which this in turn implies a coevolved relationship is much more controversial and difficult to determine (Janzen & Martin 1982; Schemske 1983; Howe 1984; Herrera 1986). The strong interactions, moreover, may play different roles in the frugivore's ecology and annual cycle: it may be useful to distinguish 'staple ' species or groups, whose fruits maintain a population over long periods and may provide energy required for reproduction, from 'keystone' resources that can tide the population over lean seasons when little other fruit is available (Terborgh 1986).

The family Melastomataceae, with over 4800 species, ranks as the seventh largest angiosperm family (Renner 1989b). Some three-quarters of the known species occur in the neotropics, where '... a dozen or more species grow naturally on every square mile ... and the various genera extend from coastal marshes at sea level to the high páramos above the tree line' (Gleason 1932). Fruits of this family are consistently mentioned as being among the most important foods for many small neotropical frugivorous birds (Snow 1962 a, b; 1965; Hilty 1980; Isler & Isler 1987; Prum & Johnston 1987; Levey 1988a; Stiles & Skutch 1989; Arango 1990; and many others). The fleshy berries of zoochorous melastomes have many tiny seeds embedded in a watery, somewhat sugary pulp. Such fruits are often considered to be 'generalized' or fed upon by 'opportunistic' frugivores (e.g. McKey 1975) although at least some small frugivores,such as certain manakins, show a high degree of dependence upon melastomes as an energy (and nutrient?) source that permits reproduction and survival over most of the year (e.g. Snow 1965; Hilty 1980). Thus, the Melastomataceae seems to represent an ideal group with which to explore questions of ecological interdependence and possible coevolution with avian frugivores.

In this paper, we commence with a detailed case study: the understory frugivorous birds and their fruits in a forest in Costa Rica. Through diet (fecal) samples, we identify those birds which interact most strongly with melastomes, and the possible role played by the melastomes in their annual cycles. Then we compare these results with several other community-level studies of birds and fruits in the neotropics, to determine whether the same groups of birds interact strongly with melastomes in these areas as well. We then apply other types of evidence, especially from distribution patterns and consumption of alternative fruit types, to judge whether a particular bird-melastome interaction is likely to be coevolved or not.

A case study: understory fruits and frugivores at 'El Plastico', Costa Rica

Study site and methods

This study was conducted from July 1987 to July 1988 at Finca 'El Plástico', 12 km SW of Las Horquetas de Sarapiquí, Heredia Prov., Costa Rica. The area is covered mainly by primary forest (Premontane Wet Forest sensu Holdridge) which is described in detail by Hartshorn & Peralta (1988) & Rosselli (1989). Our specific study site was a 12 ha plot of mature forest between 500 and 600 m elevation, crisscrossed by a grid of trails 50 m apart. The plot contained a variety of microhabitats including natural clearings, streams, ridges and ravines.

Rainfall data for El Plástico are lacking, but the overall pattern is presumably similar to that of 'La Selva' Biological Station (4,015 mm annual rainfall), 12 km north of 'El Plástico', with a dry season between January and April and a rainy season the rest of the year with peaks in July and November (Loiselle & Blake, 1991). The total annual rainfall is probably close to 5000 mm (cf. Hartshorn & Peralta, 1988). During the year of our study the first half of 1988 was exceptionally wet, interrupted by unusually low rainfall in April and June.We made monthly censuses of fruit abundance (except in December 1987) along 1700 m of trails in the study plot, counting all mature and immature fruits up to 10 m, to a fixed distance on either side of the trail. This distance was 2 m for most fruits, but was longer for conspicuous species and shorter for species harder to see. The total area in which fruits were counted varied between species according to conspicuousness, but was about 7000 m^2 for most. We used fruit dry weight (including pulp and seeds) as our overall measure of fruit abundance. Dry pulp weight would be a better measure of the resources available to dispersers (see also Blake *et al.* 1990), but we were unable to separate pulp from seeds in tiny-seeded fruits of families like Melastomataceae, Gesneriaceae, Ericaceae and some Rubiaceae. Thus, our estimates of overall fruit abundance are probably somewhat biased towards large-seeded species. Botanical specimens of all fruit species were deposited in the Herbario Nacional de Costa Rica. All fruits were measured; the seeds were drawn and described in detail. Both dry and wet (FAA solution) fruits were saved as a reference collection for fecal sample analysis.

We sampled birds using 20–25 mist nets (12 × 2.5 m, 30 mm mesh) during 3 days each month except December 1987. Each day nets were set in a different part of the study plot and opened between 5:30 and 13:00, except during heavy rains. Nets were checked once or twice hourly; all frugivores trapped were held in special cloth bags with a round vinyl bottom covered with a clean piece of filter paper, for about 20 min, during which time they were likely to defecate (Levey 1986). For the White-ruffed Manakin (*Corapipo leucorrhoa*) we sometimes injected saline solution through a thin plastic tube slowly to their stomachs until the birds defecated or vomited (Moody 1970; Herrera & Jordano, 1981). The samples obtained (including regurgitated and defecated materials) were dried on filter paper and stored in plastic bags. We experienced no deterioration of samples over periods of up to 3 years by this method.

For dense forest habitats, fecal sample analysis is one of the most effective ways to document bird diets, although it may present several biases: only birds that fly as low as the nets are caught, some soft fruits may have shorter digestion times than some insect parts and so the latter may be overestimated (Jordano & Herrera, 1981); on the other hand, different seed sizes may have different passage times, smaller seeds taking longer than larger ones, which may be regurgitated shortly after ingestion (Levey, 1986, 1987). To reduce these biases we introduced a weighting factor according to the size of the sample, ranging from 1 for very small samples (≤ 1 ml with few seeds) to 5 (volume of about 10 ml). Accordingly, we assigned greater weights to large fecal samples than to small ones, on the basis that the former were thus more representative of the diet.

For analysis we multiplied each percentage value of fruit types in the sample by these weighting factors. For analysis, fecal samples were soaked in water for 10–30 min and examined under a dissecting microscope. We estimated the percentage of total sample volume composed by fruit and insects; for the fruit portion, we estimated the percentage of total volume contributed by each species of fruit. Identification was based on seed characteristics, supplemented by skin and color details. Each seed was compared with the reference collection, that included seeds from the study area and other sites in Costa Rica. We nearly always were able to identify the seeds to family and in most cases to species or genus; in the unknown category we included pulp and fruit skins with no associated seeds, as well as seeds with no near equivalent in the reference collection.

We divided the understory frugivores included in the study into 7 groups that we found to have similar feeding habits and close taxonomic affinities:

1. Manakins (Pipridae): *Corapipo leucorrhoa* (87% of captures), *Pipra mentalis* (5%), *Pipra pipra* (6%) and *Schiffornis turdinus* (2%).
2. Tanagers (Thraupidae): *Chlorothraupis carmioli* (71%), *Tangara icterocephala* (1%), *Ramphocelus passerini* (1%), *Tachyphonus luctuosus* (4%), *Tachyphonus delatrii* (23%).
3. Euphonia (Thraupidae): *Euphonia anneae*
4. Thrushes (Turdidae): *Catharus ustulatus* (60%) and *Turdus obsoletus* (40%).
5. Solitaire (Turdidae): *Myadestes melanops*.
6. Flycatchers (Tyrannidae): *Mionectes olivaceus* (32%) and *Mionectes oleagineus* (68%).
7. Finch (Emberizidae): *Arremon aurantiirostris*.

Total captures of these groups are given in Table 1. These frugivores processed fruit in two different ways, termed 'mashing' and 'gulping' by Levey (1987). The former technique, employed by the tanagers, *Euphonia*, and *Arremon*, involves mandibulating the fruit, crushing it, squeezing out the juice (Moermond 1983), and discarding the husk and many seeds. Gulping, employed by all the other groups, involves swallowing the fruit whole and voiding the seeds by regurgitation or in the feces. This difference has impor-

Table 1. Composition of diet of understory frugivores in El Plástico, Costa Rica, from July 1987 to July 1988. All percentage values are corrected means according to fecal sample size (see text).

Groups	No. samples	% Fruit in diet	% that were Melastomes	Other tiny seeded fruits[1]	Large seeded fruits[2]	Fruit matter of unknown seed size	% Samples with melastomes
Manakins	302	94.3	57.8	7.6	27.3	7.3	80
Tanagers	111	72.0	65.7	12.7	21.6	0.0	97
Arremon	9	33.7	97.9	0.5	1.6	0.0	67
Myadestes	27	97.7	31.3	13.8	51.2	3.7	93
Thrushes	20	72.8	32.4	5.2	53.2	9.2	75
Flycatchers	36	74.4	17.9	8.6	71.8	1.7	50
Euphonia	50	99.5	7.2	3.9	88.7	0.2	52

[1] Including Ericaceae, Gesneriaceae, some Rubiaceae among others.
[2] Including Araceae, *Psychotria*, Clusiaceae, among others.

tant implications for the arguments we develop below.

Other possible frugivores captured, but not included here, either consumed little or no fruit (*Piprites griseiceps*, *Catharus mexicanus*) or had very low capture numbers combined with very different fruit types from those considered here (*Zimmerius vilissimus*, mistletoes; *Attila spadiceus*, *Guarea* sp.).

Results

The fruits

We identified 90 species of fruits from fecal samples, 66 of which were censused or recorded in the study area. Families best represented in the phenological censuses were Rubiaceae (33 spp., mainly *Psychotria* spp.), Melastomataceae (17 spp.), Araceae, Palmae and Gesneriacae (complete list of species available from authors). Overall abundance of these fruits was highest between January and March and lowest in August 1987, and between April and July 1988. The melastomes contributed only 11.5% of the overall abundance (dry weight per hectare) and no more than 15% in any month except July 1988, when they comprised 94% of available fruit dry weight (Fig. 1). Melastome abundance was highest in this month (144 g/ha) and lowest in March (1.0 g/ha), and was not correlated with overall fruit abundance (Rs = −0.54, p = 0.09, n = 11).

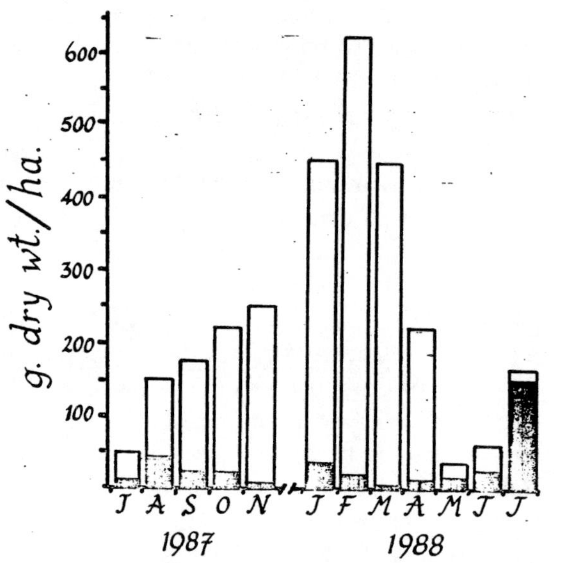

Fig. 1. Monthly fruit abundance (dry fruit weight per hectare) of understory plants of melastomes and other families in El Plástico, Costa Rica. (Only species for which we had evidence of consumption by the birds in the study are included).

Seven species contributed over 97% of the total melastome abundance through the study period: *Henriettella tuberculosa* (48.0%), *Ossaea brenesii* (19.5%), *Conostegia cooperi* (18.0%), *Ossaea macrophylla* (4.8%), *Miconia* aff. *smaragdina*

(3.6%), *Ossaea laxivenula* (1.8%) and *Miconia centrodesma* (1.7%). These species all produce small (4.8–7.3 mm diameter), tiny-seeded berries but differ greatly in abundance, productivity, fruit characteristics and spatial distribution in the area (details will be presented elsewhere).

H. tuberculosa was by far the most abundant melastome in the study area, bearing fruit nearly year-round with a very high peak in July 1988. Its seeds were found in 32% of all fecal samples, from all bird groups except *Arremon aurantiirostris*; it was most heavily consumed in September, January and July 1988, when it was most abundant. Of the samples with *H. tuberculosa* seeds, 59% were from manakins. The association between this species *C. leucorrhoa* is more than a local phenomenon, as we have observed the latter taking fruits of this melastome in the Chocó region of Colombia.

Seeds of *Ossaea brenesii*, *robusta* and *macrophylla* were indistinguishable; together they comprise ca. 25% of the annual supply of melastome fruit dry weight, most of it being *O. brenesii*. Abundance of these fruits was fairly high in late 1987, declining to low levels betwen March and July 1988; 28% of all fecal samples contained *Ossaea* seeds. Although 52% of these samples were from manakins, the proportion of *Ossaea* spp. was highest for tanagers, which contributed 34% of the samples. *Ossaea* spp. were consumed most heavily by all groups in August, October and November, when they constituted a high proportion of available melastome fruit. *O. laxivenula* produced ripe fruits between September and February; although never abundant, it contributed an important part of the diet of some bird groups during these months, notably of manakins in November.

Conostegia cooperi fruited only between May and July 1988, when it attained its peak of abundance, at which time its fruits were highly favored by manakins, tanagers, thrushes, and *A. aurantiirostris*, even though this coincided with the fruiting peak of *H. tuberculosa*. Its seeds were found in 15% of all samples, but in 43% of those from July 1988.

M. aff. *smaragdina* had ripe fruits between April and July 1988, with a peak in April; 12% of all samples contained seeds of this species. Diets of manakins, tanagers, thrushes and Mionectes spp. contained high percentages of *M.* aff. *smaragdina* during the months of high abundance. *M. centrodesma* fruits were seen in all months except April and June, but were rare except at the start and end of the study, when they constituted an important part of the diet of thrushes in August (22%) and of the flycatchers in July 1988 (28%).

The birds

Numbers of fecal samples per diet group varied from 302 (manakins) to 9 for *A. aurantiirostris*; only manakins and tanagers were represented in all months, but the flycatchers and *Euphonia* were represented in nearly all, the thrushes in 7, and *Myadestes* and *Arremon* in only 5 and 4 months, respectively. All groups except *Arremon* were highly frugivorous ($\geq 70\%$ fruit in the diet) (Table 1).

The manakins and tanagers were consistently highly frugivorous, and the melastomes comprised over half of the fruit consumed by each. For both groups, *Henriettella tuberculosa* and/or *Ossaea* spp. were the melastomes most consumed in most months, but both shifted their preferences markedly towards *M.* aff. *smaragdina* and *C. cooperi* when fruit of these species reached their respective peaks of abundance.

Euphonia annae was almost entirely frugivorous, but the fruits most consumed were from the family Araceae, especially *Anthurium* spp. (38.8% of the diet; see also Stiles & Skutch 1989 and Loiselle & Blake 1990). Consumption of melastomes was variable but relatively low in most months; only in May did melastomes (an unknown, possibly canopy species) comprise 90% of the fruit taken.

Although in only 2 of the 7 months were fruits of Melastomataceae the most consumed by the thrushes, the family was the most important overall (32% of all fruits consumed). *Miconia* aff. *smaragdina*, *Conostegia cooperi* and an unknown species were most consumed. The Solitaire (*My-*

adestes melanops) was an altitudinal migrant present only in 5 months, taking melastomes and Rubiaceae in similar proportions (30.6 and 27.6% respectively). *Ossaea* spp. were most consumed in most months, but in July 1988, *H. tuberculosa* and *Conostegia cooperi* were preferred along with *Clidemia epiphytica*, a noncensused epiphyte.

The *Mionectes* flycatchers, reported to prefer arillate fruits (Stiles & Skutch 1989; Loiselle & Blake 1990) in fact consumed Clusiaceae more than any other family (32% of fruit portion), followed by the Melastomataceae (18%), especially *H. tuberculosa*, *M.* aff. *smaragdina* and *M. centrodesma*.

Finally, *A. aurantiirostris* was poorly represented in the samples and fruit constituted only 33.7% of its diet, but nearly all of this was melastomes, especially *C. cooperi* and *Ossaea* spp.

In general, at El Plástico the importance of melastomes in the diet of the understory frugivorous birds is outstanding, even though melastomes contributed only a small portion of available fruits. Although in both gulpers and mashers the melastomes contributed an important portion of the diet, it is noteworthy that in two of the masher-type groups (tanagers and *Arremon*) the percentage value contributed by melastomes to the total fruit portion was higher than in any other (Table 1). The third masher group is *Euphonia anneae*, which as noted above, has a specialized diet based on Araceae fruits. The gulper-type groups had the higher values of large-seeded fruits and with the exeption of manakins, this type of fruits comprised over 50% of the fruit for each group. We should note, however, that this may be in part an artefact of fecal samples, as mashers typically discard large seeds while mandibulating fruit.

In most months, the melastome whose fruit was most abundant, was also the species most consumed by most or all of the frugivorous birds. The moderate tendency for some groups to prefer certain melastomes over others (e.g. manakins preferred *M.* aff. *smaragdina* to a greater extent than did tanagers, but the reverse was true for *Ossaea laxivenula*) was far outweighed by the tendency most or all species to prefer the most abundant melastome at any given time. The most important exception, lack of consumption of *H. tuberculosa* by *A. aurantiirostis* may be an artefact of the very few samples for this frugivore. Otherwise, *H. tuberculosa*, the melastome most available on the area most of the time, is a dietary staple for most of the frugivores.

The period of greatest abundance of melastome fruits (June-July 1988) coincided with the latter part of the breeding season and the start of the molt for most of the frugivore groups at El Plástico (Rosselli 1989; Stiles & Skutch 1989; cf. also Levey 1988a). This should be the most energy-demanding time for these birds, with the population swelled by many nestlings and fledglings and feather replacement beginning (cf. Stiles 1985). This, plus the moderate abundance of melastome fruits and their heavy use by understory frugivores over a 9–10 month period of each year, certainly justifies considering melastome fruits as a staple resource for these birds.

The Melastomataceae at El Plástico depend heavily on understory frugivores (chiefly manakins and tanagers) for their dispersal; these birds generally consume whatever melastomes are most abundant at a given time. Thus, one might expect selection to favor the staggering of fruiting seasons among these species, thereby reducing competition for dispersers (Snow 1965). However, this does not seem to occur at El Plástico, at least among the most abundant species. In fact, the fruiting peaks of 3 of the commonest melastomes in the area (*Henriettella*, *C. cooperi* and *M. smaragdina*) coincided between April and July 1988 (Rosselli 1989). In the areas studied by Snow (1965) & Hilty (1980), one genus (Miconia) included most of the bird-dispersed melastomes. The melastomes that contribute most fruits at El Plastico belong to three different genera; assuming that the year of observations is representative, these species evidently have very different strategies for offering their fruits. *H. tuberculosa* presents thousands of small fruits over an extended fruiting period, while *O. brenesii* invests in smaller numbers of larger fruits all year long. *C. cooperi* restricts its fruiting period to a short interval of high abundance, which in 1988 largely coincided with a major peak of fruiting of *H. tuberculosa*. The three species also differ in habitat and perhaps germination requirements; each shows a characteristic spatial distribution in the study area. All of these factors, plus the presence of numerous other fruiting plants in the study area, may dilute selection for staggering of fruiting peaks among these melastomes.

Melastome consumption by birds in other neotropical communities

For comparing data from other bird-plant communities, we consider melastomes as a group, and ask which bird species interact most intensely with melastome fruits. To obtain such 'mutual bird-melastome dependence values' we multiply the proportion of melastomes in the total number of records (visual or fecal) of frugivory for bird species x, by the proportion of all records of melastome frugivory contributed by species x (cf. Jordano 1987). In the studies of Wheelwright *et al.* (1984) and Velásquez (1991) interactions were scored as 'common', 'uncommon' and 'occasional', for which we assigned arbitrary values of 5, 3 and 1 respectively in order to calculate proportions.

In addition to El Plástico, data are available for four other bird-fruit communities in the northern neotropics (Table 2): Monteverde, Costa Rica (Wheelwright *et al.* 1984), an upper middle-elevation cloud forest and nearby open areas and edges; Reserva de Carpanta, Colombia (Arango 1990), high-elevation Andean forest and adjacent second growth; the Arima Valley, Trinidad (Snow & Snow 1971), an area of somewhat disturbed forest and tall second growth at low elevations; and San Carlos de Antioquia, Colombia, an area of wet but fairly seasonal and somewhat disturbed foothill forest (Velásquez 1991). For the Trinidad study, data were presented for fruit consumption by tanagers and honeycreepers only. Extensive fruit-frugivore community data have been taken at La Selva and Parque Nacional Braulio Carrillo, Costa Rica by Loiselle, Blake & Levey, and in SW Colombia by Hilty, but have yet to be published in detail and thus were not available for the present analysis. Because of differing methodologies and calculation procedures, the absolute values in Table 2 are not comparable between studies, nor particularly meaningful in themselves, but they do provide a reasonable relative measure of association between melastomes and different birds in a given study. The low values for most bird-plant combinations reflect the relative rarity of the birds (i.e., they contribute few feeding records), or a low preference for melastomes; few species in any community are both abundant and highly dependent upon melastome fruits (cf. Jordano 1987).

At El Plástico, the two families showing very high association with the melastomes are the manakins and tanagers. Being primarily birds of wet lowlands and foothills, the manakins are poorly represented at Monteverde and absent from Carpanta. The data set of Snow & Snow (1971) from Trinidad does not treat manakins, but other studies of these birds (Snow 1962 a,b) highlight their high degree of association with melastome berries, a conclusion also reached in studies in Costa Rica (Levey 1988a; Loiselle & Blake 1990) and in Ecuador (Prum & Johnston 1987). Many of the species and most of the genera studied by the Snows in Trinidad were also present in the Colombian study area of Velásquez, where again tanagers (notably *Ramphocelus* and *Tangara)* and manakins were the main melastome consumers.

Among the tanagers, different genera are important in different areas. The genus *Tangara* nearly always figures prominently (Table 2; see also Hilty 1980). Birds of this genus are primarily denizens of forest canopy and edge, however, and were thus poorly represented in our El Plástico study of understory frugivores. Other genera of major importance in one or more areas included *Ramphocelus*, *Tachyphonus*, *Chlorothraupis*, *Anisognathus* and *Chlorospingus*; and such 'honeycreeper' genera as *Chlorophanes*, *Dacnis*, and *Diglossopsis* (the blue *Diglossa* species, cf. Bock 1985, now usually considered to be tanagers). Members of the *Chlorophonia-Euphonia* group, by contrast, primarily consume sticky-

Table 2. The principal birds associated with melastome fruits [1] in five neotropical communities.

El plástico, Costa Rica [2,4] This study, all species	
Corapipo leucorrhoa	0.258
Chlorothraupis carmioli	0.121
Tachyphonus delattrii	0.036
Pipra pipra	0.026
P. mentalis	0.024
Euphonia annae	0.019
Myadestes melanops	0.017
Tachyphonus luctuosus	0.008
Catharus ustulatus	0.008
Arremon aurantiirostris	0.006
Schiffornis turdinus	0.006
Mionectes olivaceus	0.005
Turdus obsoletus	0.004
Mionectes oleagineus	0.004
Arima Valley, Trinidad [1,2,5] Snow and Snow 1971 (Tanagers only)	
Ramphocelus carbo	0.166
Tangara gyrola	0.060
Chlorophanes spiza	0.052
Tangara mexicana	0.033
T. guttata	0.031
Dacnis cayana	0.026
Tachyphonus rufus	0.012
Thraupis palmarum	0.011
Cyanerpes cyaneus	0.010
San Carlos, Antioquia, Colombia [3,5] Velásquez 1991: all species	
Ramphocelus dimidiatus	44/13
Manacus manacus	34/8
Tangara cyanicollis	32/8
Machaeropterus regulus	31/7
Tangara inornata	30/6
Catharus ustulatus	28/8
Chlorophanes spiza	28/6
Saltator maximus	27/7
Myiozetetes cayennensis	26/8
Tangara larvata	24/6
Hemithraupis flavicollis	24/6
Dacnis cayana	24/6
Chlorospingus flavigularis	22/6
Thraupis episcopus	22/6
Piranga olivacea	20/6
Dacnis lineata	19/5
Pipra erythrocephala	18/4
Monteverde, Costa Rica [3,5] Wheelwright *et al.* 1984: all species	
Tangara dowii	0.098
T. icterocephala	0.062
Chlorospingus ophthalmicus	0.027
Myadestes melanops	0.019
Chiroxiphia linearis	0.012
Semnornis frantzii	0.004
Chlorophonia callophrys	0.004
Columba fasciata	0.004
Turdus assimilis	0.004
Phainoptila melanoxantha	0.003
Reserva Carpanta, Colombia [2,5] Arango 1990: all species	
Tangara vassorii	0.362
Anisognatus igniventris	0.116
Tangara nigroviridis	0.006
Pipreola riefferi	0.003
Diglossa cyanea	0.003
Chlorospingus ophthalmicus	0.002
Turdus fuscater	0.001
Thraupis cyanocephala	0.001
Diglossa caerulescens	0.001

[1] Values for all studies except that of Velásquez (1991) are 'mutual dependence values' (c.f. Jordano 1987), calculated as: proportion of melastomes out of total number of frugivory records of bird species X, times proportion of all records of melastome frugivory contributed by bird species X. For Velásquez (1991), data on total melastome use (total score) and number of species of melastome visited, are given.

[2] Quantitative study.

[3] Semiquantitative data.

[4] Data from fecal samples.

[5] Mostly or entirely observational data.

seeded epiphytes such as mistletoes and *Anthurium* and are of secondary importance as melastome consumers.

Among the other bird families noted as melastome consumers, none are as closely associated with melastomes as are the manakins and tanagers. The Turdidae (especially *Myadestes)*, Emberizidae (*Arremon*, *Atlapetes*), Cotingidae (*Pipreola*), Capitonidae (*Semnornis*), Ptilogonatidae (*Phainoptila*), and Columbidae (*Columba*) may be mentioned, approximately in order of importance. Tyrannidae of the genus *Mionectes* consumed some melastomes at El Plástico but were primarily consumers of arillate seeds. However, in more open areas in Costa Rica and Colombia we have noted many larger flycatchers (e.g., *Tyrannus*, *Pitangus*, *Myiozetetes*, *Elaenia*, and Legatus) taking many melastome berries (see also Velásquez 1991).

Coevolution and distribution of melastomes and their principal consumers

The two groups primarily associated with melastome fruits in these studies, the manakins and tanagers, process these fruits by gulping and mashing, respectively (Moermond & Denslow 1985; Levey 1986, 1987). From the plant's point of view, the major difference between these techniques is that gulpers ingest all seeds, and are likely to transport them away from the parent plant before they are defecated or regurgitated. Mashers, on the other hand, usually drop many seeds below the parent plant. The number of seeds dropped is a direct function of seed size: most seeds larger than 2 mm are discarded in situ, whereas virtually all seeds smaller than 1 mm are swallowed with the pulp (Levey 1987). So far as known, this type of fruit-handling behavior is confined to the New World nine-primaried oscines, principally the tanagers and their close relatives (Moermond & Denslow 1985).

The fact that larger seeds are discarded more frecuently by mashers without being dispersed implies that, to the extent that frugivore behavior may be a major selective force on seed size, dispersal by mashers should select for very small seeds, probably embedded in a soft or liquid pulp. Dispersal by gulpers, on the other hand, should not produce strong selection on seed size beyond the obvious limit that the fruit (hence the seed) must be small enough to fit into the bird's gape. This is not to imply that frugivore behavior will be the only, or even the major determinant of seed size (see below); it does enable one to predict that, other things being equal, mashers may select for much smaller seeds than will gulpers. The typical melastome berry, with its many tiny seeds, might thus have coevolved primarily with masher-type frugivores such as tanagers and emberizine finches. The strong ecological association currently observed between melastomes and certain gulper-type frugivores, notably manakins, would not have had an appreciable impact on seed size or fruit morphology.

This hypothesis will be evaluated by examining the geographic distributions and patterns of species richness of bird-dispersed melastomes, and those of masher (especially tanagers) and gulper (especially manakins) frugivores. We would expect the distributions of bird-dispersed melastomes to agree much more closely with that of mashers than with that of the gulpers. We also examine possible evolutionary history of the interaction melastomes with these groups.

Three New World areas were chosen, based upon the fact that extensive treatments of bird and melastome distribution exist: Costa Rica, Venezuela, and the Department of Cundinamarca, Colombia. All three areas also contain a wide range of elevations and ecological conditions (see Tables 3–5 for sources). Because melastomes may also have capsular fruits and wind-dispersed seeds, we compared distribution of capsular vs. berry fruits within the Melastomataceae, on the assumption that if distribution of these two groups were to coincide, one would be less likely to place much confidence in any association of berry-fruited melastomes with the distribution of any particular group of potential dispersers.

In Costa Rica (Table 3), distributions of capsular and berry-fruited melastomes differed significantly, with the former being more diverse in dry to moist lowland and subtropical habitats, the latter in wet lowlands to subtropical elevations of the Caribbean slope; dry lowlands and very high elevations showed few melastome species (Table 3). Species from several other plant families with melastome-type berries (fleshy, many tiny seeds) show a similar distribution, albeit centered at slightly higher elevations of the Caribbean slope. The situation of the Rubiaceae is especially interesting: among the species with bird-dispersed fruits, those with many small seeds are most diverse at middle elevations, whereas those with few large seeds occur mostly at lower elevations, and more in drier areas. Among the birds, small

Table 3. Numbers of species of melastomes and other small bird-dispersed fruits, and of small avian frugivores in different geographic areas and elevations in Costa Rica.

	Pacific slope		Upper elevations			Caribbean		Totals
	Dry lowlands	Moist subtrop.	Lower montane	Montane	Paramo	wet lowlands	wet suptrop.	
Plants [1, 3]								
Melastomes								
Capsular fruits	17	18	10	3	1	2	6	57
Berry fruits	16	38	40	15	3	64	58	234
Rubiaceae								
Few-seeded	13	37	32	6	1	40	54	183
Many-seeded	2	8	17	5	2	9	8	51
Small many-seeded berries (Rubiaceae, Ericaceae, Gesneriaceae, etc.)	5	39	69	36	10	56	32	247
Smal few-seeded berrie (Rubiaceae, Myrsinaceae, Chloranthaceae, etc.)	16	47	42	10	2	55	66	238
Birds [2, 3]								
Small mashers (Tanagers, Emberizines)	7	12	17	9	5	34	31	115
Small gulpers (Manakins, flycatchers, thrushes, etc.)	26	23	21	11	1	36	47	165

[1] Plant data form Standley (1937–38).
[2] Bird data from Stiles & Skutch (1989).
[3] Results of chi-square goodness-of-fit tests between different groups (6 d.f.): capsular vs. berry-fruited melastomes: $X^2 = 42.74$, $p < 0.001$; few vs. many seeded Rubiaceae: $X^2 = 33.46$, $p < 0.001$; small mashers vs. gulpers $X^2 = 8.08$, $0.25 > p > 0.10$; small mashers vs. melastomes, $X^2 = 3.76$, $0.9 > p > 0.5$; small gulpers vs. melastomes, $X^2 = 10.78$, $0.1 > p > 0.05$.

gulpers (of the groups that take melastomes – manakins, thrushes, flycatchers, barbets, etc.) are most diverse in wet lowlands, decreasing in dry lowlands and especially at middle to high elevations; manakins, in particular, mainly occupy wet lowlands (Table 3). Small mashers (tanagers and frugivorous finches) on the other hand, are most diverse at subtropical elevations; they are notably more diverse than gulpers at high elevations, and less diverse in dry lowlands. Distribution patterns of melastomes and small mashers agree closely; melastomes and small gulpers differ more strongly, though the difference is not quite significant.

Venezuela was divided into regions and elevation belts according to the criteria of Phelps & de Schauensee (1981). The Guyana region presents special problems due to the extensive plant speciation (but much less bird speciation) associated with the tepuis and the complex savanna-forest mosaic. However, whether or not this region is included in the analysis, a highly significant difference exists between distribution patterns of capsular vs. berry-fruited melastomes. The former are much more characteristic of savannas or savanna-like habitats (including the tepuis), the latter of wet forests of low to middle elevations. Capsular fruits are also better represented in the paramos (Table 4). Among the birds, small mashers differ significantly from small gulpers, whether or not the Guyana region is included: the mashers are better represented in wet subtropical and

Table 4. Numbers of species of melastomes and frugivorous birds recorded in different regions and elevation zones in Venezuela.

Bird/melastome group	Arid Carribbean lowlands	Zulia-lago Maracaibo basin	Carribbean lowlands & foothills	Andes & northern ranges			Llanos	Guyana region	
				Subtrop.	Temperate	Paramo		Lowlands, savannas-forest	Tepuis
Melastomes [1,3]									
Capsular fruits	0	5	8	20	18	16	25	70	84
Berry fruits	3	27	60	63	27	15	34	128	64
Birds [2,3]									
Small mashers (tanagers, emberizines)	8	23	40	55	24	4	20	53	20
Small gulpers (manakins, tyrannidae, thrushes, etc.)	12	41	50	45	8	2	38	79	27

[1] Melastome data from Wurdack (1973).
[2] Bird data from Phelps and de Schauensee (1981).
[3] Chi-square values-tests for goodness of fit between different groups: capsular vs.berry-fruited melastomes, $X^2 = 51.61$, 9 d.f., $p < 0.001$; berry-fruited melastomes vs. small mashers, $X^2 = 9.98$, 7 df.: $0.25 > p > 0.10$ without including Guyana region, or $X^2 = 28.01$, 9 df., $p < 0.001$ if these are included; berry fruited melatomes vs. small gulpers, $X^2 = 30.30$, $p < 0.001$ without, or 45.03, $p < 0.001$ with Guyana region; small mashers vs. small gulpers $X^2 = 20.85$ without, 26.22 with Guyana region, in both cases $p < 0.001$.

temperate elevations, the gulpers in both dry and wet lowlands (Table 4). The distribution of the mashers accords far better than that of the gulpers with the distribution of bird-dispersed melastomes, particularly if the Guyana region is excluded.

In Cundinamarca, Colombia, altitudinal distributions of berry and capsular-fruited melastomes differed highly significantly, with capsular species more numerous in temperate and paramo elevations, berry-fruited species at tropical and subtropical elevations. Distributions of mashers show close agreement with bird-dispersed melastomes, but the gulpers differ highly significantly (Table 5).

Thus, in all three areas studied, the distribution patterns of bird-dispersed melastome species agree much more closely with those of the tanagers and their relatives than with those of the manakins and other small gulpers. Both melastomes and tanagers are most diverse at foothill and subtropical elevations and are well represented at higher elevations. In contrast, gulpers (especially manakins) are most diverse at low elevations and scarce much above the subtropics. In Costa Rica, this difference is paralleled by the difference in distribution between large- and small-seeded fruits in the Rubiaceae, and in the overall distribution of a variety of small-seeded 'masher-type' fruits, which are centered at middle elevations.

Further evidence that the melastome berry is a masher-fruit to which manakins have recently turned, comes from the use of other such fruits by tanagers (but much less often by manakins and other gulpers) in many areas. At El Plástico, the berries of various species of Ericaceae and other tiny seeded fruits (Table 1) are commonly eaten by various tanagers, but almost never by manakins (though *Corapipo* ascends freely into the canopy for other fruits, and during its displays, cf. Prum 1986; Rosselli *et al.*, pers. obs.). After the melastomes, the fruits most heavily eaten by manakins in most months were large-seeded, e.g. *Psychotria* and *Cephaelis* of the Rubiaceae. In drier areas, the staples of *Chiroxiphia* manakins are large-seeded or arillate fruits such as *Ardisia* and *Trichilia* (cf. Foster 1977). On the other hand,

Table 5. Numbers of species of frugivorous birds and melastomes recorded in different elevation zones in Dpto. Cundinamarca, Colombia.

Group	Elevation zone				
	Tropical (0–1300 m)	Subtropical (1400–2300 m)	Temperate (2400–3200 m)	Paramo (> 3200 m)	Total
Melastomes [1, 3]					
Capsular fruits	10	29	39	15	93
Berry fruits	31	57	35	10	133
Birds [2, 3]					
Small mashers	32	58	36	16	142
Small gulpers	38	40	10	3	91

[1] Melastome data from Uribe (1970).
[2] Bird data from Olivares (1969) and Hilty and Brown (1986), supplemented by personal observations.
[3] Values of Chi-square tests of goodness of fit between altitudinal distributions of different groups (3 d.f.): Capsular vs. berry fruited melastomes, $X^2 = 12.84$, $p < 0.01$; berry fruited melastomes vs. small mashers $X^2 = 1.13$, $0.9 > p > 0.5$; berry fruited melastomes vs. small gulpers, $X^2 = 13.96$, $p < 0.005$).

in many areas tanagers and their relatives feed heavily upon other small-seeded fruits such as Ericaceae (Wheelwright *et al.* 1984; Arango 1990). Perhaps chemical differences exist between fruits of melastomes, and the very similar (in size, texture, seed size) fruits of, say, ericads and that manakins have been able to adapt to the former but not the latter. Certainly the present-day feeding ecology of the manakins is often strongly associated with that of melastomes but not with other small masher-type fruits of which we are aware.

What might be the consequences to the melastomes of forming an association with the manakins in wet tropical lowlands? A notable feature of the manakins in wet lowland forests is their abundance: one or more manakins are virtually always among the birds most often captured in mist nets in the understories of these forests (Karr *et al.* 1982, 1990; Prum 1985; Martin & Karr 1986, Levey 1988a; Andrade & Mejía 1988; Rosselli 1989). Melastomes are also abundant, being especially prominent components of the early stages of gap regeneration and edge vegetation (streambanks, etc.). An increase in abundance of both parties might have become possible through their ecological association: certainly melastome seeds must reach more suitable germination sites through manakin dispersal, and manakin population density must be greater than would otherwise be the case, assuming that food is a limiting factor. Loss of weight, changes in diet and displaying activity during periods of shortage, and higher captures of manakins in high fruit production habitats are evidence of manakins being limited by fruit availability (Foster 1977; Levey 1988a, 1988b; Rosselli 1989; Schemske & Brokaw 1981; Snow 1962a). The situation appears to be different in dry forests, where only the genus *Chiroxiphia* usually occurs and does not appear to be highly dependent upon melastome fruits (cf. Foster 1977), which are relatively scarce in such forests (see above).

Is there any indication that the ecological association between manakins and melastome fruits might have evolutionary consequences? In particular, might manakins exercise some selection on the melastomes they disperse? For example, if larger seeds with greater nutrient reserves enjoy higher probabilities of germination, as has been found in many groups (Foster & Janson 1985), the acquisition of a gulper as its major disperser might relax selection for small seed size in a masher-type fruit. Interestingly, Rockwood (1985) found that only in melastomes of the families he studied in Panama and Costa Rica, seed size was significantly larger in lowland areas than above 1000 m. Another area in which the abundant manakins could exercise selective pressure is in the evolution of fruiting seasons (cf. Snow 1965). Certainly the reverse seems to be true: that abundance of melastome fruits plays a major role in the annual cycle of some manakins (e.g. Snow 1962 a,b; Levey 1988, Rosselli 1989).

The historical scenario

We will now review this coevolutionary scenario in terms of what is known about the evolution of the melastomes, their fruit characteristics, and their dispersers. The Melastomataceae belong to the order Myrtales, the oldest family of which, the Myrtaceae, was definitely in existence in the mid-Cretaceous (Raven & Axelrod 1974). The differentiation of the Melastomataceae as such probably occurred during the mid-to late Cretaceous, between the time Australia separated from Gondwanaland and the period when South America and Africa had drifted sufficiently far apart that widespread floral interchange between them was no longer possible (roughly between 100 and 60 million years ago, cf. Raven & Axelrod 1974; Renner 1989a). Raven & Axelrod (1974) proposed that the original radiation of the angiosperms occurred in West Gondwanaland, which prior to the break up of Pangaea had a dry climate. Most authorities consider the capsular fruit with small, wind-dispersed seeds to be the primitive condition in the Melastomataceae (Renner 1989b), which would accord with their original differentiation in dry or open areas. The earliest appearance of the family in the fossil record is Paleocene pollen from Colombia (Raven & Axelrod 1974); fossil leaves of the family occur widely in Eocene deposits in the Northern Hemisphere (cf. Renner 1989a). It is probable that the melastomes reached Asia via Africa and India early in the Cenozoic, but subsequently died out in Africa as conditions there became increasingly arid from the Miocene onward (Raven & Axelrod 1974). The present-day centers of diversity of the melastomes are SE Asia and South America, but the long separation of their melastome floras is indicated by the fact that no genera or tribes (if the Memecylaceae is considered a separate family) are shared between the two areas (Wurdack 1973, Renner 1989b).

According to the most widely-accepted phylogenies of the melastomes (Wurdack in van Vliet *et al.* 1981; Ter Welle & Koeck-Norman 1981), the bird-dispersed berry probably evolved indepedently in two lines of New World melastomes and once in the Old World. Although most tribes in both hemispheres are characterized by capsular fruits, the largest tribes in both hemispheres have berries (Table 6). The shift from wind to bird dispersal certainly seems to have been associated with a high degree of speciation, especially in the New World tribe Miconiae, in which the genus

Table 6. Taxonomic diversity of the Melastomataceae.

Tribes of	
New world	Old world
Blakeae (B): 2 genera, 162 spp.	Bertolonieae (C): 10 genera, 75 spp.
Cyphostyleae (C): 3 genera, 10 spp.	Dissochaeteae (B): 16 genera, ca.550 spp.
Merianeae (C): 16 genera, 214 spp.	Osbeckieae (C): 19 genera, 360 spp.
Microlicieae (C): 15 genera, 237 spp.	Oxysporeae (C): 13 genera, 150 spp.
Miconieae (B): 28 genera, ca.1700 spp.	Sonnerlileae (C): 24 genera, 360 spp.
Rhexieae (C): 3 genera, 478 spp.	
Tibouchinae (C): 26 genera, 468 spp.	
(B): berry-type fruit; (C): capsular fruit.	
Memecylaceae: 2 genera, 90 spp.	Memecylaceae: 4 genera, ca. 340 spp.
(Fruit a berry with fewer, larger seeds than in Melastomataceae).	

Miconia alone has some 1000 species (Renner 1987, 1989a). The shift to bird-dispersal was probably fundamental to the success of these groups in wet tropical forests (Renner 1989a), and probably occurred early in the Tertiary.

It is possible that having capsular fruits with numerous, small wind-dispersed seeds was a pre-adaptation for developing the small-seeded 'masher-type' fruit in New World melastomes. This logic may not hold for the Old World Dissochaeteae, whose members are mainly epiphytes and have bird-dispersed berries apparently similar to those of the New World Miconiae, albeit usually with slightly larger seeds (Regalado 1990). However, we know of no detailed studies of fruit handling and dispersal of Old World melastomes by birds; neither have we seen any detailed comparisons of fruit characteristics or seed sizes of Old and New World melastomes. Certainly the epiphytic habit in itself is often strongly associated with bird dispersal among the melastomes (Renner 1986, Regalado 1990), and can result in selection for small seeds (Rockwood 1985). This might have occurred among the New World Blakeae, which are primarily epiphytic, but not among the Miconiae, which are not (Renner 1986). More data on Old World melastomes are clearly required before their ecological and possible coevolutionary relationships with avian frugivores can be understood.

Although some data are available on the times of origin of groups of avian frugivores most closely associated with the melastomes, any inferences regarding possible coevolution of birds and melastomes must be highly speculative. There is a virtual absence of passerine fossils from the early Tertiary; most of the small arboreal land birds known from the Eocene (the first epoch from which numerous such fossils are known) represent extinct families of coraciiform or piciform affinities and unknown food habits (Feduccia 1980; Houde and Olson 1988). On the other hand, evidence from DNA-DNA hybridization places the initial divergence of the manakins (Pipridae) in the Eocene and of the tanagers and relatives (Emberizinae) only slightly later (Sibley & Ahlquist 1985). The dating procedure associated with these molecular methods has been questioned (Houde 1987), and in any case these data tell us nothing about adaptations to frugivory (if any) of the proto-manakins and proto-tanagers. It is noteworthy that the rise to dominance of the passerines in the terrestrial avifauna, and the rapid increase in prominence of many modern angiosperms, including several ornithochorous groups, both occurred in the Miocene and might have well been related (Regal 1977). The original association of the angiosperms with zoochory doubtless dates from much earlier, probably the upper Cretaceous (Tiffney 1984). Certainly there is nothing in the fossil or DNA data to preclude a tanager-melastome association extending back at least to the Miocene. In any case the uplift of the Andes in Pliocene-Pleistocene times undoubtedly resulted in extensive speciation in both tanagers and melastomes; nearly a third of the tanagers are restricted to the Andean slopes (Isler and Isler 1987).

Given the different lifespans of plant and bird species (Herrera 1985), the lack of concordance of geographical and ecological ranges of most species that interact in any given community, and the low specificity of most bird-fruit interactions (Wheelwright and Orians 1982; Schemske 1983; Howe 1984; Herrera 1986; Jordano 1987) it follows that bird-fruit coevolution must nearly always be diffuse (cf. Herrera 1982). It seems realistic to view such coevolution as occurring between evolutionary lineages – groups of phyletically related species within which the particular species vary in time and space. The open-ended nature of the species composition of communities, combined with gene flow between populations that interact with different constellations of mutualists (Horvitz & Schemske 1990), will tend to produce parallel or convergent adaptations to the mutualism among the interacting taxa. The particular species involved in the mutualism will thus have similar characteristics and will be to a considerable degree interchangeable. The bill morphology and fruit handling behavior of all of the tanagers and their relatives that interact strongly with melastomes are similar, and the fruit characteristics of melastomes and other families that interact strongly with these birds have also converged. Which members of these groups actually coexist in a given community may often be the result of random or non-evolutionary processes such as species sorting (cf. Herrera 1988). The result of such coevolution will be general 'syndromes' (e.g., 'masher-type' and 'gulper-type' fruits), rather than more specific coadaptations.

The asymmetry in plant and bird species longevity, combined with the behavioral plasticity of the birds, make it likely that the plants will be the more conservative partner in any coevolved relationship. Conversely, the acquisition of new partners in such a relationship should be via changes in foraging behavior and fruit choice in the birds, with fruit morphology being perhaps the last feature to change. This implies that strong ecological associations may occur well before precise morphological coevolution (if the latter occurs at all), just as we presently see in the association of manakins with melastome fruits. A corollary is that detailed consideration of the morphology, behavior, and distribution of the frugivores in relation to corresponding features of the fruit may lead to more realistic hypotheses concerning the limits of diffuse coevolution than simply documenting the strength of present-day interactions.

Acknowledgements

We are most grateful to those who patiently answered our queries and supplied published and unpublished materials on the Melastomataceae that made this paper possible: Frank Almeda, Clara Inés Orozco, Peter Raven, Patricia Velásquez and especially Susanne Renner. The manuscript was improved by comments from P. Charles-Dominique, Ted Fleming & Pedro Jordano. Our work at El Plástico was supported by the Organization for Tropical Studies and the Vicerrectoría de Investigación de la Universidad de Costa Rica. The staff of Rara Avis S.A., especially Amos Bien & Roberto Villalobos, helped with logistics and in many other ways. Bette Loiselle supplied several identifications of fecal samples. We thank Alejandro Estrada, Ted Fleming, the Universidad Nacional de Colombia and the Empresa de Acueducto de Bogotá for making possible our attendance to the Symposium, and to the National Science Foundation for financial support.

References

Andrade, G.I. & Mejía C. 1988. Cambios estacionales en la distribución de la avifauna terrestre en el Parque Nacional. Natural Macuira, Guajira, Colombia. Trianea 1: 147–169.

Arango, S. 1990. Aspectos morfológicos y del comportamiento de las aves frugívoras y su efectividad como dispersores en la Reserva de Carpanta, Colombia. B. Sc. Thesis, Pontífica Universidad Javeriana, Bogotá.

Blake, J.G., Loiselle B.A., Moermond T.C., Levey D.J. & Denslow J.S. 1990. Quantifying abundance of fruits for birds in tropical habitats. Studies in Avian Biol. 13: 73–79.

Bock, W.J. 1985. Is Diglossa (?Thraupinae) monophyletic? pp. 319–332 in Buckley, P.G., Foster M.S., Morton E.S., Ridgely R.S. & Buckley F.G., (eds.) Neotropical Ornithology. Ornithol. Monog. no. 36.

Feduccia, A. 1980. The age of birds. Harvard Univ. Press, Cambridge, MA.

Foster, M.S. 1977. Ecological and nutritional effects of food scarcity on a tropical frugivorous bird and its fruit source. Ecology 58: 73–85.

Foster, S.A. & Janson C.H. 1985. The relationship between seed size and establishment conditions in tropical woody plants. Ecology 66: 773–780.

Gleason, H.A. 1932. A synopsis of the Melastomataceae of British Guiana. Brittonia 1: 127–184.

Hartshorn, G.S. & Peralta R. 1988. Preliminary description of primary forests along the La Selva-Volcán Barva altitudinal transect, Costa Rica. pp. 281–296 in F.Almeda and C.M. Pringle, (eds). Tropical rainforests: diversity and conservation. Calif. Academy of Sciences, San Francisco, Calif.

Herrera, C.M. 1982. Seasonal variation in quality of fruits and diffuse coevolution between plants and avian dispersers. Ecology 63: 773–785.

Herrera, C.M. 1985. Determinants of plant-animal coevolution: the case of mutualistic dispersal of seeds by vertebrates. Oikos 44: 132–141.

Herrera, C.M. 1986. Plant adaptations for vertebrate dispersal: why they don't behave the way they should. pp 5–18 in Estrada, A. & T.H. Fleming (eds.) Frugivores and Seed Dispersal. W. Junk, Dordrecht.

Herrera, C.M. 1988. Avian frugivory and seed dispersal in mediterranean habitats: regional variation in plant-animal interaction. pp. 509–517 in H. Ouellet (ed.). Acta XIX

Congressus Internationalis Ornithologici, vol. I. Ottawa Press, Ottawa, Canada.

Herrera, C.M. & Jordano P. 1981. Prunus mahaleb and birds: the high efficiency seed dispersal of a temperate fruiting tree. Ecol. Monogr. 51: 203–218.

Hilty, S.L. 1980. Flowering and fruiting periodicity in a premontane rain forest in Pacific Colombia. Biotropica 12: 292–306.

Hilty, S.L. & Brown W.L. 1986. A Guide to the Birds of Colombia. Princeton University Press, Princeton N.J.

Horvitz, C.C. & Schemske D.W. 1990. Spatiotemporal variation in insect mutualists of a neotropical herb. Ecology 71: 108–1097.

Houde, P. 1987. A critical evaluation of DNA hybridization studies in avian systematics. Auk 104: 17–32.

Houde, P. & Olson S.L. 1988. Small arboreal nonpasserine birds from the Early Tertiary of North America. Proceedings of the 19th International Ornithological Congress: 2030–2036.

Howe, H.F. 1984. Constrains on the evolution of mutualisms. Am. Nat. 123: 764–777.

Howe, H.F. 1977. Bird activity, and seed dispersal of a tropical wet forest tree. Ecology 58: 539–550.

Isler, M.L. & Isler P.R. 1987. The tanagers: natural history, distribution, and identification. Smithsonian Institution Press, Washington D.C.

Janzen, D.H. & Martin P.S. 1982. Neotropical anachronisms: the fruits the gomphotheres ate. Science 215: 19–27.

Jordano, P. 1987. Patterns of mutualistic interactions in pollination and seed dispersal: connectance, dependence asymmetries, and coevolution. Am. Nat. 129: 657–677.

Jordano, P. & Herrera C.M. 1981. The frugivorous diet of Sylvia atricapilla wintering in Southern Spain. Ibis 123: 502–507.

Karr, J.R., Schemske D.W. & Brokaw N. 1982. Temporal variation in the understory bird community of a tropical forest. pp 441–453 in Leigh, E.G., A.S. Rand & D.M. Windsor, (eds). The ecology of a tropical forest: seasonal rhythms and long-term changes.

Karr, J.R., Robinson S.K., Blake J.G. & Bierregarrd R.O. 1990. The birds of four neotropical forests pp 237–269 in A.H. Gentry, (ed.) Four Neotropical Rainforests. Yale University Press, New Haven, Ct.

Levey, D.J. 1986. Methods of seed processing by birds and seed deposition patterns. pp 147–158 in A. Estrada & T.H. Fleming (eds.) Frugivores and Seed Dispersal. W. Junk, Dordrecht.

Levey, D.J. 1987. Seed size and fruit handling techniques of avian frugivores. Am. Nat. 129: 471–485.

Levey, D.J. 1988a. Spatial and temporal variations in Costa Rican fruit and fruit-eating bird abundance. Ecol. Monog. 58: 252–269.

Levey, D.J. 1988b. Tropical wet forest treefall gaps and distribution of understory birds and plants. Ecology 69: 1076–1098.

Loiselle, B.J. & Blake, J.G. 1990. Diets of understory fruit-eating birds in Costa Rica. Studies in Avian Biol. 13: 91–103.

Loiselle, B.J. & Blake, J.G. 1991. Temporal variation in birds and fruits along an elevational gradient in Costa Rica. Ecology 72: 180–193.

Martin, T.E. & Karr, J.R. 1986. Temporal dynamics of neotropical birds with special reference to frugivores in second-growth woods. Wilson Bull. 98: 38–60.

McKey, D. 1975. The ecology of coevolved seed dispersal systems. pp 159–191 in L.E. Gilbert & P.H. Raven (eds.) Coevolution of Animals and Plants. Univ. of Texas Press, Austin, Texas.

Moermond, T.S. 1983. Suction-drinking in tanagers and its relation to fruit handling. Ibis 125: 545–549.

Moermond, T.S. & Denslow, J.S. 1985. Neotropical avian frugivores: patterns of behavior, morphology, and nutrition with consequences for fruit selection. pp 865–897 in P.A. Buckley, M.S. Foster, E.S. Morton, R.S. Ridgely & F.G. Buckley, (eds). Neotropical Ornithology. Ornithological Monographs No. 36.

Moody, D.T. 1970. A method for obtaining food samples from insectivorous birds. Auk 87: 579.

Olivares, A. 1969. Aves de Cundinamarca. Universidad Nacional de Colombia, Bogotá.

Phelps, W.H. & Meyer de Schauensee, R. 1981. Las Aves de Venezuela. Gráficas Armitaño, Caracas.

Prum, R.O. 1985. Observations of the White-fronted Manakin (*Pipra serena*) in Suriname. Auk 102: 384–387.

Prum, R.O. 1986. The displays of the White-throated Manakin (*Corapipo gutturalis*) in Suriname. Ibis 128: 98–102.

Prum, R.O. & Johnston A.E. 1987. Display behavior, foraging ecology, and systematics of the Golden-winged Manakin (*Masius chrysopterus*). Wilson Bull. 99: 521–539.

Raven, P.H. & Axelrod, D.I. 1974. Angiosperm biogeography and past continental movements. Ann. Missouri Bot. Gard. 61: 539–673.

Regal, P.J. 1977. Ecology and evolution of flowering plant dominance. Science 196: 622–629.

Regalado, J.C. 1990. A revision of the *Medinilla* (Melastomataceae) of Borneo. Blumea 35: 5–70.

Renner, S.S. 1986. Tne neotropical epiphytic Melastomataceae: phytogeographic patterns, fruit types, and floral biology. Selbyana 9: 104–111.

Renner, S.S. 1987. Special topics: seed dispersal. Progress in Botany 49: 413–432.

Renner, S.S. 1989a. A survey of reproductive biology in neotropical Melastomataceae and Memecylaceae. Ann. Missorui Bot. Gard. 76: 496–518.

Renner, S.S. 1989b. The current classification of the Melastomataceae. Flora Malesiana Symposium, pp. 52–54.

Rockwood, L.L. 1985. Seed size and plant habit in seven families from Panamá and Costa Rica. pp. 197–205 in D'arcy, W.G. & M.D. Correa-A. (eds.) The botany and Natural History of Panamá. Missouri Bot. Gard., St. Louis, MO.

Rosselli, L. 1989. El ciclo anual de un ave frugívora migratoria altitudinal, *Corapipo leucorrhoa* (Pipridae) y los frutos que consume. M.Sc. Thesis, Universidad de Costa Rica, San José.

Schemske, D.W. 1983. Limits to specialization and coevolution in plant-animal mutualisms. pp. 67–109 in Nitecki, M.H. (ed.). Coevolution. Univ. of Chicago Press, Chicago, Ill.

Schemske, D.W. & Brokaw, N.V.L. 1981. Treefalls and the distribution of understory birds in a tropical forest. Ecology 62: 938–945.

Sibley, C.G. & Alquist, J.E. 1985. Phylogeny and classification of New World suboscine passerine birds (Passeriformes: Oligomyodi: Tyrannides) pp. 396–430 in P.A. Buckley, M.S. Foster, E.S. Morton, R.S. Ridgely & F.G. Buckley, (eds). Neotropical Ornithology. Ornithological Monographs No. 36.

Snow, B.K. & Snow, D.W. 1971. The feeding ecology of tanagers and honeycreepers in Trinidad. Auk 88: 291–322.

Snow, D.W. 1962a. A field study of the Black and White Manakin, *Manacus manacus*, in Trinidad, Zoologica 47: 65–104.

Snow, D.W. 1962b. A field study of the Golden-headed Manakin, *Pipra erythrocephala*, in Trinidad, Zoologica 47: 183–189.

Snow, D.W. 1965. A possible selective factor in the evolution of fruiting seasons in tropical forest. Oikos 15: 274–281.

Standley, P.C. 1937–1938. Flora of Costa Rica, Parts I-IV. Field Mus. Nat. Hist., Bot. Ser. 18, Chicago.

Stiles, F.G. 1985. Seasonal patterns and coevolution in the hummingbird-flower community of a Costa Rican subtropical forest. pp. 757–787 in P.A. Buckley, M.S. Foster, E.S. Morton, R.S. Ridgely & F.G. Buckley, (eds). Neotropical Ornithology. Ornithological Monographs No. 36.

Stiles, F.G. & Skutch, A.F. 1989. A guide to the birds of Costa Rica. Cornell University Press, Ithaca, N.Y.

Terborgh, J. 1986. Keystone plant resources in the tropical forest. pp. 330–344 in Soulé, M.E. (ed.). Conservation Biology: the science of scarcity and diversity. Sinauer Associates, Inc., Sunderland, Mass.

Ter Welle, B.J.H. & J. Koek-Norman. 1981. Wood anatomy of the neotropical Melastomataceae. Blumea 27: 365–394.

Tiffney, B.H. 1984. Seed size, dispersal syndromes, and the rise of angiosperm dominance: evidence and hypohtesis. Ann. Missouri Bot. Gard. 71: 551–576.

Uribe-U., L. 1972. Catálogo ilustrado de las plantas de Cundinamarca, V: Passifloraceae, Begoniaceae, Melastomataceae. Instituto de Ciencias Naturales. Bogotá.

van Vliet, G.J.C.M., Koek-Norman J. & Ter Welle B.J.H. 1981. Wood anatomy, classification, and phylogeny of the Melastomataceae. Blumea 27: 463–473.

Velásquez, M.P. 1991. Aves frugívoras y su relación con la flora en el Municipio de San Carlos, Antioquia, Colombia. B.Sc. Thesis, Universidad de Antioquia.

Wheelwright, N.T., Haber W.A., Murray K.B. & Guindon C. 1984. Tropical fruit-eating birds and their food plants: a survey of a Costa Rican lower montane forest. Biotropica 16: 173–192.

Wheelwright, N.T. & Orians G.H. 1982. Seed dispersal by animals: contrasts with pollen dispersal, problems of terminology, and constraints on coevolution. Am. Nat. 119: 402–413.

Wurdack, J.J. 1973. Flora de Venezuela, vol 8: Melastomataceae. Instituto Botánico, Caracas. 819 pp.

Vegetatio **107/108**: 75–84, 1993.
T. H. Fleming and A. Estrada (eds). Frugivory and Seed Dispersal: Ecological and Evolutionary Aspects.

Speciation and coevolution: an interpretation of frugivory phenomena

P. Charles-Dominique
URA 1183/C. N. R. S., Laboratoire d'Ecologie Générale, Muséum National d'Histoire Naturelle – 4, Avenue du Petit château, 91800 Brunoy, France

Keywords: Frugivory, Coevolution, Specific richness, Tropical forest

Abstract

Evolutionary considerations of frugivory must account for patterns of diversity and abundance of both frugivorous vertebrates (especially pulp-eaters) and of plants producing zoochorous fruits. Analysis of the diet and behavior of representative species of frugivores from the tropical forests of French Guiana (seven species of bird and eight mammals) allowed their degree of specialization for frugivory to be assessed. Analysis of the flora associated with these species showed that genera associated with specialized vertebrates had the greatest species richness. This pattern suggests that plant animal co-evolution occurs in a step-wise fashion with short transitory phases of animal and plant evolution followed by long periods of stability.

Introduction

Seed dispersal mechanisms of plants producing fleshy fruits are based upon a plant-animal mutualism (zoochory) that is common in tropical rainforests (Corner 1949; Snow 1966, 1971; Van der Pijl 1969; Hladik & Hladik 1969; McKey 1975; Howe 1977; Janzen 1983; Howe & Smallwood 1982; Janson 1983; Gautier-Hion *et al.* 1985 a). In these associations, plants provide nutritious tissues closely associated with the seed(s), and frugivorous animals eat these fruits, then regurgitate, defecate or drop viable seeds at a variable distance. Theoretical aspects of animal specializations and fruit evolution have been widely discussed (Corner (1949), Van der Pijl (1957, 1972), Snow (1966, 1971), McKey (1975), Howe (1977), Howe & Estabrook (1977), Janzen (1980), Herrera (1982, 1985), Fleming (1986) and many others). Vertebrate specializations have been correlated with pulp richness and seed size (McKey 1975; Snow 1981), fruit morphology (Van der Pijl 1957, 1972; Beehler 1983), fruit coloration (Van der Pijl 1957, 1972; Snodderly 1979; Weelwright & Janson 1985; Cooper *et al.* 1986), fruit phenology (Snow 1966, 1971; Fleming 1985, 1986; Fleming *et al.* 1985; Weelwright 1986), fruit positions (Moermond & Denslow 1983), but rarely with plant taxonomy. This last aspect has been discussed for some communities of frugivores (Snow 1966; Fleming 1986; Beehler 1989), or at a very general level for the historical (evolutionary) aspects of flowering plants and fleshy fruit formation (Crepet 1984; Doyle 1977; Fleming 1991; Herrera 1989; Heywood 1978; Regal 1977; Stebbins 1981; Tiffney 1986). Most often, fleshy fruit evolution and plant phylogeny and taxonomy are considered and discussed separately. This anomaly is due to the difficulties to establish a classification of the fruits, reflecting both functional aspects, and plant taxonomy and phylogeny (Van der Pijl 1972; Sabatier 1983). As a matter of fact, in various independent phyla, tissues from different embryological origins can give, by convergence, analogous structures resulting in very similar fruit anatomy. Conversely, in the same phylum, homologous tissues of the embryo can diverge (Sabatier 1983). In this paper, plant-animal mutualisms will be analyzed using, on one hand, different examples of frugivory, and on the other, species diversity of the plant genera involved, all coming from the same tropical forest ecosystem in French Guiana. An interpretation of the possible evolutionary mechanisms supporting such patterns will be proposed.

Feeding choices in frugivorous vertebrates

Sufficient botanical and zoological data have been collected in French Guiana during the past ten years to examine the fruit categories and the taxonomic aspects of plants associated with representative categories of frugivorous birds and mammals. The eight following categories of vertebrates will be considered in order to illustrate the main mechanisms of frugivory: six species of manakins, *Manacus manacus*, *Pipra pipra*, *P. erythrocephala*, *P. serena*, *Corapipo gutturalis*, *Tyranneutes virescens* (Pipridae); the cock-of-the-rock, *Rupicola rupicola* (Cotingidae); the short-tailed fruit bat, *Carollia perspicillata* (Phyllostomidae); three bat species of the *Artibeus* group, *A. lituratus*, *A. Jamaicencis*, *A. fuliginosus* (Phyllostomidae); the woolly opossum, *Caluromys philander* (Didelphidae); the kinkajou, *Potos flavus* (Procyonidae); the howler monkey, *Alouatta seniculus* (Cebidae); and the spider monkey, *Ateles paniscus* (Cebidae). This last species is present in French Guiana but most data come from the study of Van Roosmalen (1980) in similar forest in neighbouring Surinam. These 15 species do not represent the entire community of frugivorous vertebrates (45 frugivorous mammals, pers. obs. and 96 frugivorous birds; Erard 1988; Tostain pers. com.), but they illustrate the principal feeding strategies occuring in Guyana.

Botanical material kept in the 'Centre ORSTOM de Cayenne' herbarium since its creation in 1956 represent 49,000 specimens of 3,748 taxa. Thanks to the computerization of this herbarium (Hoff & Brisse 1986; Cremers *et al.* 1988), all relevant data can be rapidly accessed. The present analysis is also based on data in the book 'Fruits of Guyanas' (Van Roosmalen 1985), which describes 1446 fruit species.

In order to explore the interactions between plants and frugivorous animals (Table 1), the fruit consumed by the animal has been classified into three main categories depending on the sensory ability of vertebrates and their ability to reach the edible parts. These categories include: 1) Contrasting colouration when ripe *vs* homochromic colouration (cryptic fruits); 2) external pulp which is directly accessible *vs* internal pulp protected by a tough tissue (pericarp); in the latter case, the fruit can be either dehiscent (spontaneously opening when ripe) or indehiscent, requiring the animal to actively tear the pericarp; and 3) seed cross section (for medium to large seeds that are habitually swallowed): flattened *vs* subcircular. In addition, the number of genera and families corresponding to the different species consumed by each species (or group) is indicated.

Diets

Pipriadae: The six species of manakins overlap in as much as 62–90% of their diets (Théry 1990). They feed only upon fruits of contrasting color with directly accessible pulp (*i.e.* external pulp or dehiscent fruits). The important characteristic of this group is that among the 54 fruit species selected, 20 are of the Melastomataceae (including 12 species of *Miconia*) and 22 are from the genus *Psychotria* (Rubiaceae). Most of these fruits contain small seeds that pass through the gut, but the largest *Psychotria* seeds are regurgitated. Most of the fruit produced by these plants is eaten by manakins (61 to 89 % for the Melastomataceae and 52 to 75 % for the *Psychotria* species). They are produced sequentially throughout the year with six to 17 species producing fruits at the same time (Théry 1990).

Rupicola rupicola: Over a five week period, the

Table 1. Main characteristics of fruits eaten by different birds and mammals in French Guiana. The taxa are presented according to the number of species, genera and families to indicate the degree of diet diversification. 'Colored fruits' are those which contrast against the background foliage, in contrast to fruits with a green coloration when ripe. Pulp accessability depends on the toughness of the pericarp. Fruits lacking a pericarp are directly accessible. When the pericarp is hard, it can either be dehiscent when ripe or indehiscent, in which case it needs to be torn apart with strong dentition. Seed shape refers to the cross-section of large seeds (relative to the size of the frugivore) that are swallowed rather than regurgitated.

	Number of taxa eaten spp/Gen/Fam (N)	Colored fruit vs green fruit (N)	PULP ACCESS direct/dehiscent/ indehiscent (N)	SEED SHAPE flat/spherical (N)
Manacus manacus *Pipra pipra* *P. erythrocephala* *P. serena* *Corapipo gutturalis* *Tyranneutes virescens*	54/12/7	54/0	47/7/0	
Rupicola rupicola	59/48/31	57/2	37/14/8	0/4
Carollia perspicillata	17/6/6	0/17	11/1/5	1/0
Artibeus lituratus *Artibeus jamaicencis* *Artibeus fulginosus*	14/6/6	0/14	13/0/1	
Caluromys philander	36/28/18	17/19	18/8/10	
Potos flavus	30/29/22	19/11	19/2/9	4/5
Ateles paniscus	147/84/47	108/39	67/32/48	67/30
Alouatta seniculus	110/57/40	97/13	45/9/56	62/12

cock-of-the-rock consumed 65 species of fruit (59 of which were identified). However, only four species (*Pseudoxandra cuspidata* (Annonaceae), *Protium apiculatum* (Burseraceae), *Oenocarpus* cf. sp nov (Arecaceae), and *Siparuna decipiens* (Monimiaceae) represented 50 % of the items (Erard *et al.* 1989). As in the Pipridae, most fruits were of contrasting colors and few had a pulp protected by an indehiscent tissue (when it occurs, this tissue is not very tough). However, despite similar anatomy, these fruits were different from those selected by Pipridae, and the 59 species belong to 48 genera and 31 families. The cock-of-the-rock may share the resources (and compete?) with many other birds and mammals (Julien-Laferrière 1989; Théry 1990; Julliot in prep.).

Bats: *Carollia perspicillata* feeds extensively on fruits of the genera *Solanum* (Solanaceae), *Piper* (Piperaceae), and *Vismia* (Hypericaceae). The genera are easy to identify by examing the seeds collected in bat feces, but determination of species is very difficult because many congeners produce very similar seeds. The 17 species censused from fecal analysis at Piste de St Elie (Charles-Dominique 1991) represent the minimum breadth of diet. In Brazil (Dos Reis & Guillaumet 1983) or in Costa Rica (Fleming 1988), fruits of other species of *Piper* and *Solanum* also constitute the staple food of this bat. The three *Artibeus spp.* (Charles-Dominique 1986 and pers. obs.) feed mostly upon *Cecropia obtusa* (Cecropiaceae) and *Ficus spp.* (Moraceae), which have small seeds that are ingested, and *Licania spp.* (Chrysobalanaceae), *Andira spp.* (Caesalpiniaceae), *Symphonia globuliphera* (Clusiaceae), *Tyrsodium guianensis* and *T. schombugkianum* (Anacardiaceae), which contain a single large seed (sometimes 2–3 seeds for *Symphonia*).

Artibeus spp. and *Carollia perspicillata* eat a completely different set of fruits, all of which are cryptic (green when ripe; see Fleming 1986 for a similar situation in Costa Rica). Some of these species were monitored in order to estimate the proportions consumed by bats and by other vertebrates (De Foresta *et al.* 1984; Charles-Dominique 1986; Charles-Dominique & Cooper 1986). Most of them were eaten at the night: for *Solanum rugosum*, *Vismia guyanensis* and *Piper hostmanianum*, respectively, 94%, 91% and 93%, principally by *Carollia perspicillata* and to a lesser extent by two other phyllostomid bats, *Sturnira lilium* and *S. tildae*. Diurnal removal was by the opportunist tanager *Ramphocelus carbo* (see Palmeirim *et al.* 1989 for a description of a similar situation in Costa Rica). For *Cecropia obtusa*, 80% of the consumption occured at night, essentially by *Artibeus spp.* and to a lesser extent, by the kinkajou and marsupials (Charles-Dominique 1986). Diurnal removal (20%) was the result of a great number of vertebrates (birds and monkeys). Fruits of *Cecropia*, *Solanum*, *Piper*, and to a lesser extent *Vismia*, reach maturity in small numbers over an extended period. The combination of such long periods of fruit production and seasonal phenological differences among congeneric plant species results in the steady production of fruit throughout the year (Fleming 1986; Charles-Dominique 1986, 1991).

Nocturnal arboreal mammals: only 11 out of the 36 species of fruit consumed by the woolly opossum *Caluromys philander* are dispersed by endozoochory (Julien-Laferrière 1989), and 15 out of 30 for the kinkajou *Potos flavus* (Julien-Laferrière 1989). Fruits that are not swallowed are generally dropped near the parent trees. No particular fruit characteristics appear to dictate these choices, and Atramentowicz (1988) concluded that fruit selection was principally determined by pulp access ability and not by pulp composition or fruit color. Both species (opossum and kinkajou) share these fruit resources with many other nocturnal and diurnal frugivorous vertebrates. These fruit species correspond to a great number of genera and families of various morphological types (Table 1).

Primates: *Ateles paniscus* is exclusively vegetarian with fruit contributing, according to the season, from 60 to 95% of its food items (Van Roosmalen 1980). Of 147 fruit species eaten by the spider monkey, 75% of its diet comes from 50 species. The howler monkey *Alouatta seniculus* has a more folivorous diet, which in French Guiana includes 114 fruit species contributing to 21.5% of its diet (Julliot & Sabatier in press). These two primates behave as opportunistic frugivores able to eat a large number of fruit species of different types, many of which are also consumed by other vertebrates. However, some fruit species are almost exclusively eaten by *A. paniscus*, *A. seniculus*, *Cebus apella* and *C. olivaceus*: many species of the genus *Inga* (Mimosaceae) and *Pouteria*, *Micropholis*, *Chrysophyllum*, *Manilkara* (Sapotaceae). All of these species have oblong seeds of medium to large size (2–3 cm long), a rather flattened cross-section, and pulp that is closely attached to the seed and is protected by a tough indehiscent pericarp. These fruits are generally inaccessible to frugivorous birds, which can only swallow directly accessible diaspores, and small mammals, whose teeth are not strong enough to crack hard tissues.

Specializations to frugivory

Diets of frugivorous vertebrates have often been described as 'specialized' or 'opportunistic' according to the proportion of fruit they ingest (Snow 1971, 1980; Erard *et al.* 1989); the diet of 'specialists' or 'total frugivores' (Morton 1973) is nearly exclusively composed of fruit. However, the term 'specialist' is used by Howe & Estabrook (1977) for vertebrates that efficiently disperse seeds, and by McKey (1975) and Snow (1981) only for vertebrates that exploit large-seeded fruits providing a pulp rich in lipid and protein. These vertebrates are generally contrasted with 'partial frugivores' (McKey 1975), and to generalist frugivores (Howe & Eastabrook 1977; Snow 1981). These definitions are useful for zoological studies (particularly those of Morton) but are not precise enough to describe plant-animal interactions because they do not take into account the selection of particular fruit categories (syndromes) by the animal. For this reason, it seems more relevant to retain the terms 'specialized frugivores' (*i.e.* those associated with plants that correspond to only one or two syndromes) and 'non specialized frugivores' (*i.e.* those associated with plants that correspond to several syndromes) to refer only to plant-animal interaction, irrespective of other independent specializations such as insectivory or folivory.

The examples of table 1 illustrate the relationships existing between fruit types and the main groups of frugivorous vertebrates. Some vertebrate species are restricted to plant species that correspond to only one of these syndromes. Examples include manakins, cock-of-the-rock and phyllostomid bats, which can be considered as 'specialized frugivores.' Other vertebrates feed upon fruits corresponding to several syndromes, if they can reach the pulp (woolly opossum, kinkajou, primates). They can be considered as 'non-specialized frugivores' irrespective of their impact on seed dispersal (depending on the plant species involved). However, specialist frugivores which feed predominantly upon one fruit category are able to consume opportunistically some other fruit categories in small quantities (Fleming 1986 for the case of the phyllostomid bat *Carollia*).

Diversity of seed dispersers

Similarly, some plants have their fruits eaten by an assemblage of many species of frugivores while other plant species are visited by a small number of vertebrates. In fact, the degree of association of plant species with different frugivorous vertebrates must be evaluated by quantifying the proportion of fruit removed by each of them. In French Guiana *Cecropia obtusa*, *Solanum spp.*, *Piper spp.*, *Vismia spp.*, *Miconia spp.*, *Psychotria spp.*, *Inga spp.*, *Pouteria spp.*, *Chrysophyllum spp.*, *Micropholis spp.*, *Manilkara spp.*, correspond to fruit species mostly associated with assemblages of frugivores composed of a small number of vertebrate species (generally belonging to the same genus or the same family), whereas other fruit species are consumed by a larger vertebrate community, often composed of species belonging to different phyla (*Schefflera paraensis* (Araliaceae), *Cecropia sciadophylla* (Cecropiaceae), *Goupia glabra* (Celastraceae), *Simarouba amara* (Simaroubaceae), and many Clusiaceae, Annonaceae, etc. (Erard *et al.* 1989, 1991; Théry 1990; Julliot & Sabatier in press). Gautier-Hion *et al.* (1985b) showed that, in Gabon, most fruits were consumed by assemblages of very diversified frugivorous vertebrates: 'bird-monkey syndrome,' 'ruminant-odent-elephant syndrome.'

Frugivory and plant species richness

In many studies devoted to the inter-relationships between frugivorous vertebrate species (or groups) and their food, more attention have been paid to the type(s) of fruits consumed rather than to plant taxonomy (Beehler 1983; Gautier-Hion *et al.* 1985b; Janson 1983; Janson *et al.* 1986; Levey *et al.* 1984; McKey 1975; Milton 1984; Moermond *et al.* 1986; Snow 1971, 1981; Wheelwright 1985; Wheelwright *et al.* 1984; Willson & Thompson 1982). However, due to convergent evolution, the same type of fruit can exist in several plant phyla. It is thus of great importance to know whether the different fruit species associated with an animal belong to the same genus, the same family or different families. For example, the manakins and the cock-of-the-rock both feed mostly upon similar fruit types (coloration and pulp access, Table 1), but of the 54 species consumed by manakins, 12 belong to the genus *Miconia* and 22 to the genus *Psychotria* (Théry 1990), whereas the 59 species eaten by the cock-of-the-rock belong to 48 different genera and 31 different families (Erard *et al.* 1989). So, among 'specialized frugivores' it is necessary to distinguish those which consume fruits that belong to a few genera or families as opposed to those eating fruits that belong to many different genera and families.

In order to evaluate the species richness of the different plant genera exploited by the frugivorous vertebrate community, 3300 woody plants collected in French Guianan forests (Hoff & Brisse 1986; Cremers *et al.* 1988) have been grouped by genus (Fig. 1). From a total of 872 genera, 753

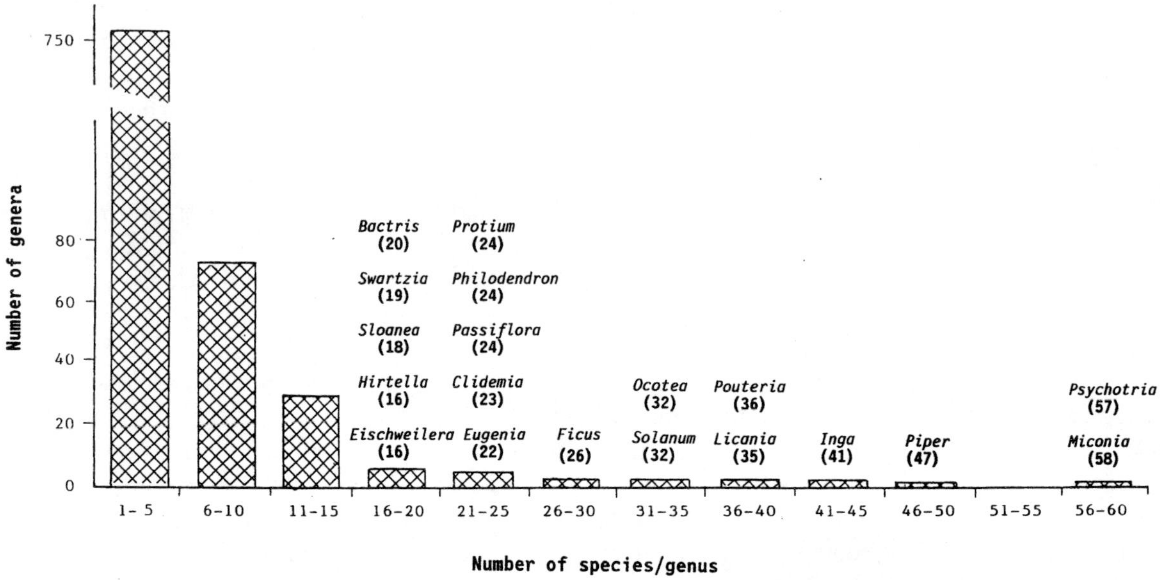

Fig. 1. Classification of woody plants grouped by genus in relation to their species richness. The number of species censused in French Guiana belonging to the most speciose genera is indicated in parentheses. These genera are associated with the most specialized frugivorous vertebrates.

(86%) have only 1 to 5 species/genus (= 45% of all 3300 species). In contrast, the 9 richest genera (= 1%, 6 last columns of the Figure) correspond to 399 (12%) species. The present analysis shows that these are the species on which specialized frugivores associated with a limited number of genera and families feed most often: Pipridae (Cf. *Miconia* and *Psychotria*), *Carollia perspicillata* (Cf. *Solanum* and *Piper*), *Artibeus spp.* (Cf. *Licania* and *Ficus* – restricted to species producing cryptic figs.) In addition, while primates consume a great number of fruit species, they behave as specialist frugivores towards the genus *Pouteria* and *Inga*. In short, plants that belong to genera with the largest number of species are principally consumed by a small number of frugivorous vertebrates with specializations which allow them first access. Similar situations have been described by Snow (1966) for manakins and the genus *Miconia*, by Fleming (1986) for *Carollia* and *Artibeus* bats associated with the genera *Piper*, *Solanum* and *Ficus*, and by Beehler (1989) for birds of paradise associated with certain Meliaceae in Papua-New Guinea. This last author concludes that 'The three genera known to be important food plants for birds of paradise (Aglaia, 104 species; *Dysoxylum*, 57 species; and *Chisocheton*, 32 species) support 193 species in the region, whereas the remaining ten genera, for which we have no records of paradisaeid, support only 18 species in total'. Such a relationship between consumption by specialized frugivores and species richness can be extended to the scale of the ecosystem, and it is important to take this into account when discussing mechanisms leading to mutualisms between plants and frugivorous vertebrates.

Discussion – Conclusion: Interpretation of plant-animal coevolution

An exclusive coevolution between one plant species and one vertebrate species could not occur for such a complex situation as frugivory. Janzen (1980) proposed a more satisfactory interpretation, that is one of 'diffuse coevolution,' which considered the selective pressures exerted between one group of animal species and one group of plant species linked by mutualistic interactions.

The principle of diffuse coevolution is generally regarded as explaining the phenomenon of frugivory, but, as pointed out by Herrera (1982, 1985), plant-animal interactions are not necessarily a direct consequence of such evolutionary mechanisms. For example, some fleshy fruit structures were present long before the vertebrates that, at present, are the main consumers and seed dispersers of these plants (plant-disperser interactions commonly observed in artificial systems recently assembled by man). In addition, the large differences between generation time of a tree and that of its associated small vertebrates (Cf. Stebbins 1949; Brodkorb 1960, 1971; Levin & Wilson 1976; Bush *et al.* 1977) leads Herrera (1985) to the conclusion that 'Slower species turnover of woody plants over geological time relative to vertebrate dispersers may have favoured a sort of very coarse diffuse coevolution, through long-lived plant species having acted as 'carriers' of information through evolutionary time.'

Numerous examples show that interactions between plants and frugivorous vertebrates concern species assemblages. For both plants and animals these assemblages are usually composed of species belonging to different families or different orders. The more an assemblage is composed of phyletically unrelated organisms, the more their respective selective pressures will work in opposition (= 'dilution of selective pressures,' Herrera 1985). Therefore, there is little likelihood that interactions between two assemblages, each one composed of unrelated organisms, will induce particular structures by coevolution. In contrast we can postulate that intimate interactions between two assemblages, each one composed of related species (of the same genus or family), can induce a coevolutionary process leading to the elaboration of particular structures. In this last situation, the selective pressure of each species of an assemblage will be very similar (*i.e.* a cumulative effect *vs* dilution). Therefore the development of a coevolutionary process between plants and frugivorous vertebrates requires that each assemblage is composed of closely related species (*i.e.* a condition necessary for preventing the 'dilution effect').

Theoretical scheme

First of all, we must recognize that phyla do not evolve uniformly (see for example Gingerich (1977) for the evolution of early mammals). During a relatively short geological period, a group of animals or plants may undergo an intense phase of speciation, while other groups are stagnant. In fact, in plant and animal communities, some genera of families are represented by a large number of species, while other genera and families are represented by very few species. In the first case, we can supose that the group has recently expanded or is going to expand, and in the second case, the group may be considered stagnant. When a plant taxon displays such an intense process of speciation, whatever characteristic this evolutionary success is based on, the species of this taxon will become abundant enough to provide a predictable and regular fruit supply over long periods, or even throughout the year. Under such conditions, and depending on the particular circumstance and ecological situation of these plants, an assemblage of frugivorous vertebrates can focus on them, becoming specialized to feed predominantly on their fruits. If this assemblage of vertebrates is composed of closely related species, the conditions required for the establishment of a coevolutionary process are met (cumulative effect), leading to various specializations of both plants and vertebrates associated with this particular mode of frugivory. However, during the course of this coevolutionary process, other vertebrates searching for new food resources (or coming by immigration) can progressively adapt to these fruits which will then be increasingly consumed by a broader range of frugivores. At a certain stage, the sum of all interactions will balance ('dilution effect'), slowing down or stopping the coevolutionary process. Fruit structures will be maintained according to the level of efficient seed dispersal achieved by this new array of vertebrates, but coevolutionary processes will stop.

In such a theoretical scheme, plant-animal coevolution will occur in steps (= step coevolution) during short transitory phases of speciation, followed by long periods of stability maintained by the multiple interactions between plants and frugivorous vertebrates. The global situation revealed by analyzing an ecosystem will in fact represent the juxtaposition of numerous old coevolutionary systems, many of which (those represented by genera poor in number of species) are in a stable phase maintained by animal partners that have not necessarily contributed to their original evolution. At the opposite, genera represented by a high number of species could be in a phase of active coevolution.

Acknowledgements

This work was based upon a part of the data collected by my colleagues from ORSTOM, Museum, CNRS and the University Paris VI engaged in the general programme on forest regeneration and plant-animal inter-relations.

I would also like to thank W. Foley for helping to translate the first manuscript, and especially T. Fleming for his numerous constructive remarks and comments and for his revision of the English text.

References

Atramentowicz, M. 1986. Etude comparée de la dynamique de population de trois Didelphidae sympatriques de Guyane. Biotropica 18: 129–136.

Atramentowicz, M. 1988. La frugivorie opportuniste de trois Marsupiaux Didelphidés de Guyane. Revue d'Ecologie (Terre et vie) 43: 47–57.

Beehler, B. M. 1983. Frugivory and Polygamy in Birds of Paradise. The Auk 100: 1–12.

Beehler, B. M. 1989. Patterns of Frugivory and the Evolution of Birds of Paradise. Symposium 10: Frugivory in Tropical Birds. 816–826.

Brodkorb, P. 1960. How many species of birds have existed? Bull. Florida State Mus. 5: 41–53.

Brodkorb, P. 1971. Origin and evolution of birds. In: Farner, D. S. & King, J. R. (eds), Avian Biology, vol. 1, pp 19–55, Acad. Press, N. Y.

Bush, G. L., Case, S. M., Wilson, A. C. & Patton, J. L. 1977. Rapid speciation and chromosomal evolution in mammals. Proc. Nat. Acad. Sci. USA 74: 3942–3946.

Carton, Y. 1988. La coévolution. La Recherche, vol. 19: 1022–1031.

Charles-Dominique, P. 1986. Inter-relations between frugivorous vertebrates and pioneer plants: *Cecropia*, birds and bats in French Guiana. In: Estrada, A. & Fleming T. H. (eds), Frugivorous and Seed Dispersal, pp 119–135, Dr W. Junk Publish., Dordrecht.

Charles-Dominique, P. 1991. Feeding strategy and activity budget of the frugivorous bat *Carollia perspicillata* (Phyllostomidae) in French Guiana. J. Trop. Ecology 7: 243–256.

Charles-Dominique, P. & Cooper, H. M. 1986. Frugivorie et transport des graines de *Cecropia* par les chauves-souris de Guyane. In: Vertébrés et Forêts tropicales humides d'Afrique et d'Amérique. Mémoires du M. N. H. N. 132: 145–157.

Charles-Dominique, P., Atramentowicz, M., Charles-Dominique, M., Gérard, H., Hladik, A., Hladik, C. M. & Prévost, M. F. 1981. Les mammifères frugivores nocturnes d'une forêt guyanaise: inter-relations plantes-animaux. Rev. Ecol. (Terre et vie) 35: 341–435.

Chivers, D. J. & Hladik, C. M. 1980. Morphology of the gastrointestinal tract in primates: comparisons with other mammals in relation to diet. J. of Morphology 166: 337-386.

Clark, D. A. & Clark, D. B. 1984. Spacing dynamics of a tropical tree: evaluation of the Janzen-Connell model. Am. Nat. 104: 769–788.

Clark, D. A. & Clark, D. B. 1985. Seedling dynamics of a tropical tree: impact of herbivory and meristem damage. Ecology 66: 1884–1892.

Cooper, H. M., Charles-Dominique, P. & Vienot, F., 1986. Etude photo-colorimétrique des fruits consommés par les oiseaux et les chauves-souris de Guyane. In: Vertébrés et Forêts tropicales humides d'Afrique et d'Amerique. Mémoires du M. N. H. N. 132: 131–143.

Corner, E. J. H. 1949. The Durian theory of the origin of the modern tree. Ann. Bot. 13: 317–414.

Cremers, G., Feuillet, C., de Granville, J. J. & Hoff, M. 1988. Liste des espèces de Phanérogames et de Ptéridophytes de Guyane française d'après l'herbier du Centre ORSTOM de Cayenne. Communication de AUBLET, éditions ORSTOM, 54 pp.

Crepet, W. L. 1984. Advanced (constant) insect pollination machanisms: pattern of evolution and implications vis-a-vis angiosperm diversity. Annals of the Missouri Botanical Garden 71: 607–630.

De Foresta, H., Charles-Dominique, P., Erard, C. & Prévost, M. F. 1984. Zoochorie et premiers stades de la régénération naturelle après coupe en forêt guyanaise. Rev. Ecol. (Terre et vie) 39: 369–400.

Denslow, J. S. & Moermond, T. C. 1982. The effect of accessibility on rates of fruit removal from tropical schrubs: an experimental study. Oecologia 54: 170–176.

Dos Reis, N. R. & Guillaumet, J. L. 1983. Les chauves-souris frugivores de la région de Manaus (Amazonie centrale brésilienne) et leur rôle dans la dissémination des espèces végétales. Rev. Ecol. (Terre et vie) 38: 147–169.

Doyle, J. A. 1977. Pattern of evolution in early angiosperms.

In: Hallam A. (ed.). Patterns of evolution, pp 509–546, Elsevier Scientific Publication Company, Amsterdam.

Ehrlich, P. R. & Raven, P. H. 1965. Butterflies and plants: a study in coevolution. Evolution 18: 586–608.

Erard, C. 1988. Bird community structure in two rainforests: Africa (Gabon) and South America (French Guiana). A comparison. In: M. L. Harmelin Vivien & F. Bourlière (eds), Vertebrates in Complex Tropical Systems, pp 89–122. Springer Verlag, New York & Berlin.

Erard, C., Théry, M. & Sabatier, D. 1989. Régime alimentaire de *Rupicola rupicola* (Cotingidae en Guyane française): relations avec la frugivorie et la zoochorie. Rev. Ecol. (Terre et vie) 44: 47–74.

Erard, C., Théry, M. & Sabatier, D. 1991. Régime alimentaire de *Tinamus major* (Timamiidae), *Crax alector* (Cracidae) et *Psophia crepitans* (Psophiidae), grands oiseaux terrestres de la forêt guyanaise. Gibier et Faune Sauvage 8: 183–210.

Estrada, A. & Coates-Estrada, R. 1986. Frugivory by howling monkeys (*Alouatta palliata*) at Los Tuxtlas, Mexico: dispersal and fate of seeds. In: Estrada, A. & Fleming T. H. (eds), Frugivores and Seed Dispersal, pp 93–104. Dr W. Junk Publish., Dordrecht.

Fleming, T. H. 1981. Fecundity, fruiting pattern, and seed dispersal in *Piper amalgo* (Piperaceae), a bat-disperser tropical shrub. Oecologia 51: 42–46.

Fleming, T. H. 1985. Coexistence of five sympatric Piper (Piperaceae) species in a tropical dry forest. Ecology 66: 688–700.

Fleming, T. H. 1986. Opportunism versus specialization: the evolution of feeding strategies in frugivorous bats. In: Estrada, A. & Fleming T. H. (eds), Frugivores and Seed Dispersal, pp 105–118, Dr W. Junk Publish., Dordrecht.

Fleming, T. H. 1988. The Short-tailed Fruit Bat – A study in Plant-Animal interactions. The University Chicago Press, 365 pp.

Fleming, T. H. 1991. Fruiting Plant-Frugivore Mutualism: the Evolutionary theater and the Ecological Play. In: Price, P. W., Lewinsohn, T. H., Wilson Fernandez, G. & Benson, W. W. (Eds.). Plant-Animal Interactions: Evolutionary Ecology in Tropical and Temperate Regions, pp 119–144, John Wiley & Sons, Inc.

Fleming, T. H., Williams, F. C., Bonaccorso, F. J. and L. H. Herst. 1985. Phenology, seed dispersal, and coloniozation in Muntinhia calabura, a Neotropical tree. Amer. J. Bot. 72: 383–391.

Gautier-Hion, A., Duplantier, J. M., Emmons, L., Feer, F., Hecketsweiler Ph., Moungazi, A., Quris, R. & Sourd, C. 1985-a. Coadaptation entre rythmes de fructification et frugivorie en forêt tropicale humide de Gabon: Mythe ou réalité. Rev. Ecol. (Terre et vie) 40: 405–434.

Gautier-Hion, A., Duplantier, J. M., Quris, R., Feer, F., Sourd, C., Decoux, J. P., Dubost, G., Emmons, L., Erard, C., Hecketsweiler, P., Moungazi, A., Roussilhon, C. & Thiollay, J. M. 1985-b. Fruit characters as a basis of fruit choice and seed dispersal in a tropical forest vertebrate community. Oecologia 65: 324–337.

Gingerich, P. D. 1977. Patterns of evolution in the mammalian fossil record. In: Hallan, A. (ed.), Patterns of Evolution, pp. 469–500. Elsevier Scientific Publish. Cie, Amsterdam.

Heithaus, E. R. 1982. Coevolution between bats and plants. In: Kunz, T. H. (ed.), Ecology of Bats, pp 327–367. Plenum Press, New York.

Herrera, C. M. 1981. Are tropical fruits more rewarding to dispersers than temperate ones? Am. Nat. 118: 896–907.

Herrera, C. M. 1982. Seasonal variation in the quality of fruits and diffuse coevolution between plants and avian dispersers. Ecology 63: 773–785.

Herrera, C. M. 1985. Determinants of plant-animal coevolution: the case of mutualistic dispersal of seeds by vertebrates. Oikos 44: 132–141.

Herrera, C. M. 1989. Seed dispersal by animals: a role in angiosperm diversification? American Naturalist 133: 309–322.

Heywood, V. H. 1978. Flowering plants of the world. Oxford University press.

Hladik, A. & Hladik, C. M. 1969. Rapports trophiques entre végétation et primates dans la forêt de Barro Colorado. Terre et Vie 1: 25–117.

Hoff, M. & Brisse, H. 1986. The botanical data bank of the Herbarium of the ORSTOM center in Cayenne (AY). Flora of the Guianas, Newsletter n°3, special Workshop Issue: 24–26.

Howe, H. F. 1977. Bird activity and seed dispersal of a tropical wet forest tree. Ecology 58: 539–550.

Howe, H. F. & Estabrook, G. F. 1977. On intraspecific competition for avian dispersers in tropical trees. Am. Nat. 111: 817–832.

Howe, H. F. & Smallwood, J. 1982. Ecology of seed dispersal. Ann. Rev. Ecol. Syst. 13: 201–228.

Howe, H. F. & Vande Kerckhove, G. A. 1981. Removal of wild nutmeg (*Virola surinamensis*) crops by Birds. Ecology 62: 1093–1106.

Janson, C. H. 1983. Adaption of fruit morphology to dispersal agents in a neotropical foret. Science 219: 187–189.

Janzen, D. H. 1966. Coevolution of mutualism between ants and acacias in centra America. Evolution 20: 249–275.

Janzen, D. H. 1980. When is it coevolution? Evolution 34: 611–612.

Janzen, D. H. 1983. Dispersal of seeds by vertebrates guts. In: Futuyma, D. J. and Slatkin, M. (Eds), pp 232–262. Sunderland, Mass: Sinauer.

Janzen, D. H. 1986. Mice, big mammals, and seeds; it matters who defecates what where. In: Estrada, A. & Fleming, T. H. (eds), Frugivores and Seed Dispersal, pp 251–271. Dr W. Junk Publish., Dordrecht.

Julien-Laferrière, D. 1989. Utilisation de l'espace et des ressources alimentaires chez *Caluromys philander* (Marsupialia, Didelphidae) en Guyane française. Comparason avec *Potos flavus* (Eutheria, Procyonidae). Thèse de l'Université Paris XIII.

Julien-Laferrière, D. & Atramentowicz, M. 1990. Feeding and

reproduction: the example of three didelphid marsupials in two neotropical forests (French Guiana). Biotropica 4: 404–415.

Julliot, C. & Sabatier, D. (in press). Diet and frugivory of the Red Howler monkey (*Alouatta seniculus*) in French Guiana. Int. J. Primatology.

Levey, D. J., Moermond, T. C. & Denslow, J. S. 1984. Fruit choice in neotropical birds: the effect of distance between fruits and preference patterns. Ecology 65: 844–850.

Levin, D. A. & Wilson, A. C. 1976. Rates of evolution in seed plantes: net increase in diversity of chromosome numbers and species numbers through time. Proc. nat. Acad. Sci. USA 73: 2086-2090.

Martin, T. E., 1985. Selection of second growth woodlands by frugivorous migrating birds in Panama: an effect of fruit size and plant density? J. Trop. Ecol. 1: 157–170.

McKey D., 1975. The ecology of coevolved seed dispersal systems. In: Gilbert, L. E. & Raven, P. H. (eds.), Coevolution of animal and plants, pp 150–191. University of Texas, Austin.

Milton, K. 1984. The role of food-processing factors in primates mates. In: Rodman, P. S. & Cant J. G. H. Adaptions for foraging in nonhuman primates. Pp 249–279. Columbia University Press.

Morden-Moore, A. L. & Willson, M. F. 1982. On the eocological significance of fruit color in *Prunus serotina* and *Rubus occidentalis*: field experiments. Can. J. Bot. 60: 1554–1560.

Moermond, T. C. & Denslow, J. S. 1983. Fruit choice in neotropical birds: effect of fruit type and accessibility on selectivity. J. Anim. Ecol. 52: 407–420.

Moermond, T. D. Denslow, J. S., Douglas, J. & Santana, E. 1986. The influence of morphology and fruit choice in neotropical birds. In: Estrada, A. & Fleming, T. H. (eds), Frugivores and Seed Dispersal, pp 137–146. Junk W. Publish., Dordrecht.

Morton, E. S. 1973. On the evolutionary advantages and disavantages on fruit-eating in tropical birds. Am. Nat. 107: 8–22.

Palmeirim, J. M., Gorchov D. L. & Stoleson, S. 1989. Trophic structure of a neotropical frugivore community: is there competition between birds and bats? Oecologia 79: 403–411.

Pratt, T. K. & Stiles, E. W. 1985. The influence of fruit size and structure on composition of frigivore assemblages in New Guinea. Biotropica 17: 314–321.

Regal, P. J. 1977. Ecology and evolution of flowering plant dominance. Science 196: 622–629.

Ridley, H. N. 1930. The dispersal of plants throughout the world. Reeve, Ashford.

Sabatier, D. 1983. Fructification et dissémination en forêt guyanaise. Thèse de 3ème cycle, Université de Montpellier, 238 pp.

Smythe, N. 1970. Relationships between fruiting seasons and seed dispersal methods in a neotropical forest. Am. Nat., 104: 23–35.

Snow, D. W. 1966. A possible selective factor in the evolution of fruiting season in tropical forest. Oikos, 15: 274–281.

Snow, D. W. 1971. Evolutionary aspects of fruit-eating by birds. Ibis 113: 194–202.

Snow, D. W. 1976. The web of adaption. Quadrangle, New York, 176 pp.

Snow, D. W. 1981. Tropical frugivorous birds and their food plants: a world survey. Biotropica 13: 1–14.

Sorensen, A. D. 1983. Taste aversion and frugivore preference. Oecologia (Berl.) 56: 117–120.

Stebbins, G. L. 1949. Rates of evolution in plant. In: Jepsen, G. L., Simpson, G. G. & Mayr, E. (eds), Genetics, paleontology and evolution, pp 229–242. Princeton Univ. Press, Princeton, NJ.

Stebbins, G. L. 1981. Why are there so many species of flowering plants? Bioscience 31: 573–577.

Stiles, E. W. 1982. Fruit flag: two hypotheses. Am. Nat. 120: 500–509.

Théry, M. 1990. Ecology et comportement des oiseaux Pipridae en Guyane: Leks, frugivorie et dissémination des graines. Thèse de Doctorat de l'Université de Paris VI, 23 janvier 1990.

Tiffney, B. H. 1986. Evolution of seed dispersal syndromes according to the fossil record. In: Murray, D. R. (Ed.). Pp 273–305. Seed Dispersal. Academic Press.

Van der Pijl, L. 1957. The dispersal of plants by bats (Chiropterochory). Acta botanica neederlandica 6: 291–315.

Van der Pijl, L. 1969. Evolutionary action of tropical animals on the reproduction of plants. Biol. J. Linn. Soc. 1: 85–96.

Van der Pijl, L. 1972. Principle of dispersal in higher plants. Springer Verlag. Berlin.

Van Roosmalen, M. G. M. 1980. Habitat preferences, diet, feeding strategy and social organization of the Black Spider Monkey (*Ateles paniscus paniscus* linnaeus 1758) in Surinam. PhD. Thesis Wageningen, 175 pp.

Van Roosmalen, M. G. M. 1985. Fruits of the guianan flora. Inst. of Systematic Botany, Utrecht Univ., Netherlands, 483 pp.

Wheelwright, N. T. 1985. Fruit size, gape with, and the diets of fruit-eating birds. Ecology 66: 808–818.

Wheelwright, N. T. 1986. A seven-year study of individual variation in fruit production in tropical bird-dispersed tree species in the family lauraceae. In: Estrada, A. & Fleming T. H. (eds), Frugivores and Seed Dispersal, pp 21–35, Dr W. Junk Publish., Dordrecht.

Wheelwright, N. T. & Janson, C. H. 1985. Colors of fruit displays of bird disperded plants in two tropical forest. Am. Nat. 126: 777–799.

Wheelwright, N. T. , Haber, W. A., Murrayn K. G. & Guindon, C. 1984. Tropical fruit-eating birds and their food plants: a survey of Costa Rican lower montane forest. Biotropica 16: 173–192.

Willson, M. F. & Thomson, J. N. 1982. Phenology and Ecology of color in bird dispersed fruit. Can. J. Bot. 60: 701–713.

Willson, D. E. & Janzen, D. H. 1972. Predation on Scheelea seeds by bruchid beetles: seed density and distance from the parent palm. Ecology 53: 954–959.

Vegetatio **107/108**: 85–104, 1993.
T. H. Fleming and A. Estrada (eds). *Frugivory and Seed Dispersal: Ecological and Evolutionary Aspects.*

Geographical ecology and variation of plant-seed disperser interactions: southern Spanish junipers and frugivorous thrushes

Pedro Jordano
Estación Biológica de Doñana, CSIC, Apdo. 1056, E-41080 Sevilla, Spain

Keywords: Annual variation, Frugivory, Geographic variation, *Juniperus*, Mediterranean scrubland, Mutualism, *Turdus*

Abstract

Spatial and temporal predictability in the mutual selective pressures of plants and frugivorous birds is a prerequisite for coevolution to occur. I examine the interaction patterns of strongly frugivorous thrushes (*Turdus* spp.) and their major winter food plants (*Juniperus* spp., Cupressaceae) and how they vary in space and time. Spatial congruency, rarely considered in seed dispersal studies, is studied at three spatial scales: 1) the total species range; 2) regional distribution; and 3) local abundance and its variation between seasons. Southern Spanish frugivorous thrushes and junipers show very low congruence in distribution patterns at each of these scales. Most juniper species show geographic distributions that are nested within the geographic ranges of thrush species. Bird species showed greater habitat breadth values than plants and were found in a greater percentage of localities. The local bird abundance was strongly correlated across years and sites with the local availability of juniper cones. Cone production varied markedly between years, but the rankings for different species in different years were statistically concordant at mid-elevation and lowland sites. Both bird abundance and cone production showed greater temporal than spatial variability. Variation of cone productions at both temporal and spatial scales was greater than variability in bird abundance. Species with strong interactions of mutual dependence showed very low values of biogeographic congruence, caused by differences in geographic range and habitat specificity. This obviously limits the possibilities for pairwise, specific coevolution to occur. However, mutual effects of species groups are possible to the extent that the component species are ecologically 'interchangeable' in their selective effects and other constraints on coevolution are not operating. The approach used here to examine the patterns of species interactions at different biogeographic scales might prove useful in comparative studies of plant-animal interactions.

Introduction

Studies documenting the geographical, regional and local variations of species interactions in plant-frugivore systems are uncommon. Mutual selective effects derived from the fruit-frugivore interaction strongly depend upon how the consequences of the interaction observed in a given locality translate and 'scale' at higher biogeographical levels. If evolutionary modification of traits involved in an interaction between mutualists depends on mutual selective pressures derived from the interaction, then the possibilities for coevolution depend to a great extent on the degree of spatial constancy of these selection pressures (Janzen 1980; Schemske & Horvitz

1984; Herrera 1988b; Horvitz & Schemske 1990). 'Spatial' is used here in a biogeographical context to include the geographic range of the interacting species, the different habitats of a given region where the species interact, and their local population sizes.

An extreme example of spatial inconsistency was given by Wheelwright (1988) who showed that at a single Costa Rican locality, emerald toucanets are the chief dispersers of *Ocotea floribunda* and that this fruit is the major food of the bird; this interaction, however, occurs in only a small portion of the ranges of the two species. This situation is probably generalizable to most avian frugivore seed dispersal systems. However, very few studies have explicitly considered spatial variation in bird-fruit interactions beyond the local scale defined as several sites within a local study area (see, e.g., Herrera 1988a; Keeler-Wolf 1988; Willson 1986). To my knowledge, no previous study has examined simultaneously the temporal and spatial variation in the species assemblages of a seed dispersal mutualism considering several years and both regional and local geographical scales.

The interaction pattern for a given pair of species in a particular community depends on the biogeographic 'attributes' (or 'forms of rarity', according to Rabinowitz 1981) of each species and how these interact (Fig. 1). More importantly, these biogeographic attributes determine how the local selective regimes sum up and translate among populations. For example, in a situation of multi-species interactions such as the frequently observed in bird-fruit communities, one can expect a priori a greater coevolutionary potential between species sharing the same combination of biogeographic attributes. Otherwise, a situation of strong asymmetry arises if one species interacts with the other in a small portion of its geographic range, or only in certain highly localized habitats or if the second species is extremely rare. Even in situations of strong asymmetry one interacting species can generate a strong selective pressure on traits of a second species whose geographic range and habitats occupied are included within those of the former.

Frugivorous thrushes (genus *Turdus*, Turdidae) are a prominent group of six species in southern Spanish frugivore communities (Santos *et al.* 1983; Herrera 1984; Jordano 1985; Tellería *et al.* 1988). All of the six species are strongly frugivorous during autumn and winter (Hartley 1954; De Juana & Santos 1981; Herrera 1981; Jordano 1981; Tejero *et al.* 1984; Guitián 1984; Debussche & Isenmann 1985a, 1985b; Obeso 1985; Obeso 1986; Snow & Snow 1988; Zamora 1990). In the highlands (scrublands above 1500 m elevation), the fleshy cones of *Juniperus* species (Cupressaceae) are a staple food of these birds (Obeso 1986; Zamora 1990). Thrushes are numerically dominant members of the frugivorous avifauna in

GEOGRAPHIC RANGE	LARGE		SMALL	
HABITAT SPECIFICITY	Wide	Narrow	Wide	Narrow
LOCAL POPULATION SIZE Large, dominant	*Turdus viscivorus* *Turdus iliacus* *Juniperus communis*	*Juniperus sabina*	*Turdus torquatus* *Juniperus thurifera* *Juniperus phoenicea*	*Juniperus cedrus* *Juniperus brevifolia*
Small, non-dominant	*Turdus merula* *Turdus philomelos* *Juniperus oxycedrus*	*Turdus pilaris*		

Fig. 1. Biogeographic patterns of *Juniperus* and *Turdus* species according to the typology established by Rabinowitz (1981). Each species can be characterized at three different spatial scales (continental, regional, and local) by evaluation of the relative size of its geographic range, the number of habitats occupied, and the average size of its local populations.

several vegetation types where junipers dominate (Santos 1982) and play a central role in the dispersal ecology of *Juniperus* species in Spain as well as elsewhere (Snow & Snow 1988; Holthuijzen & Sharik 1985).

The purpose of the paper is (1) to document between-habitat and between-year patterns of interactions between junipers and thrushes and their variability in several southern Spanish localities; (2) to assess the degree of congruence between the interaction 'strengths' recorded at each site and year and the distributions of each pair of species at the regional and continental scales; and (3) to explore some consequences for plant-frugivore coevolution of the non-coincidence between interacting species at different biogeographical scales.

Plant and bird species

Junipers are a dominant element of the southern Spanish highland vegetation (Ern 1966), with five species reaching the Iberian Peninsula. Only *J. communis* has a wide holarctic distribution and the remaining four species (*J. oxycedrus*, *J. phoenicea*, *J. sabina*, and *J. thurifera*) are restricted to the circum-Mediterranean region and certain parts of western Asian mountains (Jalas & Suominen 1985) (see Fig. 2). The six species of the genus *Turdus* (*T. iliacus*, *T. merula*, *T. viscivorus*, *T. philomelos*, *T. pilaris*, and *T. torquatus*) have a wider distribution on a continental scale (Cramp 1988). Their breeding distributions extend throughout the western Palearctic but the wintering areas are concentrated in the western part of the Mediterranean basin, specially in the eastern half of the Iberian Peninsula (Soc. Esp. Ornitología 1985) (Fig. 2).

The female strobili of *Juniperus* species (fleshy cones or 'fruits' or 'berries', hereafter) are functionally analogous to angiosperm fleshy fruits. Fresh mass of *J. oxycedrus* cones averages 0.70 g, 0.36 g of dry 'pulp' and 2.2 seeds/cone. *J. phoenicea* has 0.32 g, 0.15 g, respectively, and 8.5 seeds/cone. *J. communis* has 0.08 g, 0.05 g, respectively, and 1.9 seeds/cone. *J. sabina* has 0.08 g, 0.05 g, respectively, and 1.9 seeds/cone; and *J. thurifera* has 0.28 g, 0.17 g, respectively, and 3.1 seeds/cone. Maximum external cone diameters for these five species are 10.9 mm, 9.1 mm, 5.8 mm, 7.4 mm, and 9.2 mm respectively. Average content of the 'pulp' is 11.5% (dry mass) lipid and 4.4% protein; the fiber content is very high (mean = 44.1%) (Herrera 1987; Obeso 1985; and P. Jordano, unpubl. data).

Both *Juniperus* and *Turdus* are characteristic genera of the southern Spanish highland vegetation and frugivorous avifauna (see below). *J. communis* and *J. sabina* are dominant species in high altitude scrublands (1900–2700 m); *J. phoenicea* dominates mid-elevation vegetation (1200–1500 m) in localized habitats on dry, exposed, rocky outcrops as well as in several coastal sites on sandy soils (e.g., Parque Nacional de Doñana, Huelva; Punta del Sabinal, Almería). *J. oxycedrus* can be found as scattered individuals and is locally dominant in forest gaps and exposed, rocky outcrops at lower elevation (700–1300 m). Finally, *J. thurifera* dominates extensive juniper woodlands in central Spain. Only *J. communis*, *J. sabina*, and *J. phoenicea* – the dominant species in the highlands – are considered in detail here.

With the exception of *T. philomelos*, a lowland species rarely found at high altitude, all *Turdus* species occur in southern Spanish highlands (above 1300 m) during winter, but their abundances vary greatly from year to year (see below). Both *T. merula* and *T. viscivorus* have resident populations that increase in winter with the arrival of migrants. *T. torquatus* is mainly a transient migrant wintering in the Moroccan Atlas mountains with a minority of its population wintering in Spain. Both *T. iliacus* and *T. pilaris* occur regularly in southern Spanish mountains but the latter only in very low numbers; the two species show important between-year changes in their winter densities. A detailed description of migration patterns and geographic and habitat distributions can be found in Santos (1982) and Soc. Esp. Ornitología (1985). A summary of the main patterns of geographic distribution, habitat range, and local abundance is presented in Fig. 1.

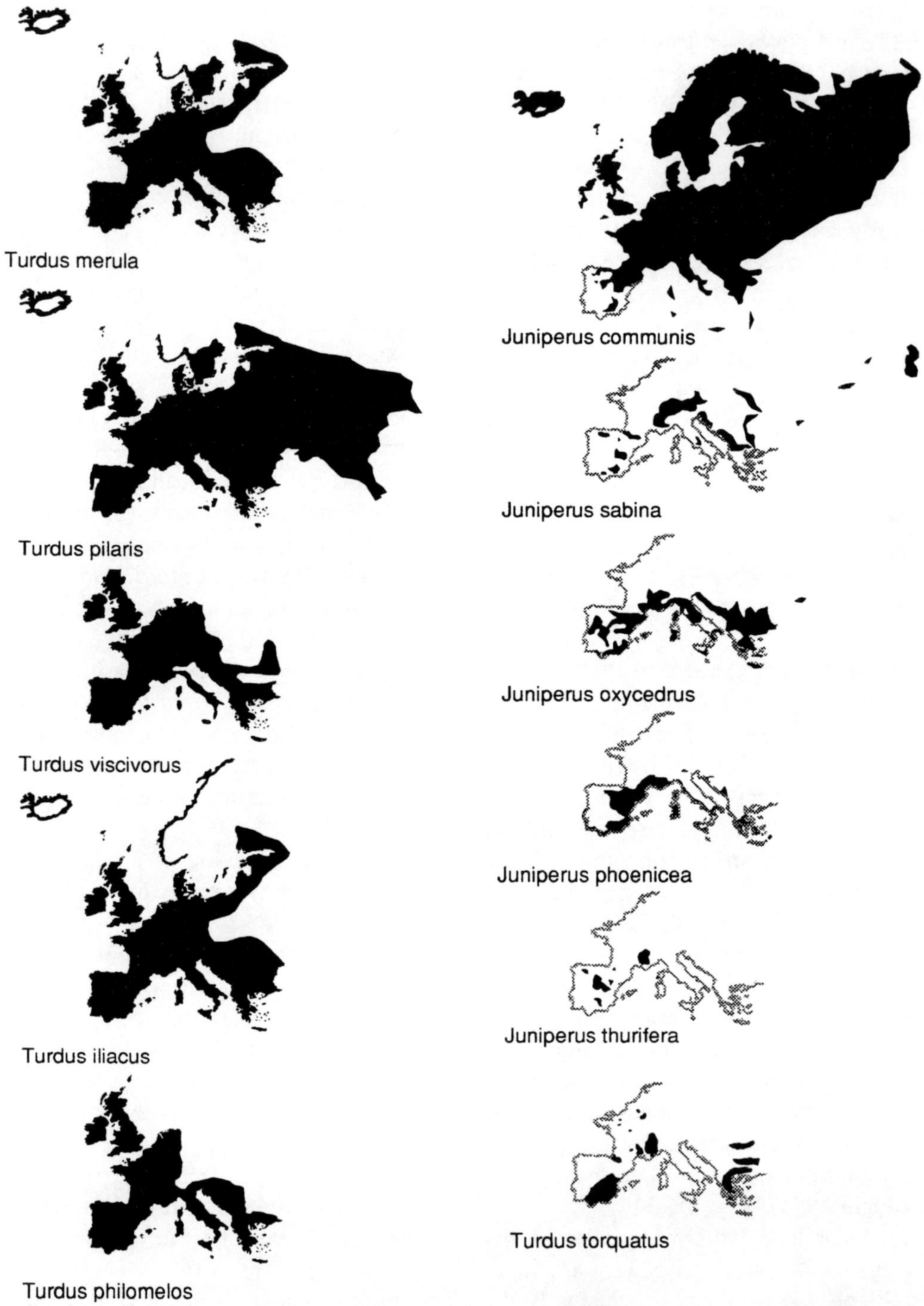

Fig. 2. Digitized outlines of the geographic distribution maps of *Juniperus* spp. and the wintering areas of *Turdus* spp. in the western Palearctic. Source: Jalas & Suominen (1985) and Cramp (1988). Note the extreme asymmetry of certain combinations of plant and bird species that show strong interactions in southern Spanish localities (e.g., *J. communis* and *T. torquatus*; *J. phoenicea* and *T. pilaris*; and *J. phoenicea* and *T. viscivorus*). Dimmed outlines of southwestern Europe have been added to provide a reference for species with restricted ranges.

Study area and methods

The study was carried out during the autumn and winter (October-March) of 1985–1989 at six localities in the highlands of Sierra de Cazorla, Jaén province (three main sites at 1550–1700 m elevation), and Sierra de Baza (2250 m) and Sierra Nevada (2550–2650 m)(Granada). The vegetation at Campos de Otero (Sierra Nevada), a north-facing slope on siliceous soil, is low scrubland dominated by *J. communis* with *Genista baetica* and large areas of open, rocky ground and pasture. Santa Bárbara (Sierra de Baza) is structurally very similar to the former site but is on calcareous rock. The vegetation is dominated by *J. sabina*, *J. communis*, *Berberis vulgaris*, *Daphne oleoides*, and *Rosa* spp. surrounded by large patches of open, rocky ground and pasture. The Cañada de las Sabinas site at Sierra de Cazorla lies at the border of pine forest (*Pinus nigra*, subsp. *salzmannii*) growing on rocky soils, with deciduous vegetation growing in the deep, cool soils of the 'navas' dominated by *Crataegus monogyna*, *Prunus mahaleb*, *Berberis vulgaris*, *Lonicera arborea*, *Daphne laureola*, *Acer monspessulanum*, and *Quercus faginea*. On dry and rocky slopes, both *J. communis* and *J. sabina* dominate the understory vegetation of the pine forest, with *J. phoenicea* being locally dominant in certain ridges and rocky outcrops. The Calarilla, Cañada del Travino and Nava Noguera sites are very similar to the preceding one in structural characteristics and vegetation composition. In the analyses below, data for the Calarilla site have been pooled with those of the nearby Cañada de las Sabinas.

Other 13 Southern Spanish localities were also surveyed at varying frequencies to determine local assemblages of *Turdus* and *Juniperus* species. The sites were: Reserva Biológica de Doñana (Huelva province, 20 m elevation); Collado del Calvario (Sierra de Cazorla, Jaén; 1400 m); Pinar Negro (Sierra de Cazorla, Jaén; 1450 m); Roblehondo (Sierra de Cazorla, Jaén; 1300 m); Guadahornillos (Sierra de Cazorla, Jaén; 1000–1250 m); Dornajo (Sierra Nevada, Jaén; 2020 m); Puerto del Boyar (Sierra de Grazalema, Cádiz; 1170 m); Sierra del Pinar (Cádiz; 1450 m); Torrecilla (Sierra de Ronda, Málaga; 1750 m); Punta del Sabinal (Almería; 14 m); Sierra de Quintana (Jaén; 950 m); Aljabaras (Sierra de Hornachuelos, Córdoba; 750 m) and La Canaleja (Sierra de Baza, Granada; 1100 m). Three highland sites (1400–1500 m) in the Haut Atlas (South Morocco), Agaiouar, near Imigdal, and Tizi-n-At-Imguer were also visited in January-February 1989.

Quantitative information on the local abundances of *Turdus* spp. at each of the main sites were obtained by counts of birds along fixed transects of variable length, usually 1–4 km. At each site, these transects were walked in at least three different months during each study year. All birds seen or heard within a 60 m band at each side of the transect were recorded. Results are given as number of birds/km census to standardize the differences in census effort. Feeding records were tallied (one record = one fruit ingested by a bird) during the census counts and also during observational watches at feeding trees. Additional information on the local abundance of *Turdus* spp. at other Southern Spanish sites was obtained from the literature (Herrera 1984; Jordano 1985; Santos 1982; Suárez & Muoz-Cobo 1984; Obeso 1987).

Counts of mature-sized fleshy fruits present in different permanent quadrats (30 m $\times$ 1.5 m; $n = 12$–30 per site) were carried out at each of the six main sites (see Blake *et al.* 1990 for further discussion of the method). Similar but non-permanent transects were used at the other sites to determine the relative cover of each *Juniperus* species and other fleshy-fruited species. Individual *Juniperus* plants at each main site were also marked to determine cone crop sizes, totalling (all sites pooled) 85 *J. phoenicea*, 70 *J. communis*, and 47 *J. sabina*.

Digitized outlines of the geographical ranges of *Juniperus* and *Turdus* species were obtained from Jalas & Suominen (1985) and Cramp (1988). Maps were divided in 5° quadrats ($n = 128$) and the presence-absence of each species was recorded. The percentage of quadrats with presence was used as an index of each species' geographic range. Similarly, the proportion of southern Spanish sites with confirmed presence of each species was used as an index of habitat breadth. Habitat occupancy patterns were derived from the presence-absence data for each species in 19 southern Spanish sites. The average quantitative importance of each species in 9 local assemblages was derived from values of relative cover (plants) or population density (birds) obtained for the study sites as well as from literature sources cited above. Then, for each species, single values were obtained to characterize the relative size of the geographic range (proportion of latitudinal quadrats where present), the proportion of habitats occupied, and the relative local abundance. Similarities between taxa at each of these biogeographical levels were determined using the Sørensen coefficient (presence-absence data for the geographic range and the types of habitats occupied) and the Gower similarity coefficient (for quantitative data on local abundances)(Legendre & Legendre 1979).

Results

Turdus *feeding on* Juniperus *fleshy cones*

Cones of *Juniperus* species made up an important part of the winter fruit diet of *Turdus* spp. (Table 1). Both *T. torquatus* and *T. viscivorus* showed the highest reliance on *Juniperus* with 37.5% and 39.3%, respectively, while *T. iliacus* and *T. pilaris* consumed mainly other fruits (e.g., *Crataegus monogyna* and *Berberis vulgaris*). *T. pilaris* is a very rare species found in very low number only in certain years, with feeding records for *J. phoenicea* at Nava Noguera and for *J. sabina* at Pinar Negro in years of high fruit production. *T. merula* inhabits denser vegetation and feeds chiefly on rosaceous fruits (*C. monogyna*, *Rosa* spp.) and infrequently on juniper berries. Finally, *T. philomelos* and *T. iliacus* are important for *Juniperus* at lower elevations, relying heavily on *J. oxycedrus* and *J. phoenicea*. For example, a mast cone production of *J. phoenicea* occurred at R. B. Doñana during the winter 1986–1987; there, 98.9% of the feeding records for *T. iliacus*, 96.5% for *T. philomelos*, and 91.3% for *T. merula* were for this species. *T. merula* adults were even feeding their young during the 1987 spring with the remaining ‘surplus’ cones.

At the study sites, *Turdus* species are quantitatively the more important *Juniperus* seed dispersers among frugivorous birds. *Turdus* records made up 97.6% of the feeding observations at these plants. With the possible exceptions of *J. phoenicea* and *J. oxycedrus*, for which carnivorous mammals are also important dispersers (Herrera 1989), *Turdus* species are therefore the most common dispersers of *Juniperus*, especially for *J. communis* and *J. sabina*. *T. torquatus* and *T. viscivorus* are the main dispersers of three species of *Juniperus* in the highlands. Data from Herrera (1984), Obeso (1985) and Santos (1983) indicate that thrush species also disperse most seeds of lowland junipers (*J. oxycedrus*, *J. thurifera*).

Table 1. Feeding records of *Turdus* species on *Juniperus* spp. and other fruits. Data for all southern Spanish highland study sites have been pooled.

Tudus species	*Juniperus* species			Other fruit species	Number of records
	communis	*phoenicea*	*sabina*		
merula	11	10	6	355	382
iliacus	20	2	10	219	251
philomelos	0	0	0	13	13
torquatus	165	189	63	694	1111
pilaris	0	8	0	105	113
viscivorus	43	115	48	318	524
Other birds	0	15	2	528	545
Number of records	239	339	129	2232	2939

For every pair of interacting *Juniperus*/*Turdus* species in Table 1, with the exclusion of *T. philomelos*, I computed the percentage of the bird's feeding records accounted for by the plant species and the percentage of the plant's records made up by the bird species. These values serve as a first quantitative evaluation of the reciprocal dependency between each species pair. Values of plant dependence on bird and bird dependence on plant were strongly correlated across pairs of species ($r_s = 0.854$, $p < 0.01$, $n = 15$), indicating symmetrical interactions. However, as a group, the birds' dependence on junipers was much lower than the plants' dependence on thrushes.

Geographic variation in local Turdus/Juniperus *assemblages: regional and local scales*

Juniper and thrush species differ markedly in their distribution patterns on a continental scale (Fig. 2). As a group, *Turdus* spp. are widely distributed during winter throughout the southwestern Palearctic. Only *T. torquatus* has a relatively localized wintering area in eastern Spain and Morocco. The wintering ranges of all six species overlap over a relatively small area encompassing the southeastern part of the Iberian Peninsula, south France, and northwestern Africa.

This area of highest concentration of wintering thrush species overlaps extensively with the area of maximum diversity of *Juniperus* species, most of which have circum-Mediterranean distributions (Fig. 2). When all possible species pairs are considered, all the ranges of *Juniperus* species are included within the ranges of each *Turdus* spp., with the exception of *T. torquatus*. Thrush species, in contrast, are distributed over vast areas where *J. communis* is the only juniper present. With the exception of *J. communis*, no juniper species is present in more than 17% of the 5° quadrats. This situation contrasts with *Turdus* spp. which, excluding *T. torquatus* (11.7%), occupy more than 36% of the quadrats (Table 2).

To examine the patterns of habitat specificity, I calculated the percentage of the 19 habitats surveyed where each *Turdus* and *Juniperus* species was present, irrespective of their local abundances (Table 2). Plants showed higher habitat specificity than birds, and were present in an average of 44.7% of the habitats surveyed. *Turdus* species, with the exception of the extremely localized *T. pilaris* (10.5%) were present in 63.7% of the sites, and as a group showed greater habitat breadth than *Juniperus* (U = 1, $p = 0.016$; Mann-Whitney U-test; Table 2).

Table 2 summarizes the quantitative composition of the local assemblages and its regional variation for the intensively-studied areas (see also Appendix 1 and 2). The total cover of *Juniperus* species reached 62% in the highland scrublands and 52% in the coastal woodland at R. B. Doñana but was only 30% at most sites. *Turdus* species represented, on average, 42% of the censused birds in highland juniper scrub, 48% in coastal sites, and decreased to 8–23% in lowland scrublands, olive groves, and oak woodlands.

The local assemblages of both species groups were usually dominated by one species, particularly at the highland sites. These habitats are dominated either by *J. communis* (62.3% horizontal cover at Campos de Otero) or *J. sabina* (38.6% at Santa Bárbara), with *J. phoenicea* being extremely scarce (Table 2). Common thrushes in these highland sites were *T. torquatus* (70–86% of censused *Turdus*), *T. pilaris* (23%) and *T. viscivorus* (8–16%). Large flocks of *T. torquatus* are characteristic in these sites during large crop years. At Campos de Otero, flocks of 3–62 individuals were recorded during the 1985–1986 winter (mean = 17 individuals, $n = 32$). Flocks of *T. viscivorus* (mean = 9, range = 3–35 individuals, $n = 21$) or *T. iliacus* (mean = 4, range = 3–8, $n = 10$) were less numerous. Local assemblages in the mixed mountain pine forests showed higher equitability, with *T. merula*, *T. viscivorus*, or *T. torquatus* as the dominant species.

Cone production was extremely variable from year to year (see below), but when data for single localities were pooled, only *J. phoenicea* showed statistically significant variation between sites in cone production ($F_{4,11} = 10.84$, $p < 0.001$; data square-root transformed; Appendix 1). The pooled, total production of *Juniperus* ranged from

Table 2. Relative values for size of geographic range, habitat specificity, and local population size of *Juniperus* and *Turdus* species. Values for local population sizes (as %) are given for the main study sites.

	Percent cover				Percent of birds censused					
	JCOM	JSAB	JPHO	JOXY	TMER	TVIS	TILI	TPHI	TTOR	TPIL
Geographic range[1]	66.4	11.7	14.1	17.2	50.8	35.2	50.8	42.9	11.7	62.5
Habitat breadth[2]	63.2	36.8	42.1	36.8	78.9	84.2	94.7	52.6	84.2	10.5
Local population size[3]	12.6	6.1	11.3	0.6	21.1	29.1	7.4	1.4	38.1	2.9
Highland juniper scrubland										
Campos de Otero (Sierra Nevada)	62.3	0.0	0.0	0.0	0.0	8.8	4.9	0.0	86.3	0.0
Sta. Bárbara (Sierra de Baza)	1.6	38.6	0.0	0.0	0.0	15.9	6.8	0.0	77.3	0.0
Pinar Negro (Sierra de Cazorla)	21.3	13.9	0.6	0.0	4.6	13.6	12.7	0.9	45.5	22.7
Mixed mountain forest										
Nava Noguera (Sierra de Cazorla)	8.3	0.0	7.3	0.0	15.2	37.3	8.2	2.1	36.9	0.4
Cañada Travino (Sierra de Cazorla)	8.3	6.2	0.9	0.0	31.2	17.7	11.3	1.6	38.2	0.0
Cañada Sabinas (Sierra de Cazorla)	21.3	13.9	0.6	0.0	15.3	63.5	1.2	0.0	20.0	0.0
Roblehondo (Sierra de Cazorla)*	0.3	0.0	0.4	4.9	49.1	43.9	0.0	7.0	0.0	0.0
Coll.Calvario (Sierra de Cazorla)*	2.1	0.0	0.0	1.3	53.3	32.2	14.2	0.3	0.0	0.0
Coastal juniper woodland										
Reserva Biológica de Doñana	0.0	0.0	51.8	0.01	12.2	2.5	18.4	66.4	0.5	0.0

[1] Percentage of 5° latitude quadrats where the species is present (wintering period for *Turdus* spp.) in the western Palearctic.
[2] Percentage of 19 sites surveyed where each species was recorded.
[3] Local population sizes are given, for each of 9 main study sites in southern Spain, as average % horizontal cover in permanent transects (*Juniperus*) and average % of censused individuals in transect counts (*Turdus*).
* Data from Obeso (1987).
Abbreviations: JCOM – *J. communis*; JSAB – *J. sabina*; JPHO – *J. phoenicea*; TMER – *T. merula*; TILI – *T. iliacus*; TPHI – *T. philomelos*; TTOR – *T. torquatus*; TPIL – *T. pilaris*; TVIS – *T. viscivorus*.

30.5 ± 21.6 cones/m^2 (mean ± s.e.) at Campos de Otero to 2.7 ± 1.9 cones/m^2 at Santa Bárbara ($F_{4,11} = 3.95$, $p = 0.024$). Among *Turdus* species, only the mid-elevation and lowland species (*merula*, *iliacus*, and *philomelos*) showed significant variations in abundance between sites ($F_{4,11} \geq 6.58$, $p < 0.005$), as expected from their greater habitat specificity; the highland species showed no significant variations. Total abundance of *Turdus* did not vary significantly between sites ($F_{4,11} = 1.00$, $p = 0.448$) and ranged from 5.2 ± 1.9 birds/km (Santa Bárbara) to 27.6 ± 21.7 birds/km (Campos de Otero).

Temporal variations in local assemblages

Temporal (between-year) variation in mutualistic assemblages has three main components: different species can be present or absent in a particular year; the rankings of species' abundances can vary between years or remain highly concordant; and, even with low qualitative variation and concordance of the abundance rankings, species can show abundance variations from one year to the next.

A characteristic feature of local *Juniperus* and *Turdus* species assemblages is their extreme between-year variation in cone production and bird abundance. However, presence/absence of particular species is relatively constant. Only *J. phoenicea* and *J. sabina* failed to set fruits in two years at Nava Noguera and Santa Bárbara (Appendix 1), but years of extremely low fruit availability (<0.1 fruits/m^2) are not infrequent and result from a low frequency of reproductive individuals. Among the birds, only *T. pilaris* and *T. iliacus*, two species with irrupting movements, are absent in certain years (e.g., at Campos de Otero, Nava Noguera) and were recorded only in years with high cone production (Appendix 2). For example, *T. iliacus* was observed at Campos de Otero only in 1985 and 1986 when crops of *J. communis* were large; the following winters, when very few cones were available (Appendix 1), birds were seen at lower elevations feeding on *Crataegus monogyna* and *Rosa canina*.

The local thrush abundance in a given year and site was strongly correlated with both the total availability of fleshy fruits ($F_{1,15} = 72.0$, $r^2 = 0.847$, $p < 0.0001$) (Fig. 3) and with the total production of *Juniperus* cones ($F_{1,15} = 53.6$, $r^2 = 0.793$, $p < 0.0001$). Between-year peaks in local abundance of *Turdus* matched the local maxima in *Juniperus* availability. *T. torquatus* reached 84 birds/km at Campos de Otero during 1985 (Appendix 2), a year with an extraordinary crop of *J. communis* (94 cones/m^2; Appendix 1). Similarly, at the Santa Bárbara site, high abundance of *T. torquatus* (6 birds/km) in 1986–1987 matched a good crop year for *J. sabina* (6.4 cones/m^2). Small flocks of *T. pilaris* at Nava Noguera in 1988 (0.15 birds/km) were feeding mostly on *J. phoenicea*, with an availability of 4 cones/m^2 that year. This species was also recorded at Pinar Negro (Sierra de Cazorla) during the 1987–1988 winter when *J. sabina* cones were abundant; birds were consuming this species and fruits of *Berberis vulgaris*. Observations at the lowland sites showed similar trends. The 1986–1987 winter, in which *J. phoenicea* had a mast year with 103.5 cones/m^2, was exceptional at R. B. Doñana. *T. philomelos* reached densities of 32.2 birds/km, and two other species, *T. torquatus* (0.4 birds/km) and small flocks of *T. pilaris* were also seen.

To determine how consistent between years were both bird numbers and fruit abundance, I compared both the abundance rankings of *Turdus* species and the availability rankings of *Juniperus* cones at each site, including also the lowland R. B. Doñana site (2 years), across study years (see Appendix 1 and 2). Significant values of Kendall's coefficient of concordance for *Turdus* were obtained for Nava Noguera, Cañada del Travino and R. B. Doñana (i.e., the two mixed woodland sites and the lowland site; $W > 0.60$, $p < 0.001$). Non-significant values were obtained for the three highland sites (Campos de Otero, Santa Bárbara, and Calarilla) that showed extreme between-year variations both in the identity of the most abundant species and the abundance of each species. The mixed woodland and lowland sites showed more constancy in the abundance rankings of each species than the highland scrublands, although bird numbers varied extremely. The rankings of juniper cone abundances were compared for the Calarilla and Nava Noguera sites at which two or three species were present. For the former, there was significant between-year constancy in the rankings of juniper species regarding cone availability ($W = 0.933$, $p = 0.05$). At Nava Noguera, annual rankings were inconsistent ($W = 0.758$, $p = 0.435$).

Other fruit crops certainly also have an influence on annual variations in the bird numbers illustrated in Appendix 2. *Crataegus monogyna*, for example, produced a good crop at several Sierra de Cazorla sites (Nava Noguera, Cañada del Travino, Cañada de las Sabinas-N. Correhuelas) in 1987, with > 10 fruits/m^2, and high densities of *T. torquatus* (> 7 birds/km) and *T. iliacus* > 5 birds/km) were recorded there that winter; individuals of all six *Turdus* species were seen feeding simultaneously at the same Crataegus tree.

Annual variations in bird abundance and cone production were greater than between-site variations (Table 3). For all the sites and years pooled, annual variation in both cone production and bird numbers were greater (mean CV = 137.3%) than variation between sites (mean CV = 96.4%) ($F_{1,76} = 11.41$, $p < 0.001$), suggesting that abundance patterns show greater temporal than spatial variability in thee taxa. On the other hand, cone production showed a significantly greater CV (for all years and sites pooled) (mean CV = 132.5%) than *Turdus* abundance (mean CV = 107.9%) ($F_{1,76} = 4.13$, $p < 0.05$). This indicates an overall smaller variation in bird numbers than cone availability.

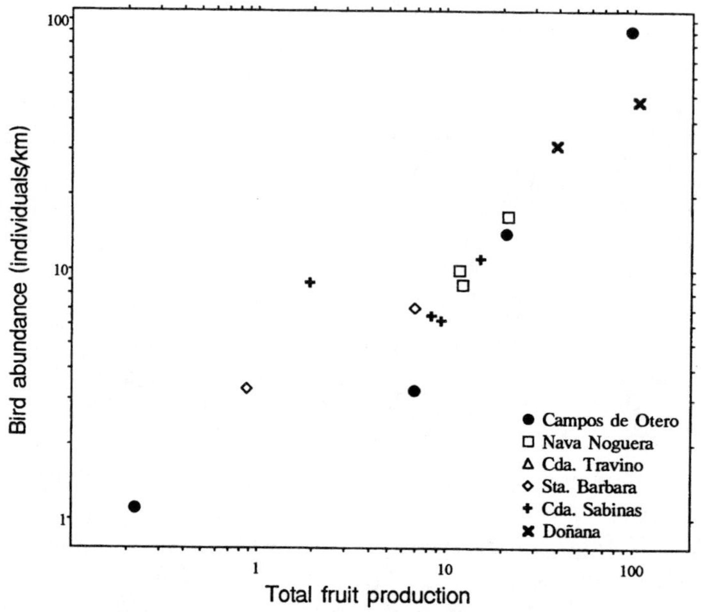

Fig. 3. Relationship between local availability of fleshy fruits (the total production of *Juniperus* cones plus fleshy fruits of angiosperm species) and the total abundance (number of birds/km census) of *Turdus* species. Symbols identify values corresponding to six localities in different study years.

Table 3. Coefficients of variation in the abundance of thrush species (number of birds/km) and the production of juniper cones (number/m^2) in southern Spanish habitats.

Species group	CV (%) Among sites			CV (%) Among years, within sites		
	N*	Mean	Range	N*	Mean	Range
Juniperus species	14	114.31	[65.6–180.4]	15	149.55	[58.3–221.7]
Turdus species	26	86.77	[1.78–181.6]	25	129.92	[58.2–224.1]

* Refers to the number of year-site and site-year combinations for the species in each group.
CV = 100 SD/mean.

Mutual congruency of distribution patterns and interaction strengths

Each species of *Juniperus* and *Turdus* can be characterized by its abundance patterns at three different spatial scales namely, geographic range, habitat use, and local population size (Fig. 1). The congruence patterns between interacting species pairs can be examined by considering their similarities at each of these biogeographic levels (Table 4).

For the species studied, no obvious relationship exists between the size of the geographic range and the degree of habitat specificity (Fig. 4A). *T. pilaris*, with the largest range among *Turdus* (Fig. 2), showed the highest habitat specificity, being found only in highland juniper scrub and mixed mountain forest. *T. torquatus*, in contrast, has a highly restricted wintering range but was recorded in most habitats surveyed (Fig. 4A). The four *Turdus* species with greatest relative importance in the seed dispersal of *Juniperus*, as derived from feeding records (*merula*, *iliacus*, *torquatus*, and *viscivorus*; Table 1), occupy a greater proportion of the habitats surveyed compared with *Juniperus* species. With the exception

Table 4. Similarity values for the geographic ranges, habitat breadth, and relative size of the local populations of *Juniperus* and *Turdus* species pairs. Interaction strength between the two species in each pair was estimated as the number of records for that pair relative to the total number of records for the two species. Dependence of the plant on the bird was estimated as the proportion of fruits consumed by the bird species. Dependence of the bird on that plant species was calculated as the proportion of feeding records on the plant species. A total of 15 pairs of species (excluding *T. philomelos*; see Table 1) were used in the computations.

Species pair	Similarity [1]			Interaction strength	Bird dependence on plant	Plant dependence on bird
	Geographic range	Habitat breadth	Local abundance			
communis-merula	0.361	0.545	0.644	0.0177	2.9	4.6
phoenicea-merula	0.425	0.609	0.868	0.0139	2.6	3.0
sabina-merula	0.389	0.714	0.756	0.0117	1.6	4.7
communis-iliacus	0.425	0.316	0.856	0.0408	8.0	8.4
phoenicea-iliacus	0.364	0.500	0.661	0.0034	0.8	0.6
sabina-iliacus	0.336	0.720	0.475	0.0264	4.0	7.8
communis-torquatus	0.349	0.375	0.728	0.1222	14.9	69.0
phoenicea-torquatus	0.360	0.588	0.707	0.1303	17.0	55.8
sabina-torquatus	0.311	0.545	0.523	0.0508	5.7	48.8
communis-pilaris	0.260	0.222	0.439	0.0000	0.0	0.0
pilaris-phoenicea	0.259	0.632	0.687	0.0117	7.1	2.4
sabina-pilaris	0.208	0.500	0.553	0.0000	0.0	0.0
communis-viscivorus	0.098	0.286	0.740	0.0564	8.2	18.0
phoenicea-viscivorus	0.279	0.636	0.894	0.1333	22.0	33.9
sabina-viscivorus	0.171	0.593	0.669	0.0735	9.2	37.2

[1] Similarities of the geographic ranges and habitats occupied (presence-absence data, see Methods) were calculated with the Sørensen coefficient. Those for local abundances (quantitative data), with the Gower similarity coefficient (Legendre & Legendre 1979).

of *J. communis*, all junipers show reduced, patchy geographic distributions and are restricted to particular habitats. Local population size and geographic range showed no clear association (Fig. 4B), but there was a trend for species with larger population sizes to be habitat generalists and to occupy a greater number of habitats than rare species ($r_s = 0.689$, $p < 0.04$) (Fig. 4C).

No plant-bird species pair emerges from Fig. 4 that would suggest a degree of biogeographic congruence between the two genera. Correlations across species pairs ($n = 15$, see Table 4) between the similarity of geographic range, habitat specificity, and local abundance and the values of interaction strength derived from feeding records were non significant (Table 5). As expected from previous analyses, bird dependence on plants was greater for those species pairs exhibiting similar abundance patterns, suggesting some type of resource tracking, but increasing similarity of geographic ranges or habitat distributions were not correlated with greater mutual dependence.

A situation of close matching between two species at all levels would result in close proximity between their points in each plane depicted in Fig. 4. I computed the inverse euclidean distance between the species in each pair of Table 4 by simultaneously considering the scales of geographic range, habitat use, and local population. This was taken as an estimate of overall biogeographic congruence for each pair, and the resulting correlation coefficient with interaction strength was not significant ($r_s = -0.24$, $p = 0.38$, $n = 15$; Fig. 5). Species pairs with strong mutual dependence such as *J. sabina-T. torquatus*, *J. sabina-T. viscivorus*, and *J. communis-T. torquatus* (Table 1) showed very low values of biogeographic congruency. Thus, *J. sabina* is a geographically restricted species with narrow habitat specificity that interacts strongly with birds species of

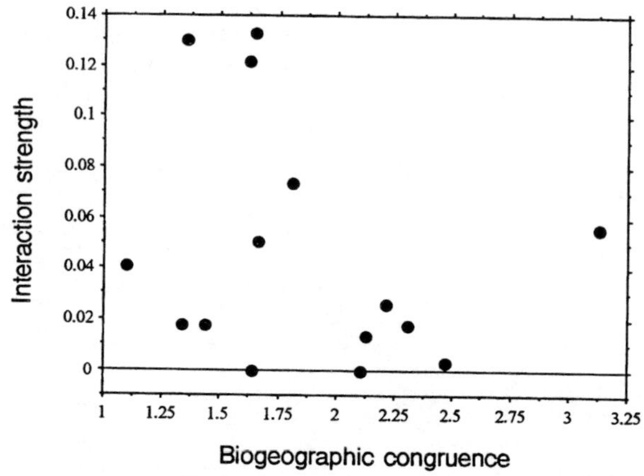

Fig. 5. Plot of biogeographic congruence values between each interacting *Turdus-Juniperus* species pair and their interaction strength in southern Spanish sites (data for the main study sites pooled). Overall biogeographic similarity was estimated as the inverse euclidean distance between the species in each pair when considering size of geographic range, habitat breadth, and local population size. Interaction strength was estimated as the percentage of all feeding records contributed by the records corresponding to each pair of species.

large geographic range and/or wide habitat specificity (*T. viscivorus*, *T. torquatus*). The *J. communis-T. torquatus* pair is also very asymmetric in biogeographic affinities, deriving mainly from differences in geographic ranges (Fig. 2). On the other hand, *J. sabina-T. iliacus* show a high biogeographical similarity but very low interaction strength. Therefore, juniper and thrush species showing strong local interactions in these habitats do not show similar geographic ranges, patterns of habitat use, or relative local popula-

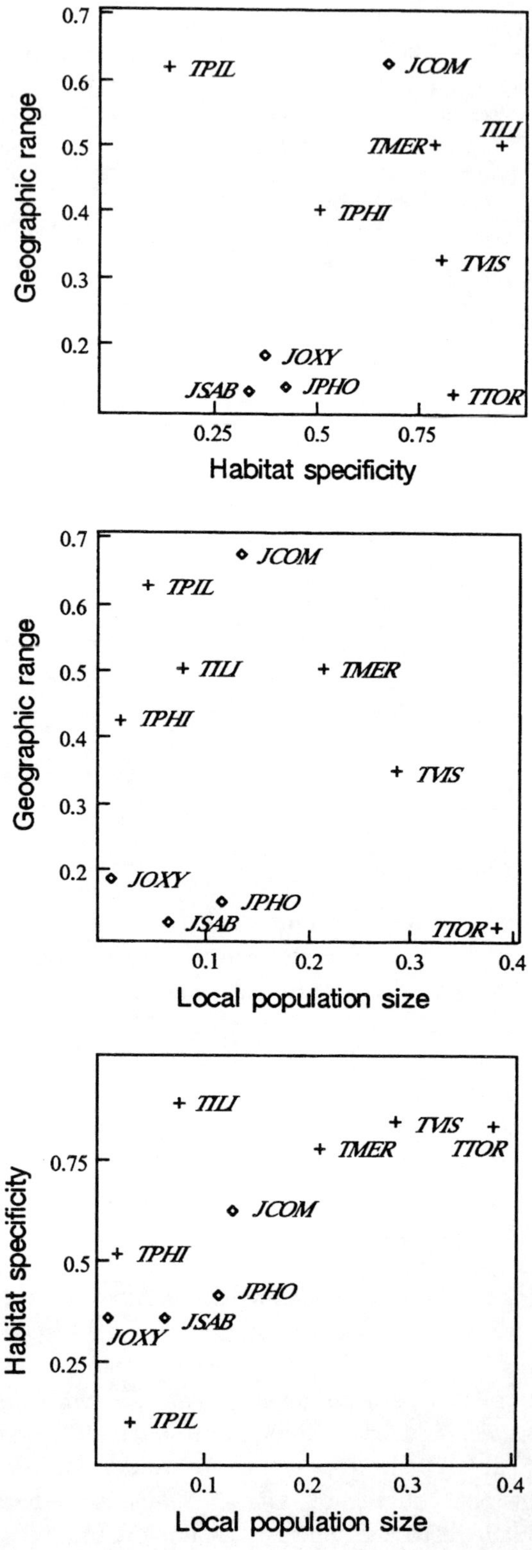

Fig. 4. Patterns of interspecific congruency between the population characteristics of *Juniperus* and *Turdus* species at different biogeographic levels. Importance values at the geographical scale were estimated as the proportion of 5° degrees quadrats where the species is present in the western Palearctic. Values for habitat specificity were calculated as the proportion of southern Spanish localities where the species is present. Figures for the local scale were calculated as the average proportion of total cover (*Juniperus* species) or total thrush abundance (*Turdus* species) in the study localities. The position of each species is indicated by symbols (plus signs, *Turdus* spp.; diamonds, *Juniperus* spp.). Abbreviations for species names as in Table 2.

Table 5. Rank correlations (Kendall's coefficient) between the similarity at different biogeographical levels for each pair of *Juniperus* and *Turdus* species and variables characterizing their interaction pattern at local study sites (see Table 4).

	Similarity between bird and plant in species pair[1]		
	Geographic range	Habitat specificity	Local abundance
Interaction strength	-0.097^{ns}	0.126^{ns}	0.298^{ns}
Plant dependence on bird	0.010^{ns}	0.058^{ns}	0.230^{ns}
Bird dependence on plant	-0.125^{ns}	0.116^{ns}	0.364*

* $p = 0.058$; ^{ns} non-significant.

[1] See Table 4.

tion sizes. Local interaction patterns of strong interdependency are not mirrored by a high degree of biogeographic congruence between the interacting species. This lack of congruence arises from the marked differences of geographic distribution and habitat breadth among *Juniperus* and *Turdus* species.

Discussion

Comparing congruence of plant and frugivore distributions at different scales

Hypotheses about the coevolutionary implications of local interaction patterns can be addressed only if different spatial scales (i.e. the geographic distribution, the regional patterns of habitat use, and the local population sizes) are considered simultaneously. Variation in the outcome of interactions is the raw material for their evolution (Thompson 1988), and we need to understand how the interaction patterns vary and how their outcomes (in terms of fitness implications for the participants) map onto this variation. Biogeographical levels or spatial scales act as nested filters such that congruence at all levels and scales is a necessary condition for an interaction to have taxa-specific evolutionary implications (Wheelwright 1988; Howe 1984). Otherwise, the selective pressures that emerge from strong localized interactions of mutual dependence, if any, are expected to be 'screened-off' by the variation in mutualistic assemblages between habitats or geographic regions.

This study provided no evidence that plant and frugivore species with more similar distribution patterns at each of the above levels showed greater mutual dependence at a local scale. Several alternative outcomes can be defined and the *Juniperus*/*Turdus* interactions described here illustrate some of them:

1. Close fitting between the geographic range, habitat affinity and local population size, together with strong interdependence. None of the interactions studied here fits this situation, but the case of *T. torquatus* and *J. thurifera* could possibly be an example. In general, I would expect this situation to arise in the interactions between two restricted endemic species or between continental species restricted to insular conditions (e.g., high-elevation habitats, etc.).
2. Close fitting at the regional and local scales but low similarity of geographic ranges. This situation is illustrated by the interaction between, e.g., *J. communis*-*T. torquatus* and *J. sabina*-*T. viscivorus*. This type of asymmetry in geographic distribution will be relevant to the evolution of strong interactions when the range of the more restricted species is included within the range of the mutualist. The selection pressures or demographic effects of the widely distributed species will be relevant to the restricted species, given the close habitat association and low variance (e.g., temporal) of the interactions. In contrast, the reciprocal effects of the restricted species are expected to be irrelevant for the widespread species, at

least when considering a large geographic scale. 3. Low congruence in geographic and regional distributions but strong local interdependence. This is a case of association by chance or opportunistic exploitation and is probably very common in plant-frugivore interactions. The interactions between junipers and the two irruptive thrush species (*T. pilaris* and *T. iliacus*) fit in here. Other situations of biogeographic unmatching could be imagined and exemplified, but most *Juniperus/Turdus* interactions fit situation 2 because of the general asymmetry of geographic distributions between the two species groups, with most *Juniperus* areas being 'nested' within the winter ranges of *Turdus*. In addition, the four *Turdus* species with greater relative importance in *Juniperus* dispersal occupied a greater range of habitat types than junipers, indicating a higher species turnover among localities for plant assemblages than for the bird assemblages (see also Herrera 1988a). Therefore, when considering simultaneously all three biogeographic levels, no concordance emerged between local interaction strengths and the congruence of abundance patterns. This result provides quantitative support to earlier interpretations of the lack of biogeographic congruence as a serious constraint for the coevolution of mutualisms (Wheelwright 1988; Herrera 1988a; Howe 1984). The analytical procedure outlined here, based on the typology of forms of rarity proposed by Rabinowitz (1981) appears to be a useful tool for comparing interaction patterns in different biogeographic areas and for testing hypotheses about biogeographic factors limiting coevolutionary interactions.

Variations in space and time: their patterns and causes

Results of this study illustrate marked spatial (between habitat types) and temporal (autumn-winter seasons in several years) variations in the abundance and species involved in interaction between *Juniperus* and *Turdus* species. Very few seed dispersal studies have explicitly addressed the question of variation in time and space and its quantification. Instead, they have considered a more restricted spatial or temporal scale than in the present study (Wheelwright 1988; Herrera 1988a; Keeler-Wolf 1988; Willson 1986, Bronstein & Hoffmann 1987; Jordano 1985; Howe 1980; Howe 1981; Levey 1988; Loiselle 1988). These studies have documented strong variations in the interactions and suggested that a spatio-temporal mosaic of sizes and compositions of seed disperser assemblages is the rule in nature.

At the regional level, most spatial variation in local *Juniperus-Turdus* species assemblages is due to differences in habitat selection.Habitat associations, defined by an elevational gradient, exist among the two species groups and the particular species of each genus present in each habitat is highly predictable. Highland habitats show a characteristic composition with *J. communis* or *J. sabina* as the locally dominant juniper species and *T. viscivorus* and *T. torquatus* as the dominant frugivores. Both *T. iliacus* and *T. pilaris* have irruptive movements associated (at particular sites) with the availability of large cone crops, and they can be found both in highland scrublands and in mid-elevation forests. All *Turdus* species enter the mountain forest and mid-elevation oak forest, but *T. merula* and *T. philomelos* are rarely seen in the highland juniper scrubland. Between-habitat differences in the composition of seed disperser assemblages at particular plant species have also been attributed to differential habitat selection patterns among frugivores and how these interact with the particular locations of individual trees (Bronstein & Hoffmann 1987; McDiarmid, *et al.* 1977; Herrera & Jordano 1981).

A clear pattern emerges for *Turdus* as a group to have a greater habitat breadth than *Juniperus* species. This supports earlier findings of Willson (1986) and Herrera (1988a), who found a greater habitat specificity for plants than for frugivorous birds. For example, Willson (1986) reported that the average percentage of habitat types occupied by wintering frugivorous birds in two major eastern North American regions was 58% and 67%; bird-dispersed plants occupied 49% of the habitats in the whole region. Southern Spanish thrushes occupy, on average, 64% of the surveyed sites ($n = 19$), while *Juniperus* species occupy 45%. This, together with the fact that the CV's for both temporal and spatial variations in *Turdus* abundance were much lower than the CV's of *Juniperus* cone abundance, indicates a greater overall predictability of the frugivore assemblage across habitats and seasons.

Between year variations in juniper cone crop production are related to their long ripening and cone growth periods (> 1.5 years for the Spanish species; Ceballos & Ruiz de la Torre 1979; Obeso, 1985; P. Jordano, pers. obs.) and masting behaviour (Silvertown 1980; Roques, *et al.* 1984). Climate, resource levels, and tradeoffs of previous reproductive events influence the reproductive behaviour in a given year (Lev-Yadun & Liphschitz 1987). Years with no reproductive individuals were frequent in the highland populations, as revealed by the present study, but also in a good crop year cones usually remained during the following winter in good condition for consumption and added to that year's crop. Thus, the greatest variability observed for the *Juniperus-Turdus* system occurred for cone availability among years within sites. The temporal variation (among years, within sites) of both *Juniperus* and *Turdus* abundances were much higher than the variation among sites, as estimated by the CV's. These facts, together with coexistence of several *Juniperus* species in the same area, result in a highly patchy winter resource distribution for the birds in both space and time.

Previous studies of wintering thrushes have pointed out their strong dependence on juniper berries for winter food (Santos, *et al.* 1983; De Juana & Santos 1981; Debussche & Isenmann 1985a, 1985b; Zamora 1990) and the association of irruptive movements of wintering flocks with the local availability of berries (Tyrväinen 1970; Simms 1978). Also, causes external to local food availability (cold spells, food availability and climatological conditions outside the wintering areas) have been proposed to explain the extreme variation and unpredictability of winter bird movements (see Santos 1982 and Soc.Esp.Ornitología 1985 for a review). The strong positive relationship between local availability of juniper and other berries and the local abundance of thrushes suggests behavioral 'resource tracking' and strong bird responses to local food conditions. This type of functional response to fruit availability is expected in this particular system since: 1) juniper cone production made up an important percentage of the total fruit production at the highlands and is frequently the only fruit resource available; 2) important differences in cone availability can be found between nearby areas; and 3) frugivorous thrushes, the main seed dispersers, show erratic or nomadic movement patterns resulting in local altitudinal migrations or regional displacements. The positive relationship that exists across juniper-thrush species pairs between the similarity of local abundance patterns and the bird dependence values on the plants supports these conclusions. It is also interesting to point out the absence of a strong asymmetry in the pairwise interactions between plants and birds in this system, in contrast to their generality in plant-disperser interactions (Jordano 1987).

Concluding remarks

In the particular case of the *Juniperus-Turdus* interaction examined here, the strong uncoupling of biogeographic attributes of plants and frugivores implies that taxon-specific coevolved interactions are extremely unlikely. In this scenario, interactions between two species are restricted to a particular subset of their respective populations and show extremely variable outcomes in time and space. This situation might prove to be common in plant-frugivore interactions in nature. There are few studies containing both quantitative descriptions of how the species identity and numbers in local assemblages vary and quantitative descriptions of the outcome of their interactions (e.g., Herrera 1988b; Horvitz & Schemske 1990) but they support the hypothesis of a diffuse selection regime by a variable assemblage of mutualists.

It could be argued that *Turdus* species exert such a diffuse selection regime on plant fruiting traits given the asymmetry and nested interaction with junipers at a large geographical scale and the strong dependence of the plants on *Turdus* spp. for seed dispersal. No information is available, however, on the relative magnitude of bird effects on plant fitness through differential seed removal and influence on seed deposition and early seedling recruitment. Large variation is to be expected among *Turdus* species due to the variability of forging patterns among the main species at the highlands (*T. torquatus*, *T. viscivorus*, and *T. iliacus*). These include territorial defence of fruiting junipers (*T. viscivorus*; Snow & Snow 1988; P. Jordano, pers. obs.), foraging in large flocks in open areas (*T. torquatus*), and foraging at more closed vegetation in small flocks. This variation could have important effects on seed deposition patterns generated by the three species. For example, potential negative consequences of territorial defence include increased rodent seed predation below or nearby juniper patches (Schupp & López-Soria, pers. com.); in contrast, survival of seedlings from seeds dispersed to open areas can be low due to competition with grasses (Johnsen 1962; Fitter & Jennings 1975).

Therefore, a critical evaluation of the diffuse selection hypothesis would need to address at least three main points: 1) variation in space and time of the size, composition and numbers of the species interacting in local areas; 2) evaluation of the outcomes of interactions and mapping their variations (for example, by focusing on key components of these outcomes such as fruit removal, probability of postdispersal seed predation as a result of seed deposition patterns, and effects on early seedling survival); and 3) test the hypothesis that the effects of the interaction outcomes due to different species are not opposed, but rather additive or compensatory.

Acknowledgements

I appreciate the help of Regino Zamora and M. Luz Hernández during my stays in Sierra Nevada and Sierra de Baza; they were a continuous source of hospitality, support, and valuable discussions. The Agencia de Medio Ambiente authorized my work in Cazorla and provided logistic and housing facilities. I am especially grateful to Myriam for her untiring help with field work. Juan A. Amat, Manolo Carrión, Carlos M. Herrera, José Ramón Obeso, Gene Schupp, and Luis López Soria deserve special mention and gratitude for their helpful discussions and assistance with field work. N.T. Wheelwright and T.H. Fleming made valuable comments and suggestions to the manuscript. Over the years, this study was supported by a postdoctoral grant from Consejo Superior de Investigaciones Científicas, C.I.C.Y.T., the Frank M. Chapman Memorial Fund, and D.G.I.C.Y.T. project PB-87–0452.

Appendix 1. Summary of data on production of fleshy cones ('fruits') by *Juniperus* species in different southern Spanish localities and years. The total production of fleshy fruits (both *Juniperus* and other species included) is also summarized for each site and year.

Site, species, year	Transect surface sampled (m^2)	Number of fruits	Fruits/ m^2	Total fruit production/ m^2
Calarilla-Cda. Sabinas (Sierra de Cazorla)				
J. communis				
1985	3750	365	0.10	1.94
1986	450	2164	4.81	9.32
1987	450	2959	6.58	15.23
1988	450	266	0.59	8.41
1989	450	10	0.02	1.39
J. sabina				
1985	3750	0	0.00	
1986	450	955	2.12	
1987	450	450	1.00	
1988	450	60	0.13	
1989	450	15	0.03	
J. phoenicea				
1985	3750	43	0.01	
1986	450	326	0.72	
1987	450	245	0.54	
1988	450	2268	5.04	
1989	450	30	0.07	
Campos de Otero (Sierra Nevada)				
J. communis				
1985	540	25360	93.93	96.91
1986	540	11345	20.90	20.9
1987	540	1273	6.83	6.89
1988	540	97	0.18	0.22
Nava Noguera (Sierra de Cazorla)				
J. communis				
1985	37500	1270	0.03	–
1986	540	1327	2.55	11.68
1987	540	2090	3.87	21.41
1988	540	165	0.31	12.17
1989	540	765	1.42	30.54
J. phoenicea				
1985	37500	10	0.00	
1986	540	0	0.00	
1987	540	158	0.29	
1988	540	1924	3.56	
1989	540	1068	1.98	
Santa Bárbara (Sierra de Baza)				
J. sabina				
1985	–	–	–	–
1986	900	5777	6.42	6.88
1987	900	417	0.46	0.87
1988	900	1012	1.12	4.09

Appendix 2. Summary of census results (number of birds/km) in different southern Spanish localities and wintering seasons.

Site and Winter	*Turdus merula*	*Turdus philomelos*	*Turdus iliacus*	*Turdus torquatus*	*Turdus pilaris*	*Turdus viscivorus*	*Number of censuses*	*Distance (km)*
Calarilla-Cda. Sabinas (Sierra de Cazorla)								
85–86	0.75	0.00	0.15	2.24	0.00	5.82	9	6.7
86–87	2.00	0.00	0.00	0.50	0.00	3.75	2	4
87–88	3.24	0.00	4.71	0.00	0.00	3.24	10	6.8
88–89	1.32	0.00	0.55	0.11	0.00	4.62	11	9.1
Campos de Otero (Sierra Nevada)								
85–86	0.00	0.00	5.50	83.50	0.00	3.00	2	2
86–87	0.00	0.00	0.42	10.97	0.00	2.64	14	7.2
87–88	0.00	0.00	0.00	0.73	0.00	2.55	8	5.5
88–89	0.00	0.00	0.00	1.11	0.00	0.00	4	2.7
Cda. Travino (Sierra de Cazorla)								
85–86	6.82	0.00	0.45	5.45	0.00	6.82	3	2.2
86–87	8.18	0.00	2.73	1.82	0.00	5.45	2	2.2
87–88	6.49	0.39	5.84	11.43	0.00	5.45	11	7.7
88–89	6.45	0.00	0.97	1.29	0.00	3.23	16	12.4
Nava Noguera (Sierra de Cazorla)								
85–86	3.87	0.97	0.65	4.52	0.00	10.65	5	3.1
86–87	4.00	0.00	0.29	0.86	0.29	4.57	5	3.5
87–88	1.62	0.18	2.43	7.12	0.00	5.14	12	11.1
88–89	2.53	0.00	0.22	2.75	0.15	3.19	32	27.3
Santa Bárbara (Sierra de Baza)								
85–86	–	–	–	–	–	–	–	–
86–87	0.00	0.00	0.48	5.48	0.00	1.13	13	6.2
87–88	0.00	0.00	0.16	0.79	0.00	2.38	9	6.3
88–89	–	–	–	–	–	–	–	–

References

Bronstein, J. L. & Hoffmann, K. 1987. Spatial and temporal variation in frugivory at a neotropical fig, *Ficus pertusa*. Oikos 49: 261268.

Ceballos, L. & Ruiz de la Torre, J. 1979. Arboles y arbustos de la España peninsular. Escuela Técnica Superior de Montes, Secc. Publ. Madrid.

Cramp, S. (ed.). 1988. The birds of the Western Palearctic, vol. V. Oxford Univ. Press. Oxford.

De Juana, E. & Santos, T. 1981. Observations sur l'hivernage des oiseaux dans le Haut-Atlas (Maroc). Alauda 48: 1–12.

Debussche, M. & Isenmann, P. 1985a. An example of Redwing diet in a mediterranean wintering area. Bird Study 32: 152–153.

Debussche, M. & Isenmann, P. 1985b. Le regime alimentaire de la Grive musicienne (*Turdus philomelos*) en automne et en hiver dans les garrigues de Montpellier (France méditerranéenne) et ses relations avec l'ornithochorie. Rev. Ecol. (Terre Vie) 40: 379388.

Ern, H. 1966. Die dreidimensionale Anordnung der Gebirgsvegetation auf der Iberischen Halbinsel. Bonn. Geographische Abhandlungen 37: 1–136.

Fitter, A. H. & Jennings, R. D. 1975. The effects of sheep grazing on the growth and survival of seedling junipers (*Juniperus communis* L.). J. Appl. Ecol. 12: 637–642.

Guitián, J. 1984. Sobre la importancia del acebo (*Ilex aquifolium* L.) en la ecología de la comunidad invernal de paseriformes en la Cordillera Cantábrica Occidental. Ardeola 30: 65–76.

Hartley, P. H. T. 1954. Wild fruits in the diet of british thrushes. A study in the ecology of closely allied species. British Birds 47: 97–107.

Herrera, C. M. 1981. Datos sobre la dieta frugívora del mirlo (*Turdus merula*) en dos localidades del sur de España. Doñana Acta Vert. 8: 306–310.

Herrera, C. M. 1984. A study of avian frugivores, bird-dispersed plants, and their interaction in mediterranean scrublands. Ecol. Monogr. 54: 1–23.

Herrera, C. M. 1987. Vertebrate-dispersed plants of the Iberian peninsula: a study of fruit characteristics. Ecol. Monogr. 57: 305–331.

Herrera, C. M. 1988a. Avian frugivory and seed dispersal in mediterranean habitats: regional variation in plant-animal interaction. In: Ouellet, H. (eds), Acta XIX Congressus Int. Ornithologici. pp: 509–517, Ottawa, Canada.

Herrera, C. M. 1988b. Variation in mutualisms: the spatio-temporal mosaic of a pollinator assemblage. Biol. J. Linn. Soc. 35: 95–125.

Herrera, C. M. 1989. Frugivory and seed dispersal by carnivorous mammals, and associated fruit characteristics, in undisturbed mediterranean habitats. Oikos 55: 250–262.

Herrera, C. M. & Jordano, P. 1981. *Prunus mahaleb* and birds: the high efficiency seed dispersal system of a temperate fruiting tree. Ecol. Monogr. 51: 203–21.

Holthuijzen, A. M. A. & Sharik, T. L. 1985. The avian seed dispersal system of eastern red cedar (*Juniperus virginiana*). Can. J. Bot. 63: 1508–1515.

Horvitz, C. C. & Schemske, D. W. 1990. Spatio-temporal variation in insect mutualists of a neotropical herb. Ecology 71: 1085–1097.

Howe, H. F. 1980. Monkey dispersal and waste of a neotropical fruit. Ecology 61: 944–959.

Howe, H. F. 1981. Dispersal of neotropical nutmeg (*Virola sebifera*) by birds. Auk 98: 88–98.

Howe, H. F. 1984. Constraints on the evolution of mutualisms. Am. Nat. 123: 764–777.

Jalas, J. & Suominen, J. 1985. Atlas Florae Europaeae. Vol. 1. Cambridge Univ. Press. Cambridge.

Janzen, D. H. 1980. When is it coevolution? Evolution 34: 611–612.

Johnsen Jr., T. N. 1962. One-seeded juniper invasion of Northern Arizona grasslands. Ecol. Monogr. 32: 187–207.

Jordano, P. 1981. Alimentación y relaciones tróficas entre los paseriformes en paso otoñal por una localidad de Andalucía central. Doñana Acta Vert. 8: 103–124.

Jordano, P. 1985. El ciclo anual de los paseriformes frugívoros en el matorral mediterráneo del sur de España: importancia de su invernada y variaciones interanuales. Ardeola 32: 69–94.

Jordano, P. 1987. Patterns of mutualistic interactions in pollination and seed dispersal: connectance, dependence asymmetries, and coevolution. Am. Nat. 129: 657–677.

Keeler-Wolf, T. 1988. Fruit and consumer differences in three species of trees shared by Trinidad and Tobago. Biotropica 20: 38–48.

Legendre, L. & Legendre, P. 1979. Ecologie numérique. La structure des données écologiques. Vols. 1 and 2. Masson. Paris.

Lev-Yadun, S. & Liphschitz, N. 1987. The ontogeny of gender of *Cupressus sempervirens* L. Bot. Gaz. 148: 407–412.

Levey, D. J. 1988. Spatial and temporal variation in Costa Rican fruit and fruit-eating bird abundance. Ecol. Monogr. 58: 251–269.

Loiselle, B. A. 1988. Bird abundance and seasonality in a Costa Rican lowland forest canopy. Condor 90: 761–772.

McDiarmid, R. W., Ricklefs, R. E. & Foster, M. S. 1977. Dispersal of *Stemmadennia donnell-smithii* (Apocyanaceae) by birds. Biotropica 9: 925.

Obeso, J. R. 1985. Comunidades de Passeriformes y frugivorismo en altitudes medias de la Sierra de Cazorla. Ph. D. Thesis. Univ. Oviedo, Oviedo.

Obeso, J. R. 1986. Alimentación del Zorzal charlo (*Turdus viscivorus*) en la Sierra de Cazorla, SE de España. Doñana Acta Vert. 13: 95–102.

Obeso, J. R. 1987. Comunidades de passeriformes en bosques mixtos de altitudes medias de la Sierra de Cazorla. Ardeola 34 37–59.

Rabinowitz, D. 1981. Seven forms of rarity. In: Synge, H. (ed.), The biological aspects of rare plant conservation. pp: 205–217, John Wiley & Sons, New York.

Roques, A., Rimbault, J. P. & Goussard, F. 1984. La colonisation des cônes et galbules des genévriers mediterranéens par les insectes et acariens et son influence sur les possibilités de régeneration naturelle de ces essences. Ecologia Mediterranea 10: 147169.

Santos, T. 1982. Migración e invernada de zorzales y mirlos (género *Turdus*) en la Península Ibérica. Ed. Univ. Complutense. Madrid.

Santos, T., Suárez, F. & Tellería, J. L. 1983. The bird communities of the Spanish juniper woodland (*Juniperus thurifera* L.). In: Purroy, F. J. (ed.), Proc. VII Int. Conf. Bird Census Work. pp: 172179, Univ. León, León.

Schemske, D. W. & Horvitz, C. C. 1984. Variation among floral visitors in pollination ability: a precondition for mutualism specialization. Science 225: 519–521.

Silvertown, J. W. 1980. The evolutionary ecology of mast seeding in trees. Biol. J. Linn. Soc. 14: 235–250.

Simms, E. 1978. British thrushes. Collins. London.

Snow, B. K. & Snow, D. W. 1988. Birds and berries. T. & A.D. Poyser. Calton.

Soc. Esp. Ornitología 1985. Estudio sobre la biología migratoria de la tribu Turdini (Aves) en España. ICONA, Monogr. no 39, Publ. Ministerio de Agricultura, Pesca y Alimentación. Madrid.

Suárez, F. & Muñoz-Cobo, J. 1984. Comunidades de aves invernantes en cuatro medios diferentes de la provincia de Córdoba. Doñana, Acta Vert. 11: 45–63.

Tejero, E., Soler, M. & Camacho, I. 1984. Alimentación del Zorzal común (*Turdus philomelos* Brehm. 1831) en olivares de la provincia de Jaén (otoño-invierno). Anal. INIA, Ser. Forestal 8: 9–32.

Tellería, J. L., Santos, T. & Carrascal, L. M. 1988. La invernada de los passeriformes (O. Passeriformes) en la Península Ibérica. In: Tellería, J. L. (ed.), Invernada de aves en la Península Ibérica. Monogr. S.E.O., no 1. pp: 153–166, Sociedad Española de Ornitología, Madrid.

Thompson, J. N. 1988. Variation in interspecific interactions. Ann. Rev. Ecol. Syst. 19: 65–87.

Tyrväinen, H. 1970. The mass occurrence of the fieldfare (*Turdus pilaris*, L.) in the winter of 1964–65 in Finland. Ann. Zool. Fenn. 7: 349–357.

Wheelwright, N. T. 1988. Four constraints in coevolution between fruit-eating birds and fruiting plants: a tropical case history. In: Ouellet, H. (ed.), Acta XIX Congressus Int. Ornithologici. pp: 827–845, Ottawa.

Willson, M. F. 1986. Avian frugivory and seed dispersal in eastern North America. In: Johnston, R. F. (ed.), Current ornithology. Vol. 3. pp: 223-279, Plenum Press, New York.

Zamora, R. 1990. The fruit diet of Ring-ouzels (*Turdus torquatus*) wintering in the Sierra Nevada (South-east Spain). Alauda 58: 67–70.

Part 2

Plant strategies

Introduction

Papers in this section focus on various aspects of the reproductive biology of plants, including the evolution of disperser specialization, the evolution of masting behavior, and the evolution of fruit and seed size, as well as the causes of mortality in a well-studied fleshy-fruited tropical tree, *Virola nobilis* (= *V. suranamensis*).

Fleming *et al.* attempt to answer two questions (1) What determines optimal disperser coterie size? and (2) Why does disperser specialization occur along taxonomic lines? by means of graphical models. They develop a graphical model to predict optimal coterie size as a function of the interaction between number of seeds removed by different disperser species and the proportion of those seeds effectively dispersed by each species. They use a fitness set approach to determine when plants should specialize on particular disperser taxa. Specialization is favored when different disperser taxa are substitutable or antagonistic in their fitness effects on plants; otherwise, mixed taxa coteries are favored. When specialization is favored, birds should be the 'targets' more often than bats and primates because of their larger species pool.

Augspurger and Franson use computer simulations to examine the effects of within-crop variation in mean seed size and its variance on seed distributions around wind-dispersed parent plants. Their results indicate that alterations in mean seed size vs. variance have different effects on seed distributions. Either an increase in variance or a decrease in mean wing loading increases the area and uniformity of density of the seed distribution. They also demonstrate that differences in wind speed on seed distributions is consistently greater than differences among parents with contrasting seed characteristics. Thus, evolutionary responses to selection for changes in seed characteristics are likely to be slow.

Mast-seeding is a conspicuous reproductive strategy in a variety of tropical and non-tropical plants. Sork addresses three non-mutually exclusive hypotheses to explain the evolution of mast-seeding in temperate trees of the genus *Quercus*. The anti-predator and large seed hypotheses appear to best explain mast-seeding in temperate oaks, and she predicts that mast-seeding in tropical oaks should be especially sensitive to the need to synchronize flowering and fruiting as well as the selective advantage of large seed size.

Howe presents an overview of the causes of mortality in seeds and seedlings of *Virola nobilis* in Panama. His detailed studies indicate that early mortality results primarily from weevil attacks, which select for long distance dispersal because they are distance-dependent. Later mortality is caused primarily by herbivorous mammals and is distance- and density-independent. Hence, mortality due to mammals confers no advantage to long distance seed dispersal.

Wheelwright examines within and between plant variation in fruit size in the bird-dispersed tree, *Ocotea tenera*, over an 11 year period. Considerable variation occurs within and between plants, and birds selectively feed at large-fruited trees and take larger-than-average fruits within trees. Phenotypic variation in fruit size is highly heritable, indicating that this plant should be able to respond to directional selection by birds. He concludes, however, that compared with selection for traits such as fecundity and phenology, selection on fruit size is likely to be weak.

Vegetatio **107/108**: 107–120, 1993.
T. H. Fleming and A. Estrada (eds). *Frugivory and Seed Dispersal: Ecological and Evolutionary Aspects.*

Opportunism vs. specialization: the evolution of dispersal strategies in fleshy-fruited plants

T. H. Fleming[1], D. L. Venable[2] & L. G. Herrera M.[1]
[1] *Department of Biology, University of Miami, Coral Gables, FL 33124, USA;* [2] *Department of Ecology and Evolutionary Biology, University of Arizona, Tucson, AZ 85721, USA*

Keywords: Dispersal coteries, Dispersal quality, Dispersal syndromes, Coevolution

Abstract

In this paper we address two questions concerning the interaction between fleshy-fruited plants and their seed dispersers: (1) What determines optimal disperser coterie size (designated as $\hat{C}$) and (2) Why does disperser specialization occur along taxonomic lines? We review factors that affect the evolution of $\hat{C}$ and conclude that seed size and disperser quality (designated as Q) are especially important. We present a simple graphical model for determining $\hat{C}$ based on Q and conclude that $\hat{C}$ will be small (i.e. a specialized disperser strategy) when Q declines rapidly with increasing coterie size; $\hat{C}$ will be large when Q declines slowly with increasing coterie size. We construct a model based on fitness set theory to predict that specialization on particular vertebrate taxa (e.g. birds or mammals) will be favored when different disperser taxa are perfectly substitutable or antagonistic (*sensu* Tilman 1982); mixed-taxa coteries are favored when different disperser taxa are complementary. Finally, we predict that when conditions favor taxonomic specialization, plants will evolve bird fruits more often than bat or primate fruits because of the greater species richness of birds compared with bats and primates.

Introduction

The similarity between the title of this paper and the title of Fleming's paper at the 1985 Frugivore and Seed Dispersal symposium ('Opportunism vs. specialization: the evolution of feeding strategies in frugivorous bats') is not coincidental. In that paper, Fleming asked (1) To what extent is diet choice in bats opportunistic or selective and (2) What ecological factors influence bat diet choice and foraging behavior? Similar questions can be asked about plants providing fleshy fruits for frugivores. Just as many frugivorous animals are selective about the fruits they eat, many plants likewise appear to be selective about the animals they 'choose' to consume their fruits and disperse their seeds. Thus, some plants (e.g. mistletoes, Loranthaceae) appear to have small disperser coteries whereas others (e.g. *Cecropia* and *Ficus*, Moraceae) have much larger disperser coteries. Here we use 'coterie' to denote a group of disperser species.

In addition to differing in the species richness of their disperser coteries, plants also differ in the taxa they use as primary or exclusive dispersers. This specialization is reflected in the existence of dispersal 'syndromes'. Beginning with van der Pijl (1957), the syndrome concept has been widely discussed in the literature (e.g. Gautier-Hion *et al.* 1985; Howe 1986; DeBussche & Isenmann 1989; Herrera 1989). Examples of vertebrate dispersal syndromes are given in Table 1. Bird fruits generally differ from mammal fruits by being small and red, black, or purple in color, lacking a scent

Table 1. Seed dispersal syndromes of vertebrate frugivores. Modified from Howe (1986).

A. Bird fruit syndrome	
Obligate frugivore fruits	Large arillate seeds or drupes; seeds > 10 mm; scentless; rich in lipid or protein; black, blue, green, purple, red
Opportunist fruits	Small or medium-sized arillate seeds, berries, or drupes; seeds < 10 mm; no scent; rich in lipid, protein, sugar or starch; black, blue, orange, red, white
B. Mammal fruit syndrome	
Arboreal frugivore fruits	Large arillate or compound; aromatic, rich in protein, sugar, or starch; brown, green, orange, yellow, white
Aerial frugivore (bat) fruits	Large or small; odorless or musky, often rich in starch or lipid; green, yellow, white or whitish; often pendant
Elephant fruits	Large, tough and indehiscent; fibrous pulp; dull colors; resistant seeds

and a protective husk, and often containing significant amounts of lipids and proteins. Fruit characteristics associated with different dispersal syndromes presumably reflect qualitative as well as quantitative differences in the morphology, physiology, and behavior of potential vertebrate dispersal agents as discussed in detail by Howe (1986) and Schupp (this volume).

It is well-known that these syndromes have many exceptions and that fruit morphological/chemical characteristics are only moderately predictive about which animals disperse particular species of fruit (Heithaus 1982; Howe 1986). Nonetheless, the existence of recognizable 'syndromes' defined by O'Dowd and Gill (1986) as the nonrandom occurrence of combinations of diaspore traits related to the nature of dispersal agents, suggests that many plants indeed specialize on different subsets of animals to disperse their seeds. Instead of producing fruits that are universally attractive to fruit-eating vertebrates, plants often appear to 'aim' their fruits at nonrandom subsets of potential dispersers. This raises two important evolutionary questions: (1) What determines optimal disperser coterie size and (2) Why does disperser specialization occur along taxonomic lines?

The purpose of this paper is to explore possible answers to these two questions using theoretical approaches that have not previously been applied to frugivory/dispersal systems. Because birds and mammals are numerically the most important biotic seed dispersers in most contemporary habitats (Howe 1986; Fleming 1991a), we will restrict our discussion to interactions between plants and those two groups of vertebrates. We will proceed by first discussing constraints on the evolution of dispersal strategies and the concept of seed dispersal quality. We will then develop simple models of optimal coterie size and taxon-specific specialization. Our overall conclusions based on this approach are that conditions favoring the evolution of generalized (= opportunistic) dispersal strategies apparently outnumber those favoring taxon-specific strategies, but that when taxon-specificity is favored, birds will be the 'targets' of specificity more often than other groups of vertebrates.

Factors influencing the evolution of fruiting strategies

We begin our discussion of the evolution of dispersal strategies of fleshy-fruited plants by noting that the goal of such a strategy is to maximize the number of surviving, reproducing offspring produced by a plant. Given this goal, an important question then becomes, What determines the coterie size, symbolized as $\hat{C}$ throughout this paper, that maximizes a plant's fitness?

Before addressing this question, we note that many factors in addition to the characteristics of potential dispersers are involved in the evolution of plant fruiting strategies. For example, the ways in which three factors – establishment site predictability, plant successional status, and plant growth form – influence seed size, a critical factor in fruit choice and fruit availability in birds and mammals (Wheelwright 1985; Dowsett-Lemaire 1988; DeBussche & Isenmann 1989; Willson *et al.* 1989), is illustrated in Fig. 1. As discussed by Fleming (1988), a seed's size critically influences its mobility and availability to vertebrate consumers. Seed size, in turn, is influenced by a plant's establishment strategy, as discussed by Grime (1979), Willson (1983), and Foster (1986), among others.

Because the number of vertebrate species capable of handling and/or ingesting fruits/seeds is inversely related to diaspore size (Wheelwright 1985; Fleming 1988), seed size and all abiotic and biotic factors influencing seed size (including the size of potential vertebrate dispersers, see below) will automatically have a major influence on $\hat{C}$ (Martin 1985). We suggest that the inverse relationship between fruit/seed size and coterie size should be the null expectation in any discussion of the evolution of $\hat{C}$ in fleshy-fruited plants. Because of their fruit/seed size, some plants will have smaller disperser coteries than others and hence will appear to be more specialized. What is of interest in discussions of the evolution of $\hat{C}$ is variation around the regression line. For example, seeds of *Virola sebifera* and *V. surinamensis* differ considerably in mass (0.7 g vs. 2.0 g, respectively) but have similar small disperser coteries of 7–8 species of birds on Barro Colorado Island, Panama (Howe 1981; Howe & Vande Kerchove 1981). Controlling for the effect of fruit or seed size, why do some plant species have a larger or smaller $\hat{C}$ than expected by chance? Is this variation random or does it conform to a predictable pattern?

Some of the biotic constraints involved in the evolution of the dispersal strategies of fleshy-fruited plants are listed in Table 2. Constraints in the plant column influence diet choice in frugivorous vertebrates and constraints in the animal column can influence the evolution of plant dispersal strategies. For example, as mentioned above, fruit and seed size strongly affect the accessibility of fruit to birds and mammals. Whereas large birds and mammals can handle a wide range of fruit sizes containing either small or large seeds, small species can only handle small fruits containing relatively small seeds. A plant selected to maximize the breadth of its disperser coterie because of its seedling establishment requirements (e.g. a pioneer plant) will be constrained to produce fruits containing many small seeds. Optimal fruit size will also be small (e.g. *Muntingia calabura*, many *Ficus* species) or the fruit must be capable of being eaten and seeds ingested in piecemeal fashion (e.g. *Cecropia* species).

Differences in the body sizes of New World and Old World tropical frugivorous birds and mammals appear to have influenced maximum fruit (and seed?) size in several families of fleshy-

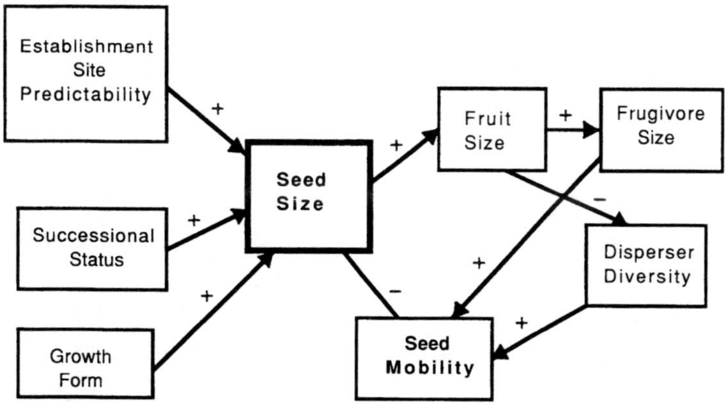

Fig. 1. Diagram indicating some of the factors influencing seed size and the effect of seed size on seed mobility and the diversity of disperser coteries. Plus and minus signs indicate positive and negative correlations, respectively, between variables; the arrow points to the dependent variable in each correlation. Redrawn from Fleming (1988).

Table 2. Biotic constraints in fruit disperser-fruit coevolution.

A. Plant-based constraints	B. Frugivore-based constraints
1. Fruit/seed size	1. Body size (gape width)
2. Degree of seed protection	2. Masticatory morphology
3. Nutritional characteristics/ secondary compounds	3. Digestive physiology
4. Fruit detectability	4. Sensory physiology
5. Fruit accessibility	5. Fruit harvesting behavior
6. Spatio-temporal availability	6. Spatio-temporal availability

Note: Additional constraints on plants but not on animals include effects of seed/seedling predators, pathogens, and competitors.

fruited plants. Fleming *et al.* (1987) reported that maximum sizes of contemporary frugivorous birds, primates, and bats are greater in the paleotropics than in the neotropics. This led A. Mack (unpubl. manuscript) to survey fruit sizes in eight families of plants widely distributed in both the New and Old World tropics. He found that mean fruit size was significantly larger in Old World Moraceae and Anacardiaceae and that the largest genera in the Burseraceae, Lauraceae, Meliaceae, Myristicaceae, Palmae, and Simaroubaceae occurred in the Old World more often than expected by chance. These results suggest that, in addition to the general factors indicated in Fig. 1, fruit size in at least some families of plants is sensitive to selection pressures provided by potential dispersal agents. This kind of animal-influenced plant evolution will influence $\hat{C}$ of tropical plants.

A model of optimal disperser coterie size

Any model of optimal disperser coterie size must ultimately explain some pattern in nature. However, with our currently limited database, we know relatively little about how actual or optimal Cs are distributed among and within species of fleshy-fruited plants in most communities. Are they normally or log-normally distributed among species? Are extremely broad- or narrow-niched species exceptional or are they common? Do patterns in tropical communities differ qualitatively from those in non-tropical communities? Izhaki (unpubl. manuscript) reviewed available data on the size of disperser coteries of bird-dispersed plants and reported that coterie size in temperate plants ($\bar{x} = 8.6$ species, range = 1–31 species) was about half as large as that of tropical plants ($\bar{x} = 14.6$, range = 1–43). Similar questions can be asked within species of fleshy-fruited plants; what is the shape of the curve relating $\hat{C}$ to lifetime fitness? Given the 'diffuse' nature of plant-frugivore interactions (Fleming 1991a), it is likely that $\hat{C}$ varies considerably in space and time (both geological and contemporary) for many plant species. Because of this variation, the answer to the question, What is $\hat{C}$ for plant species *i*?, must often (always?) be location- and time-dependent.

With this caveat in mind we suggest that within a subset of potential vertebrate dispersers (determined by factors in column A of Table 2), $\hat{C}$ will be determined by patterns of variation in seed dispersal quality (Q) among dispersers. For purposes of this paper, we will define Q for disperser species *j* and plant species *i* as $Q_{ij} = r_{ij} \times p_{ij}$ where r_{ij} = the proportion of seeds of *i* removed by *j* and p_{ij} = the proportion of seeds of dispersed by *j* that establish as seedlings. These will be called 'removal' and 'dispersal effectiveness', respectively, for short. Note that r_{ij} depends on the abundance of species *j* and the attractiveness or availability of fruits of *i* to *j*. It may also vary with the species composition of the coterie due to competitive interactions among dispersers. Dispersal effectiveness, p_{ij}, depends on the foraging movements and seed handling behavior of *j*.

Thus, the total number of seeds of species *i* effectively dispersed (a measure of *i*'s fitness) is

$$W_i = F_i \sum_{j=1}^{C} Q_{ij} = F_i Q_i(C)$$

where F_i is the total number of seeds produced per plant of species *i* per season and $Q_i(C)$ is the dispersal quality of the entire coterie consisting of C members.

Another way of expressing a disperser's Q_{ij} is to determine the proportion of established seedlings of species *i* that it dispersed. Whichever way it is measured, Q_{ij} will be very difficult, if not impossible, to measure for most species (but see Reid 1989). It should be noted that our terminology differs somewhat from that of Schupp (this volume). Our Q_{ij} is called 'effectiveness' and our p_{ij} is called 'quality' by Schupp.

Just as $\hat{C}$ is likely to vary among plant species, Q_{ij} is also likely to vary among vertebrate frugivores in a community. We currently have little information on how Q_{ij} varies among species (Is it normally-distributed or right- or left-skewed?). Are some birds or mammals likely to be consistently high (or low) quality dispersers across plant species? Current information (Levey 1986 and Schupp, this volume) suggests that within disperser species, Q_{ij} is likely to vary considerably across plant species. For instance, Chestnut-mandibled toucans (*Ramphastos swainsonii*) are excellent dispersers of the seeds of *Virola surinamensis* but are very poor dispersers of seeds of *Tetragastris panamensis* in Panama (Howe 1990). Some seeds ingested by the howler monkey *Alouatta palliata* (e.g. *Agonandra macrocarpa*, *Muntingia calabura*) are killed by gut passage whereas others (e.g. *Bursera simaruba*, *Cupania guatemalensis*) are not (Chapman 1989).

Plant fitness can also be expressed as the product of the total number of seeds removed times the mean proportion effectively dispersed, both of which are functions of coterie size, C (and coterie membership). The number of seeds of species *i* removed by C disperser species equals F_i times the total proportion of seeds removed, $r_{i\,tot}(C)$, which equals $\sum_{j=1}^{C} r_{ij}$. The mean proportion of these seeds that are effectively dispersed is

$$\bar{p}_i(C) = \sum_{j=1}^{C} \frac{r_{ij}\, p_{ij}}{\sum_{j=1}^{C} r_{ij}}$$

This proportion can be thought of as the probability that a removed seed will be effectively dispersed, averaged over all dispersers. Plant fitness can then be expressed as

$$W_i = F_i \Sigma r_{ij} \cdot \sum_{j=1}^{C} \frac{r_{ij}\, p_{ij}}{\sum_{j=1}^{C} r_{ij}} = F_i r_{i\,tot}(C) \cdot \bar{p}_i(C)$$

where

$$Q_i(C) = r_{i\,tot}(C) \cdot \bar{p}_i(C)$$

is the seed dispersal quality of the entire coterie.

Fitness is composed of multiplicative components which, if converted to logarithms, can be expressed as a sum for the purpose of making a simple graphical model. Thus,

$$\ln W_i(C) = \ln F_i + \ln r_{i\,tot}(C) + \ln \bar{p}_i(C)\,.$$

Since F_i is not determined by coterie size or membership, we can find the fitness-maximizing coterie size by concentrating on the components of dispersal success, i.e. the combined dispersal quality of the entire coterie,

$$\ln Q_i(C) = \ln r_{i\,tot}(C) + \ln \bar{p}_i(C)\,.$$

If disperser species are added to the coterie in their order of decreasing quality (Q_{ij}), then $\bar{p}_i(C)$ will usually decline with C whereas the total proportion of seeds removed ($r_{i\,tot}(C)$) will generally increase with C. When the logs of $\bar{p}_i(C)$ and $r_{i\,tot}(C)$ are plotted against log C, log dispersal success of plant species *i* can be calculated graphically as the sum of the two curves (Fig. 2). The coterie size favored by selection ($\hat{C}$) will be the one with the highest sum. $\hat{C}$ can be thought of as the coterie size for which the fitness increase from having more seeds dispersed equals the fitness decrease from lowering the effectiveness of their dispersal. That is, another species should be added to the coterie as long as the increase in the proportion of seeds removed is greater than the reduction in the mean effectiveness of dispersal ($\Delta \ln r_{i\,tot}(C) > -\Delta \ln \bar{p}_i(c)$; cf. Venable 1992; Lloyd & Venable 1992). An easy way to think of logarithms here is as proportional changes; thus $\hat{C}$ is the coterie size at which the proportional increase in seed removal that could be achieved by adding more species to the coterie is exactly balanced by the proportional decrease in the effectiveness of dispersal that would result.

The model predicts that $\hat{C}$ will be small (a specialized disperser strategy) when the mean

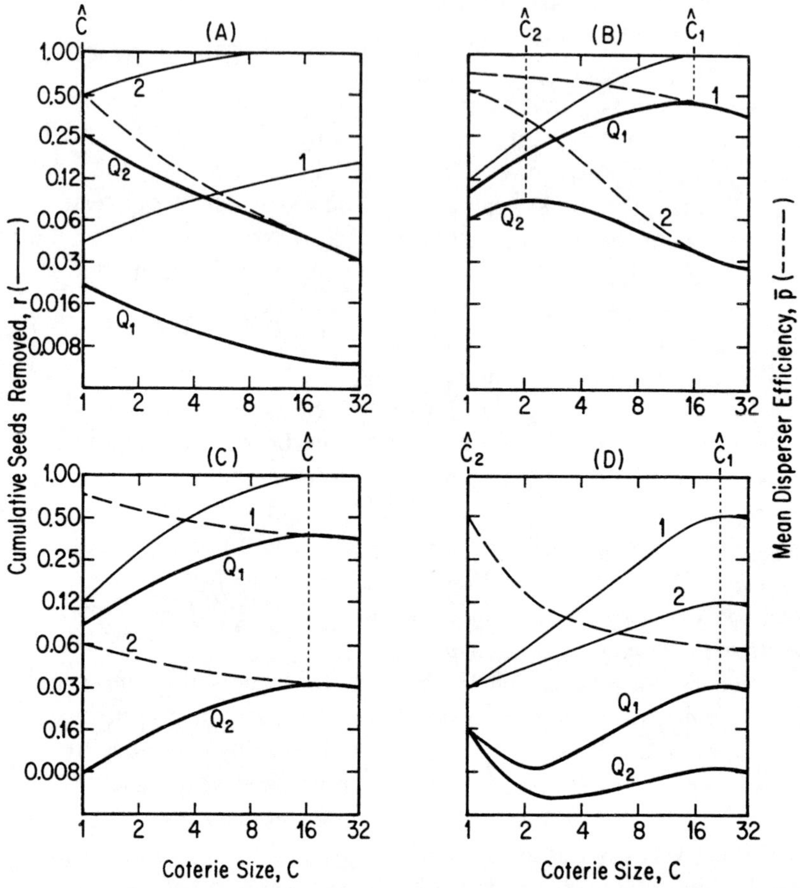

Fig. 2. Graphical models for predicting optimal disperser coterie size ($\hat{C}$) as a function of the cumulative proportion of seeds removed (r) and the proportion of seeds effectively dispersed (p) by different disperser species ranked by their seed dispersal quality. Curves marked Q indicate overall plant fitness. Optimal coterie size ($\hat{C}$) is that coterie at which Q is greatest. Numbers = 1 & 2 refer to specific cases. See text for further explanation.

dispersal effectiveness tends to decline faster than seed removal increases with increasing C. Conversely, $\hat{C}$ will be large (a generalized or opportunistic disperser strategy) when removal increases faster than the mean dispersal effectiveness declines.

The former case could occur if seeds are generally unattractive or dispersers rare so that the proportion of the fruit crop removed ($\ln r_{i\,tot}(C)$) increases only slowly with C (Fig. 2A, case 1). The mean dispersal effectiveness curve would decline rapidly if only one or a few species are very effective dispersers, and $\hat{C}$ would be small. The same prediction of disperser specialization would hold if seeds were very attractive and the dispersers abundant, but again with only a few species being very effective at dispersing the seeds they remove (Fig. 2A, case 2). In the first case, adding species to the coterie does not improve the proportion of seeds removed much because the potential dispersers are not very attracted (or abundant) and the increment in seed removal is low relative to the reduced effectiveness of dispersal. In the second case, adding species does not improve removal very much because the first species already removes most of the seeds.

It is important to note that we have added the two curves in a downward direction treating 1.0 as the origin. This is because both $\bar{p}_i(C)$ and $r_{i\,tot}(C)$ are proportions with values between O and 1. Thus their logarithms are negative and adding them gives a more negative value (just as multiplying a seed removal fraction of 0.5 by an effective dispersal fraction of 0.5 gives a smaller proportion, 0.25, of the seeds that are both removed and effectively dispersed).

Also notice that, in this model, a specialized dispersal coterie can only be favored if dispersers interfere with each other's seed removal or dispersal effectiveness. Even if a second species had a zero dispersal effectiveness, it could not lower total dispersal quality of the coterie unless it had a negative effect on the seed removal or dispersal efficiencies of the better quality dispersers.

Coterie size should vary inversely with the variance in dispersal effectiveness. This is because variation in dispersal effectiveness steepens the mean dispersal effectiveness curve (Fig. 2B, cf. cases 1 and 2). If all dispersers are similar in effectiveness, the mean dispersal effectiveness curve will be flat. Dispersers can be either ineffective or highly effective and have low variance and a flat mean effectiveness curve (Fig. 2C, cases 1 and 2). This should select for large dispersal coteries as long as the seed removal curve is steeper.

Consider one last example that illustrates several important points not touched on above. Suppose the potential disperser community contains one species that does not remove very many seeds but is highly effective at depositing those removed in favorable sites. All of the other potential dispersers remove even fewer seeds, have a negative effect on seed removal by the first species, and have very low effectiveness at depositing seeds they do remove in favorable sites (Fig. 2D). For a plant with this set of potential dispersers, there may be two coterie sizes that are local optima, depending on parameter values. One local optima will be to specialize on the effective disperser; the other will be to be a generalist. The latter is a local optimum as long as some coterie sizes exist for which the increments in proportion of seeds dispersed are greater than the continued reduction in mean dispersal effectiveness. Whether the generalist strategy is a global optimum or not depends on how much the large coterie can increase the proportion of seeds effectively dispersed (Fig. 2D, cf. cases 1 and 2).

Regardless of whether the generalist or specialist strategy is the global optimum, there is likely to be a fitness valley between them. This creates a role for phylogenetic conservatism and rapid punctuated evolution in fruit biology. A specialist species would not be favored to become somewhat more generalist, even if the highly generalized strategy were the global optimum (Fig. 2D, case 1). Likewise, selection would not favor a slightly less generalized generalist, even if the specialist were the global optimum (Fig. 2D, case 2). Only macro-mutations affecting traits that influence fruit attractiveness or availability to frugivores could permit plants to cross the fitness valley.

This model provides a graphical tool for thinking about the evolution of coterie size. It predicts how coterie size should evolve with different communities of dispersers having different distributions of removal proportions and dispersal effectivenesses. Many scenarios not discussed here could be analyzed, yielding predictions that might not be easy to derive otherwise. The key idea is to think in proportional changes. We have presented some simple examples of predictions that can be generated with this model but have not explored it exhaustively here.

Taxon-specific disperser specialization

The existence of fruit syndromes implies that selection has favored the evolution of specialization along taxonomic lines. Why should fruit syndromes exist? Are the qualitative and quantitative characteristics listed in Table 2 different enough between birds and mammals to favor the evolution of taxonomically specialized dispersal strategies?

Here we develop a graphical model for determining the dispersal strategy that maximizes a plant's fitness when it has two (or more) vertebrate taxa to 'choose' from as its seed dispersers. First we plot the proportion of a plant's seeds removed by each of two groups of dispersers (e.g. birds and mammals) for different fruit designs (Fig. 4a). Seed removal is defined as in the previous section, so that r_{ib} and r_{im} are the proportions of the F_i seeds of plant species i that are removed by birds and mammals, respectively. These points are bounded by a *removal constraint curve*. Each point in the stippled area represents the bird and mammal removal fractions that result from a particular feasible fruit design. For each point below the removal constraint curve, some aspect of fruit design could be improved to give greater seed removal by one disperser without any cost in terms of seed removal by the other. All such changes should be favored by natural selection, leading to fruit designs that fall along the constraint curve. Points on the constraint curve are at the design limit where no further improvement in fruit removal by either taxon can be attained without a reduction in removal by the other taxon, thus the negative slope. Such tradeoffs are due to differences between birds and mammals in size, food-gathering morphology, digestive physiology, and sensory physiology (e.g. Table 2; Shupp, this volume; Martinez del Rio and Restrepo, this volume). That is, fruits that are more attractive to birds owing to their color, nutrient composition, time of ripening (e.g. day vs. night), etc. will be less attractive to mammals and vice versa.

This model is similar to the one presented in the previous section. There we dealt with several to many disperser species (without regard to taxonomy) but only discussed in a general way the fact that adding an additional disperser may reduce seed removal or effectiveness of animals already in the coterie. In the present model, we will explicitly graph these interactions and consider how changes in fruit design increase or decrease seed removal by various actual or potential dispersers (though our 2 dimensional graphical approach only deals with 2 disperser species or 2 taxa at a time).

As mentioned previously, seed removal is determined partly by the attractiveness of the fruit, and thus by the constraints listed in Table 2, and partly by the relative abundance of the potential dispersers (Fig. 3). If, for example, it is difficult to design a fruit that attracts birds but is easy to design one that attracts mammals, then the removal constraint curve will tend to have a shal-

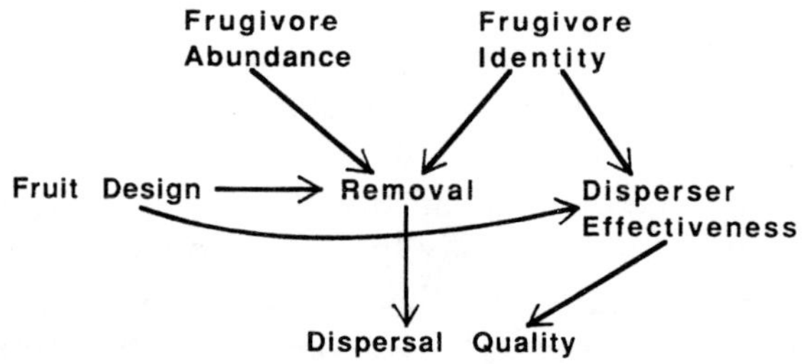

Fig. 3. Schematic plot of the interaction of factors affecting the dispersal quality experienced by a fleshy-fruited plant attracting vertebrate frugivores.

low slope. If birds are more abundant than mammals, it may be easier to obtain higher removal proportions with birds than with mammals. This would generate a steep removal constraint curve (e.g. Fig. 4A).

The removal constraint curve is not necessarily straight, and its degree of curvature has an important effect on the superiority of a specialist vs. a generalist dispersal syndrome. If the curve bows inward, a generalist syndrome will have a lower total seed removal ($r_{i\,tot} = r_{ib} + r_{im}$) than could be achieved by specializing on one (or perhaps either) of the disperser taxa. The curvature is determined by such factors as the competitive interactions between birds and mammals for fruit and by how different the specialized bird and mammal dispersal syndromes are. For example, if birds and mammals undergo interference competition, the line will bow toward the origin. Following the terminology of Tilman (1982) and others, such a curve can be called an 'antagonistic removal constraint'. If the dispersers experience some kind of mutualistic interaction, perhaps by inadvertently warning each other about predators while feeding, the line would bow out (i.e. a 'complementary removal constraint'). If an attractive mammal fruit is very different in design from an attractive bird fruit, it may be that an intermediate fruit design will not result in much seed removal by either group. This would result in an antagonistic curve bowed in towards the origin. If the two syndromes are not very different, intermediate designs may result in considerable removal by both groups and a complementary removal curve.

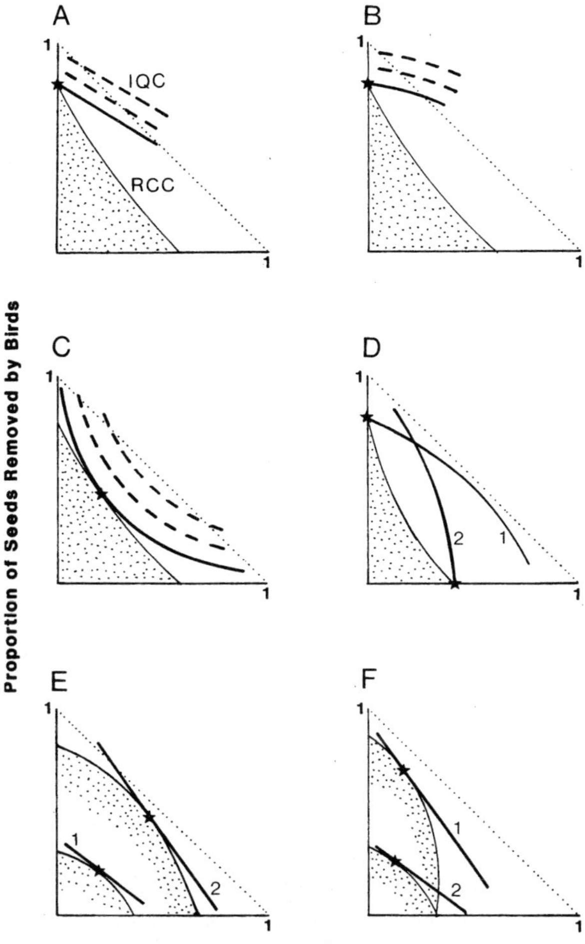

Fig. 4. Graphical models for predicting the optimal disperser strategy (*i.e.* specialization on one taxon or on a mixture of taxa) in fleshy-fruited plants. Axes indicate the proportion of a plant's seed crop removed by either birds or mammals. The solid light lines represent removal constraint curves (RCC). The solid heavy and dashed lines represent lines of constant disperser effectiveness (i.e. isoquality curves, IQC). The dotted line indicates maximum seed removal. Stars indicate the optimal disperser strategy. Numbers 1 & 2 refer to specific cases. See text for further explanation.

As discussed in our previous model, seed removal does not necessarily translate directly into dispersal quality because taxa may differ in their effectiveness at transporting seeds to sites favorable for establishment (Shupp, this volume). Specifically, the dispersal quality of a particular disperser *j* for plant species *i* has been defined as $Q_{ij} = r_{ij} \times p_{ij}$ and the dispersal quality of plant species *i* due to the entire coterie is given by $Q_i(C) = \sum_{j=i}^{C} r_{ij} \times \bar{p}_{ij}$.

We can translate removal proportions into dispersal quality by adding to our graph a plot of *isoquality curves* which connect the removal proportions that yield equal dispersal quality. In the case of bird vs. mammal dispersal, such an isoquality curve is given by $Q_i(C) = r_b p_b + r_m p_m = K$ where K is a given constant dispersal effectiveness and which can be related to fitness by $W_i = F_i Q_i(C) = F_i K$. There will be a family of isoquality curves parallel to one another, each with a different dispersal quality K. As total seed removal increases, the K curves will move farther from the origin (Fig. 4A). Such curves are sometimes called *indifference curves* because a plant should be indifferent to the set of removal proportions specified by particular isoquality curves (and corresponding fruit design) because they all generate the same fitness.

The fundamental question for the evolution of dispersal syndromes is, Which fruit design gives the highest K and thus the highest fitness? This can be ascertained by determining the highest K isoquality curve that intersects the removal constraint curve (the heavy solid line in Fig. 4A). The one with highest K will always be tangent to the removal constraint curve (or touch it at the boundary where it meets one of the axes), and the best fruit design will be the one associated with the seed removal proportions at this point. This type of graphical analysis is similar to fitness set analysis (Levins 1968) and to the approach used by Tilman (1980, 1982) and others (e.g. Rapport 1971; Covich 1972) in foraging theory. The graphical and algebraic techniques for such models are well developed (cf. Venable & Brown 1988; Lloyd & Venable 1992).

If the dispersal effectiveness of birds and mammals is invariant with changes in fruit design, the isoquality curves described above will form straight lines given by $r_b = K/p_b - (p_m/p_b)r_m$ when plotted in r_b, r_m space. In this situation, birds and mammals can be considered to be perfectly substitutable dispersers (*sensu* Tilman 1982). The slope will be determined by the relative effectiveness of birds and mammals. If birds are more effective than mammals, the slope of the curve will be shallow and will intersect the bird removal axis at a low point determined by K/p_b. The opposite situation will hold if mammals are the more effective dispersers.

The intuition behind isoquality curves may be clearer if it is remembered that they tend to function in an inverse fashion to the removal constraint curves. Removal values farther from the origin are 'better', so a steep removal curve means that birds remove more seeds than mammals. In contrast, an isoquality line closer to the origin on the bird axis means that birds are more effective than mammals. This is because the same dispersal quality can be achieved with fewer seeds removed by birds. Thus a slope more shallow than -1 implies that birds are more effective dispersers than mammals. Indeed, the fitness maximum occurs where the 2 curves touch, i.e. where the removal curve is out far enough and an isoquality curve is in far enough that they coincide. Thus, if birds removed more seeds than mammals (giving a steep removal curve such as in Fig. 4A) and birds were also more effective at dispersing removed seeds (yielding a shallower isoquality curve as in Fig. 4A), then the 2 curves will intersect at the left of the graph at a point representing a specialized bird fruit, just as our intuition tells us it should.

Isoquality curves will not be straight lines if disperser effectiveness varies with fruit design. Recall from the previous model that the dispersal quality of an entire coterie can be thought of as the product of the total proportion of seeds removed and the average effectiveness of dispersers, i.e. $Q_i\,(C) = r_{i\,tot} \times \bar{p}_i$. Thus, if the average dispersal effectiveness is lower at intermediate fruit designs, removal proportions must be higher to maintain the constant dispersal quality of an isoquality curve. In this case, the isoquality curves would bow away from the origin (an 'antagonistic isoquality curve'; Fig. 4B). If average dispersal effectiveness were higher at intermediate fruit designs, then removal proportions would have to be lower to maintain a constant dispersal quality, and the curves would bow in (a 'complementary isoquality curve'; Fig. 4C). Technically, the condition for an antagonistic isoquality curve is that $p_b + p_m$ is greater for specialized bird and mammal plants than for generalists (the inequality is reversed for complementary isoquality curves, and a straight isoquality curve is generated by constant values of p_b and p_m). Again, the intuition is the reverse of that for the removal constraint curves: an outward bowing isoquality curve is an antagonistic one signifying that it is hard to have high quality dispersal with an intermediate, generalist fruit dispersal strategy.

As an example, consider the following scenario. As a species becomes more specialized for dispersal by birds, not only is the attractiveness to birds and seed removal by birds likely to increase, but also birds may become more effective dispersers of removed fruit. This might result from evolutionary changes in fruit design affecting, e.g. hardness of the endocarp, the production of mucilage on the seed coats, or changes in the laxative properties of the pulp. If other modifications occur when the fruit specializes more on mammals, the effectiveness of mammals might also increase. Under these circumstances, p for specialized bird or mammal fruits would be higher than p for generalized fruits, which would result in an outward bowing (antagonistic) isoquality curve (Fig. 4B).

It is theoretically possible, though unlikely, that increased specialization and increased removal by a disperser taxon might be accompanied by a declining effectiveness of that disperser. For example, if birds and mammals dispersed seeds to different types of safe sites for germination and both types were limited in availability, density dependent mortality might occur among seedlings within each type of safe site. This could result in lower effectiveness as specialization towards one disperser increased. The density effects might be alleviated somewhat by utilizing both types of dispersers in a generalized syndrome that would have a higher $\bar{p}$, which would result in a complementary isoquality curve (Fig. 4C).

Curved isoquality lines could also result if the effectiveness of two disperser taxa interacted. For example, the curves might bow in (a complementary effect; Fig. 4C) if dispersal by either taxon alone was not very effective, but successive dispersal by both was highly effective. In this scenario, $\bar{p}$ would be highest with intermediate, more generalized fruits (or perhaps with dual dispersal mechanisms). Possible examples of this include combined bird and ant dispersal (Byrne & Levey, this volume) and combined ballistic and ant dispersal (Beattie 1985).

We have discussed how the biology of frugivore-plant systems outlined in Fig. 3 gives rise to differently shaped removal constraint and isoquality curves. We can now put these curves together to predict how the different factors in Fig. 3 select for taxon-specific or generalized dispersal syndromes. One scenario that we consider likely is that fruit traits favoring attraction and removal by birds may often be sufficiently different from those favoring removal by mammals that intermediate fruits are not very attractive to either and thus experience low seed removal. That is, birds and mammals are antagonistic dispersers. In many habitats birds are more abundant than mammals, which could result in the removal constraint curve intersecting the bird axis at a high removal proportion and a steep removal slope (Fig. 4D). Since specialization on either taxon is likely to increase dispersal effectiveness, the isoquality curve will be antagonistic (bowed out) with its slope depending on the relative effectiveness of birds and mammals. This scenario probably describes many dispersal systems, and it always favors specialization on one taxon over generalization. Which taxon depends on the quantitative interplay of removal and effectiveness. If birds, by virtue of their abundance, tend to remove more seeds than mammals, then specialization on birds will often be favored (Fig. 4D, case 1). Mammals would have to be significantly more effective than birds for mammal fruits to be favored by selection (Fig. 4D, case 2).

Consider what would happen with two different bird groups such as mashers (tanagers) and gulpers (manakins) (Levey 1986; Stiles and Roselli, this volume). The optimal fruit design for these taxa is likely to be similar enough that many fruits may be removed if they are of intermediate design. If seed removal is generally low, either because dispersers are uncommon or fruits have low attractiveness (perhaps due to repellant chemistry), then total seed removal ($r_{i\ tot}$) is likely to be higher with intermediate designs, which would result in a strongly complementary (bowed out) removal curve (Fig. 4E, case 1). If seed removal by these 2 taxa is very high, they are likely to compete resulting in a straighter removal curve (total removal cannot be greater than 1 so the removal curve cannot bow out past the dashed line in Fig. 4E). Dispersal effectiveness of these 2 bird taxa is not likely to change much as fruit design changes which leads to a fairly straight isoquality curve. In this scenario, our model predicts that the optimal fruit will be generalized and attractive to both disperser taxa. An exception to this (Fig. 4F) would be if one group had significantly higher removal (case 1) or effectiveness (case 2), in which case specialization might still be favored.

Up to this point we have considered birds and mammals to be homogeneous groups, but this is clearly an oversimplification. In addition to specialization on birds or mammals (e.g. Janson 1983), we expect to see specialization on ecologically and behaviorally similar taxonomic groups within birds and mammals (e.g. Gautier-Hion *et al.* 1985). Groups of birds with unusual ecologies such as oil birds (Steatornithidae) may represent special opportunities involving peculiar fruit designs and high potential returns for specialization. Likewise, plants should be specialists rather than generalists on mammals as different as elephants and bats.

Many other scenarios could be explored. To do this, one must first determine how different ecological and behavioral factors lead to different removal constraint and isoquality curves. These curves can then be used to predict the optimal dispersal syndrome.

These concepts can be related to actual dispersal systems by determining whether or not seed removal and seed wasteage increase, remain the same, or decrease when dispersal strategies contain a mixture of taxa compared with when they are specialized on one taxon. We predict that, because of their morphological, physiological, and behavioral differences, birds and mammals are more likely to have antagonistic removal and isoquality curves than complementary ones. Because of its importance in determining optimal coterie size and degree of taxonomic specialization and its implications for conservation biology, this is an important topic for future research.

Taxon-specific specialization and fruit-frugivore coevolution

Because birds greatly outnumber bats, primates, and other kinds of mammals in number of species and families (Howe 1986; Fleming *et al.* 1987; Fleming 1991a), the removal constraint curve should be steep and should intersect the bird removal axis at a high point (e.g. Fig. 4A). Yet as more and more bird-dispersed species migrate into or evolve in a community, new plants will be less likely to attract sufficient numbers of birds, and eventually other vertebrates will be able to remove as many or more of their seeds. There should come a point where disperser species are utilized in proportion to their availability (appropriately weighted by their effectiveness). This process should function at the interspecific level much as ideal free distribution theory suggests that individuals should behave within species (e.g. Milinski & Parker 1991). If this idea is correct, we should expect to find plants specializing on birds as their exclusive dispersers more frequently than on mammals whenever taxon-specific dispersal syndromes are favored. Data summarized in Willson *et al.* (1989) for several tropical and subtropical localities generally support this prediction. Except for Gabon, bird or bird-bat fruits outnumber mammal fruits by a factor of at least 2 in South Africa, Australia, and the neotropics. A similar bias towards bird fruits exists in temperate habitats (Stiles 1980; Skeate 1987; Herrera 1989).

This bias towards birds rather than mammals in specialized seed dispersal systems raises an important evolutionary question: To what extent have the speciation rates of fleshy-fruited plants and their vertebrate dispersers been coupled or uncoupled? Some theoretical models of speciation between mutualistic plants and animals (e.g. Kiester *et al.* 1984) indicate that coupled speciation is likely. The general feeling among researchers studying fruit-frugivore interactions, however, is that the interaction is too diffuse to promote diversity via coupled speciation (Howe 1984, 1986; Herrera 1985). In the case of one well-studied fruit-frugivore system (*Piper* fruits and *Carollia* bats, Fleming 1988), plant speciation has clearly proceeded independently of bat speciation. In lowland Central America, *Piper* species richness increases from five species in tropical dry forest to nearly 50 species in tropical wet forest whereas *Carollia* diversity only increases from two to three species (Fleming 1985, 1991b; Marquis 1991). Charles-Dominique (this volume) suggests that the association between certain specialized frugivorous birds and mammals and species-rich genera of plants has promoted plant speciation, but he provides no coevolutionary model to explain this process.

Conclusions

We have used a variety of theoretical approaches to address questions of optimal disperser coterie size and degree of disperser-taxon specialization in fleshy-fruited plants. $\hat{C}$ depends on a variety of factors of which fruit/seed size and the seed dispersal quality (Q) of different disperser species appear to be especially important. The way in which Q varies among potential dispersers can influence the evolution of specialized or generalized disperser coteries. We suggest that the degree of taxonomic specialization of a plant's disperser coterie depends on whether different disperser taxa (e.g. birds, mammals) are perfectly substitutable, complementary, or antagonistic (*sensu* Tilman 1982). Taxonomic specialization, as reflected in dispersal 'syndromes', is likely to evolve when disperser taxa are perfectly substitutable or antagonistic. Otherwise, evolution will favor taxonomically-mixed disperser coteries. Finally, we suggest that plants evolving taxonomically-specialized dispersal strategies should 'target' birds more often than bats or primates because of their greater species richness. Important future work in this area includes determining (1) Q values of frugivores within and among species of plants, (2) whether different taxonomic or ecological groups of dispersers are perfectly substitutable, complementary, or antagonistic, and (3) the extent to which rates of speciation in fleshy-fruited plants and their dispersers are interdependent. In addition to their theoretical importance, each of these topics has important implications for the conservation of biodiversity.

Acknowledgements

We thank Gene Schupp for his helpful comments on a draft of this paper. Contribution No. 391 from the Program in Ecology, Behavior, and Tropical Biology, Department of Biology, University of Miami.

References

Beattie, A. J. 1985. The Evolutionary Ecology of Ant-plant Mutualisms. Cambridge University Press, Cambridge.

Chapman, C. A. 1989. Primate seed dispersal: the fate of dispersed seeds. Biotropica 21: 148–154.

Covich, A. 1972. Ecological economics of seed consumption by *Peromyscus* – a graphical model of resource substitution. Trans. Conn. Acad. Arts and Sci. 44: 71–93.

DeBussche, M. & Isenmann, P. 1989. Fleshy fruit characters and the choices of bird and mammal seed dispersers in a Mediterranean region. Oikos 56: 327–338.

Dowsett-Lemaire, F. 1988. Fruit choice and seed dissemination by birds and mammals in the evergreen forests of upland Malawi. Rev. Ecol. 43: 251–285.

Fleming, T. H. 1985. Coexistence of five sympatric *Piper* (Piperaceae) species in a tropical dry forest. Ecology 66: 688–700.

Fleming, T. H. 1988. The Short-tailed Fruit Bat. University of Chicago Press, Chicago.

Fleming, T. H. 1991a. The fruiting plant-frugivore mutualism: the evolutionary theater and the ecological play. In: Plant-animal Interactions: Evolutionary Ecology in Tropical and Temperate Regions (eds. P. W. Price, T. W. Lewinsohn, G. W. Fernandes, & W. W. Benson): 119–144. John Wiley & Sons, New York.

Fleming, T. H. 1991b. The relationship between body size, diet, and habitat use in frugivorous bats of the genus *Carollia* (Phyllostomidae). J. Mammal. 72: 493–501.

Fleming, T. H., R. L. Breitwisch & G. H. Whitesides. 1987. Patterns of tropical vertebrate frugivore diversity. Ann. Rev. Ecol. Syst. 18: 91–109.

Foster, S. A. 1986. On the adaptive value of large seeds for tropical moist forest trees: a review and synthesis. Bot. Rev. 52: 260–299.

Gautier-Hion, A., Duplantier, J. -M., Quiris, R., Feer, F., Sourd, C., Decoux, J. -P., Subost, G., Emmons, L., Erard, C., Heckestweiler, P., Moungazi, A., Roussilhon, C., & Thiollay, J. -M. 1985. Fruit characters as a basis of fruit choice and seed dispersal in a tropical forest vertebrate community. Oecologia 65: 324–337.

Grime, J. P. 1979. Plant strategies and vegetation processes. John Wiley & Sons, New York.

Heithaus, E. R. 1982. Coevolution between bats and plants. In: Ecology of Bats (ed. T. H. Kunz): 327–367. Plenum Press, New York.

Herrera, C. M. 1985. Determinants of plant-animal coevolution: the case of mutualistic dispersal of seeds by vertebrates. Oikos 44: 132–141.

Herrera, C. M. 1989. Frugivory and seed dispersal by carnivorous mammals and associated fruit characteristics in undisturbed Mediterranean habitats. Oikos 55: 250–262.

Howe, H. F. 1981. Dispersal of a neotropical nutmeg (*Virola sebifera*) by birds. Auk 98: 88–98.

Howe, H. F. 1984. Constraints on the evolution of mutualisms. Am. Nat. 123: 764–777.

Howe, H. F. 1986. Seed dispersal by fruit-eating birds and mammals. In: Seed Dispersal (ed. D. R. Murray): 123–189. Academic Press, Sydney.

Howe, H. F. 1990. Seed dispersal by birds and mammals: implications for seedling demography. In: Reproductive Ecology of Tropical Forest Plants (eds. K. S. Bawa & M. Hadley): 191–218. UNESCO & Parthenon Publ. Group, Paris.

Howe, H. F. & Vande Kerchkove, G. A. 1981. Removal of wild nutmeg (*Virola surinamensis*) crops by birds. Ecology 62: 1093–1106.

Janson, C. H. 1983. Adaptation of fruit morphology to dispersal agents in a Neotropical forest. Science 219: 187–189.

Kiester, A. R., Lande, R., & Schemske, D. W. 1984. Models of coevolution and speciation in plants and their pollinators. Am. Nat. 124: 220–243.

Levey, D. J. 1986. Methods of seed processing by birds and seed deposition patterns. In: Frugivores and Seed Dispersal (eds. A. Estrada and T. H. Fleming): 147–158. W. Junk, Dordrecht, Netherlands.

Lloyd, D. G. & Venable, D. L. 1992. Some properties of natural selection with single and multiple constraints. Theor. Pop. Biol. 41: 90–110.

Levins, R. 1968. Evolution in changing environments. Monogr. Pop. Biol. Princeton Univ. No. 2.

Martin, T. E. 1985. Resource selection by tropical frugivorous birds: integrating multiple interactions. Oecologia 66: 563–573.

Marquis, R. J. 1991. Herbivore fauna of *Piper* (Piperaceae) in a Costa Rican wet forest: diversity, specificity, and impact. In: Plant-animal Interactions: Evolutionary Ecology in Tropical and Temperate Regions (eds. P. W. Price, T. M- .. Lewinsohn, G. W. Fernandes & W. W. Benson): 179–208. John Wiley & Sons, New York.

Milinski, M. & Parker, G. A. 1991. Competition for resources. In: Behavioural Ecology, 3rd. ed. (eds. J. R. Krebs & N. B- . Davies): 137–168. Blackwell Sci. Publs., Oxford.

O'Dowd, D. J. & Gill, A. M. 1986. Seed dispersal syndromes in Australian *Acacia*. In: Seed Dispersal (ed. D. R. Murray): 87–121. Academic Press, Sydney.

Pijl, L. van der. 1957. Seed Dispersal Mechanisms. Springer-Verlag, Berlin.

Rapport, D. J. 1971. An optimization model of food selection. Am. Nat. 105: 575–587.

Reid, N. 1989. Dispersal of mistletoes by honeyeaters and flowerpeckers: components of seed dispersal quality. Ecology 70: 137–145.

Skeate, S. T. 1987. Interactions between birds and fruits in a northern Florida hammock community. Ecology 68: 297–309.

Stiles, E. W. 1980. Patterns of fruit presentation and seed dispersal in bird-disseminated woody plants in the eastern deciduous forest. Am. Nat. 116: 670–688.

Tilman, D. 1980. Resources: a graphical-mechanistic approach to competition and predation. Am. Nat. 116: 362–393.

Tilman, D. 1982. Resource competition and community structure. Monographs in Pop. Biol. Princeton Univ. No. 17.

Venable, D. L. 1992. Size-number tradeoffs and the variation of seed size with plant resource status. Am. Nat. 140: 287–304.

Venable, D. L. & Brown, J. S. 1988. The selective interactions of dispersal, dormancy, and seed size as adaptations for reducing risk in variable environments. Am. Nat. 131: 360–384.

Wheelwright, N. T. 1985. Fruit size, gape width, and the diets of fruit-eating birds. Ecology 66: 808–818.

Willson, M. F. 1983. Plant Reproductive Biology. John Wiley & Sons, New York.

Willson, M. F., Irvine, A. K. & Walsh, N. G. 1989. Vertebrate dispersal syndromes in some Australian and New Zealand plant communities, with geographic comparisons. Biotropica 21: 133–147.

Vegetatio **107/108**: 121–132, 1993.
T. H. Fleming and A. Estrada (eds). Frugivory and Seed Dispersal: Ecological and Evolutionary Aspects.

Consequences for seed distributions of intra-crop variation in wing-loading of wind-dispersed species

Carol K. Augspurger[1] & Susan E. Franson[2]
[1]*Department of Plant Biology, University of Illinois, Urbana, Il 61801, USA;* [2]*Department of Ecology, Ethology, and Evolution, University of Illinois, Urbana, IL 61801 USA (present address: US Environmental Protection Agency, Las Vegas, NV 89193, USA)*

Keywords: Area of distribution, Dispersal distance, Fruit area, Seed mass, Uniformity of distribution

Abstract

We used a computer simulation to quantify how intra-crop variation in wing-loading in a wind-dispersed species affects the seed distribution around a parent plant. We used a data set of seed distributions generated from a previous field study using artificial fruits varying in seed mass or fruit area dispersed from a tower into a tropical forest. For this study, the spatial distribution from each hypothetical parent's fruit crop of 1000 was calculated by randomly drawing locations of dispersed fruits from the previous data base. Three parents with contrasting fruit crops were used to test two hypotheses: 1) Increasing within-parent variance in wing-loading (= weight/area), while maintaining the mean, will lead to an increase in the area and uniformity of the seed distribution, without changing the mean dispersal distance. 2) Decreasing within-parent mean wing-loading, (which also decreases variance), will lead to an increase in mean dispersal distance, area, and uniformity of the seed distribution. The hypotheses were tested under four wind speeds.

Increasing variance in wing-loading resulted in increasing the area and uniformity of density of the seed distribution without changing mean dispersal distance. Decreasing mean and variance in wing-loading resulted in increasing the area and uniformity of density of the seed distribution, as well as increasing the mean dispersal distance. Similar results occurred whether the differences in wing-loading arose by altering seed mass or fruit area. The effect of wind speed was consistently greater than the effect of parent. Generally, the same pattern of parent effects on seed distributions occurred, regardless of wind condition.

The effects on seed distributions differed for alterations in mean versus variance, specifically in whether mean dispersal distance was increased. How selection may act on intra-crop mean and variance in wing-loading will depend on additional factors, e.g. the relative importance of distance, area, and density on seedling recruitment and the relative costs for crop size and seedling establishment of making fruit crops of a given mean and variance.

Introduction

Seed dispersal offers several potential advantages to a parent. Dispersal moves offspring away from the parent and spreads them in lower densities over a greater area. Seedling recruitment may be enhanced by this dispersion pattern because density-dependent and/or or distance-dependent mortality of seeds/seedlings are lowered (Augspurger 1984a; Clark & Clark 1984). Also, dispersing seeds over a greater area, thereby increasing the sampling of the habitat, e.g., light-gaps or other favorable microhabitats, may increase seedling recruitment by reducing risk of mortality due to fungal pathogens (Augspurger 1984a) or because almost all seedlings survive better or grow faster under higher light intensities (Augspurger 1984b).

For wind-dispersed species, a number of factors have been demonstrated to affect dispersal and the resulting seed distribution. Some of these factors are under direct control of the parent, e.g. mass, area, and morphology of the diaspores (Siggins 1933; McCutchen 1977; Green 1980; Greene & Johnson 1990), height of release (Isaac 1930), and force required to sever the diaspore from the parent plant. Other factors, including variability in wind speed and direction (Isaac 1930; Mair 1973), and surrounding topography and vegetation (Cremer 1965), are external to the parent tree.

The mass and/or area of seeds or fruits within a parent's crop are never uniform (Janzen 1977; 1978, Michaels 1988). This variability can arise due to differences in seed number per fruit (Janzen 1978; Augspurger & Hogan 1983; Augspurger 1986) or perhaps, in one-seeded fruits, to position effects, heterogeneity of resource availability, or inbreeding effects. Regardless of cause, there is intra-crop variation in wing-loading, defined as diaspore weight/area, a major variable determining rate of descent and dispersal of any wind-dispersed seed or fruit. As wing-loading increases, due either to greater weight or smaller area of the diaspore, rate of descent increases and dispersal distance decreases.

Janzen (1977), having found a 3.6 fold variation in seed mass within one parent's crop, hypothesized that this variability could enhance a parent's fitness by generating a larger and more homogeneous seed distribution than would result were seeds of very similar mass. By homogeneous he was referring to an overall more thorough coverage of the habitat. He made this argument first for a water-dispersed species (Janzen 1977) and later for a wind-dispersed species (Janzen 1978). For *Crinum erubescens*, a water-dispersed perennial with highly variable seed size, Manasse (1990) argued that the possibility that its different sized seeds are adapted to different dispersal ranges is unlikely. Instead, the large variation in seed size appears to be due to inbreeding (Manasse & Stanton 1991). No explicit test of Janzen's hypothesis has been made for wind-dispersed species. However, for multi-seeded wind-dispersed fruits within one parent's crop, fruits with more seeds have higher wing-loading and are dispersed shorter distances (Augspurger & Hogan 1983; Augspurger 1986). Thus each fruit type with a given number of seeds has its unique dispersion pattern. How the overall dispersion pattern of the parent is affected by considering all fruit types with different numbers of seeds and hence different, but to some degree overlapping, seed distributions has not been quantified.

Previously, we determined experimentally the effect of differences in wing-loading of fruits on the spatial distribution of artificial wind-dispersed fruits that mimicked fruits of a tropical tree species, *Tachigalia versicolor* Standley & Williams (Leguminosae) (Augspurger & Franson 1987). The results from that field study indicated that, among five fruit models that varied in either seed mass or fruit area, those with lower wing-loading were dispersed greater distances, over larger areas, and in more uniform density than were fruits of larger wing-loading. Thus, that study identified the single model that maximized those three dispersal measures. The differences in wing-loading among the five fruit models can be viewed either as differences among five contemporary parents or species or as five sequential changes within one species occurring over evolutionary time. In either case, each parent (or species) was considered to make a uniform fruit crop having a specified mean wing-loading with no variance.

The present study takes the data base accumulated in the previous study and uses it in a computer simulation to investigate the effect of within-parent mean and variance in wing-loading on the spatial distribution of dispersed fruits. It is designed to test two hypotheses:

1. *Increased variance hypothesis*. Increasing within-parent variance in wing-loading, while maintaining the mean, increases the area and uniformity of the seed distribution without changing the mean dispersal distance. This is a modification of the Janzen hypothesis (1977, 1978). We added the constancy of mean dispersal distance and expanded his homogeneity concept to include uniformity of density within the seed distribution.
2. *Decreased mean hypothesis*. Decreasing within-parent mean wing-loading, (which also decreases its variance), increases mean dispersal distance, area and uniformity of the seed distribution. This hypothesis deals with how the seed distribution is affected by a parent having a monomorphic crop, relative to one with maximal variance in wing-loading. It represents directional selection away from a variable fruit crop toward a monomorphic crop made up of the model of lowest seed mass (or greatest fruit area) identified in our earlier study as achieving the maximum values for all dispersal measures. Through directional selection a decrease in mean wing-loading is achieved, but variance in wing-loading is lost. The two hypotheses differ in that Hypothesis 1 tests the effect of intra-crop variance, while Hypothesis 2 tests the effect of a crop comprised of only the best model for seed distributions.

The hypotheses are tested by altering first, seed mass and second, fruit area within a parent's fruit crop. We also investigate the influence of different wind speeds on the above hypotheses. Finally, we compare the differences in seed distributions resulting from increasing variance in wing-loading versus decreasing mean wing-loading of a parent's fruit crop.

Methods

In our previous study (Augspurger & Franson 1987), we built artificial fruits mimicking the natural fruit of *Tachigalia versicolor*. This wind-dispersed fruit has one flat seed centrally located in a flat symmetrical and elongated samara. A typical fruit is 43.1 cm^2 in area and has a mass of 1.45 g at dispersal; its seed alone has a mass of 0.61 g. To construct the artificial fruits, we used a semi-stiff paper of the same density (16 $mg \cdot cm^{-2}$) and approximate rigidity as the natural fruit of *T. versicolor*. To mimic the seed, we folded a thin pliable sheet of paraffin ('Parafilm'), with a density of 11.9 $mg \cdot cm^{-2}$, into a rectangle of the desired mass and shape of a natural seed. One seed was glued to each fruit center.

Alterations in wing-loading in the previous study were accomplished in two ways: by changing seed mass or fruit area. Our 'mass series' consisted of five artificial fruits with constant area but different seed mass. Model 3 was built as a replica of the natural fruit, while models 1, 2, 4 and 5 had seed masses 20%, 60%, 140% and 180% of the mass of Model 3. Our 'area series' consisted of five artificial models with constant seed mass but different fruit areas. Model 3 again served as the standard, while Models 6, 7, 8 and 9 had areas 20%, 60%, 140% and 180% of the area of Model 3. These artificial fruits were released from 40 m on a meterological tower into the semi-deciduous forest on Barro Colorado Island, Panama. In all, we had five different trials to quantify seed distributions resulting from dispersal in different wind speeds. The artificial fruits were recovered after each trial and the location of each fruit in each trial was recorded by identifying the 2.5×2.5 m quadrat from which it was recovered. Descriptive statistics and drawings of these models as well as results of that study are in Augspurger & Franson (1987).

For the present study, we first determined the number of each of the five artificial fruit models to be included in parent crops so as to have the mean and variance required to test each of the two hypotheses. One series of parent crops was created using models varying in seed mass; a second series used models varying in fruit area. In particular, to test the Increased Variance Hypothesis for the mass series, Parent 1 consisted entirely of Model 3 fruits, Parent 3 consisted of equal numbers of each of the five models, and Parent 2 was designed as an intermediate between Parents 1 and 3 (Fig. 1). These three parent crops have the same mean mass but variance in mass increases through the sequence. Parent 1

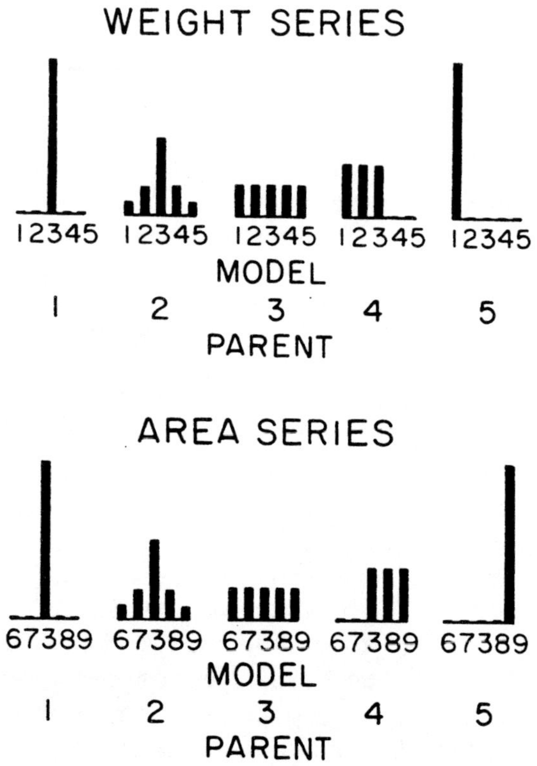

Fig. 1. The numerical composition of fruit crops for each of five parents in both the mass and area series. See Table 1 for descriptive statistics of each parent's fruit crop. Model 1 (6) represents the lowest value, model 5 (9) the highest value of mass (area). See Augspurger and Franson (1987) for descriptive statistics (Table 1) and a drawing of each model (Fig. 1).

displays no variance, Parent 2 approximates a normal distribution of fruit mass, while Parent 3, with its equal numbers of five distinct models, represents a uniform distribution of fruit mass.

To test the Decreased Mean Hypothesis, we used Parent 3 of the above hypothesis as the initial condition. Parent 5 consisted entirely of Model 1 fruits (the lightest and thus with lowest wing-loading), and Parent 4 represented an intermediate condition with equal numbers of Models 1, 2 and 3 (Fig. 1). These three parent crops represent a progression of directional selection toward a monomorphic crop with the lowest mean seed mass and no variance. Mean and variance values for wing-loading in these parent crops are presented in Table 1.

Five artificial parent crops also were compiled for the area series, exactly as described for the mass series above. For the Increased Variance Hypothesis, Parent 1 consisted entirely of Model 3, Parent 3 of equal numbers of Models 6, 7, 3, 8 and 9, with Parent 2 intermediate (Fig. 1). For the Decreased Mean Hypothesis, Parent 3 consisted of equal numbers of five models, Parent 5 consisted entirely of Model 9 fruits (those with the largest area and thus lowest wing-loading), while Parent 4 consisted of equal numbers of Models 3, 8 and 9 (Fig. 1). Parent 5 had a monomorphic crop with the largest mean fruit area and no variance. Mean and variance values for wing-loading in these parent crops occur in Table 1.

From the previous study, we had recorded the quadrat location of each fruit of each model dispersed from the tower. We used that data base for this study by randomly drawing from it the requisite number of each type of model required for each parent's crop. In total, fruit locations were drawn randomly from each of three separate trials of the previous study. This allowed an assessment of the relative effect of wind speed versus parent fruit crop on seed distributions. We chose Trials 3, 1 and 4 to represent dispersal in Low, Medium, and High Winds (mean velocity from release height to ground estimated to be 1.8, 3.7, and 5.3 km $\cdot$ hr^{-1}, respectively). In addition, dispersal by each parent in a variety of windspeeds was simulated by combining the locations of all fruits of a model in all five trials of the previous study before randomly choosing the locations of the proper numbers of the model for each artificial parent crop. This is referred to as the Combined Wind.

Initially, five replicate crops of 200 fruits each were created randomly for each parent in each series and in each of four winds. Because the seed distributions from these replicates did not differ significantly within a parent for a given wind, these five replicates were lumped. All further analyses were performed on these lumped parent crops of 1000 fruits each and are reported here.

Next, the seed distributions of the parent crops dispersed in different winds were compared using these three measures as described in Augspurger & Franson (1987):

Table 1. Descriptive statistics of mass (fruit and seed combined), fruit area, (wing-loading) 1/2, and rate of descent in still air for fruit crops of five parents. See Fig. 1 for the numerical composition of models in each parent's fruit crop. Values represent mean and (variance). n = 200 total for each parent's fruit crop.

	Parent				
	1	2	3	4	5
Mass series					
Mass (g)	1.43	1.45	1.45	1.21	0.98
	(0)	(0.06)	(0.11)	(0.03)	(0)
Area (cm^2)	40.1	40.3	40.5	40.6	41.0
	(0)	(0.1)	(0.1)	(0.2)	(0)
$(\text{Wing-loading})^{1/2}$ $(\text{mPa})^{1/2}$	186.91	187.11	186.48	170.49	152.91
	(0)	(247.68)	(507.32)	(194.16)	(0)
Rate of descent (cm/s)	199.8	195.8	192.2	175.1	160.4
	(0)	(411.4)	(628.9)	(306.8)	(0)
Area series					
Mass (g)	1.43	1.47	1.48	1.76	2.03
	(0)	(0.09)	(0.18)	(0.06)	(0)
Area (cm^2)	40.1	41.2	42.0	58.4	75.9
	(0)	(269.4)	(565.3)	(215.4)	(0)
$(\text{Wing-loading})^{1/2}$ $(\text{mPa})^{1/2}$	186.91	197.25	209.43	174.33	161.91
	(0)	(1203.68)	(2679.42)	(104.47)	(0)
Rate of descent (cm/s)	199.8	209.4	225.2	171.2	157.8
	(0)	(2846.3)	(6163.4)	(404.2)	(0)

Mean dispersal distance

$$= \frac{\sum^{NF} \sqrt{x_{i^2} + y_{i^2}}}{NF\, i = 1},$$

Mean distance to centroid

$$= \frac{\sum^{NF} \sqrt{(x_i - \bar{x})^2 + (y_i - \bar{y})^2}}{NF\, i = 1},$$

Mean crowding

$$= \frac{\sum^{NQ} n_q (n_q - 1)}{NF\, q = 1}$$

$$\text{Note: } \sum_{q=1}^{NQ} n_q = NF$$

where:

xi, yi = Cartesian coordinates of center of quadrat to which i^{th} fruit was dispersed (parent is at origin).

NF = Total number of fruits dispersed.

NQ = Total number of occupied quadrats.

n_q = Number of fruits in q^{th} quadrat.

Distance to centroid (center of gravity) reflects the relative area of the seed distribution. Greater mean distances to centroid are associated with distributions of greater area. Mean distance to centroid does not distinguish among different spatial patterns, e.g., a homogeneous distribution cannot be distinguished from certain patterns of clumping. Therefore, we used mean crowding to measure the homogeneity of the spatial distributions of fruits. Crowding (Lloyd 1967) is the number of fruits of the same parent with which each fruit shares its quadrat. It reflects the relative uniformity of density within seed distributions. For example, a fruit that was the sole occupant of its quadrat has a crowding value of zero, while each fruit in a quadrat with 10 fruits would have a crowding value of nine. Lower crowding values result from a parent dispersing fruits over a greater area, and thus to more quadrats, as well as from

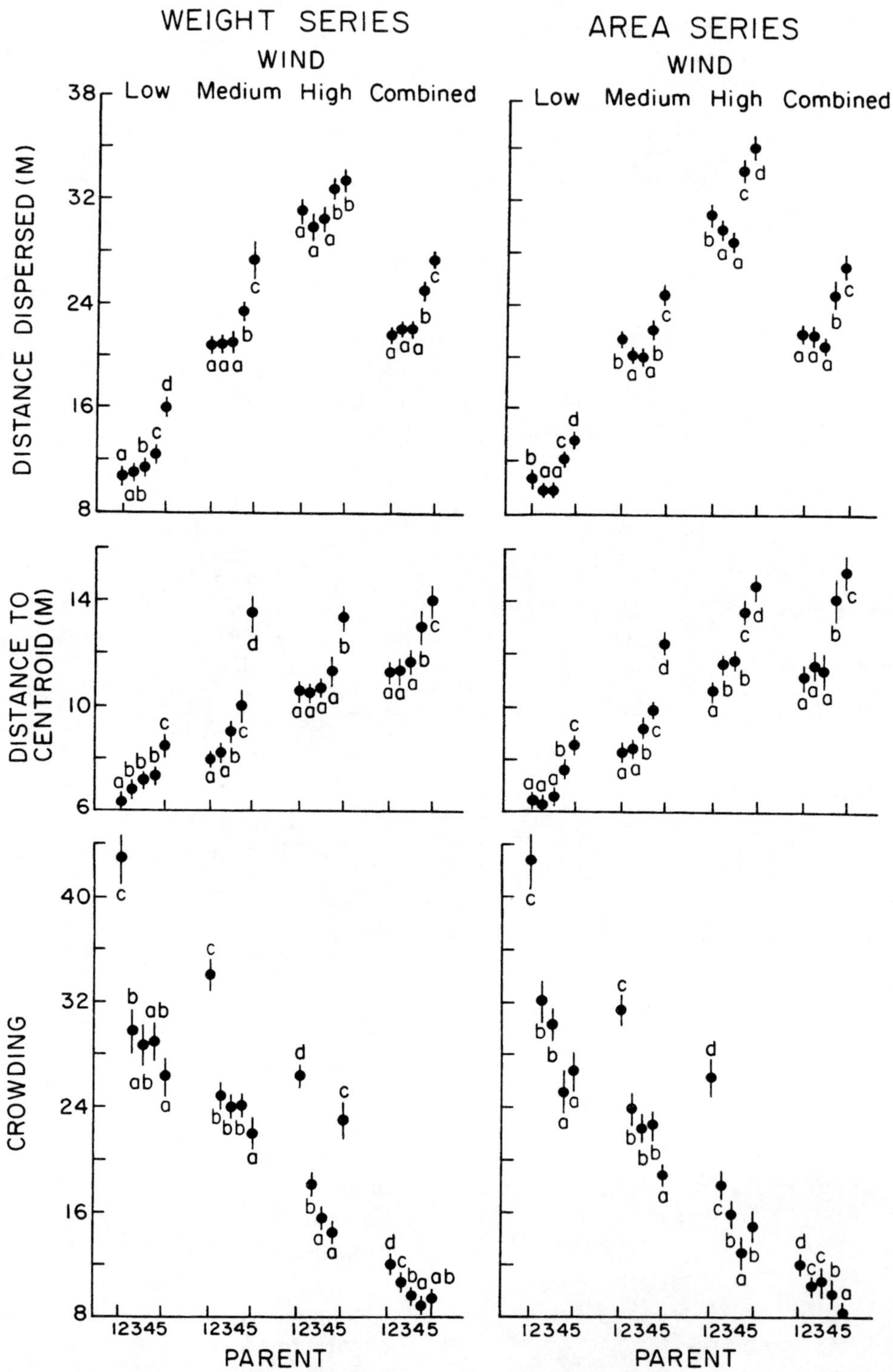
WEIGHT SERIES
AREA SERIES
WIND
Low
Medium
High
Combined
DISTANCE DISPERSED (M)
DISTANCE TO CENTROID (M)
CROWDING
PARENT
12345

a more even distribution of fruits among those quadrats. Hereafter, lower crowding will be taken to infer a greater uniformity of density within a distribution.

Finally, we compared the relative effects of parent and wind on dispersal. We defined the parent effect in the manner of the model effect of the previous study (Augspurger & Franson 1987). Calculated for each dispersal measure in each series within each hypothesis, the parent effect was the absolute deviation of each parent's mean value from the mean for all parents within each wind trial, expressed as a proportion of that latter mean. Similarly, wind effect was found by expressing the absolute deviation of each wind's mean value from the mean of all winds within each parent as a proportion of that latter mean. Expression of the absolute deviation as a proportion allowed us to compare parent and wind effects among different dispersal measures with different units. We then subjected these values to standard statistical tests for comparisons of parent effect and wind trial effect within series, between series, between hypotheses, and among dispersal measures.

Results

Increased Variance Hypothesis – Parents 1, 2 and 3

Simulations for parents in the mass series supported the hypothesis that increasing within-parent variance in wing-loading, without changing the mean wing-loading, results in increasing area and uniformity of the seed distribution without changing mean dispersal distance (Fig. 2). The mean dispersal distances of Parents 1, 2 and 3 did not differ in Medium, High, and Combined Winds, while in Low Wind, mean dispersal distance slightly increased with increasing variance in wing-loading. Mean distance to centroid increased, indicating an increase in the area of the fruit distribution, with increasing variance in wing-loading in all but High Winds, with significant differences among the parents exhibited in Low and Medium Winds. The uniformity of the seed distribution showed a pronounced increase with increasing variance in wing-loading, as indicated by a decrease in crowding, with significant differences among Parents 1, 2 and 3 in all winds.

Simulations for parents in the area series for dispersed distance were somewhat less compelling in their support of the Increased Variance Hypothesis than were those of the mass series. Mean dispersal distance tended to decrease slightly with increasing variance in wing-loading, and significant differences between Parents 1 and (2 and 3) were found in all but the Combined Wind (Fig. 2). Mean distance to centroid was greater, indicating that fruits were dispersed over a greater area, for Parent 3 (with greatest variance in wing-loading) than for Parent 1, significantly so in Medium and High Winds. The uniformity of the seed distribution was greater for parents with variance in wing-loading, as indicated by lower values in crowding of Parents 2 and 3 as compared to Parent 1, with significant differences in all winds.

Decreased Mean Hypothesis – Parents 3, 4 and 5

Simulations for parents in the mass series supported the hypothesis that decreasing both mean and variance in wing-loading increases mean dis-

Fig. 2. For the simulated mass and area series, distance dispersed from the tower, distance to centroid of the distribution, and crowding (see methods) in four types of winds. Values, representing mean ± 2 standard errors, are shown for five parents with contrasting fruit crops. See Table 1 for descriptive statistics and Fig. 1 for numerical composition of each parent's fruit crop. Identical letters (within each series and each wind for each measure) indicate parent's fruit crops that are not significantly different ($p > 0.05$ for Student-Newman-Kuels test following Analysis of Variance). Statistical results for Increased Variance Hypothesis are indicated by three letters above mean values for Parents 1–3. For Decreased Variance Hypothesis by three letters below mean values for Parents 3–5. Parent 3 is included in both hypotheses.

persal distance, area, and uniformity of the seed distribution (Fig. 2). Mean dispersal distance increased with decreasing wing-loading, with significant differences among Parents 3, 4, and 5 in all winds. The area of the seed distribution increased with decreasing wing-loading, as indicated by significant increases in distance to centroid from Parent 3 through Parent 5 in all winds. Likewise, the uniformity of the seed distribution increased by decreasing wing-loading. The greatest uniformity, as indicated by the lowest values of crowding, occurred for Parent 5 in all four winds.

Conclusions drawn from simulations for parents in the area series were very similar to those of the mass series. Mean dispersal distance and area of the seed distribution, as reflected by mean distance to centroid, both increased significantly in all winds with decreasing wing-loading (Fig. 2). Likewise, the uniformity of seed distribution, as indicated by crowding values, increased with decreasing wing-loading. Crowding was lowest for either Parent 4 or 5, relative to Parent 3, in all winds; which of the two parents was least crowded varied with wind.

Comparisons among measures, between hypotheses, and between series

Seedling recruitment may be differentially affected by changes in distance from parent, area of habitat sampled, and crowding with siblings. Therefore, we asked whether, within each series for each hypothesis, the parent effect was the same for the three dispersal measures. For the Increased Variance Hypothesis, differences in wing-loading among parents led to greater differences in mean crowding than in either mean dispersal distance or mean distance to centroid (Table 2). For the parents in the Decreased Mean Hypothesis, the three dispersal measures were affected equally.

Next, we asked whether the two hypotheses differed in the parent effect on each dispersal measure. If some aspect of the seed distribution confers a greater potential fitness advantage, then it is important to know whether or not this aspect is altered equally by a change in the mean or variance in wing-loading. For both the mass and area series, the parent effect on dispersal distance and distance to centroid was greater for the Decreased Mean Hypothesis than for the Increased Variance Hypothesis, while the parent effect on crowding was greater for the Increased Variance Hypothesis than for the Decreased Mean Hypothesis (Table 2).

Finally, changes in mean or variance in wing-loading can be effected by changes in either seed mass or fruit area. Changes in these two variable

Table 2. Parent effect and wind effect on each dispersal measure for each series in each hypothesis. Value presented is the mean of the absolute deviation from the mean as a proportion of the mean (see methods for further explanation). Results for the parent effect are given for: 1) analysis of variance among dispersal measures within each series and hypothesis (see a, b; identical letters indicate groups that are not significantly different, $p > 0.05$ from Student – Newman – Kuels tests); 2) t-test between hypotheses for each dispersal measure in each series (z indicates the significantly larger mean); and 3) t-test between series within each hypothesis (r indicates the significantly larger mean). Absence of letters indicates no significant difference was found ($P > 0.05$). The wind effect was significantly greater than the parent effect for each dispersal measure for each series in each hypothesis (t-test, $p < 0.05$).

	Distance	Distance to centroid	Crowding
Increased Variance Hypothesis (Parents 1, 2, 3)			
Mass Series			
Parent effect	0.015 a	0.031 a	0.156 b, z
Wind effect	0.240	0.184	0.332
Area Series			
Parent effect	0.029 a, r	0.030 a	0.137 b, z
Wind effect	0.255	0.205	0.323
Decreased Mean Hypothesis (Parents 3, 4, 5)			
Mass Series			
Parent effect	0.083 z	0.097 z	0.077
Wind effect	0.219	0.161	0.335
Area Series			
Parent effect	0.091 z	0.098 z	0.078
Wind effect	0.255	0.198	0.336

may have quite different consequences in terms of construction costs, crop size, and seedling survival. Therefore, we asked whether the effects on the seed distribution were the same whether the desired wing-loading was achieved by changing seed mass or fruit area. Comparisons of the parent effect on each dispersal measure within each hypothesis showed that only for dispersal distance in the Increased Variance Hypothesis was there a difference between the mass and area series, with the larger parent effect occurring in the area series (Table 2).

Comparisons between parent and wind effects

Tests of the hypotheses were run under four wind conditions. Analyses revealed that the wind effect was significantly greater than the parent effect for each dispersal measure for each series in each hypothesis (*t*-test, $p < 0.05$, Fig. 2). Comparisons among winds alone indicated that as wind velocity increased, both mean dispersal distance and mean distance to centroid, an indication of area, increased, while crowding decreased (Fig. 2). In combined winds, mean dispersal distance was comparable to that from a medium wind, distance to centroid equaled that from a high wind, while crowding was lower than in all three other winds.

Discussion

The simulations generally support the Increased Variance Hypothesis. Parents with greater intra-crop variance in wing-loading disperse their seeds over a greater area and in greater uniformity of density, relative to parents with uniform crops. Such an outcome occurs from intra-crop variance in either seed mass or fruit area. Mean dispersal distance in all winds does not differ among Parents 1 to 3 with intra-crop variance in mass. Likewise, in Combined Winds in the area series, there is no difference among Parents 1 to 3 in mean dispersal difference. An unexpected result arose in the area series for the three non-combined wind trials in which mean dispersal distance decreases slightly with increasing variance in wind-loading. This result may have arisen, in part, because mean mass, and hence mean wing-loading, increased slightly from Parents 1 to 3 (Table 1). This increase in mean wing-loading would have opposed the effect of increasing variance in wing-loading.

Likewise, the simulations support the Decreased Mean Hypothesis. Parents producing a monomorphic crop with lowest mean seed mass (or greatest mean fruit area), and hence lowest mean wing-loading, disperse their seeds the greatest mean distance over the largest area with the greatest uniformity in density, even though they have no variance in wing-loading.

The three dispersal measures were not affected equally by changes designed in the Increased Variance Hypothesis. Crowding was altered the most and dispersal distance the least. In contrast, in the Decreased Mean Hypothesis the three measures were affected equally. These results support both of our initial hypotheses. The comparative analyses pinpoint the specific aspects of the seed distribution altered most by shifts in the mean or variance of the fruit crop. Depending on which attribute of the seed distribution most strongly influences seedling recruitment will determine whether it is advantageous for a parent to have intra-crop variance in wing-loading. Such an argument does not apply to a decreased mean in wing-loading because all measures were affected equally.

Comparisons between the two hypotheses revealed that the three dispersal measures were affected differently. For both the mass and area series, the parent effect on dispersal distance and distance to centroid was greater for the Decreased Mean Hypothesis than for the Increased Variance Hypothesis. The opposite was true for uniformity of density (crowding). These results indicate that if greater distance and greater area enhance seedling recruitment, then selection will favor the parent with decreased mean wing-loading. If uniformity of the seed distribution is important, however, selection will favor an increase in variance in wing-loading. Changing the mean versus the variance have different implications and are interrelated with whether the changes are made in the mass or area series (see below).

Overall, comparisons of the parent effect between the mass and area series showed that comparable distributions result by altering seed mass or fruit area. Which of the two variables exhibits more natural variance is unknown (but see below). The ecological implications for altering seeds or fruits may be quite different. In the Decreased Mean Hypothesis, lowering mean seed mass may diminish the likelihood of successful seedling establishment, but may be accompanied by an increase in crop size. Increasing mean fruit area per se does not alter seedling establishment, but may be accompanied by a decrease in crop size. This may depend upon the photosynthetic contribution of the fruit itself. Likewise, in the Increased Variance Hypothesis, different ecological implications arise from altering mass and area. The effects of increasing variance in seed mass without changing the mean mass are likely to be complex. Some seeds are smaller and may have lower probability of survival. Alternatively, they may be better able to utilize a new subset of habitats. Some seeds become larger and presumably have a higher probability of survival. In contrast, variability in fruit area has no effect on seed mass and it is not obvious how it would affect seedling establishment. Therefore, whether it is advantageous to alter wing-loading by changes in seed mass or fruit area depends upon relative construction costs of seeds and fruits, plasticity of size, and overall seedling ecology.

Of the four wind tests, the Combined Wind is the most realistic. A parent never experiences one constant wind speed from one direction throughout its entire dispersal period. It is important to note that seeds cover the greatest area in the greatest uniformity of density in Combined Winds; distance is not exceptional. This distribution undoubtedly arises because of slight differences in wind direction that accompanied the different wind trials. Generally, the patterns among parents were similar in different wind types. The wind effect was consistently greater than the parent effect on the distributions. This implies that evolutionary responses to selection for changes in intra-crop characteristics may be slow, due to the strong environmental component of wind determining seed distributions.

It is relevant to consider how these simulations using artificial fruits relate to the amount of variance in natural fruit crops. Variation in seed mass within an individual can be as high as 16-fold (Thompson & Pellmyr 1989). Variation in seed mass within and among individuals is common in temperate trees and herbs (Michaels *et al.* 1988; Willson *et al.* 1990). In their study of 39 species, Michaels *et al.* (1988) found the within-individual component of variance was larger (= 62%) than the among-individual component (= 38% of total variance). The coefficient of variation of seed mass commonly exceeded 20%.

This substantial variation in seed mass within individuals is inconsistent with the predictions of models for the evolution of a constant seed size (McGinley *et al.* 1987) and with empirical observations of directional selection on seed mass (Stanton 1984; Winn 1988). However, individuals may be unable to produce a uniform crop because resource availability changes throughout the season, because of selection for variation in traits correlated with seed mass, or because of a true genetic constraint on the evolution of uniform offspring size (Winn 1991). Hence, variation in qualitites affecting wing-loading have both genetic and environmental components. Which of the two is the stronger component is likely to be species-specific and will affect the rate of any evolutionary response.

More relevant to the current study, the observed variation in wing-loading in a crop of natural wind-dispersed tropical fruits is somewhat lower than that experimentally created. For the artificial fruits there was a 1.4× and 1.9× maximum difference in wing-loading in a crop varying in seed mass and fruit area, respectively. These values were intentionally made somewhat extreme so as to offer a rigorous test of the hypotheses. A logical next step is to design parent crops to determine the minimum variance required to produce an effect. Little is known about variance in wing-loading within natural fruit crops of *Tachigalia versicolor*, the species mimicked by the artificial fruits. The major variance for *T. versicolor* arises when a tree produces some 2- and even 3-seeded fruits in addition to the majority 1-seeded fruits. Such multi-seeded fruits are rare in this species. For other wind-dispersed tree species, intra-crop seed mass varies 2 × in *Ateleia herbert-smithii*, (Janzen 1978), fruit mass varies 2 × when multi-seeded fruits are included for *Lonchocarpus pentaphyllus* (Augspurger & Hogan 1983), and fruit mass varies 1.7 × in 1- and 2-seeded fruits of *Platypodium elegans* (Augspurger 1986). Wing-loading varies 1.2 × and 1.3 × for the last two species.

How these hypotheses relate to non-wind-dispersed species is less clear. They may relate best to some water-dispersed species as Janzen (1977) first suggested. Animal-dispersed tropical tree species appear to have intra-crop variance comparable in seed mass to that of these wind-dispersed species. In *Virola surinamensis* there is a 1.6 × intra-crop comparable difference in seed mass (Howe & Richter 1982). In *Gustavia superba* there is a 2.0 × intra-fruit difference in seed mass (V. Sork, unpublished data). Although it is clear that animal dispersal agents can make choices among *Virola* trees differing in mean seed mass (Howe & Vande Kerckhove 1981), there is no evidence yet that they make feeding choices based on intra-crop variance in some attribute. In studies of temperate shrubs, birds in feeding trials selectively choose larger berries (E. W. Stiles, pers. comm.) or those with fewer seeds per fruit (G. Murray, this volume). Hence intra-crop variances is expected to affect seed distributions for animal- as well as wind-dispersed species.

Acknowledgements

This research was supported by an NSF grant (BSR 8219856) to CKA and was made possible by logistical support from the Smithsonian Tropical Research Institute.

References

Augspurger, C. K. 1984a. Seedling survival among tropical tree species: interactions of dispersal distance, light-gaps, and pathogens. Ecology 65: 1705–1712.

Augspurger, C. K. 1984b. Light requirements of neotropical tree seedlings: a comparative study of growth and survival. J. Ecol. 72: 777–795.

Augspurger, C. K. 1986. Double- and single-seeded indehiscent legumes of *Platypodium elegans*: consequences for wind dispersal and seedling growth and survival. Biotropica 18: 45–50.

Augspurger, C. K. & Franson. S. E. 1987. Wind dispersal of artificial fruits varying in mass, area, and morphology. Ecology 68: 27–42.

Augspurger, C. K. & Hogan K. P. 1983. Wind dispersal of fruits with variable seed number in a tropical tree (*Lonchocarpus pentaphyllus*: Leguminosae). Amer. J. Bot. 70: 1031–1037.

Clark, D. A. & Clark D. B. 1984. Spacing dynamics of a tropical rain forest tree: evaluation of the Janzen-Connell model. Amer. Nat. 124: 769–788.

Cremer, K. W. 1965. Dissemination of seed from *Eucalyptus regnans*. Aust. For. 30: 33–37.

Green, D. S. 1980. The terminal velocity and dispersal of spinning samaras. Amer. J. Bot. 67: 1218–1224.

Greene, D. F. & Johnson E. A. 1990. The dispersal of winged fruits and seeds differing in autorotative behavior. Can. J. Bot. 68: 2693–2697.

Howe, H. F. & Vande Kerckhove G. A. 1981. Removal of wild nutmeg (*Virola surinamensis*) crops by birds. Ecology 62: 1093–1106.

Howe, H. F. & Richter W. M. 1982. Effect of seed size on seedling size in *Virola surinamensis*; a within and between tree analysis. Oecologia 53: 347–351.

Isaac, L. A. 1930. Seed flight in the Douglas fir region. J. For. 28: 492–499.

Janzen, D. H. 1977. Variation in seed size within a crop of a Costa Rican *Mucuna andreana* (Leguminosae). Amer. J. Bot. 64: 347–349.

Janzen, D. H. 1978. Inter- and intra-crop variation in seed weight of Costa Rican *Ateleia herbert-smithii* Pitt. (Leguminosae). Brenesia 14–15: 311–323.

Mair, A. R. 1973. Dissemination of tree seed: Sitka spruce, western hemlock and Douglas fir. Scot. For. 27: 308–314.

Manasse, R. S. 1990. Seed size and habitat effects on the water-dispersed perennial, *Crinum erubescens* (Amaryllidaceae). Amer. J. Bot. 77: 1336–1341.

Manasse, R. S. & Stanton M. L. 1991. The influence of mating system on seed size variation in *Crinum erubescens* (Amaryllidaceae). Evolution 45: 883–890.

McCutchen, C. W. 1977. The spinning rotation of ash and tulip tree samaras. Science 197: 691–692.

McGinley, M. A., Temme D. H. & Geber M. A. 1987. Parental investment in offspring in variable environments: theorectical and empirical considerations. Amer. Nat. 130: 370–398.

Michaels, H. J., Benner B., Hartgerink A. P., Lee T. D., Rice S., Willson M. F. & Bertin R. I. 1988. Seed size variation: magnitude, distribution, and ecological correlates. Evol. Ecol. 2: 157–166.

Stanton, M. L. 1984. Seed size variation in wild radish: effect of seed size on components of seedling and adult fitness. Ecology 65: 1105–1112.

Thompson, J. N. & Pellmyr O. 1989. Origins of variance in seed number and mass: interaction of sex expression and herbivory in *Lomatium salmoniflorum*. Oecologia 79: 395–402.

Siggins, H. W. 1933. Distribution and rate of fall of conifer seeds. J. Agri. Res. 47:119–128.

Willson, M. F., Michaels H. J., Bertin R. I., Benner B., Rice S., Lee T. D., Hartgerink A. P. 1990. Intraspecific variation in seed packaging. Amer. Mid. Nat. 123: 179–185.

Winn, A. A. 1988. Ecological and evolutionary consequences of seed size in *Prunella vulgaris*. Ecology 69: 1537–1544.

Winn, A. A. 1991. Proximate and ultimate sources of within-individual variation in seed mass in *Prunella vulgaris* (Lamiaceae). Amer. J. Bot. 78: 838–844.

Vegetatio **107/108**: 133–147, 1993.
T. H. Fleming and A. Estrada (eds). Frugivory and Seed Dispersal: Ecological and Evolutionary Aspects.

Evolutionary ecology of mast-seeding in temperate and tropical oaks (*Quercus* spp.)

V. L. Sork
Department of Biology, University of Missouri-St. Louis, St. Louis, MO 63121, USA

Keywords: Acorn production,Mast-fruiting, Mast-flowering, Pollination efficiency, Predator-satiation hypothesis, *Quercus*

Abstract

Mast-seeding is the synchronous production of large seed crops within a population or community of species every two or more years. This paper addresses three non-mutually exclusive hypotheses explaining the evolution of mast-seeding in temperate tree species, especially the genus *Quercus*: (1) mast-seeding is a consequence of mast-flowering which evolves to increased pollination efficiency in mast-flowering years; (2) mast-seeding has evolved as an anti-predator adaptation by which large seed crops during mast years satiate the seed predators and allow survival of some of the seeds; (3) selection on seed size by habitat can indirectly affect the evolution of masting if trees with large seeds require more time to accumulate reserves to mature those seeds. I find support for the pollination hypothesis in several wind-pollinated temperate tree species but not oaks. However, oaks show evidence favoring the predation and seed size hypotheses. I then develop a model to illustrate the relationships among the three hypotheses in their effects on the evolution of masting. Finally, using data from herbaria and Floras, the influence of selection via flowering, fruiting, and seed size in the evolution of masting in tropical oaks is discussed. I conclude that the need for a supra-annual cue to synchronize flowering and fruiting as well as the larger seed size found in many tropical oak species should contribute to the evolution of masting to a greater extent than seen among temperate oaks.

Introduction

Mast-seeding is the synchronous production of large seed crops within a population or community of tree species every two or more years (Janzen 1969, 1971, 1974; Silvertown 1980; Waller 1979). This phenomenon is most likely to occur in trees which, as long-lived plants, are less affected by the costs of delayed reproduction (Waller 1979). Mast-seeding has been best documented in north temperate tree species (Silvertown 1980; Waller 1979) but is also well-known for tropical dipterocarpaceae (Janzen 1974; Ashton *et al.* 1988) and possibly other tropical species (Janzen 1978; Wheelwright 1986). In general, it appears that mast-seeding has evolved by postponing seed production during some years to accumulate the resources necessary to produce a large crop in other years. Thus, there are two aspects of this reproductive phenomenon: size of seed crop during a mast year and length of intermast interval, that is, the number of years between mast years.

The most parsimonious explanation for year-to-year fluctuations in seed production is weather (Sork *et al.* in press; Norton & Kelly 1988). Plant populations may simply be responding to variation in climatic conditions that are most suitable to the production of large seed crops (Norton & Kelly 1988). In fact, several studies have shown various weather variables to be associated with large seed crops (Mathews 1955; Wenger 1957; van Vredenburch and la Bastide 1969; Sork *et al.* in press). When a large seed crop is part of a mast-seeding reproductive strategy, there should be a reallocation of resources rather than only a matching of the seed crop size to current resource levels (Norton & Kelley 1988; Sork *et al.* in press). Evidence of resource allocation would include a negative correlation between crop size one year and crop size during one or more previous years, or a trade-off between vegetative and reproductive growth. For many species, vegetative growth is reduced during mast years (Kozlowski 1971; Kozlowski & Keller 1976). In Missouri oaks, we found that size of acorn crop was inversely correlated with prior acorn production, suggesting that individual trees must accumulate new resources after a mast year before they can produce another large crop (Sork *et al.* in press). Moreover, we also found that weather clearly influences the size of the crop, but internal cues determine when the next mast crop will occur (Sork *et al.* in press). It is likely that this pattern of interaction between weather cues and internal physiological cues will be observed in many masting species.

There are several non-mutually exclusive hypotheses that may account for the evolution of masting. The first hypothesis states that masting in wind-pollinated species has evolved owing to the proportional increase in fertilization (especially by outcross pollen) and seed set during mast years (Nillson & Wastljung 1987; Norton & Kelly 1988; Smith *et al.* 1990). A second hypothesis proposes that mast-seeding is an anti-predator adaptation that allows survival of seeds by periodically satiating seed predators (Janzen 1969, 1971, 1974; Smith 1970; Silvertown 1980). A third hypothesis concerns the potential impact of selection on seed size, a factor that may indirectly affect the evolution of masting. This hypothesis proposes that selection for increased seed size could indirectly select for greater intermast interval owing to the physiological need to accumulate resources for a longer period of time to produce a large seed. A likely selection pressure is habitat selection for increased seed size to enhance survival of seedlings under low light conditions (Grime & Jeffrey 1965; Gross 1984; Winn 1985; Mazer 1989). Support for this possibility comes from observations like those of Mazer (1989) of an Indiana dune community and Foster and Janson (1985) of moist forest communities who found that large seeded species are more likely to occur in closed canopy conditions. Thus, the evolution of masting may have been influenced by selection at any or all of the flower, fruit, and seedling stages of the species' life history.

In this paper, I will discuss the potential role of selection at three life-history stages in the evolution of mast-seeding in temperate tree species with special emphasis on oaks. I will then develop a model illustrating the interactions of these factors for temperate oaks. Oaks also occur in tropical habitats where the climate can be much less seasonal than found in the temperate zone. Consequently, the reproductive biology may not be identical to that of temperate oaks. While little is known about tropical oaks, I will use information gathered from herbaria and regional Floras to summarize current knowledge on their reproductive biology. These findings will be used to make predictions about which oak species are most likely to have evolved a mast-seeding reproductive pattern. I hope that these predictions will pave the way for future research on tropical oaks.

Evolutionary ecology of masting in temperate trees

Pollination efficiency in temperate tree species

In this paper, pollination efficiency will be defined as the percentage of flowers that are fertilized. Because most wind-pollinated tree species are highly outcrossing (Loveless & Hamrick 1984), we would expect fertilization success to be higher from outcross rather than self-pollination owing to the effects of inbreeding depression. Within a season, selection for pollination efficiency should favor synchronization of flowering by trees within a population to maximize opportunity for cross-pollination. For temperate tree species, pollination usually occurs during early spring over a relatively short time period probably using a combination of daylength and temperature as a cue.

Across years, pollination efficiency may result in the synchronization of flowering so that all trees within a population produce their larger flower crops during the same year. Thus, pollination may play a role in the evolution of masting. The pollination-efficiency hypothesis states that masting has evolved to favor the increased fertilization success that occurs when more pollen is available (Norton & Kelley 1988; Smith *et al.* 1990). Since the majority of mast-fruiting species in temperate forests are wind-pollinated, it may be that selection has acted through the flowering stage in addition to or instead of other stages. In other words, masting may have evolved through selection for mast-flowering due to increased efficiency of pollination during mast-flowering years. Alternatively, masting may be common in wind-pollinated species not because of selection for increased flower production but because wind-pollination, in contrast to animal pollination, does not loose effectiveness with increasing flower production. For example, animal-pollinated tree species may have an upper limit on size of the flower crop caused by pollinator limitation while size of flower crop in wind-pollinated species is limited only by the internal resources needed to mature a seed crop.

Smith *et al.* (1990) developed a model to evaluate the potential advantages of mast-flowering for pollination efficiency in wind-pollinated species. The two most important assumptions of their model are that (1) fluctuations in male and female flower production are positively correlated across years; and (2) that cost of sexual reproduction through female function is the same regardless of fertilization. This second assumption refers to the fact that many plant species produce the reproductive structure (e.g., a cone) whether or not the ovules within it are fertilized or matured. Because the cost of producing a flower includes the production of fruit, selection should favor increased pollination efficiency through mast-flowering to minimize wasted reproductive effort. They conclude that mast-flowering is particularly advantageous in gymnosperms because (a) many species invest in production of cones regardless of fertilization and (b) gymnosperm species often occur in monodominant stands where mast pollen production can dilute self-pollen and increase cross-fertilization.

Several types of evidence would provide support for the pollination-efficiency hypothesis: (1) percentage fertilization of a flower crop should be greater during a mast-flowering year; (2) male and female flower production should be temporally correlated across years; and (3) high percent fertilization is correlated with mast-fruiting years. The first type of evidence is required to demonstrate that pollination efficiency could select for mast-flowering. The second type of evidence also would result if masting is a consequence of selection for mast-flowering because we would expect that male and female flower production should be synchronized to the same year. The third type of evidence is necessary but not sufficient to support the pollination efficiency hypothesis. If the other two types of evidence are absent, then high fertilization success only demonstrates that large crop size can result from flowering success; it would not indicate increased pollination efficiency resulting from mast flowering (Sork & Bramble, manuscr.).

Data from several gymnosperm and angiosperm species were reviewed to test the pollination-efficiency hypothesis (Table 1). As predicted by the Smith et al. (1990) model, in all cases where mast flowering was found to enhance fertilization success, the plant produces a reproductive structure regardless of fertilization. For example, *Pinus*, *Dacryidium*, *Betula*, and *Fagus* all mature a cone or fruit even when the seeds are absent or undeveloped. Because the reproductive structure is produced regardless of seed set, it is easy for the investigators to estimate percentage fertilization by counting hollow seeds or fruit. Al-

Table 1. Studies examining the three kinds of evidence in support of the pollination efficiency hypothesis to account for the evolution of masting in temperate tree species. See text for discussion of criteria.

Species	Reference	Increased fertilization success during years with high flowering?	Correlation of male and female flowering during mast year?	Increase seed set in mast years?
Dacrydium cupressinum	Norton & Kelly 1988	yes	–	yes
Fagus sylvatica	Nilsson & Wastljung 1987	yes	–	yes
Betula pubescens & B. verrucosa	Sarvas 1952	yes	yes	yes
Picea abies	Sarvas 1957	yes	–	yes
Pinus silvestris	Sarvas 1955	no	–	yes
Quercus alba, Q. rubra, & Q. velutina	Sork & Bramble, in prep.	no	no	yes

– = data not available.

though only two of the studies which showed fertilization success also measured male and female flowering, in only one case male and female flowering correlated temporally (Table 1).

The pollination-efficiency hypothesis was also examined using collections of aborted flowers and fruit from individual trees of *Quercus alba*, *Q. rubra*, and *Q. velutina* from spring through autumn of 1984–1988 (Sork & Bramble, manusc.). To test the hypothesis that percent fertilization is greater during mast years, we estimated number of flowers initiated and number of flowers fertilized and then calculated percent fertilized. Unlike the species discussed above, it is not possible to determine fertilization success by counting hollow fruit. Oaks often have two peaks of premature abscission. The first one occurs in June, which coincides with a period after fertilization should have taken place, and the second one occurs in midsummer just prior to fruit enlargement (Shellburne 1983; Sork *et al.* in press). Assuming that the number of abscised flowers falling prior to late June represents unfertilized flowers, we used the first early peak as an estimate of the unfertilized flowers. To test the hypothesis that male and female reproduction is synchronized across years, we measured the catkin biomass that fell into traps (cone-shaped plastic bags with a 0.5 m^2 opening) placed beneath the canopy of individual trees. This provides us a relative estimate of pollen production, which we correlated with total number of flowers. Finally, we assessed the size of the mature acorn crop by counting the mature acorns in the fall.

Our results do not support the pollination-efficiency hypothesis (Table 1). Percent fertilization was not greater in years of high pistillate flower production or years of high pollen production (Sork & Bramble, manuscr.). Total number of female flowers was not correlated with male catkin biomass. Although the pollination-efficiency hypothesis was supported in beech (Nilsson & Wastljung 1987) another member of the Fagaceae (Table 1), we cannot draw the same conclusions for oaks. A major difference between the two genera is that oaks do not allocate energy to the fruit if the seed is unfertilized. Indeed, large proportions of the flower crop are aborted quite early. In spite of our observation that mature acorn crop size correlated with percent fertilization, this finding alone without the other two types of evidence should not be interpreted as support for the pollination-efficiency hypothesis.

Role of seed predation in temperate tree species

The predation hypothesis proposes that mast-seeding is an anti-predator adaptation that allows survival of seeds by satiating their predators (Janzen 1969, 1971, 1974; Smith 1970; Silvertown 1980). In Janzen's view, both the ability to satiate seed predators during mast years and to suppress predator populations during nonmast years are important factors in the evolution of masting. In his review of a wide range of species, Silvertown (1980) found evidence supporting predator satiation, but was unable to find support for Janzen's hypothesis about the importance of the intermast interval for the reduction of predator populations. Silvertown (1980) suggested that masting is most likely to evolve in populations subject to intense predation. His principal support for this was a significant correlation between the coefficient of variation of crop size (an index of masting) and the highest observed percent predispersal mortality (an index of intensity of predation). An alternative explanation for the intermast interval is that it is simply a physiological consequence of producing a large crop and is unrelated to suppression of predator populations (Silvertown 1980; Sork *et al.* in press).

The best evidence demonstrating that high crop size reduces seed predation is increased percent seed survival from either insect and/or vertebrate predators during mast years. Several mast-fruiting species show evidence of predator satiation. Silvertown (1980) reanalyzed data from mast-seeding species to determine those cases in which probability of seed survival is correlated with log crop size for data sets with five or more years of crop size data. He examined 59 cases for 25 tree species, but the majority (n = 39) of the cases did not report the cause of mortality (whether animal or something else), thus preventing any test of the predator-satiation hypothesis. In Table 2, I list the significant cases in which either insect or vertebrate predation was identified as a source of mortality. These seven species provide an indication of the range of species in which masting results in predator satiation. In Table 2, I also list six species in which the predator-satiation hypothesis was specifically examined. Of those studies, only Nilsson and Wastljung (1987) measured seed predation by both insect and vertebrate predators. In fact, this study is the only one that I am aware of in which both the pollination and predator hypotheses are tested. They concluded that both pollination and

Table 2. Evidence addressing the predator-satiation hypothesis in temperate mast-fruiting tree species. The studies cited by Silvertown (A) presented data that he reanalyzed to test for a relationship between percent survival from predation and crop size. The second group of studies (B) were specifically designed to address the predator satiation hypothesis.

Species	Reference	Increased seed survival from insect predators during mast years?	Increased seed survival from vertebrate predators during mast years?
A. Studies cited in Silvertown (1980)[a]			
Abies concolor	Fowells & Schubert 1956	yes	no
Fagus grandifolia	Gysel 1971	yes	no
Fagus sylvatica	Nielsen 1977	yes	–
Pinus banksiana	Kraft 1968	yes	–
Pinus palustris	McLemore 1975	yes	–
Pinus ponderosa	Fowells & Schubert 1956; Schubert 1970	yes	yes
Pinus resinosa	Matson 1971	yes	–
B. Studies directly testing predator satiation			
Fagus sylvatica	Nillson and Wastljung 1987	yes	yes
Hammamelis virginiana[b]	De Steven 1983	yes	–
Carya glabra	Sork 1983	no[c]	no
Quercus alba	Sork & Bramble, manusc.	no[c]	–
Quercus rubra	Sork & Bramble, manusc.	yes	–
Quercus velutina	Sork & Bramble, manusc.	yes	–

[a] References cited by Silvertown 1980 who reanalyzed their data.
[b] Shrub species.
[c] More seeds escaped seed predation during mast years but there was not a proportional advantage of mast years as found in other studies in this table.
– = not available.

predation contribute to the evolution of masting in beech.

In a study of three temperate oak species, we examined the relationship between crop size and insect infestation in 1981–1988 (Sork & Bramble, manusc.). We found that percent infestation shows a negative exponential relationship with log crop size in *Quercus rubra* and *Q. velutina*, members of the black oak subgenus. For *Quercus alba*, a member of the white oak subgenus, a proportional advantage of large acorn crops did not occur, but numerically more acorns survived during a mast year than a non-mast year. Thus, *Quercus alba* is less efficient in satiating its seed predators than the other two species but in all three species large acorn crop enhances escape from seed predators. When seed predation is more intense, as in *Quercus alba*, it may cause stronger selection for increased masting (Silvertown 1980). So, current directional selection may be acting on *Quercus alba* to increase crop size even further until proportional satiation of predators occurs. It is interesting to note that in some areas, the masting interval of *Quercus alba* can be as high as ten years (Schopmeyer 1974) rather than the three years observed in one Missouri population.

Because the two subgenera of oaks have such different patterns of reproduction in North America, I decided to evaluate the extent to which masting has evolved in each. The differences in reproductive patterns include the following: 1) members of the white oak subgenus mature fruit in one season while the North American members of the black oak subgenus require two seasons; 2) acorns of the white oak subgenus germinate in the fall immediately upon falling while black oak acorns are dormant until the following spring; and 3) acorns of the white oak subgenus tend to have greater moisture content (Korstian 1927). To examine the differences between the two subgenera, we utilized data from Schopmeyer (1974) on length of the intermast interval in common eastern North American deciduous oak trees. The length of the intermast interval provides an index of masting. When Schopmeyer reported a range of years, the midpoint of the range was used.

The eighteen tree species in Schopmeyer's review includes 13 species which show intermast intervals of two or more years (Table 3). Our analysis showed that intermast interval varied across species but that, on the average, white oak species have a greater interval than black oak species (Table 3). This suggests that selection has been stronger for greater masting in the white oak subgenus than in the black oak subgenus among North American oaks.

In conclusion, the temperate deciduous oaks show evidence that masting succeeds in predator satiation of insect seed predators. In general, species of the white oak subgenera show greater masting, as indicated by length of intermast interval, than species of the black oak subgenus, which may be due to greater selective pressure by seed predators. Further research is needed to understand how phylogenetic constraints contribute to the evolution of masting. Nonetheless, the predator-satiation hypothesis is supported by several North American oak species.

Seed size, habitat selection and masting in temperate forest trees

Seed size can play an important role in the successful establishment of seedlings: larger acorns produce more vigorous seedlings than small acorns (Korstian 1927; Beaman 1981). It has been shown for a range of species that larger seeds are favored under conditions of shading (Grime & Jeffrey 1965; Gross 1984; Winn 1985), drought (Baker 1972; Wulff 1986) and in high latitudes characterized by short growing seasons (McWilliams *et al.* 1968). Aizen and Patterson (1990) found seed size in temperate deciduous oaks to be correlated with geographical range and concluded that large seed size enables those species to colonize a broader range of habitats. Selection on seed size could affect masting indirectly because size of the seeds may influence the length of time to accumulate reserves for seed production. A large seed may require more local reserves to develop than a small seed.

Using data from Schopmeyer (1974), Waller

(1979) reported that, within six genera (*Acer*, *Betula*, *Fraxinus*, *Pinus*, *Prunus*, and *Quercus*), species that had an intermast interval greater than 1.5 years had significantly larger seed size than those producing seeds every year. Using Schopmeyer's (1974) data for oak tree species, I estimated seed size by transforming the average number of cleaned acorns per pound to mean grams per acorn. I found a positive correlation between log of intermast-interval and log seed size (Fig. 1, $r = 0.617$, $n = 18$, $P < 0.01$). Moreover, white oak species have significantly larger acorns than black oak species (Table 3).

In spite of the seed size differences between the two North American subgenera, some members of the black oak subgenus do produce large acorns and show long masting intervals. For example, *Quercus rubra*, a large seeded species, tends to occupy late successional habitats that typically are associated with larger seed size. It appears that phylogenetic constraints alone do not explain the evolution of seed size or masting in *Quercus*. To examine whether habitat correlates with seed size, I categorized seed size into small (less than 2.5 g) and large (greater than 2.5 g) after examining the size distribution of seed weights. Then I used information from Fowells (1965) to categorize each species according to successional status (mid- or late succession). I found that large seeded species are more likely to be classified as late successional species $\chi^2 = 5.84$, $df = 1$, $P < 0.05$) than small-seeded species. *Quercus alba* and *Q. rubra* in Missouri have larger seeds than *Quercus velutina* and show significantly greater indices of masting (Sork *et al.* in press). Those former two species are mature forest co-dominants whereas *Q. velutina* is distributed in the more xeric and open sites.

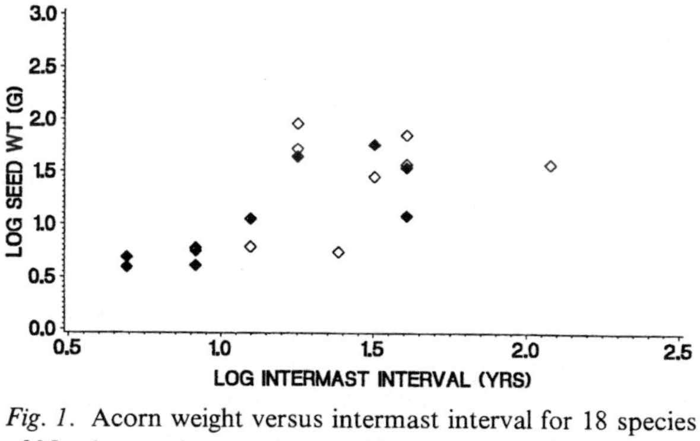

Fig. 1. Acorn weight versus intermast interval for 18 species of North American eastern deciduous oaks in the white oak subgenus (hollow diamond) and the black oak subgenus (solid diamond). Data were taken from Schopmeyer (1974).

What we cannot determine with current information is whether (a) selection for increased seed size puts increased selective pressure on masting because a crop of the same magnitude but with larger seeds requires a longer intermast interval to store the necessary resources; or (b) whether factors which have selected for masting such as seed predation have also favored large seed size. Normally, seed predation should favor a tree to produce more and presumably smaller seeds (Janzen 1969). However, if masting has evolved, the tree may have the resources to produce both large crops and large seeds. Given that masting species produce fewer crops than non-masting species, there may also be selection to increase the probability of seedling survival through large seed size. Consequently, selection for large crop size with a long mast interval may also select for larger seeds. Further research is needed to understand the relationship between seed size, masting, and habitat selection. Moreover, because white oak species have larger seeds than black oak species, future work should include an analysis of possible phylogenetic constraints on the evolution of seed size and masting (e.g. Donohue 1989).

General model for temperate deciduous oaks

This discussion has shown that selection could have acted during several life history stages. These possible effects are not mutually exclusive and probably include interactions among them. To summarize the relationships among flower phenology, seed predation, selection on seed size and mast-fruiting, I have developed a schematic diagram (Fig. 2).

Selection via pollination efficiency could contribute to the evolution of masting either through its effect on the intermast interval or on acorn crop size. Selection to maximize pollination effi-

Table 3. Successional stage, intermast interval, and seed size in selected species of eastern Nort American oaks.

Subgenus	Species	Successional stage	Mast interval (yrs)	Seed size size (g)
White oaks	*Q. alba*	mid	7.0	3.28
	Q. bicolor	late	4.0	3.46
	Q. lyrata	late	3.5	3.53
	Q. macrocarpa	mid	2.5	5.24
	Q. micheauxii	mid	4.0	3.99
	Q. prinus	late	2.5	3.67
	Q. stellata	mid	2.0	1.10
Black oaks	*Q. coccinea*	mid	4.0	1.80
	Q. falcata	mid	1.5	0.83
	Q. imbricaria	mid	3.0	0.83
	Q. laurifolia	late	1.0	0.85
	Q. niger	mid	1.5	0.98
	Q. nuttalli	mid	3.5	4.61
	Q. palustris	mid	1.5	1.07
	Q. phellos	mid	1.0	0.95
	Q. rubra	late	4.0	2.77
	Q. shumardii	late	2.5	4.45
	Q. velutina	mid	2.0	1.75
White oak	MEAN (S.E.)		3.64 (0.64)	3.42 (0.47)
Black oak	MEAN (S.E.)		2.32 (0.35)	1.90 (0.43)
	W.O. vs B.O.[a]		$P<0.05$	$P<0.05$

[a] T-test for unequal variances using log-transformed data.

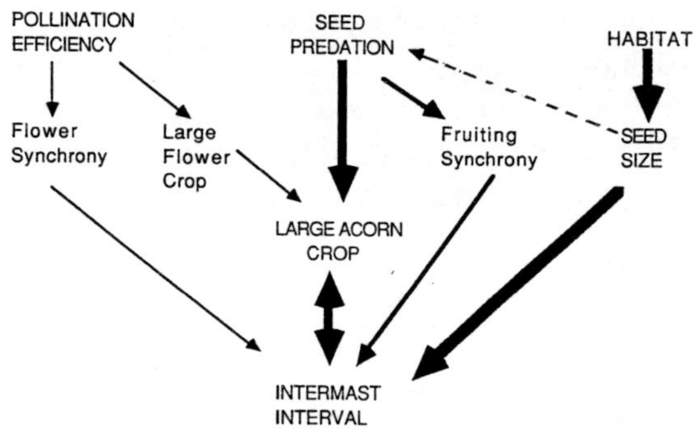

Fig. 2. General model for the evolution of mast-fruiting in temperate deciduous oaks. Thickness of line indicates hypothesized relative importance of the relationship between the two variables. See text for description.

ciency via outcrossing may act on flowering synchrony across years so that year-to-year variation in flower production is synchronized. This later process could potentially influence the intermast interval if such synchronization requires some specific cue that occurs every so many years. However, because synchronization of flower production within a year is strongly cued to early spring in North American oaks, I do not expect that the need for flower synchrony would be the most important factor contributing to the intermast interval. I have indicated this with thin arrows (Fig. 2). Pollination efficiency could contribute to the evolution of masting by selecting for larger flower crop which could result in an increased crop size. However, my research on three temperate oak species does not provide strong support for this path, so I have drawn thin arrows (Fig. 2) to indicate the extent of importance.

For predator satiation to be effective, acorn maturation should be synchronized within a season and among years. In the temperate zone, oaks mature acorns during early fall. Among years, some cue is necessary to synchronize acorn production within a population to the same year. My research has shown that each of the three species (a) produce large and small crops synchronously across years within a species, and (b) has its own internal cue for the production of a mast year (Sork *et al.* in press). Selection via seed predation should affect both the crop size and mast interval components of masting. However, I suggest that the major impact of selection by seed predators is on increased crop size, as indicated by the thick line (Fig. 2). Then, as selection continues on crop size, the need for greater resources to produce a mast crop could cause a concomitant increase in the intermast interval. Moreover, if the intermast interval is due to physiological needs to accumulate sufficient resources for another large crop, then selection on intermast interval should be a function of crop size times seed size or total biomass allocated to acorn production.

The significant correlation between habitat and seed size, combined with the significant correlation of seed size and intermast interval, suggests that habitat also may influence the evolutionary ecology of masting. I suggest that seed size influences the intermast interval component rather than the crop size component of masting as indicated by the line between seed size and intermast interval (Fig. 2). The effect on the mast interval would occur because as seed size increases more resources are needed to mature a seed which could in turn affect the masting interval if simultaneously selection favors increased crop size. I speculate that large seed size could indirectly affect the crop size component of masting (indicated by dashed line in Fig. 2) by increasing the vulnerability of seeds to predators, which then causes selection for increased crop size. Clearly, the relationship among these factors requires direct investigation.

In summary, selection through predation and seed size but not pollination efficiency seem to be quite important in the evolutionary ecology of masting in temperate deciduous oaks.

Applications to tropical oaks

Although more than half of the world's species of oaks occur in the tropics, we know very little about them. Like their temperate zone counterparts, tropical oaks, especially in Central America, are ecologically significant in that they often dominate the forests in which they occur and they provide a major food source for wildlife. Montane tropical oak forests are disappearing as fast as lowland rain forest enhancing the need to understand the factors contributing to the production of seeds so important for tree regeneration. In this section, I will utilize existing, albeit scanty, information about Central American tropical oak species to modify the temperate oak model of the evolutionary ecology of masting for application to tropical oaks. My prediction is that masting has evolved in many of these species.

Pollination and masting

Unlike the temperate zone where spring provides a sharp cue for flowering, tropical oaks, especially those in less seasonal climates, do not have such a strong seasonal cue. In this way, the tropical oaks may be similar to the dipterocarps in Malaysia, which have a large, high-lipid fruit like *Quercus* (Janzen 1974; Ashton *et al.* 1988). Although *Quercus* spp. are wind-pollinated and dipterocarp species are insect-pollinated, both types of pollination would require a cue to synchronize flowering to achieve fertilization, especially in aseasonal environments. Apparently, in Malaysia mast-fruiting is more common in the aseasonal dipterocarp species than in the seasonal species from which they probably evolved. Ashton *et al.* (1988) proposed that because the aseasonal species have no clear seasonal cue, they must have evolved to respond to a supra-annual cue that can be detected by a large number of individuals. Flowering in the aseasonal species is hypothesized to be cued by El Nino years (Ashton *et al.* 1988) but flower synchrony has probably evolved through selection for mast-fruiting (Janzen 1974; Ashton *et al.* 1988) rather than mast-flowering. Because dipterocarps are insect-pollinated, mast-flowering is unlike to improve fertilization success (Smith *et al.* 1990).

We have no data on the relationship between supra-annual weather cues and mast-flowering or mast-fruiting in tropical oaks. To evaluate whether oaks use a reliable seasonal cue, I first wanted to determine whether they flower at the same time each year. For this purpose, herbarium specimens were examined for the oaks of Costa Rica and Panama from herbaria at the Instituto Nacional de Biodiversidad in Costa Rica, the Missouri Botanical Garden, and the California Academy of Science. Although herbarium data do not have the same reliability as data collected systematically in the field, they are the best data currently available on flowering and fruiting phenology of neotropical oaks. I found that wet-forest species in Costa Rica were collected in flower in a range of months throughout the year (Table 4A). For most species, 75% of the flowering record fell in one of the six potential months of the dry season (December through May) with March being the most common month of flowering. Breedlove (pers. comm.) hypothesizes that for flowering of many oaks are keyed to a three day dry spell followed by rain, which may explain why flowering occurs during so many months because that dry spell could occur anytime in the dry season. For one species in Costa Rica, (*Quercus bumelioides*, I found specimens with flowers in nine months of the year (Table 4A). This species occurs in high elevation cloud forest with no clear dry season. In contrast to temperate oaks in which flowering occurs consistently within the same month each year, tropical oaks seem to flower in different months across years. If within year cues for flowering are not highly predictable, which is probably true for species in less seasonal habitats, then mast-flowering which is synchronized to a strong supra-annual cues might evolve to improve pollination efficiency. In spite of the fact that temperate oaks show little evidence in support of the pollination-efficiency hypothesis (Sork & Bramble, manusc.), tropical oaks that occupy wet habitats (even in the 'dry' season) may have strong selective pressure to flower synchronously during years when environmental conditions are unusually good. Probably the first cue that should be tested is the El Nino years.

Predator satiation and masting

Only one study that I am aware of has examined seed predation in tropical oaks. Boucher (1981) demonstrated that the synchronous production of

Table 4. Observations of herbarium collections recording flowering and fruiting times for six wet-forest species of Costa Rica oaks. Each observation represents a different plant.

Species	Dec	Jan	Feb	Mar	Apr	May	Jun	July	Aug	Sep	Oct	Nov	Total
A. Flowering observations													
Q. benthamii	1	1		3									5
Q. bumelioides	1	1	1	5	2		1	1		1		2	15
Q. costaricensis		2	2	3	1		1						8
Q. insignis			1	2								2	5
Q. lancifolia				1									1
Q. salicifolia	4		4	2			1		2	1			14
Total flowering	6	4	7	16	3	0	3	1	2	2	0	4	48
A. Fruiting observations													
Q. benthamii						3			1				4
Q. bumelioides			2	3					1			1	8
Q. costaricensis		2			2					1			5
Q. insignis						1		1	1				3
Q. lancifolia					1		1			1			3
Q. salicifolia				1		2		2	1	1	1		8
Total fruiting	0	2	2	4	3	6	1	3	4	3	1	2	31

acorn crops by several individuals within a population was necessary to satiate mammalian seed predators of *Quercus oleiodes*. While other tropical trees can exhibit a mast-fruiting reproductive pattern (e.g. Wheelwright 1986), only a few cases of predator satiation have been demonstrated. Research on other large-seeded tropical tree species, including the understory tree *Gustavia superba* (Sork 1987) and the canopy species *Dipteryx panamensis* (De Steven & Putz 1984; Clark & Clark 1984), demonstrate that high seed density is critical for escape from predation. Consequently, we should expect predation by seed predators to play as much a role in the evolution of mast-fruiting in tropical oaks as it does in temperate oaks.

Synchronization of fruit maturation in tropical oaks should be extremely important to achieve predator satiation. As with flowering synchrony, we might expect species without a clear seasonal cue to rely on a supra-annual cue in order for individuals within a species to synchronize to the same year and time of year. To evaluate fruiting phenology, I used herbarium specimens for Costa Rica and Panama oaks to determine whether there was any seasonal pattern of acorn production. I recorded date for a mature fruit only if I had evidence that it was attached to the specimen at the time of sampling or the label stated that mature fruit were present in the trees. Although the sample sizes are not always large, it is apparent that the timing of fruit maturation is highly variable from year to year (Table 4B). Thus, one might predict that if predator satiation is necessary for seed survival, we should see the evolution of mast-fruiting by synchronizing fruit production to supra-annual cues as observed in the Malaysian dipterocarps in such a way that tropical oak populations produce large acorn crops in the same year and during the same time of year.

Given that we have no reason to believe that tropical oaks are any less vulnerable to seed predators than temperate oaks, I predict that selective pressure through predation has contributed to the evolution of masting in these species. At this point, we need data on insect infestation, predation by vertebrates, dispersal agents as well as patterns of acorn production in tropical oak species to evaluate their vulnerability to predation and the extent of masting. It will be interesting to see whether the tendency for members of the white oak genus to show greater masting intervals than those of the black oak genus as has been observed among North American oaks.

Seed size and habitat

Tropical oaks show tremendous variation in seed size both within and across species. Because moist conditions sometimes select for smaller seeds (Grime & Jeffrey 1974), one possible hypothesis is that wet-forest species should have smaller seed size than dry-forest species. However, if differences in forest type also affect light conditions, then because low light environments usually are associated with larger seed size (Grime & Jeffrey 1974; Foster & Janson 1985; Mazer 1989), one could hypothesize that larger seeds would be associated with wet forest which tend to have more closed canopy.

I examined three flora – (1) Costa Rica, (2) Nicaragua, and (3) State of Chiapas, Mexico – to evaluate whether seed size correlated with habitat in tropical oaks and whether there were differences between subgenera. Using data from herbaria and Floras, I measured length and width of mature acorns and recorded the date, location, and elevation of each observation. The dependent variable of acorn size (log transformed) was estimated by calculating the volume from length and width measurements. Most oak species could be classified as inhabitants of either wet forest or dry forest. The nomenclature is that of D. Breedlove (pers. comm.).

For the Costa Rican oaks, I utilized the individual observations from herbaria on six wet-forest species of oaks to test for differences across species as well as subgenera. Seed size differed significantly among species (ANOVA, $F(4,33) = 13.4$, $P < 0.0001$) and was significantly greater for species in the white oak subgenus than those in the black oak subgenus (ANOVA, $F(1,33) = 11.0$, $P < 0.01$). Because elevation provides an environmental gradient, I included this variable in the ANOVA model as a covariate and found that acorn volume co-varied inversely with elevation (ANOVA, $F(1,33) = 4.8$, $P < 0.05$). Thus, seed size may be affected by environmental factors associated with elevation (e.g. temperature, precipitation, length of growing season). If moisture increases with elevation then selection on smaller seeds in mesic environments may account for the trend. Alternatively, some evidence suggests that length of growing season may decrease with elevation. Using data on date of collection on specimens with young staminate or pistillate flowers and mature attached acorns where December 1 was day 1 and November 30 was day 365, I found that mean day of fruiting within a year was negatively correlated with elevation ($r = -0.388$, $n = 27$, $P < 0.05$) even though day of flowering was not correlated with elevation ($r = -0.125$, $n = 55$, $P > 0.10$). These two observations together suggest that higher-elevation trees have a shorter period of time to mature their acorns and thus produce smaller acorns. Regardless of the explanation, the relationship between elevation and seed size in Costa Rican oaks indicates that habitat may also put selective pressure on seed size.

For Nicaragua oaks, I estimated acorn volume using the midpoint of the range of length and width measurements reported in the Flora of Nicaragua (Breedlove, pers. comm.). Many of the species are also found in Costa Rica but additional dry-forest species are also present. Again, I found that the white oak subgenus had significantly larger acorns than the black oak subgenus (dependent variable = log transformed acorn volume: ANOVA, $F(1,9) = 7.17$, $P < 0.05$; Fig. 3A). I also found that wet-forest species had significantly larger acorns than dry-forest species (ANOVA, $F(1,9) = 9.17$, $P < 0.05$, Fig. 3B). This finding runs counter the observation that mesic environments select for small seed size. However, because other studies have found large seeded species to be associated with closed habitats (Mazer 1989) or more mature forest (Foster & Janson 1985), perhaps light availability is lower in wet forests than in dry forests.

As an additional test of the habitat hypothesis, I examined data from the Flora of Chiapas, Mexico which contains a larger number of species than Nicaragua. Breedlove (pers. comm.) has identified 24 species, but I chose to use only the 17 species occupying wet and dry forest and to omit those found in the cold, high elevation forest. Based on information from Breedlove (pers. comm.), species were assigned to either the small (volume less than 5 cm^3) or large (volume greater than 5 cm^3) seed size class. Using a goodness of fit test, I found that more species within the black oak subgenus had small acorns than did members of the white oak subgenus $\chi^2 = 7.244$, $df = 1$, $P < 0.01$; Fig. 4A) and that dry-forest species were more likely to have small acorns than wet-forest species $\chi^2 = 4.735$. $df = 1$. $P < 0.05$; Fig. 4B).

In conclusion, tropical oaks show a similar phylogenetic constraint observed in temperate

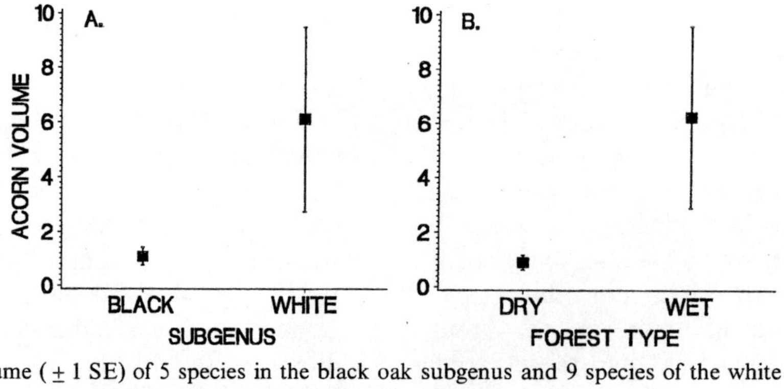

Fig. 3. Mean acorn volume (± 1 SE) of 5 species in the black oak subgenus and 9 species of the white oak subgenus for Nicaraguan oaks (A) and for 6 dry-forest and 6 wet-forest species from Nicaragua (B).

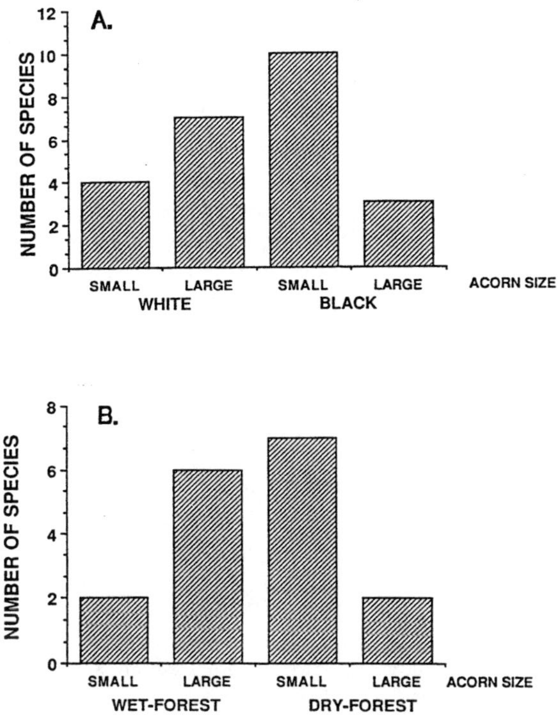

Fig. 4. Number of species from Chiapas, Mexico with small and large acorns which belong to the two subgenera (A) and occur in dry-forest and wet-forest.

oaks that the white oak subgenus has larger acorn size, on the average, than the black oak subgenus. They also show that acorn size is associated with habitat. Moreover, some tropical oaks have much larger acorns than any among the temperate deciduous oaks. For example, the large-seeded *Quercus insignis* in Costa Rica has a mean seed volume of 25.6 cm^3 which is much greater than that of a large-seeded North American oak, *Quercus rubra*, which has an acorn volume of 10.4 cm^3. If seed size directly or indirectly selects for mast crop size and intermast interval in tropical oaks as it appears to do in North American oaks, then we should expect the tropical large-seeded species to have even longer intermast intervals than observed among temperate masting species. Thus, I predict that one should look for the evolution of masting in species that are in the white oak subgenus, occupy wet forests habitats, or are known to have large acorns.

General model for tropical oaks

The general model of the evolutionary ecology of masting in temperate oaks should probably be modified for tropical oaks. As before, I have indicated important relationships with thick lines (Fig. 5). While the nature of the relationships should be similar between temperate and tropical oaks, I think certain differences in the strengths of the relationships are likely to occur. The major difference concerns the role of pollination. As stated above, the need for increased pollination efficiency in less seasonal climates should select for a response to a supra-annual cue. Because of the importance of tight synchronization within a year, this cue must be strong enough to trigger flowering within the same year and time of year. Thus, the scale of the intermast interval will be influenced by the type of cue that is used. Unlike temperate oak species, I predict that selection for pollination efficiency should have a strong effect on the evolution of mast-flowering. Such mast-flowering will result in mast-fruiting.

Like temperate oaks, predator satiation will play a strong role in the evolution of mast-fruiting. Again, because tropical climates tend to be less seasonal, there may be selection to respond to a supra-annual cue in order to ensure that individuals within a population mature fruit in the same year and season. Because of the need for sufficient reserves to produce a mast crop, especially in the large-seeded tropical oaks, a mast year will

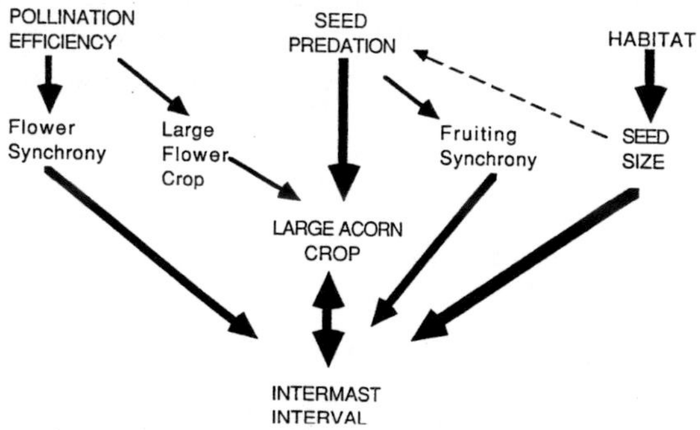

Fig. 5. General model for the evolution of mast-fruiting in Central American tropical oaks. Thickness of line indicates hypothesized relative importance of the relationship between the two variables. See text for description.

require a number of elapsed years since the last mast crop as well as a strong external cue. In temperate oaks, a good acorn crop was a function of both internal cues and weather conditions (Sork *et al.* in press). In tropical oaks, I expect this interaction between internal physiological and external cues to be at least as strong.

In summary, I wish to make four predictions concerning tropical oaks. Like temperate oaks, I do not anticipate that all species will have evolved the same degree of mast-fruiting. However, I do believe it is going to be found to be quite common. First, improved pollination efficiency through mast-flowering will be more prevalent among tropical oaks than temperate oaks owing to the less seasonal nature of tropical environments creating a greater need for a supra-annual cue for flowering. Second, I predict that predator satiation will also be important in tropical oaks thus creating a need for mast-fruiting to be tied to a supra-annual cue to insure the synchronous production of acorn crops for the less seasonal species (e.g. Ashton *et al.* 1988). In fact, if a supra-annual cue is found to be associated with masting in tropical oaks, it might be difficult to distinguish whether it evolved through the flowering or fruiting life-history stages. Third, I predict that white oak species will have greater intermast intervals than black oak species, on average. Determining whether this is true because of their larger seed size or phylogenetic tendency toward longer intermast interval can only be sorted out with a phylogenetic analysis of character-state evolution (e.g. Donoghue 1989). Fourth, I predict that because wet-forest species have larger seed size, they will also show longer intermast intervals than dry-forest species.

Acknowledgements

I wish to thank all the undergraduate and graduate students who have helped me over the years collect, measure, and weigh acorns. I thank Nancy Schellhorn for assistance in collecting herbarium data. Carol Augspurger, Diane De Steven, Carol Kelly, Bette Loiselle, Christopher Smith, and Mary Willson provided valuable comments on this manuscript. I am also grateful to Dennis Breedlove who has shared data on oaks and improved my understanding of the ecology and systematics of tropical oaks. Finally, I want to acknowledge Dan Janzen who piqued my curiosity about masting years ago while I was a student at the University of Michigan. My research has been supported by funds from the Missouri Department of Conservation, the U.S.D.A. Forest Service Experimental Station Cooperative Grant Program, the National Science Foundation, the University of Missouri-St. Louis Improved Research Quality grant program, and the University of Missouri Weldon Spring Fund.

References

Aizen, M. A. Patterson III, W. A. 1990. Acorn size and geographical range in the North American oaks (*Quercus* L.). J. Biogeog. 17: 327–332.

Ashton, P. S., Givnish, T. J. & Appanah, S. 1988. Staggered flowering in the Dipterocarpaceae: new insights into floral induction and the evolution of masting in the aseasonal tropics. Amer. Natur. 132: 44–66.

Baker, H. G. 1972. Seed weight in relation to environmental conditions in California. Ecology 53: 997–1010.

Beaman, B. A. 1981. Factors affecting the establishment, growth, and survival of white oak (*Quercus alba* L.) in an upland hardwood forest. Ph. D. thesis, Duke University, Durham, NC.

Boucher, D. H. 1981. Seed predation by mammals and forest dominance by *Quercus oleoides*, a tropical lowland oak. Oecologia 49: 409–419.

Clark, D. A. & Clark, D. B. 1984. Spacing dynamics of a tropical rain forest tree: evaluation of the Janzen-Connell model. Amer. Natur. 124: 769–788.

De Steven, D. 1983. Reproductive consequences of insect seed predation in *Hammamelis virginiana*. Ecology 64: 89–98.

De Steven, D. & Putz, F. E. 1984. Impact of mammals on early recruitment of a tropical canopy tree, *Dipteryx panamensis*, in Panama. Oikos 43: 207–216.

Donohue, M. J. 1989. Phylogenies and the analysis of evolutionary sequences, with examples from seed plants. Evolution 43: 1137–1156.

Foster, S. A. & Janson, C. H. 1985. The relationship between seed size and establishment conditions in tropical woody plants. Ecology 66: 773–780.

Grime, J. P. & Jeffrey, D. W. 1965. Seedling establishment in vertical gradients of sunlight. J. Ecol. 53: 621–642.

Gross, K. L. 1984. Effects of seed size and growth form on seedling establishment of six monocarpic perennial plants. J. Ecol. 72: 369–387.

Janzen, D. H. 1969. Seed-eaters versus seed size, number, dispersal, and toxicity. Evolution 23: 1–27.

Janzen, D. H. 1971. Seed predation by animals. Ann. Rev. Ecol. Syst. 2: 465–492.

Janzen, D. H. 1974. Tropical blackwater rivers, animals, and mast fruiting by the Dipterocarpaceae. Biotropica 6: 69–103.

Janzen, D. H. 1978. Seeding patterns of tropical trees. In: P. B. Tomlinson and M. H. Zimmerman (eds.), Tropical trees as a living systems. Cambridge Univ. Press, New York.

Korstian, C. F. 1927. Factors controlling germination and early survival in oaks. Yale Univ. Sch. For. Bull. 19: 5–118.

Kozlowski, T. T. 1971. Growth and development of trees. Vol II. Academic Press, NY

Kozlowski, T. T. & Keller, T. 1976. Food relations of woody plants. Bot. Rev. 32: 293–382.

Loveless, L. & Hamrick, J. 1984. Ecological determinants of genetic structure in plant populations. Ann. Rev. Ecol. Syst. 15: 65–95.

Mathews, J. D. 1955. The influence of weather on the frequency of beech mast years in England. Forestry 28: 107–115.

Mazer, S. J. 1989. Ecological, taxonomic, and life history correlates of seed mass among Indiana dune angiosperms. Ecol. Monogr. 59: 153–175.

McWilliams, E. L., Landers, R. Q. & Mahlstede, J. P. 1968. Variation in seed weight and germination in populations of *Amaranthus reflexus* L. Ecology 49: 290–296.

Nilsson, S. G. & Wastljung, U. 1987. Seed predation and cross-pollination in mast-seedling beech (*Fagus sylvatica*) patches. Ecology 68: 260–265.

Norton, D. A. & Kelly, D. 1988. Mast seeding over 33 years by Dacyrdium cupressinum Lamb. (rimu) (Podocarpaceae) in New Zealand: the importance of the economies of scale. Func. Ecol. 2: 399–408.

Sarvas, R. 1952. on the flowering of birch and the quality of seed crop. Communicationes instituti Forestalis Fenniae 40: 1–35.

Sarvas, R. 1957. Studies on the seed setting of norway spruce. Meddelelser fra det Norske skogskogforsoksvesen 48: 533–556.

Sarvas, R. 1955. Investigations into the flowering and seed quality of seed crop. Communicationes Instituti Forestalis Fenniae 45: 1–37.

Schopmeyer, T. 1974. Seeds of Woody Plants in the United States. Agric. Handb. 450, U.S.D.A. For. Serv., Washington, D.C.

Shellburne, O. R. 1983. Patterns in the production, development, and maturation of acorns in *Quercus alba* (Fagaceae). M. S. Thesis, University of Missouri, St. Louis.

Silvertown, J. W. 1980. The evolutionary ecology of mast seeding in trees. Biol. J. Linnean Soc. 14: 235–250.

Smith, C. C. 1970. The coevolution of pine squirrels (*Tamiasciurus*) and conifers. Ecol. Monogr. 40: 349–371.

Smith, C. C., Hamrick, J. L. & Kramer, C. L. 1990. The advantage of mast years for wind pollination. Amer. Natur. 136: 154–166.

Sork, V. L. 1983. Mast-fruiting in hickories and availability of nuts. Amer. Midl. Natur. 109: 81–88.

Sork, V. L. 1987. Effects of predation and light on seedling establishment in *Gustavia superba*. Ecology 68: 1341–1350.

Sork, V. L., Sexton, O. & Bramble, J. Ecology of mast-fruiting in Missouri oaks. Ecology. (in press).

van Vredenburch, C. L. H. & la Bastide, J. G. A. 1969. The influence of meteorological factors on the cone crop of Douglas-fir in the Netherlands. Silvae Genetica 18: 182–186.

Waller, D. M. 1979. Models of mast fruiting in trees. J. Theor. Biol. 80: 223–232.

Wenger, K. F. 1957. Annual variation in the seed crops of loblolly pine. J. Forestry 55: 567–569.

Wheelwright, N. T. 1986. A seven-year study of individual variation in fruit production in tropical bird-dispersed tree species in the family Lauraceae. In: A. Estrada and T. H. Fleming (eds.), Frugivores and seed dispersal. Dr. W. Junk Pub., Dordrecht.

Winn, A. A. 1985. Effects of seed size and microsite on seedling emergence of *Prunella vulgaris* in four habitats. J. Ecology 73: 831–840.

Wulff, R. D. 1986. Seed size variation in *Desmodium paniculatum*. Il. Effects on seedling growth and physiological performance. J. Ecology 74: 99–114.

Vegetatio **107/108**: 149–162, 1993.
T. H. Fleming and A. Estrada (eds). Frugivory and Seed Dispersal: Ecological and Evolutionary Aspects.

Aspects of variation in a neotropical seed dispersal system

Henry F. Howe
Department of Biological Sciences (M/C 066), University of Illinois at Chicago, Box 4348, Chicago, IL 60680, USA

Keywords: Ecology, Frugivory, Herbivory, Seed dispersal, Seedling Ecology, *Virola*

Abstract

This paper explores the causes and consequences of seed and seedling mortality of the tree *Virola nobilis* (Myristicaceae) Central Panama in order to understand the advantage to local seed dispersal by birds and monkeys. Post-dispersal mortality due to insects (primarily *Conotrachelus* spp., Curculionidae) accounts for 30–35% of seed and seedling death during the first 12 weeks after seed fall. Because more seeds and seedlings are killed under and near fruiting trees than 15–45 m away, seed dispersal confers a 20–40 fold advantage on seeds carried 45 m from fruiting adults. In contrast, >60% of seed and seedling death during the first year is due to seed predation by mammals, with >90% due to mammals among the <2% seeds that survive until maternal endosperm is exhausted ± 12 weeks after seed fall. Mortaliy due to mammals is independent of distance from parent trees, confering no advantage to seed dispersal. Insects account for variation in mortality attributable to distance effects, mammals to between site effects.

Early weevil infestations put a premium on seed removal by large birds (*Ramphastos swainsonii*, *R. sulfuratus*, *Penelope purpurascens*), which carry >50% of the seeds that they eat >40 m, as compared with smaller birds (*Baryphthengus martii*, *Tityra semifasciata*, *Trogon massena*) and monkeys (*Ateles geoffroyi*), which leave most or all of the seeds that they eat under or near the tree crown.

Introduction

This paper re-evaluates causes and consequences of seed and seedling mortality in *Virola nobilis* A.C. Smith (Myristicaceae; Duke 1962) (misnamed for South American *V. surinamensis* in Croat (1978)) in central Panama (Howe 1986a) in order to refine understanding of a classic 'dispersal system,' defined by a tree population and the animals that enhance or diminish seed and seedling recruitment (Howe 1977). Here I dissect the relative contribution of insect and mammal predation on seeds and seedlings to the previously-reported 'escape advantage' of seed dispersal away from parent trees of this species (Howe *et al.* 1985). One goal is to evaluate the relative importance of dispersal agents that drop *Virola* seeds different distances from fruiting trees.

This analysis also suggests an important unfilled role of frugivory and seed dispersal in understanding dynamics of tropical tree communities (Janzen 1970; Connell 1971; Hubbell 1979, 1980). Tests of density-dependence that might detect forest equilibrium usually rely upon interactions of adult trees and saplings (Hubbell & Foster 1987; Hubbell *et al.* 1991). Seed and seedling studies may well important if, as in the case of *Virola nobilis* and other trees of moderate or low abundance, seed and seedling mortality is so high that few saplings exist near parents. Strong *early* density-dependence would not be detected by adult interactions with saplings, despite a profound effect on adult tree dispersion.

Background and system

Snow (1971) and McKey (1975) identified what appear to be extremes of adaptation in patterns fruit production, fruit quality, and fruit use by animals, while Janzen (1970) offered a constellation of possible consequences of dispersal for plants. McKey in particular argued that some large-seeded tree species that invest heavily in both seeds and nutritional enticements for frugivores should attract dedicated specialists that consistently disperse seeds; trees that invest less in fruits might depend on opportunistic species that are individually less predictable and often less effective dispersal agents. Janzen pointed out that even very local seed dissemination might help seeds and seedlings avoid devastating mortality from insects, mammals, or disease in high densities near parent trees.

Study of the *Virola nobilis* population and dispersal assemblage on Barro Colorado Island, Panama between 1979 and 1987 was conceived as an exploration of the degree of ecological dependence of the tree for dispersal and of the community of 80 fruit-eating birds and mammals for this plant (Howe 1983, 1986a, 1990a). The tree was selected as a possible exemplar of a 'specialist' tree that bears moderate numbers of bulky seeds, made attractive to animals by a showy, lipid-rich aril. The *V. nobilis* study remains one of the few long-term attempts to understand patterns of fruit production, use by animals, and consequences of animal use for tree recruitment.

The dispersal system

Virola nobilis is a dioecious canopy tree of moderate abundance and fecundity (see Howe 1986a). Five years of sampling with fruit traps and animal censuses at 17–23 focal trees show that this is a species with low fecundity; approximately 50% of the crop is removed by a limited disperser assemblage dominated by 8 of 68 diurnal and 1 of 12 nocturnal frugivores which consistently use the fruits for food (Table 1). An additional 320 hours of extended watches at 8 trees show that one toucan (*Ramphastos swainsonii)* takes 35% of the seeds removed by animals; 4 other birds (black-crested guan, *Penelope purpurascens*; keel-billed toucan, *Ramphastos swainsonii*; rufous motmot, *Baryphthengus martii*; and slaty-tailed trogon, *Trogon massena*) each take >3% of the seeds dispersed. Crude estimates of dispersal distances due to different animals show that, among principal visitors, large toucans and guans drop >50% of the seeds carried far (>40 m) from fruiting adults, while smaller trogons and motmots drop most under or near (<20 m) the canopy of fruiting plants.

Seed and seedling survival

Of special interest here are manipulations carried out in 1982, 1983, and 1984 to test the hypothesis that seeds dropped under and near the crown would suffer higher mortality than those dropped further away (Howe *et al.* 1985; Howe 1986a). This was designed as a test of the Janzen (1970) 'escape hypothesis,' as elaborated by Howe and Smallwood (1982). Because the original objective was to provide a sample that would survive to one year of age, and because initial mortality was extraordinarily high, three cohorts of seeds or seed-

Table 1. Summary of *Virola nobilis* (once *V. surinamensis*) dispersal system, based on on 4–5 years of observation at 17–23 fruiting trees.

Item	Years	Summary data
Large aril and seed	5	Wet: 1.0–1.6 g and 1.4–4.7 g
High nutrition	5	Aril 63% lipid by dry mass
Low fecundity	5	Median crop 2,200–8,500/tree
Efficient dispersal	5	45–65% seeds taken/tree
Small assemblage	4	6 of 80 common frugivores

Source: summarized from Howe (1983, 1986a, unpublished).

lings, each representing a different stage of development, were planted in consecutive years.

Fresh seeds, germinating seeds, or established seedlings were planted in concentric rings around 17, 13, and 7 fruiting adults, respectively. Radii at 5 m were under the crown, 15 m just beyond the crown edge, and 25 m, 35 m, and 45 m well beyond the crown of focal plants. In 1982, 40 fresh seeds were placed at randomly located points on each of the 5 circles (= rings) around 17 female trees (total N = 3,400) and monitored for 6 weeks. Screenhouse space limited later plantings to 25 germinating seeds 6 weeks old (radicles protruding; 1983) and 25 established seedlings 12 weeks old (endosperm depleted, 2–4 leaves flushed; 1984) on circles around 13 trees and 7 trees (N of 1,625 and 875, respectively).

Plantings of freshly fallen seeds (1982) and germinating seeds (1983) showed a clear early advantage for seeds dropped at least a few meters from fruiting trees (Fig. 1). Plantings of established seedlings (≥ 12 weeks old) in 1984 showed 96% mortality in one year, independent of distance from fruiting trees. Plantings at 5 m from males in 1982 and 1983 produced indistinguishable mortality from that of seeds 45 m from females. Mortality of established seedlings around males in 1984 was indistinguishable from mortality around females at any distance.

Causes of mortality vary. From 0–12 weeks,

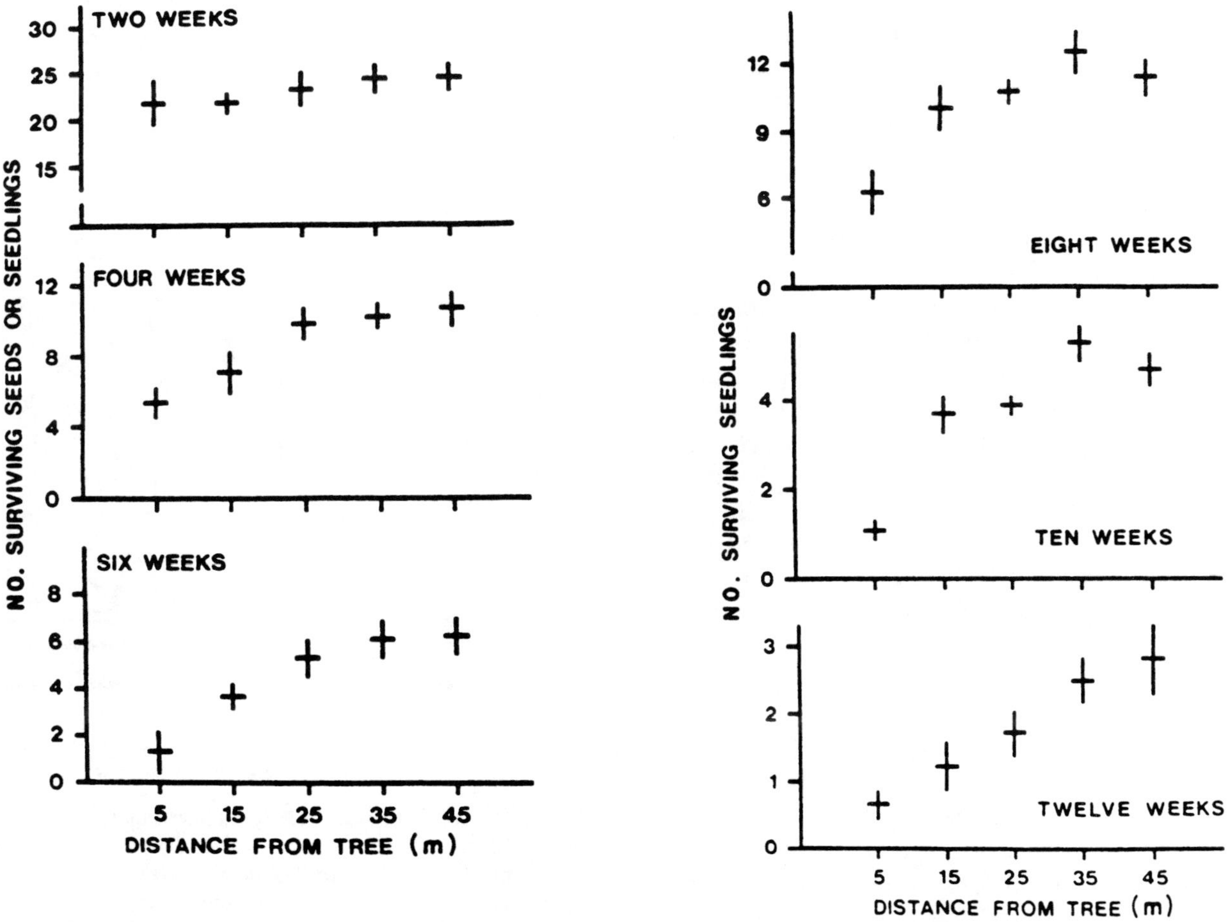

Fig. 1. Survival as a consequence of distance for two cohorts of seeds and seedlings planted in concentric rings around *Virola nobilis* trees. Samples were 40 seeds per ring for each of 5 rings at 17 trees (1982; total N = 3,400), and 25 seedlings per ring for each of 5 rings at 13 trees (1983; total N = 1,625). Means ± 1 SE. From Howe *et al.* (1985).

virtually all mortality can be attributed to weevils (*Conotrachelus* spp.), bark beetles (*Nitidulidae*), or mammals (*Agouti paca, Dicotyles tajacu, Odocoilius virginiana*). Agoutis (*Dasyproctus punctata*) do not eat freshly fallen (1–2 day old) seeds of this species (Larson & Howe 1987), but indirect evidence in this and other studies suggest that agoutis probably eat older seeds and seedlings (Forget & Milleron 1991). *V. nobilis* seed powder at many planting sites suggested predation by pacas, which press the seeds against the ground and gnaw them to bits, rather than burying by these or other rodents. Ants bury seeds still encased in arils (Howe, *pers. observation*), and in 45 hr of observation one agouti was observed to bury a *V. nobilis* seed where it found it (no seeds were carried by agoutis >2 m from the site of discovery; Larson & Howe 1987). It is unlikely that burying is a factor unless seeds are moved substantial distances, since by far the greatest mortality by either mammals or insects is of erect seedlings >2 weeks after seed fall. After 12 weeks, rodents and browsing mammals (e.g. tapirs, *Tapirus bairdii*) are the primary sources of established seedling and small (<1 m high) juvenile mortality, whether under *Virola* trees or not.

Insect damage to seeds can be seen directly by exit holes and frass; mammalian damage to seedlings by clipping, defoliation, pulling-up, or disappearance from sites not subject to wash-out. Mortality from seasonal drought, which potentially could have an enormous impact on this population (Howe 1990b), is not directly relevant because the vast majority of seedlings and young juveniles are killed by insects or mammals before the first dry season when drought is an issue.

The case for a specialized dispersal system

V. nobilis, consistent with predictions, invests heavily in small numbers of fruits, relying on a limited assemblage of dispersal agents which appear to depend on this species over a 4 month fruiting season, year after year (Table 1). This approaches the extreme of specialization suggested by Snow (1971), McKey (1975) and Howe and Estabrook (1977); *V. nobilis* shows every indication of being at the center of a specialized dispersal system playing to a small gallery of fruit-eating connoisseurs.

Moreover, dispersal counts. Seeds dropped near have a lower probability of survival than those dropped well beyond the crown edge of focal trees. This can be seen in either half of the period during which seedlings depend on parental endosperm (Fig. 1), or in the number of seedlings surviving the full 12 weeks on concentric rings around 17 fruiting trees in 1983 (Fig. 2). Summary calculations from products of proportions surviving at each distance during the 1982 and 1983 years (Fig. 2, inset) suggest as much as a 44-fold advantage to seed dissemination 45 m from fruiting trees (Howe *et al.* 1985; see Howe & Westley 1988, p. 154). Direct calculations from the cohort of 3,400 seeds planted in 1982 that survived to 12 weeks suggest a 20-fold ad-

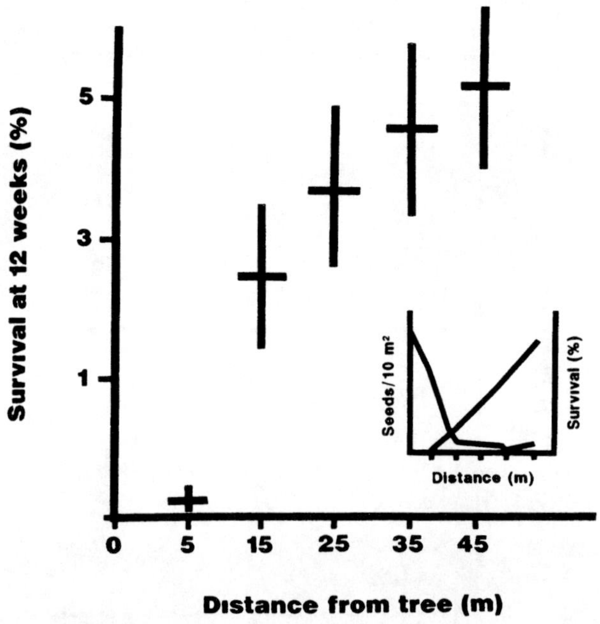

Fig. 2. Seedling survival to 12 weeks after fruit fall as a function of distance from 17 fruiting *Virola nobilis* trees for 3,400 seeds planted in 1982. Twelve weeks marks the end of dependence on parental endosperm, and of vulnerability to insect seed predators. Inset: summary survival calculated by multiplying the probabilities of survival at 6 and 12 weeks. Means ± 1 SE.

Table 2. Chestnut-mandibled Toucans as dispersal agents of *Virola nobilis* on Barro Colorado Island, Panama.

30 × better than Rufous Motmots & Slaty-tailed Trogons
15 × better per seed handled
2 × as many seeds taken
8 × better than Spider Monkeys
2 × better per seed handled
4 × as many seeds taken
4 × better than Keel-billed Toucans
As good per seed handled
4 × as many seeds taken
3 × better than Crested Guans
As good per seed handled
3 × as many seeds taken

vantage to dispersal 45 m from fruiting trees (Fig. 2).

Finally, different dispersal agents count. As predicted by McKey (1975) and others, the birds responsible for the most effective dispersal were large species that commonly fed at the trees; smaller trogons or motmots left most seeds directly under or very near the crown edge, where virtually all seeds or seedlings (99.94%) died before 3 months of age. Seed and seedling survival data from manipulations (Howe *et al.* 1985) coupled with those from extended watches at fruiting trees (Howe & Vande Kerckhove 1981) permit a crude ranking of dispersal agents (Table 2). The two toucans and the guan were roughly equivalent in the quality of *Virola* dispersal *per seed carried*, but differed in the extent to which they consumed fruits. Motmots and trogons were also more or less equivalent from the *Virola* perspective per seed carried, but left seeds much closer to fruiting trees than the larger species.

Seed rings: a much closer look

Seed and seedling rings described above (Fig. 1) document the strong likelihood of an immediate and profound advantage to local seed dispersal. Seeds dropped right under the crown or at the crown edge (as consistently done by trogons and motmots) are more likely to die during the first three months than those taken further away (as consistently accomplished by toucans and guans).

As first analyzed and presented, these data do not speak to the possible consequences of variation among trees in mortality and survival, nor do they distinguish alternative sources of mortality that might bear upon variation in recruitment between tree neighborhoods. Additionally, the initial picture of a linear increase in survival with distance from parent trees (Fig. 2, inset) provided by summary calculations does not identify possibly important biological effects of density-dependent, distance-dependent (indistinguishable from density-dependent in Hubbell's program), and density-independent sources of mortality.

Seeds and seedlings: one-way anovas

Survival data presented above were originally analyzed with single classification anovas for each census, with the number of survivors from each circle used as a sample for each of the 5 distances from each tree (Table 3). The results convincingly show an effect of distance on either seed (0–6 weeks) or seedling (6–12 weeks) survival, but not on established seedling survival (12–24 weeks). These analyses suggest an early influence of insect, but not mammalian, seed and seedling predation on the escape advantage. That is, early seed predation is sometimes strongly density or distance related, but later mortality is not.

Table 3. One-way anovas of survival of *V. nobilis* seedlings as a function of distance (5, 15, 25, 35, 45 m) from fruiting trees.

Source	df	Ms	F-statistic
	0–6 weeks (1982)		
Distance	4	68.42	7.46***
Error	80	9.17	
	6–12 weeks (1983)		
Distance	4	12.86	6.11***
Error	60	2.10	
	12–24 weeks (1984)		
Distance	4	4.88	0.31
Error	30	15.98	

*** $P < 0.001$.

Seeds and seedlings: two-way anovas

One-way anovas may be insufficient for either of two reasons. Seeds or seedlings on one circle might influence the survival of those on other rings around the same trees; this would result in a strong interaction of distance with tree in a two-way analysis. This can be tested with variables representing crop history. Alternatively, different trees might represent neighborhoods that strongly influence seedling survival (e.g. an agouti does or does not live there, high or low soil moisture, etc.). This might lead to a strong site advantage or disadvantage, independent of distance.

Two-way anovas, using 'tree' as a fixed variable representing neighborhood, provide an evaluation of these alternatives (Table 4). There cannot be interaction between tree and distance in this analysis because trees do not share the same circle, so the interaction term is not used. During germination and the early seedling stage (0–6 weeks), when developing plants depend entirely on parental endosperm, there is an overwhelming effect of distance and a substantial effect of tree, accounting for 60% of the variation in the model. Variance components further allocate most explicable variance to distance, much less to tree, and much to 'error,' which is variation within rather than between sites. In these two-way ANOVAs, such effects are obscur in slightly older seedlings (6–12 weeks). Established seedlings (> 12 weeks) show highly significant variation in survival among tree neighborhoods, but not distances. In 12–24 week samples, over half of the explicable variation can be attributed to variation within samples (error). The general message is that distance counts during seedling establishment, site counts after establishment.

Analyses with 'tree' as a fixed variable explain substantial variance, but meaning of the variable is unclear. Given the importance of insects and mammals on mortality, it is more than likely that a history of annual infestation or foraging habits cause the crops produced by individual trees to be a factor in 'neighborhood' defined here by 'tree.' Exploratory analyses using both the crop of the year and the number of seeds *not* dispersed each year explains substantial variance; the most significant influence is with the crop or undispersed seed means of the year in question with its previous value the year before (crop or undispersed seeds for each tree for 1981 and 1982 for the 1982 experiment; 1982 and 1983 for the 1983 experiment; Table 5). Tests of survival with crops and the numbers of undispersed seeds of the year are not significant or are marginal (often $P \geq 0.05$). Analyses using two-year means show a substan-

Table 4. Two-way anovas: distance and 'tree' as fixed effects on survival.

Source	df	Ms	F	% Variance
		0–6 weeks; R sq = 0.60)		
Distance	4	68.42	10.92***	64%
Tree	16	20.77	3.31***	4
Error	64	6.27		32
		6–12 weeks; R sq = 0.48)		
Distance	4	12.87	6.68***	52%
Tree	12	2.82	1.46	1
Error	48	1.93		46
		12–24 weeks; R sq = 0.77)		
Distance	4	4.89	1.02	0%
Tree	6	60.76	12.70***	45
Error	24	4.79		55

*** P < 0.001.

Table 5. Two-way anovas: distance and 2-year total and undispersed crop means on seedling survival.

Source	df	Ms	F	% Variance
		0–6 weeks; R sq = 0.35)		
Distance	4	25.87	8.51*	22%
Crop mean	1	65.16	7.45**	21
Dist * crop	4	3.04	0.35	1
Error	75	8.75		56
		0–6 weeks; R sq = 0.30)		
Distance	4	19.63	6.46*	7%
Undispersed	1	16.10	1.71	4
Dist * undisp	4	3.04	0.32	0
Error	75	9.40		89

* P < 0.05; ** P < 0.01.

tial power of mean crop size in explaining much of the tree neighborhood effect on survival, though the undispersed seeds (which should be more powerful) are of less use for the 0–6 week period. For 6–12 weeks (not shown), two-way anovas show no influence of distance on survival, while influences of a two-year history of undispersed seeds ($F = 3.5$, $df = 55$, $P = 0.07$) are suggestive. In these analyses, two-year historical variables are roughly as significant in early seedling survival as 'distance,' and more significant than crop values of the year.

Non-significant interactions are presented because crop variables are continuous, not fixed; inclusion of continuous variables with the fixed effect 'distance' constitutes a Model II or mixed model anova. In such a case the F-statistic for the main effect (distance) is derived from the interaction mean squares rather than the error mean squares, as computer programs (which virtually always assume a Model I anova) automatically calculate. This may either reduce or increase the F-statistic. Absence of interaction between distance and 'tree' effects represented by crop variables suggest that interaction among circles is negligible.

Finally, the rules clearly change for survival and mortality after independence from parental endosperm at 12 weeks of age. Two-way anovas exploring the effects of distance and crop history on surival of established seedlings planted in 1984 at 24 and 52 weeks showed no effects ($P > 0.9$) of distance, as reported in discussion of single classification anovas above. However, in the two-way analyses historical crop variables were highly significant, with the highest F-ratios for mean undispersed seeds per tree (mean of 1982 and 1983 values) at 24 weeks ($F = 27.195$, $df = 25$, $P < 0.001$) and at 52 weeks ($F = 13.064$, $df = 25$, $P = 0.001$). Notably, correlations of surviving seedlings with historical crop variables *at each distance* for the 7 focal fruiting trees in the 1984 sample showed consistently *positive* relationships, with the strongest correlations between survival and the mean undispersed seeds per tree (all but one r from 0.72 to 0.96, $P \leq 0.05$ to 0.001). Whereas mortality during establishment is negatively density-dependent, mortality after that stage appears to be density-independent or even *positively density-dependent*. Of these seedlings, planted well after peak mortality, 0 to 8 survived 52 weeks on rings in the vicinity of highly fecund trees, far fewer near trees of lower fecundity (overall means per annulus ≤ 1 seedling).

Insects and mammals: one-way anova

Simple enumeration shows that insects account for 30–35% of seeds and establishing seedling mortality, while mammals account for most remaining mortality until the onset of the dry season in late December (Table 6). Enumeration by distance shows that insects kill far more seeds near than far from fruiting trees by 6 weeks ($F = 10.26$, $df = 4$, $P < 0.001$), and that some of this effect continues through the second half of the establishment period between 6–12 weeks ($F = 2.36$, $df = 4$, $P = 0.06$). One-way analyses show no influence of mammals with distance (Fig. 3). It follows that the advantage to seed dispersal is entirely due to the influence of insects, mostly during the first 6 weeks after seed fall. Extensive burying of *V. nobilis* seeds by rodents has not been observed in this or other studies (cf. Forget & Milleron 1991), and is unlikely to be important because most mortality occurs after germination (see Background and System above).

Insects and mammals: two-way anovas

Inspection of apparent patterns suggest that the one-way anovas are insensitive to possible de-

Table 6. Seed and seedling fates: number (%) of three cohorts killed by insects, mammals, or washout.

Weeks	Survived	Killed by		Lost:
		Insects	Mammals	Washout
0–6	396 (12)	1026 (30)	1783 (52)	195 (6)
6–12	118 (7)	583 (36)	893 (55)	31 (2)
12–52	32 (4)	0 (0)	± 800 (91)	2 (0)

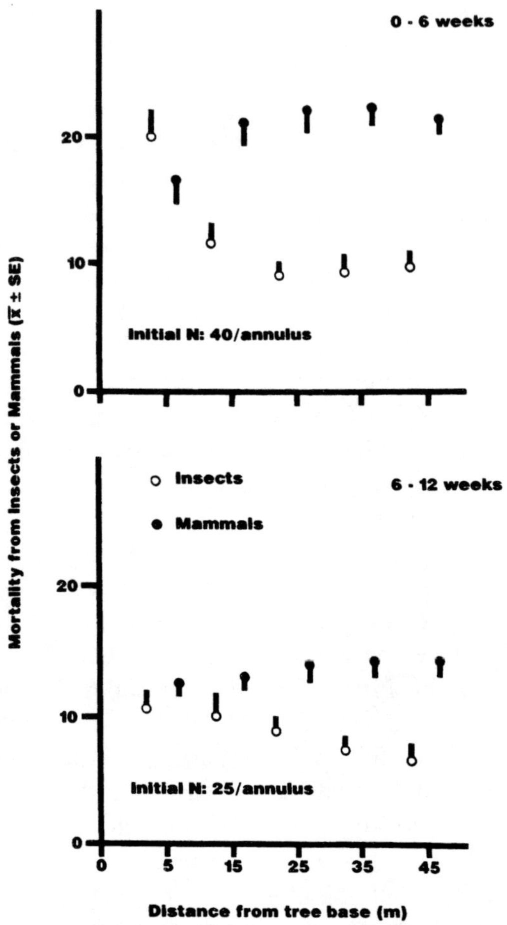

Fig. 3. Mortality of *Virola nobilis* seeds and seedlings due to insects or mammals as a function of distance from trees, using the above design (Fig. 1 and 2). Means $\pm$ 1 SE.

Table 7. Summary of effects of distance, two-year crop mean, and two-year undispersed crop means on mortality at 6 weeks, caused by insects or mammals

Source	df	F	Source	df	F
			Insects		
Distance	4	8.46*	Distance	4	8.26*
Crop mean	1	3.42	Undispersed	1	4.18*
Dist * crop	4	0.56	Dist * undisp	4	0.46
Error	75		Error	75	
R sq = 0.38			R sq = 0.39		
			Mammals		
Distance	4	2.90	Distance	4	2.63
Crop mean	1	14.57***	Undispersed	1	8.91**
Dist * Crop	4	0.69	Dist * undisp	4	0.72
Error	75		Error	75	
R sq = 0.26			R sq = 0.22		

* $P<0.05$; ** $P<0.01$; *** 0.001.

pression of survival near fruiting trees (Fig. 3). Repetition of each analysis with a two-way anova, using 'tree' as a fixed effect, produces powerful indications of effects of *both* distance and tree neighborhood. Indeed, R squares for distance and tree effects on insect and mammal predation are 0.80 and 0.76 for 0–6 weeks and 0.71 and 0.65 for 6–12 weeks, respectively. For established seedlings planted at 12 weeks of age and scored at 24 (before serious dry-season mortality but subject to mammalian herbivory), an overall R square of 0.77 ($P<0.001$) indicates a profound influence of tree neighborhood, but none for distance.

Exploration with historical crop variables shows distance but weak or absent crop effects for insect seed predation during the first 6 weeks (Table 7). Crop effects predominate for mammals. Again, historical characters are better predictors than contemporary estimates for the 1982 experiment. The small number of juveniles remaining by 24 or 52 weeks suggests that the *positive* relationship between survival of these older seedlings and increasing fecundity is less likely to be due to satiation of mammals than to edaphic characteristics of the site.

Discussion

The *V. nobilis* seed dispersal system offers a number of insights into general patterns of frugivory and seed dispersal, and into the methodological problems associated with exploring those patterns. A flood of empirical observations have shown that the extreme ranges of fruit and frugivore specialization exist, as predicted by Snow (1971), McKey (1975), and Howe and Estabrook (1977), but that most species fall between dichotomies (e.g. Howe & Smallwood 1981; Wheelwright & Orians 1982). Janzen's (1970) many possible relations between seed dispersal and seedling survival also seem plausible in light of recent evidence (Clark & Clark 1984; Howe 1986b), but documenting actual patterns and sorting out causes of different patterns has lagged far behind his original insights, and in fact are rarely attempted.

Lessons from V. nobilis

Results of this relatively comprehensive study of a tropical tree and its dispersal assemblage are consistent with the general hypothesis that specialized frugivores are more effective than others in seed dispersal and ultimately in seedling establishment of dependent plants (McKey 1975; Howe & Estabrook 1977; Howe & Vande Kerckhove 1981; Howe *et al.* 1985). Here I explore the reasons why seeds dispersed by large toucans and guans are more likely to survive than those dispersed by smaller trogons and motmots. As important, the *V. nobilis* system offers insights into methodological issues required in studying frugivory and its consequences in tropical trees.

First, *V. nobilis* provides insights into the 'distance' or 'density' effects predicted by Janzen (1970) and Connell (1971). The anomaly of a linear relation between seedling survival and distance from fruiting trees (Fig. 2, inset) appears to be an artifact of two classes of confounded processes, mortality due to insects and mortality due to mammalian seed predation. Within each class, different species and abundances of insects and different species and abundances of mammals will surely influence the distribution of surviving seedlings (e.g. seedling mortality due to some mammals may be more density-dependent than that due to others). On Barro Colorado Island in 1982 and 1983, these tended to cancel each other out to produce something like a straight line in summary calculations (Fig. 2, inset). A plot of seedling survival from the cohort 3,400 seeds planted in 1982 that lived to 12 weeks of age fails to suggest this linearity, and in fact indicates an asymptote to a plateau as originally predicted by Janzen (1970; Fig. 2). In short, the 'escape advantage' exists and in this species is substantial, but varies in extent and form in time. A one year sample is a snapshot of a dynamic process, not a constant state.

Secondly, variation in dispersal advantage from one tree to another is obvious and probably important. A strong 'tree effect,' indicates – as Hubbell (1980) predicted – that non-zero survival occurs under some trees even in the face of extremely high overall mortality. This should lead, as Hubbell (1980), Becker *et al.* (1985), and I (Howe 1989) have suggested, to aggregated and patchy distributions of recruitment at varying densities. The actual densities achieved by a species may reflect general dispersal and establishment strategies (Howe 1989) and local conditions of density-independent mortality (e.g. Howe 1990a).

In *V. nobilis*, excessive initial mortality eliminates almost all seedlings from the vicinity of fruiting trees. Even positive correlations of surviving seedlings (planted at 12 weeks, after peak mortality) with tree fecundity, at all distances to 45 m, left a average of no more than one seedling per ring at 52 weeks. Hence, adult densities are low and declining. A strong advantage to seed escape from parents on Barro Colorado Island exists against a backdrop of high density-*in*dependent mortality in a declining population.

These results suggest that a number of factors may operate independently, or in a hierarchichal fashion. For instance, quantitative differences in dispersal ability, from the plant perspective, exist among *V. nobilis* dispersal agents, but many differences are strictly qualitative (Table 2). One could imagine communities in which smaller motmots and trogons outnumber larger toucans or guans, or in which the larger birds have other preferred foods. Either case would alter seed and seedling distributions. The obvious potential for independent variation in insect, mammal, and seasonal drought on *Virola* seedling survival likewise opens the possibility that recruitment patterns could vary widely in space and time, with varied consequences for seedling recruitment.

In the case of *V. nobilis*, some sources of mortality are not independent, suggesting a hierarchy of effects set by insect seed predation. On Barro Colorado Island, insects set the stage for later mortality. A relatively minor source of overall mortality (<35%) from weevil infestations has a major effect on all that follows by eliminating most seeds and seedlings near adults; density-independent mortality from mammals only may occur among survivors of density-dependent earlier effects. Continuing density-independent mortality then maintains the general pattern of dispersion away from fruiting adults, but at lower densities.

It is fair to ask whether, with a different complement of mammals or in the absence of a large mammal population, insects would kill as many seeds and seedlings? If they did, would insects strongly accentuate the advantage of seedlings dispersed away from parent trees? Would enough seeds and seedlings be killed to avoid recruitment of numerous saplings under and around fruiting *V. nobilis* trees? We can guess that insects in the absence of mammals would accentuate spacing, perhaps completely eliminating recruitment near adults. But this is only conjecture. It remains to be seen whether, in the absence of mammals, density-independent effects of seasonal drought would kill as many seeds (Howe 1990a; Fisher *et al.* 1991). One might expect drought mortality to leave more seedlings in gaps and moist ravines than mammals, but this too is only an educated guess.

In short, differences in recruitment almost certainly occur in time or space if mammals are more or less common, or if insects with different search patterns and life histories infest the seed crops. There is every reason to believe that important variations in *V. nobilis* seed dispersal and seedling recruitment will occur on Barro Colorado Island over time, as they should in any forest. Useful investigations for the future would include comparative studies of density-dependent and density-independent seedling mortality at different sites in Panama and in other localities in Central and South America. A wider comparative analysis would lay the groundwork for defining the limits of dispersal advantage and the animals that mediate it in the *V. nobilis* dispersal system. Extension to dispersal systems of focal congeners, or of other members of the Myristicaceae, would provide an even more comprehensive understanding of the boundaries of dispersal advantage in a pantropical family of ancient lineage, and undoubtedly ancient relationships with frugivores.

The *V. nobilis* study also gives insights into the methods by which one approaches tropical dispersal systems that involve large canopy trees. Most students of frugivory have been preoccupied with monitoring potential dispersal agents (e.g. see Howe 1980; Howe & Vande Kerckhove 1981); most people who justify their interest in frugivory as a medium of seed dispersal do not study patterns and consequences of seed dispersal. As a consequence, most cannot answer the central question, 'Do different frugivores influence plant recruitment in different ways?' The question cannot be answered by observations of animals without examining their effects on seed and seedling distribution, and ultimately on seed and seedling survival.

Moderate to low abundance and extremely high natural mortality leave too few small juveniles and saplings alive to permit comprehensive demographic analyses (seed to adult) of causes of mortality for many or even most tropical tree species (Howe 1989, 1990a, b). That is, the paradigm of cross-sectional demographic studies promoted by Harper (1977) for abundant pasture plants, and adopted virtually throughout plant population biology, is not feasible for sparse species in which seeds or seedlings are very hard to find. A danger exists that workers attempting to conform to the Harperian paradigm will select species with especially resistent seeds or seedlings, thereby biasing insights towards characteristics of a small minority of unusual trees (e.g. Sarukhan 1980; Howe 1990b). The danger is evident (see below), that simply ignoring seed and seedling demography in favor incomplete demographies of more easily monitored saplings and adults could lead an entire discipline to ignore fundamental processes that determine plant distribution and abundance.

If statistically robust demographies are often infeasible, reductionism has its cost. Manipulation tailored to the particular system is a far more efficient means of *identifying* sources of mortality than cross-sectional demographic studies for all but a small minority of common tropical tree species. But only a fortunate investigator can determine the relative importance of sources of mortality or make any but the most general predictions of future population changes. In *V. nobilis*, for instance, the planting manipulations discussed here permit a definitive statement that an escape advantage exists, but can only begin to sort out the influences of different animals on seedling recruitment. More extensive controlled experiments could probably detect differences in the effects of different mammals on recruitment, but it would be fortuitous if these effects were independent enough to allow projections of *V. nobilis* population changes based on such experiments. If deer were absent, would pacas produce the same effect? Likewise, careful determination of effects of drought on *V. nobilis* seedling mortality show that it could be an important source of density-independent mortality, and quite likely an important determinant of tree distributions (Howe 1990b; Fisher *et al.* 1991), if only more seedlings survived mammalian herbivory long enough to experience seasonal drought.

A clear message from the *V. nobilis* analysis is that every empirical challenge in this infant discipline should be met with as many approaches as possible. Where feasible, annual enumerations of naturally dispersed seeds and seedlings will yield insight into the status of a population, and allow projection of its population changes. Experimental manipulations permit one to sort out causes of demographic patterns that are lost in simple counts. In the *V. nobilis* case, field manipulations designed to test effects of distance (Howe *et al.* 1985), gap proximity (Howe 1990b), and water balance (Fisher *et al.* 1991) on seed or seedling survival give different insights into the factors that affect recruitment of this tree, and permit one to offer educated guesses about its responses to future changes in environmental conditions. Together, enumeration and experiment permit one to describe the present and predict the future as if it resembled the present, and to have some rough insight into the future if factors change radically from the present.

In hindsight, possible improvements in the *V. nobilis* study are obvious. Clear refinements of both the zoological and botanical aspects are in order. The weakest link is in determining the actual seed distributions produced by disperser activity; estimates are biased in favor of short visits, a small but undoubtedly measurable frequency exists with which a bird drops *Virola* seeds under another *Virola* tree, and the number of individual birds of each species using the trees are not known. Radio telemetry, or some other method of following individual birds as they carry seeds (e.g. additional assistants who could quietly follow birds, as an observer at a fruiting tree directs), would permit a far better determination of the actual distributions of seeds produced by each species of dispersal agent (see Fleming *et al.* 1977). Banding or wing tags would allow one to assess the number of birds using *Virola* trees, as well as the extent of their use of fruit resources produced by other tree species.

When possible, cohorts of naturally dispersed seeds, seedlings, and juveniles should be monitored. Because this will often be impractical, manipulations must be planned with care, substantial replication, and with close attention to sampling designs that allow rigorous analysis. This statement would not be necessary in most of ecology, but it is necessary where many workers infer species-wide patterns from single individual trees, or unreplicated measures around focal trees.

Many ambiguities in determining the effects of seed dispersal on seedling survival could have been removed with attention to the scale of mortality obvious from casual inspection; hardly any seedlings existed near *V. nobilis* trees. In this species, 3,400 seeds planted in the first cohort was too small to produce much of a sample after six weeks. A greater effort, involving placement of perhaps two to three times as many seeds around a larger sample of female trees (high densities induce seed predation; the number of seeds on each ring must not be 'too' high) would have eliminated the awkward necessity of evaluating mortality with different samples for each half of the seedling period, in different years. Greater replication per gap and gap vicinity in the exclosure studies (Howe 1990a) might have allowed a within and between site analysis; a substantially larger number of *unprotected* seedlings might have permitted a direct comparison of herbivory in and outside of gaps. As it was, the number of survivors was too small to show whether seedlings survived better in or out of gaps.

Other sampling schemes might be developed, with trial and error and careful inspection of existing seed and seedling distributions contributing to a reasonable design (e.g. 3,400 seeds would have been quite unnecessary with *Tetragastris panamensis,* which has far lower seedling mortality than *V. nobilis* [Howe 1990b]). The natural history of each species will determine the adequacy of different sampling methods.

Broader issues: tests of density-dependence

Tropical trees might reach stable equilibrial species abundances if, as Janzen (1970) and Connell (1971) originally proposed, density-dependent mortality were high enough to ensure that the site occupied by a tree of each species was replaced by a member of another species. Alternatively, if mortality is only weakly density-dependent, or density-independent, every tree might replace itself with an offspring (or conspecific brought to it by frugivores) (see Hubbell 1979, 1980). In this extreme scenario species will always be increasing or decreasing in abundance, a process that can be roughly assessed with counts of seedlings, saplings, and adults in each species. Increasing species will have a surfeit of juvenile plants, decreasing species a deficit.

This contentious issue in contemporary ecology has been all but ignored by students of frugivory and seed dispersal. *V. nobilis*, as discussed here, has very few conspecific saplings as neighbors, exists in low densities, and is clearly in decline (also Howe 1990b). Beyond a few other supporting cases (e.g. Augspurger 1983, 1984; Schupp 1988a, b), casual observation suggests that Janzen and Connell might be right; trees produce 10^3 to 10^6 seeds per year, and one never finds 10^3 to 10^6 seedlings of the year near adult fruiting trees. On the other hand, some evidence indicates that Hubbell may be right for some species, which replace themselves as frequently or more frequently than expected on the basis of adult abundance (Hubbell & Foster 1987). Seedling demographies sometimes confirm Hubbell's prediction of exploding populations. *Tetragastris panamensis* adults are often surrounded by juveniles on Barro Colorado Island, and themselves occur in dense aggregations that suggest a population boom (Howe 1980, 1990b). In general, however, determinations of proportions of seeds that are dispersed away and live or die, or that fall under the crown and live or die, are simply not attempted by field biologists who are best suited to resolve these issues.

Given the paucity of directly relevant information about seed and seedling distributions and mortality, Hubbell and his colleagues (1987, 1991) have tested for density-dependent effects of adult trees on conspecific saplings. Such studies allow a firm understanding of replacement rates, but they do not allow an evaluation of the processes that *produce* sapling, pole, and adult distributions. Hubbell and his colleagues find few trees with sufficient neighboring saplings for robust tests, and fewer yet with clear negative density-dependence as shown by decreased survival of saplings near adults. From this base they conclude that there is little evidence of density-dependence that might produce equilibrial tropical communities.

Such demographic studies of saplings and trees do not explain why so few trees have large numbers of neighboring conspecific saplings, or why so few trees replace each other. The answer may be that extreme density-dependent mortality of seeds and seedlings near adults, as originally predicted by Janzen (1970) and Connell (1971) and later argued by Becker *et al.* (1985), coupled with greater or lesser levels of subsequent density-independent mortality, reduce each cohort far below the densities needed for biologically meaningful tests of adult influence on saplings. In plants, in which almost all mortality occurs in the seed or seedling stages (Harper 1977), tests of density-dependence for all but the most common species in a forest should begin with seed and seedling distributions. As in *Virola*, the stage may be set very early for subsequent patterns of mortality, and ultimately may strongly influence adult spatial dispersions.

Acknowledgements

I am grateful to C. Brandaris and E. W. Schupp for field assistance. Other help was provided by O. Finke, A. Herre, J. Idol, D. Larson, R. Manasse, W. Richter, E. Rodriguez, and L. C. Westley. This work was primarily supported by the National Science Foundation, with additional assitance from the Smithsonian Tropical Research Institute and the University of Iowa.

References

Augspurger, C. K. 1983. Offspring recruitment around tropical trees: changes in cohort distance with time. Oikos 40: 189–196.

Augspurger, C. K. 1984. Seedling survival of tropical tree species: interactions of dispersal distance, light-gaps, and pathogens. Ecology 65: 1705–1712.

Becker, P., Lee, L., Rothman, E. D., & Hamilton, W. D. 1985. Seed predtion and the coexistence of tree species: Hubbell's models revisited. Oikos 44: 382–390.

Clark, D. B., Clark, D. A. 1984. Spacing dynamics of a tropical rain forest tree, evaluation of the Janzen-Connell model. Am. Nat. 124: 769–788.

Connell, J. H. 1971. On the role of natural enemies in preventing competitive exclusion in some marine animals and in rain forest trees. In: Dynamics of Populations, den Boer, P. J. & Gradwell, G. R. (eds.) Pudoc, Wageningen, pp. 298–312.

Croat, T.B. 1978. The Flora of Barro Colorado Island. Stanford University Press, Stanford.

Duke, J.A. 1962. Myristicaceae. Anals Missouri Bot. Garden 49: 214–225.

Fisher, B. L., Howe, H. F. & Wright, S. J. 1991. Survival and growth of *Virola surinamensis* yearlings: Water augmentation in gap and understory. Oecologia 86: 292–297.

Fleming, T. H., Heithaus, E. R. & Sawyer, W. B. 1977. An experimental analysis of the food location behavior of frugivorous bats. Ecology 58: 619–627.

Forget, P-M., Milleron, T. 1991. Evidence for secondary seed dispersal by rodents in Panama. Oecologia 87: 596–599.

Howe, H. F. 1980. Monkey dispersal and waste of a neotropical fruit. Ecology 61: 944–959.

Howe, H. F. 1983. Annual variation in a neotropical seed dispersal system. In: Tropical rain forest: ecology and management, Sutton, S. L., Whitmore, T. C., and Chadwick, A. C. Blackwell (eds.) Scientific, Oxford. pp 211–277.

Howe, H. F. 1986a. Consequences of seed dispersal by birds: a case study from Central America. J. Bombay Nat. Hist. Society 83: 19–42 (supplement).

Howe, H. F. 1986b. Seed dispersal by fruit-eating birds and mammals. In: Seed Dispersal. Ed. Murray, D. R. Academic Press, Sydney, pp. 123–189.

Howe, H. F. 1989. Scatter- and clump-dispersal and seedling demography: hypothesis and implications. Oecologia 79: 417–426.

Howe, H. F. 1990a. Survival and growth of juvenile *Virola surinamensis* in Panama: effects of herbivory and canopy closure. J. Trop. Ecology 6: 259–280.

Howe, H. F. 1990b. Seed dispersal by birds and mammals: implications for seedling demography. In: Reproductive Ecology of Tropical Forest Plants, K. S. Bawa & M. Hadley (eds.) UNESCO & Parthenon Publ. Group, Paris. pp. 191–218.

Howe, H. F. & Estabrook, G. F. 1977. On intraspecific competition for dispersal agents in tropical trees. American Naturalist 111: 817–832.

Howe, H. F., Schupp, E. W. & Westley, L. C. 1985. Early consequences of seed dispersal for a neotropical tree (*Virola surinamensis*). Ecology 66: 781–791.

Howe, H. F. & Smallwood, J. 1981. Ecology of seed dispersal. Ann. Rev. Ecol. Syst. 13: 201–228.

Hubbell, S. P. 1979. Tree dispersion, abundance, and diversity in a tropical dry forest. Science 203: 1299–1309.

Hubbell, S. P. 1980. Seed predation and the coexistence of tree species in tropical forests. Oikos 35: 214–229.

Hubbell, S. P. & Foster, R. 1983. Diversity of canopy trees in a neotropical forest and implications for conservation. In: Sutton, S. L., Whitmore, T.C. & Chadwick A. (eds.) The tropical rain forest: ecology and management. Blackwell Scientific, Oxford. pp. 25–41.

Hubbell, S. P., Foster, R. 1987. The spatial context of regeneration in a neotropical forest. In: Gray, A., Crawley, M. J. & Edwards, P. J. (eds.) Colonization, succession, and stability. Blackwell Scientific, Oxford. pp. 395–412.

Hubbell, S. P., Condit, R., & Foster, R. B. 1991. Presence and absence of density dependence in a neotropical tree community. Proc. Royal Soc. London B 330: 269–281.

Janzen, D. H. 1970. Herbivores and the number of tree species in tropical forests. Am. Nat. 104: 501–528.

Larson, D., Howe, H. F. 1987. Seed dispersal and destruction of *Virola surinamensis* seeds by agoutis: Appearance and reality. J. Mammalogy 68: 859–860.

McKey, D. 1975. The ecology of coevolved seed dispersal systems. In: Coevolution of animals and plants (eds.) L. E. Gilbert and P. Raven. U. Texas Press, Austin. pp. 159–191.

Moermond, T., Denslow, J. 1983. Fruit choice in tropical

frugivorous birds: effects of fruit type and accessibility on selectivity. J. Animal Ecology 52: 407–421.

Sarukhan, J. 1980. Demographic problems in tropical systems. In: O. Solbrig (ed.) Demography and evolution in plant populations. Univ. California Press, Berkeley. pp. 161–188.

Schupp, E. W. 1988a. Factors affecting post-dispersal seed survival in a tropical forest. Oecologia 76: 525–530.

Schupp, E. W. 1988b. Seed and early seedling predation in the forest understory and in treefall gaps. Oikos 51: 71–78.

Snow, D. W. 1971. Evolutionary aspects of fruit-eating in birds. Ibis 113: 194–202.

Wheelwright, N. T., Orians, G. 1982. Seed dispersal by animals: contrasts with pollen dispersal, problems of terminology, and constraints on coevolution. Am. Nat. 119: 402–413.

Vegetatio **107/108**: 163–174, 1993.
T. H. Fleming and A. Estrada (eds). Frugivory and Seed Dispersal: Ecological and Evolutionary Aspects.

Fruit size in a tropical tree species: variation, preference by birds, and heritability

Nathaniel T. Wheelwright
Department of Biology, Bowdoin College, Brunswick, ME 04011, USA

Keywords: Ocotea tenera, Lauraceae, Seed dispersal, Frugivory, Plant reproduction

Abstract

Ocotea tenera (Lauraceae), an understory bird-dispersed tree, produces single-seeded fruits that vary in diameter from 1.4 to 2.4 cm. Much of the variation within a population at Monteverde, Costa Rica occurred within individual trees. The relative size of fruits produced by different trees remained generally constant over an 11-year period despite slight differences between years in the average size of fruits produced by a given tree.

Fruit-eating birds could thus express their preferences for particular fruit size characteristics by choosing among trees that have distinct distributions of fruit diameters, and between individually variable fruits within trees. In a field study of individually marked fruits, birds removed 46.2% of fruits; the rest of the fruits were destroyed by invertebrate (25.3%) and vertebrate (4.3%) pulp-feeders or aborted by the plant after remaining ripe but uneaten for as long as 100 days (24.2%). The four major avian seed dispersers of *O. tenera* each have gape widths exceeding all but the largest fruits. Birds preferred plants with greater-than-average-sized fruits; within trees, they favored larger fruits, apparently because net pulp mass increases with fruit diameter. Fruits that ripened early in the season were more likely to be removed and were removed more quickly than late-ripening fruits.

Based on mother-offspring regressions of mean fruit size, the phenotypic variation in fruit diameter in *O. tenera* is highly heritable, indicating the potential for an evolutionary response to selection by birds. Nonetheless, directional selection on fruit size or shape is likely to be inconsistent, constrained by genetic correlations, and weak compared to selection on traits like fecundity or phenology.

Introduction

When flocks of emerald toucanets (*Aulacorhynchus prasinus*) forage through the lower montane forests of Costa Rica, they encounter a diversity of fruits. On a given day, the birds might stop to feed on the clustered orange fruits of *Urera elata*, which measure only 3 mm in diameter, then move on to eat a few 22-mm diameter black drupes of *Beilschmeidia pendula* before swallowing 6-mm *Sapium oligoneuron* seeds covered by red arils (Wheelwright *et al.* 1984). In a temperate-zone forest the variety of fruits seen by birds is less than in the tropics, but nonetheless birds such as American robins (*Turdus migratorius*) consume fruits as distinct as the 3-mm furry scarlet drupes of *Rhus copallina* and the 10-mm watery fruits of *Pyrus* sp. (Wheelwright 1986a; White & Stiles 1991).

Fruit variation within species, though less extreme than variation between species, can also be striking. Individual plants of the same species often produce fruits that differ from those of other members of the population (Gorchov 1985). Even within a plant, fruits may vary. Fruit-eating birds must therefore make a series of choices at different scales: between plant species, between individual plants of a single species, and between individual fruits within the canopy of a particular tree.

Most recent research on fruit-eating birds and fruiting plants has tried to make sense of these choices by describing variation in plant reproductive traits and evaluating how different traits affect fruit selection and handling by birds (Willson 1986; Jordano 1992). The problems of fruit variation and avian preferences have implications for understanding the diversity of diets among fruit-eating birds, particularly dietary specialization versus generalization, as well as the degree of mutual dependence between birds and plants, and the strength of their interaction. Birds' choices translate into different rates of fruit removal from plants and influence seed dispersal patterns. Therefore, birds may affect plant fitness and direct the evolution of certain plant traits, but only when the following conditions are met. There must be variation in a particular reproductive trait within plant populations, birds must pick or handle fruits non-randomly with respect to that trait, their preferences must influence the probability of successful seed dispersal and seedling establishment, and at least some of the variation in the plant reproductive trait must be heritable. Once we determine to what extent these conditions apply in natural populations, we will have a much better understanding of the diversity of plant reproductive features and the degree to which they have been influenced evolutionarily by interactions with seed dispersers.

Of these conditions, recent work has clearly documented variation in fruit traits at various levels – between species, between populations, between individuals, and within individuals (Herrera 1981; Howe & Vande Kerckhove 1981; Janson 1983; Wheelwright *et al.* 1984; Gorchov 1985; Johnson *et al.* 1985; Pratt & Stiles 1985; Dowsett-Lemaire 1988; Foster 1990). We have also learned a great deal about the bases of birds' preferences, specifically the role of plant features such as fruit size, coloration, structure and fecundity, as well as extrinsic, non-heritable features such as habitat or the characteristics of neighboring plants (Howe 1983; Gautier-Hión *et al.* 1985; Wheelwright & Janson 1985; Denslow 1987; Jordano 1987a; Debussche & Isenmann 1989; Sargent 1990). We know far less about where birds drop seeds and how seed dispersal patterns influence seedling survival and establishment (Estrada & Fleming 1986, Murray 1988), much less plant fitness or the interaction between preferences and patterns of dispersal. With respect to the genetic basis of variation in fruit traits, we know next to nothing about non-commercial, long-lived plant species.

Nonetheless, there is abundant (albeit indirect) evidence that plant reproductive traits have responded to natural selection by birds and other animals. In this paper I concentrate on one such trait, fruit size, in an attempt to illustrate phenotypic variation in natural populations at different scales and show how seed dispersers potentially influence the evolution of fruit traits. The justification for focusing on fruit size is that, of the variety of traits that have been investigated, fruit size seems to affect handling and profitability directly and to bear a relatively predictable relationship to the feeding behavior and morphology of birds. Birds in nature and in captivity vary in how they select or handle fruits of different sizes. The reasons are various, but they include the physical difficulty of picking or swallowing large fruits because of gape limitations (Pratt & Stiles 1985; Wheelwright 1985a; Jordano 1987b; Lambert 1989), differences in profitability between large and small fruits due to differences in pulp:seed ratios (Herrera 1981; Howe & Vande Kerckhove 1981; Howe 1983; Jordano 1987; Snow & Snow 1988) or handling rates (White & Stiles 1991), and digestive problems associated with large versus small seeds (Levey 1987; Murray *et al.* 1992). Most of the studies cited above examined birds' fruit size preferences between plant species, but a few have shown that birds respond to fruit size variation between populations (Herrera 1981) or individual plants (Howe 1983; Wheelwright

1985a; Piper 1986; see also Bonaccorsco 1979). In several temperate zone studies the relationship between a bird's gape or body size and the number or size of fruits that it ate was weak or non-existent (Johnson *et al.* 1985; Malmborg & Willson 1988, but see Jordano 1987b), presumably because temperate zone fruits tend to be small relative to birds' gapes, and because the range of fruit sizes is limited compared to tropical habitats (Wheelwright 1988a). In general, though, most studies agree that fruit size alone is one of the surest predictors of whether or not a bird will consume a fruit and disperse its seeds (Jordano 1992).

Within habitats, the distribution of gape widths of fruit-eating birds often mirrors the size distribution of the fruits on which they depend for food, further evidence that fruit size is an important feature of plants, one that could direct co-evolution between species and determine community-wide patterns (Wheelwright 1985a; Pratt & Stiles 1985; Dowsett-Lemaire 1988). Moreover, the allometry of fruit form – specifically, the tendency for larger fruits to be more elongated than smaller fruits (Mazer & Wheelwright, 1993) – indicates that the evolution of fruit shape may be constrained by the ability of potential seed dispersers to swallow fruits.

The common interpretation of such patterns of size and shape in the fruits of bird-dispersed plants is that they reflect mutual adaptation with contemporary seed dispersers. Such patterns could be influenced by selection at the level of species as well as genotype. For example, parallel distributions of fruit sizes and gape widths of fruit-eating birds, as well as upper limits on fruit size, could be the result of a process of species selection producing broad patterns. Contemporary fruit forms may have arisen in response to historical dispersers (Janzen & Martin 1982; Wheelwright 1988b). Furthermore, fruit size and shape conceivably could have evolved in response to selection on genetically correlated characters such as flower or leaf traits (Primack 1987).

To understand the evolution of traits such as fruit size, it is important to 1) quantify phenotypic variability among and within individual plants; 2) measure correlations between fruit size and other traits of relevance to foraging birds, such as seed characteristics or crop size; 3) evaluate the constancy of the trait in different seasons and throughout the life of individual plants; 4) determine if fruits of distinct size differ in their profitability to potential seed dispersers; 5) show that birds can distinguish and actually choose among fruits on the basis of size at the level of species, individual plants, or individual fruits within a canopy; 6) demonstrate that birds' preferences result in differential survival and recruitment of the seeds of fruits of distinct sizes; and 7) determine how much of the variation is heritable versus induced by environmental factors such as climate or microhabitat. In other words, we need to ask whether fruit size really matters to birds, and if so, whether their choices influence plant fitness.

This paper addresses some of these questions by describing the nature of fruit variation in *Ocotea tenera*, a tree species which I have studied in Costa Rica since 1979. Diameter is used as the measure of fruit size most meaningful to avian seed dispersers (Pratt & Stiles 1985; Wheelwright 1985a; Lambert 1989). I describe variation at two levels: within a natural population and within individual plants. I also present the results of a study of fruit removal as a function of mean fruit size within plants and individual fruit size. Finally, I provide estimates of the heritability (h^2) of fruit diameter based on parent-offspring regressions.

Methods

Study species

Ocotea tenera (Lauraceae) is a small, dioecious, bird-dispersed tree which reaches a maximum height of 12 m and diameter-at-breast-height (DBH) of 20 cm. *O. tenera* occurs in clumped distributions along forest edges and in small light gaps, although once established, saplings can survive for many years in dense shade in the understory (Fig. 1). Endemic to Costa Rica, the species' range extends from an elevation of 1700 m

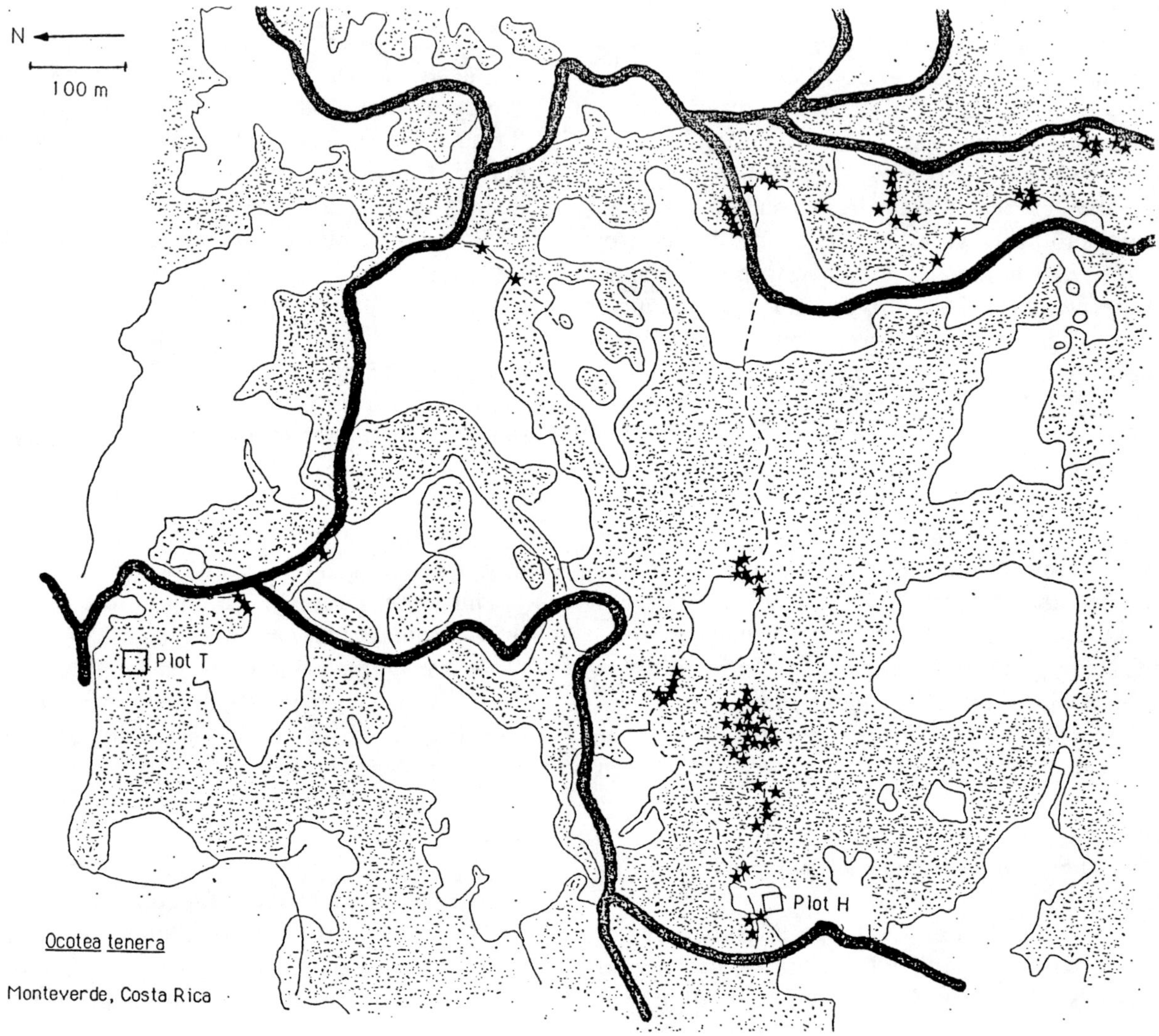

Fig. 1. Map of the study site. Stars indicate the locations of all reproductively mature *Ocotea tenera* trees known in 1980. Dark lines indicate dirt roads or major trails; dashed lines indicate minor trails. Open areas designate pastures, and stippled areas, forests. Plot T and Plot H were established in 1981 and 1984, respectively; in each plot, 32 trees of known maternity survived to reproductive maturity.

in the central highlands to as low as 50 m on the Caribbean slope; it is commonest between 1200 and 1350 m (Burger & van der Werff 1990). In earlier publications (Wheelwright *et al.* 1984, Wheelwright 1986b) the species was referred to as *O. bernouliana* (sic), now considered by Burger & van der Werff (1990) to be a distinct species.

O. tenera begins to flower at a median age of 5 years, height of 1.5 m, and DBH of 2–3 cm (Wheelwright & Bruneau, 1992). Its major pollinators are small, generalist wasps, bees, and flies. The fruits, borne in shallow receptacles at the end of enlarged, scarlet pedicels, become full-size about 6 months after pollination. As they ripen, they change in color from green to black over a 2–3 day period. The principal dispersers of *O. tenera* are resplendent quetzals (*Pharomachrus mocinno*), three-wattled bellbirds (*Procnias tricarunculata*), emerald toucanets, keel-billed toucans (*Ramphastos sulfuratus*) and black guans

(*Chamaepetes unicolor*) (Wheelwright *et al.* 1984). Birds swallow the fruits whole and regurgitate the large single seeds after about 50 min (Wheelwright 1991).

Study site and procedures

In 1980, as part of a study of the phenology of the Lauraceae (Wheelwright 1986b), I marked every reproductively mature individual of *O. tenera* encountered in 15 km^2 of lower montane wet and rain forests around Monteverde, Costa Rica (10° 18 N, 84° 48′ W; see Lawton and Dryer (1980) & Wheelwright (1985b) for a more complete description of the site) (Fig. 1). Thereafter, all trees have been measured and their fruits counted individually at least once a year during the peak of fruit production in mid-dry season (mid-February to early March). For most of the female trees, haphazard samples of ripe fruits, sometimes comprising all that were produced by the tree, were collected in 1981, and their dimensions (length and maximum diameter of fruits and seeds) were measured to 0.1 mm using dial calipers; mass was determined to 0.1 g using portable balances. At least 10 fruits were measured per plant, except for those with crop sizes less than 10 fruits. I measured the fruits of several plants in at least 5 different years spanning an 11-year period.

To determine the factors influencing fruit removal by birds, I monitored the disappearance of ripe fruits from 29 trees in 1980–1. For trees with crop sizes less than the median for the population (30 fruits), I observed all fruits; for trees with larger crop sizes, I followed a subset of accessible fruits (those borne at heights lower than 3 m). Fruits were identified by writing a number in black permanent marker on the pedicel. For 524 fruits the diameter at the widest point was measured with calipers, as was the length of the exposed fruit (from the margin of the receptacle to the tip). From late December, 1980 until late May, 1981, 597 fruits were censused at 3-day intervals. The dates of ripening and removal (or damage caused by animals or desiccation) were determined. Because the stages of ripening were signaled by gradual color changes, 3-day censuses were frequent enough to determine the exact day when fruits became fully ripe. Removal was obvious because of the empty receptacle left on the tree; I checked beneath plants for evidence that fruits had been dropped rather than dispersed. Pulp-feeding animals left characteristic marks of their activity (pericarp disks excised or totally naked seeds left by leaf-cutter ants (*Acromyrmex* spp.), patchy holes in the pericarp excavated by weevils (Curculionidae), and incisor scars produced by arboreal rodents. Virtually all removal of fruits was by birds, as indicated by direct observations of feeding behavior by birds (Wheelwright 1991), negligible fruit removal rates during the night, absence of seeds in mammal feces or beneath bat roosts, and examination of isolated dispersed seeds showing that the pulp had been cleanly removed.

In 1981 and 1984 I established experimental populations consisting of 185 *O. tenera* seedlings grown from seeds of known maternity, representing 11 maternal families; 64 plants survived until 1991. Of those, most proved to be males and a few had not reached reproductive maturity by 1991. For 25 female F_1 plants whose maternal phenotypes were known, fruits were collected and measured. The mean fruit diameters of siblings were then averaged within sibships and regressed against maternal mean fruit diameters to estimate heritability (h^2). Because pollen donors were unknown, the slope of the single parent-offspring regression equation was multiplied by 2 to calculate h^2 (Falconer 1989). Note that this method potentially overestimates h^2 for several reasons. First, it overlooks possible maternal genetic effects (e.g., cytoplasmic genes; long-term effects caused by maternal provisioning of the seed are less likely). Second, it assumes that F_1 trees grown from seeds taken from the same parent tree were half-sibs, when their relatedness may actually be greater if they shared the same pollen donor. Another potential problem is that fruits produced by F_1 trees may have resulted from mating between siblings in the experimental plots. In spite of these potential confounding factors, a strong parent-offspring resemblance in fruit traits almost certainly would indicate an appreciable genetic basis to phenotypic variation. Regressions and other statistical analyses were performed on Statview (Abacus 1988) on a Macintosh computer. Unless noted otherwise, descriptive statistics are presented as $\bar{x}$ (± 1 SD).

Results

Variation in fruit diameter

Within the natural *O. tenera* population, there was wide variability in fruit size. The smallest fruits (1.41 cm in diameter) were only 60% the size of the largest (2.37 cm; $\bar{x} = 1.82 \pm 0.14$ cm, coefficient of variation [CV] = 7.7%, N = 674 fruits). Although none of the fruits was too small to be eaten profitably by the major seed dispersers of *O. tenera*, the bigger fruits were too broad to be swallowed by bellbirds and perhaps toucanets (Wheelwright 1985a).

Fruit diameter was positively correlated with other traits, including fruit length, fruit mass, seed diameter, seed length, and seed mass (linear regression; $r^2 = 0.21$, 0.72, 0.77, 0.21, and 0.60, respectively; $P < 0.001$). From the perspective of fruit profitability (Herrera 1981), fruit diameter was positively correlated with pulp mass (fruit mass − seed mass) ($r^2 = 0.56$; N = 415 fruits; $P < 0.001$) (Fig. 2), although it was a relatively

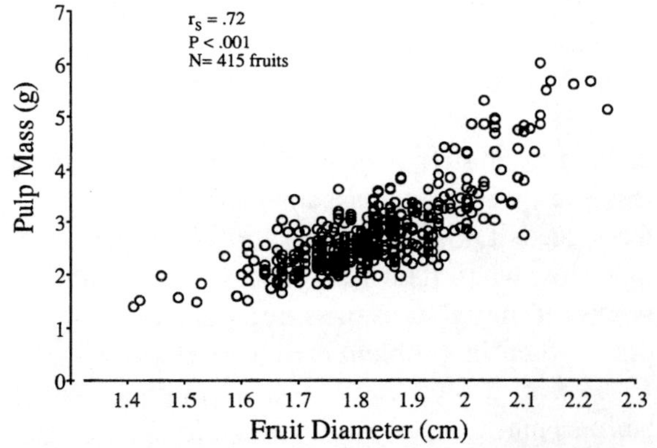

Fig. 2. Relationship between fruit diameter and pulp mass (fresh fruit mass − seed mass) in *Ocotea tenera*.

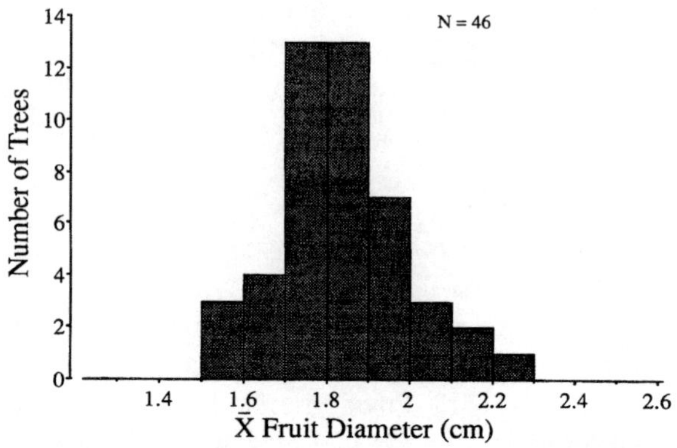

Fig. 3. Frequency distribution of the mean fruit diameters for 46 *Ocotea tenera* trees.

poor predictor of seed mass/fruit mass ($r^2 = 0.01$; $P < 0.05$). A relatively small increase in diameter was associated with a relatively large gain in edible pulp. For example, a 6% difference in fruit diameter (e.g., 1.8 cm to 1.9 cm in diameter) meant an 18% increase in pulp mass (2.61 to 3.08 g).

Trees differed substantially in the size of the fruits they produced. Some trees had characteristically small fruits ($\bar{x} = 1.64$ cm in diameter) whereas others produced large fruits ($\bar{x} = 2.22$ cm; $\bar{x}$ of tree means = 1.88 ± 0.15 cm; one-way ANOVA comparing trees: $P < 0.001$) (Fig. 3). Fruit diameter within trees varied significantly among years; the magnitude of year-to-year differences varied between trees (two-way ANOVA: among years: P = 0.004; among trees: $P < 0.001$; year-by-tree interaction: P = 0.002). Nonetheless, absolute average fruit size difference within trees between years were slight (Fig. 4), and a tree's

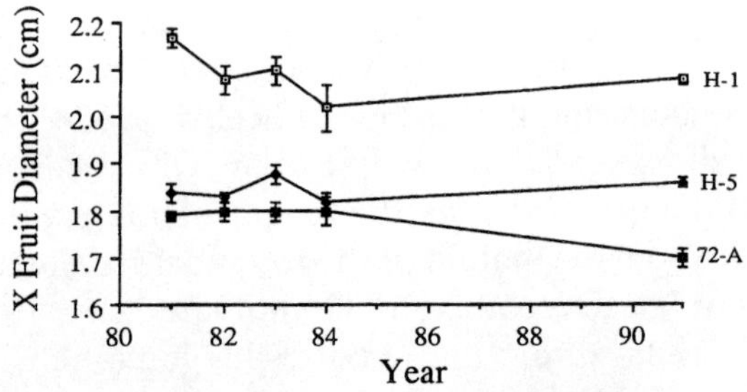

Fig. 4. Mean fruit diameter in three *Ocotea tenera* trees whose fruits were measured in at least five years between 1981 and 1991. Error bars represent ± 1 SE.

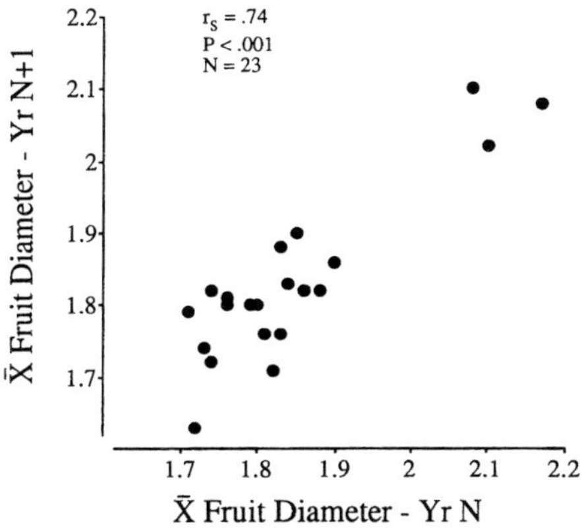

Fig. 5. Relationship between mean fruit diameter in a given year and mean diameter in subsequent years for eight trees whose fruits were measured in at least four years.

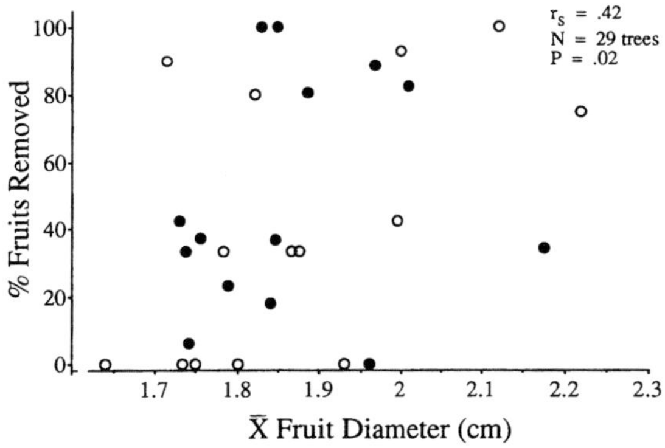

Fig. 6. Relationship between mean fruit diameter and the proportion of the fruit crop removed by avian seed dispersers within individual trees. Open circles represent trees with crop sizes less than the median for the population (30 fruits); closed circles represent trees with crop sizes exceeding the median.

mean fruit diameter in a given year was a good predictor of its fruit diameter in subsequent years (Fig. 5).

There was also appreciable variation in fruit diameter within individual trees. In trees 72a and H-1, for example, the largest fruits had diameters 22–29% broader than the smallest fruits (range = 1.55–2.00 and 1.95–2.37 cm; $\bar{x} = 1.79 \pm 0.08$ and 2.11 ± 0.10 cm; CV = 4.5 and 4.7%, respectively). Thus, if fruit size influenced birds' choices of fruits, their preferences could be expressed at the level of trees or individual fruits within trees.

Fruit diameter and removal

Ocotea tenera fruits ripened over a 5-month period from late December into May, with 60.7% of fruits ripening during a 4-week period. Counting all marked fruits, only 46.2% were removed by avian seed dispersers, as judged by empty receptacles and no evidence of damage, predation, or dropped fruits. About an equal proportion of the remaining fruits was damaged by herbivores or fungi, or stayed uneaten on the plant until they desiccated or were aborted.

The proportion of a tree's fruit crop that was removed and presumably dispersed by birds varied from 0 to 100% (Fig. 6). Overall, fruits that were removed were slightly larger in diameter than fruits that were ignored or preyed upon (1.85 ± 0.14 cm vs. 1.81 ± 0.14 cm; N = 213 vs. 311; ANOVA: $P < 0.001$). Much of this was due to differences in removal rates between plants differing in mean fruit diameter, and did not necessarily indicate selectivity by birds within a canopy. Thus, there was a positive correlation between mean fruit diameter and the proportion of the fruit crop that was removed (Fig. 6). Nonetheless, in some of the trees, birds preferentially took larger fruits (ANOVA: $P < 0.05$).

For all fruits that were eventually removed, there was no correlation between diameter and the number of days between ripening and removal ($r_s = 0.07$; N = 210 fruits; $P = 0.31$). Nor was mean fruit diameter within a tree related to the mean duration of fruits on the tree ($r_s = -0.08$; N = 29 trees; $P = 0.72$) (Fig. 7).

The fact that there was a positive correlation between fruit diameter and the probability of removal, but there was no correlation between fruit diameter and the duration of ripe fruits on plants, indicates that other factors influence removal by birds and points out the necessity of analyzing several variables simultaneously. Therefore I performed a multiple regression with the proportion of the fruit crop removed as the dependent variable and each tree's mean fruit diameter, mean pulp mass, mean seed mass/fruit mass, crop size, and mean ripening date as independent variables.

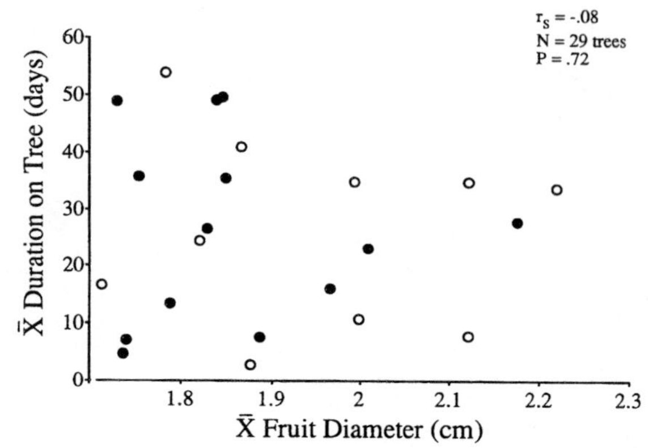

Fig. 7. Relationship between mean fruit diameter and the mean number of days between ripening and removal within individual trees. Open and closed circles as in Fig. 6.

Controlling for other variables, mean diameter continued to be positively correlated with the proportion of fruits removed (P = 0.018). The only other variable that had a statistically significant effect on the proportion of fruits removed was ripening date (P = 0.005): trees whose fruit crops ripened early in the season had a greater proportion of their fruits removed than those whose fruits ripened relatively late, regardless of crop size (Fig. 8). Overall, the multiple regression model including all independent variables was statistically significant (P = 0.005).

When a multiple regression was performed with the same independent variables and mean dura-

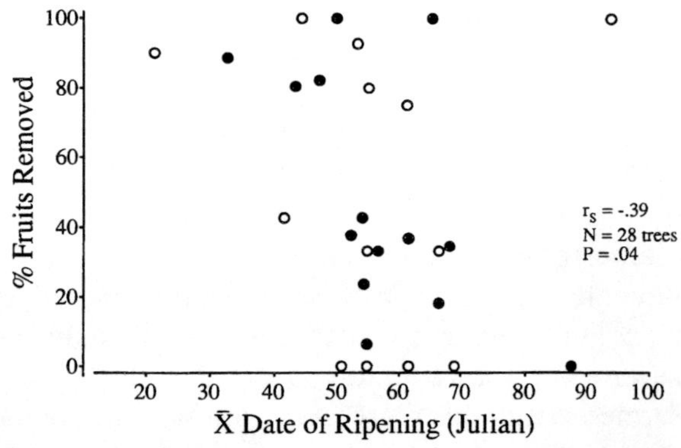

Fig. 8. Relationship between mean date of ripening and the proportion of the fruit crop removed by avian seed dispersers within individual trees. Open and closed circles as in Fig. 6.

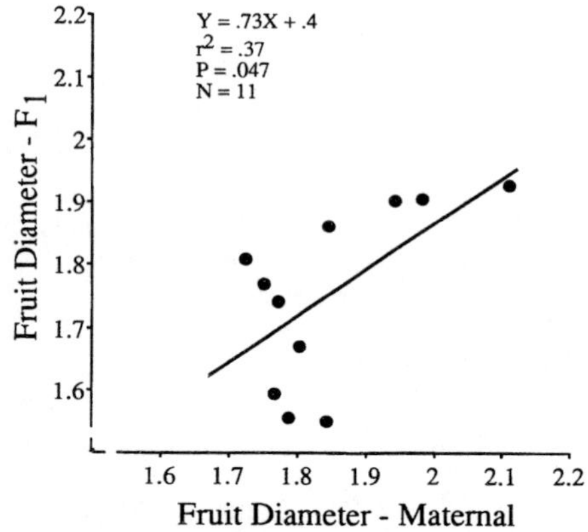

Fig. 9. Regression of mean maternal fruit diameter against mean offspring fruit diameter (averaged across siblings). N = 11 maternal lineages. Twice the slope in single-parent-offspring regressions provides an estimate of h^2 (c. 1.0).

tion of fruits as the dependent variable, only ripening date was related to mean duration (P = 0.04), and overall the model was not statistically significant (P = 0.31).

Heritability of fruit diameter

Upon reaching reproductive maturity, trees grown from seeds of known maternity produced fruits that closely resembled their mothers' fruits in diameter. Regression of the mean fruit diameters of maternal trees against those of their offspring yielded a significant positive correlation with a slope of 0.73 ($\pm$0.32; N = 11 maternal lineages; r^2 = 0.30; P = 0.047) (Fig. 9). Twice the slope gives an estimate for h^2 that exceeds 1.0, indicating that a large proportion of the variation in fruit diameter in *O. tenera* is genetic. In general fruit and seed dimensions in this species show high resemblance between female parents and their offspring (h^2 = 0.61 → 1.0; Wheelwright unpubl. data), as do non-reproductive traits such as leaf shape (Wheelwright *et al.* unpubl.).

Discussion

To the extent that a fruit's characteristics affect its detection by potential seed dispersers, influences its probability of being eaten or the time between ripening and removal, or determines how it is handled externally or processed internally, such traits may be the target of natural selection. The implications of such a conclusion are that the day-to-day interactions between fruit-eating animals and fruiting plants that biologists commonly document can have evolutionary significance. Moreover, much of the richness of reproductive features among animal-dispersed plants may legitimately be considered 'adaptations,' shaped by generations of interactions with animals acting as seed vectors. This requires, of course, that fruit traits have a genetic basis. Research in the field of seed dispersal and plant-animal interactions has explicitly or implicitly assumed that fruit characters influence the probability of dispersal, and that much of the phenotypic variation in those characters is heritable, but the longevity of most trees and the brevity of most studies have prevented a full assessment of such assumptions.

This study demonstrates that in at least one tree dispersed by birds, a fundamental fruit character, diameter, shows wide variation between and within individuals, in contrast to earlier studies which emphasized the constancy of seed (and by implication, fruit) size within species (Harper *et al.* 1970). Other species in the Lauraceae are equally variable in fruit morphology (Mazer & Wheelwright, 1993). Faced with a hierarchy of choices between trees that have distinct fruit traits, and within trees that produce fruits of varying sizes, avian seed dispersers appear to be discriminating foragers. Birds favored fruits of larger-than-average diameter, presumably because such fruits offered more pulp per fruit (although not necessarily a more profitable pulp-:seed ratio) (Herrera 1981; Howe and Vande Kerckhove 1981; Piper 1986; Levey 1987; Murray *et al.* 1992). Birds' fruit-size preferences mean that plants with less-favored fruits disperse few or no seeds, and that, within canopies, some fruits are ignored and exposed for longer periods to pulp- or seed-damaging herbivores. If birds' preferences are relatively stable, they should have a compounding effect on a plant's seed dispersal success year after year, because mean fruit diameter changes little throughout the course of a tree's life, at least in *O. tenera*.

Mean fruit diameter also changes little between generations within a lineage; in other words, it is highly heritable in *O. tenera*. Heritability expresses the percentage of phenotypic variation of a trait that is attributable to genotype (the additive genetic variance). Although my estimate of h^2 may be biased upward because of maternal effects and the assumption that F_1 trees were only half-sibs (see Methods), the results clearly show a strong genetic basis to fruit morphology. The significance of demonstrating that fruit diameter in *O. tenera* is heritable is that relatively small selective pressures can quickly drive evolutionary change in such traits. Selection on traits that are genetically correlated with fruit diameter could have effects similar to direct selection on fruit diameter (Primack 1987). Traits that are most closely related to fitness, as fruit diameter might be expected to be, are ordinarily expected to have relatively low h^2 because most of the additive genetic variance has already been fixed by natural selection (Falconer 1989), but sufficient variability may be introduced by mutation or maintained by fluctuating selection to result in high heritabilities of traits such as fruit diameter.

If *Ocotea tenera* is representative of other species in the Lauraceae, which have fruits of similar size and form, share the same avian seed dispersers, and are an important component of neotropical forests (Wheelwright *et al.* 1984), there may be the potential for evolutionary responses on the part of those plants to birds' fruit-size preferences. Whether other plant taxa show equivalent variation in reproductive traits, interact with their seed dispersers in comparable ways, and exhibit high heritabilities in such traits is still unknown, but it is not unreasonable to expect qualitatively similar patterns (cf. Mazer & Wheelwright 1993).

The results of this study should not be construed to imply that directional selection on fruit size or any other reproductive character is likely to be strong or constant. Depending upon whether one views the amount of variation in fruit size found in this study as wide or narrow, the high heritability of fruit traits could be interpreted as reflecting stabilizing selection (Gorchov, pers. comm.). In any case, the maintenance of extensive phenotypic variation in fruit size within the *O. tenera* population appears to be inconsistent with strong directional selection, and it may be explained by a variety of factors. First, this study has not answered a crucial question: does speedy fruit removal in fact result in measurably improved plant fitness? If it does not, directional selection on fruit diameter would be weak, despite apparent fruit size preferences by birds. Second, there is clearly ample phenotypic variation in fruit form in *O. tenera*, and regressions of maternal traits against that of their daughters established that the variation is heritable, but to what extent is variation within fruit crops caused by differences in pollen donors, and what is the genetic basis of those effects? A recent study using molecular techniques to examine population genetic structure and investigate paternity and maternity in seedlings and saplings in *O. tenera* may be able to fill in many important details about the relationship between fruit removal by birds and plant fitness, and shed light on the paternal effects on fruit form (Gibson and Wheelwright, unpubl.). Third, even though this study shows that fruit diameter hardly fluctuates between years, and other studies suggest that interactions with seed dispersers, or at least assemblages of dispersers, remain quite constant in the face of wide annual variation (Howe 1983), other parameters change extensively, such as a plant's fecundity (Wheelwright 1986b, unpubl.) and, perhaps more importantly, the fecundity of its neighbors (Howe 1983; Sargent 1990). Fourth, for many reasons plants may be constrained in their evolutionary response to selection by avian seed dispersers. Among these are fluctuating selection in space and time, opposing selection, asymmetries in generation lengths of animals and the plants whose seeds they disperse, and genetically correlated characters (Wheelwright & Orians 1982; Herrera 1985; Wheelwright 1988b; Mazer & Wheelwright 1993). Slight preferences on the part of birds for a trait such as fruit size, for example, may be overshadowed by their response to differences in fecundity between plants (Howe 1983; Murray 1987; Jordano 1987a).

General rather than specific adaptation of plants to their seed dispersers is the more likely outcome of evolutionary interactions, and may explain both diversity and convergence in fruit traits within habitats and between taxonomically distinct plant communities. The general premises of models of coevolution between mutualists look more credible in light of this study because, in at least one neotropical tree species, reproductive traits are highly variable, they matter to birds, and they have a genetic basis.

Acknowledgements

I thank Monteverde residents, especially D. Lee, S. Perly, the Hoges, Hollanders, Stuckeys and Trostles, for permitting me to study on their land; F. Joyce, Jr., S. Sargent, J. Wolfe, and W. Zuchowski for helping to establish and monitor the experimental plots; D. Gorchov, T. Fleming and A. Estrada for helpful comments on an earlier draft; V. Sork for illuminating discussions about maternal genetic effects; and the Organization for Tropical Studies, the Douroucouli Foundation, the National Science Foundation, the Council for International Exchange of Scholars, and Bowdoin College for support of my research.

References

Bonaccorsco, F.J. 1979. Foraging and reproductive ecology in a Panamanian bat community. Bull. Florida State Mus. 24: 359–408.

Burger, W. & H. van der Werff. 1990. Lauraceae. Flora Costaricensis. Fieldiana (Botany New Series) 23: 1–121.

Debussche, M. & P. Isenmann. 1989. Fleshy fruit characters and the choices of bird and mammal seed dispersers in a mediterranean region. Oikos 56: 327–338.

Denslow, J.S. 1987. Fruit removal rates from aggregated and isolated bushes of the red elderberry, *Sambucus pubens*. Can. J. Bot. 65: 1229–1235.

Dowsett-Lemaire, F. 1988. Fruit choice and seed dissemination by birds and mammals in the evergreen forests of Upland Malawi. Revue d'Ecologie (Terre Vie) 43: 251–286.

Estrada, A. & T.H. Fleming (eds.). 1986. Frugivores and Seed Dispersal. W. Junk, Amsterdam.

Falconer, D.S. 1989. Introduction to quantitative genetics. Longman, Essex.

Fleming, T.H. 1986. Opportunism versus specialization: the evolution of feeding strategies in frugivorous bats. In: Estrada, A. & T.H. Fleming (eds.), Frugivores and Seed Dispersal. W. Junk, Amsterdam. Pp. 105–118.

Foster, M.S. 1990. Factors influencing bird foraging preferences among conspecific fruit trees. Condor 92: 844–854.

Gautier-Hión, A., Duplantier J.M., Quris R., Feer F., Sourd C., Decoux J.P., Dubost G., Emmons L., Erard C. & Hecketsweiler P. 1985. Fruit characters as a basis of fruit choice and seed dispersal in a tropical forest vertebrate community. Oecologia 65: 324–337.

Gorchov, D.L. 1985. Fruit ripening asynchrony is related to variable seed number in *Amelanchier* and *Vaccinium*. Amer. J. Bot. 72: 1939–1943.

Harper, J.L., Lovell P.H. & Moore K.G. 1970. The shapes and sizes of seeds. Ann. Rev. Ecol. Syst. 1: 327–356.

Herrera, C.M. 1981a. Fruit variation and competition for dispersers in natural populations of *Smilax aspera*. Oikos 36: 51–58.

Herrera, C.M. 1981b. Are tropical fruits more rewarding to dispersers than temperate ones? Amer. Natur. 118: 896–907.

Herrera, C.M. 1985. Determinants of plant-animal coevolution: the case of mutualistic dispersal of seeds by vertebrates. Oikos 44: 132–141.

Herrera, C.M. 1988. The fruiting ecology of *Osyris quadripartita*: individual variation and evolutionary potential. Ecology 69: 233–249.

Herrera C.M. 1989. Frugivory and seed dispersal by carnivorous mammals and associated fruit characteristics in undisturbed mediterranean habitats. Oikos 55: 250–262.

Hoppes, W.G. 1987. Pre- and post-foraging movements of frugivorous birds in an eastern deciduous forest woodland, USA. Oikos 49: 281–290.

Howe, H.F. 1983. Annual variation in a neotropical seed-dispersal system. In: Sutton, S.L., Whitmore T.C. & Chadwick A.C. (eds.) Tropical rainforest ecology and management. Blackwell Scientific Publications, London. Pp. 211–227.

Howe, H.F. & Richter W.M. 1982. Effects of seed size and seedling size in *Virola surinamensis*: a within and between tree analysis. Oecologia 53: 347–351.

Howe, H.F. & Vande Kerckhove G.A. 1981. Removal of wild nutmeg (*Virola surinamensis*) crops by birds. Ecology 62: 1093–1106.

Janson, C.H. 1983. Adaptation of fruit morphology to dispersal agents in a neotropical forest. Science 219: 187–189.

Johnson, R.A., Willson M.F., Thompson J.N. & Bertin R.I. 1985. Nutritional values of wild fruits and consumption by migrant frugivorous birds. Ecology 66: 819–827.

Jordano, P. 1987a. Avian fruit removal: effects of fruit variation, crop size, and insect damage. Ecology 68: 1711–1723.

Jordano, P. 1987b. Frugivory, external morphology and digestive system in mediterranean sylviid warblers *Sylvia* spp. Ibis 129: 175–189.

Jordano, P. 1989. Pre-dispersal biology of *Pistacia lentiscus* (Anacardiaceae): cumulative effects on seed removal by birds. Oikos 55: 375–386.

Jordano, P. 1992. Fruits and frugivory. In: Fenner, M. (ed.). Seeds: the ecology of regeneration in plant communities. CAB International, London. *In press.*

Katusic Malmborg, P. & Willson M.F. 1988. Foraging ecology of avian frugivores and some consequences for seed dispersal in an Illinois woodlot. Condor 90: 173–186.

Lambert, F. 1989. Fig-eating birds in a Malaysian lowland rain forest. J. Trop. Ecol. 5: 401–412.

Lee, W.G., Wilson J.B. & Johnson P.N. 1988. Fruit color in relation to the ecology and habit of *Coprosma* spp. (Rubiaceae) in New Zealand. Oikos 53: 325–331.

Levey, D.J. 1987. Seed size and fruit-handling techniques of avian frugivores. American Naturalist 129: 471–485.

Levey, D.J. & Moermond T.C. 1984. Fruit choice in neotropical birds: the effect of distance between fruits on preference patterns. Ecology 65: 84850.

Loiselle, B.A. & Blake J.G. 1990. Diets of understory fruit-eating birds in Costa Rica: seasonality and resource abundance. Studies Avian Biol. 13: 91–103.

Mazer, S.J. & Wheelwright N.T. Fruit size and shape: allometry at different taxonomic levels in bird-dispersed plants. Evolutionary Ecology. In press.

Murray, K.G. 1987. Selection for optimal fruit crop size in bird-dispersed plants. Amer. Natur. 129: 19–31.

Murray, K.G. 1988. Avian seed dispersal of three neotropical gap-dependent plants. Ecol. Monogr. 58: 271–298.

Murray, K.G., Winnett-Murray K., Cromie E.A. & Minor M. 1993. Feeding preferences of frugivorous birds: the influence of fruit color, seed packaging, and sugar concentration. Vegetatio. *In press.*

Obeso, J.R. 1989. Fruit removal and potential seed dispersal in a southern Spanish population of *Berberis vulgaris* (Berberidaceae). Acta Oecol. Oecol. Plant 10: 321–328.

Piper, J.K. 1986. Seasonality of fruit characters and seed removal by birds. Oikos 46: 303–310.

Piper, J.K. 1986. Effects of habitat and size of fruit display on removal of *Smilacina stellata* (Liliaceae) fruits. Can. J. Bot. 64: 1050–1054.

Pratt, T.K & Stiles E.W. 1985. The influence of fruit size and structure on composition of frugivore assemblages in New Guinea. Biotropica 17: 314–321.

Primack, R. 1987. Relationships among flowers, fruits, and seeds. Ann. Rev. Ecol. Syst. 18: 409–430.

Sargent, S. 1990. Neighborhood effects on fruit removal by birds: a field experiment with *Viburnum dentatum* (Caprifoliaceae). Ecology 71: 1289–1298.

Snow, B.K. & Snow D.W. 1988. Birds and berries. T. & A.D. Poyser, Calton.

van der Pijl, L. 1982. Principles of dispersal in higher plants. Springer-Verlag, Berlin.

Wheelwright, N.T. 1985a. Fruit size, gape width, and the diets of fruit-eating birds. Ecology 66: 808–818.

Wheelwright, N.T. 1985b. Competition for dispersers, and the timing of flowering and fruiting in a guild of tropical trees. Oikos 44: 465–477.

Wheelwright, N.T. 1986a. The diet of American Robins: an analysis of U.S. Biological Survey records. Auk 103: 710–725.

Wheelwright, N.T. 1986b. A seven-year study of individual variation in fruit production in tropical bird-dispersed tree species in the family Lauraceae. In: Estrada, A. & Fleming T.H. (eds.) Frugivores and Seed Dispersal. W. Junk, Amsterdam. Pp. 19–35.

Wheelwright, N.T. 1988a. Fruit-eating birds and fruiting plants in the tropics and temperate zone. Trends in Ecology and Evolution 3: 270–274.

Wheelwright, N.T. 1988b. Four constraints on coevolution between plants and their seed dispersers: a tropical case history. Proceedings of the XIX International Ornithological Congress, pp. 827–845.

Wheelwright, N.T. 1991. How long do fruit-eating birds stay in the plants where they feed? Biotropica 23: 29–40.

Wheelwright, N.T. & Bruneau, A., 1992. Population sex ratios and spatial distribution of *Ocotea tenera* (Lauraceae) trees in a tropical forest. Journal of Ecology 80: 425–432.

Wheelwright, N.T., W.A. Haber, K.G. Murray, and C. Guindon. 1984. Tropical fruit-eating birds and their food plants: a survey of a Costa Rican lower montane forest. Biotropica 16: 173–191.

Wheelwright, N.T. & Orians G.H. 1982. Seed dispersal by animals: contrasts with pollen dispersal, problems of terminology, and constraints on coevolution. American Naturalist 119: 150–164.

White, D.W. & Stiles E.W. 1990. Co-occurrences of foods in stomachs and feces of fruit-eating birds. Condor 92: 291–303.

White, D.W. & Stiles E.W. 1991. Fruit harvesting by American Robins: influence of fruit size. Wils. Bull. *In press.*

Willson, M.F. 1986. Avian frugivory and seed dispersal in eastern North America. Current Ornithology 3: 223–279.

Worthington, A.H. 1989. Adaptations for avian frugivory: assimilation efficiency and gut transit time of *Manacus vitellinus* and *Pipra mentalis*. Oecologia 80: 381–389.

Part 3

Frugivore strategies

Introduction

Papers in this section deal primarily with the animal side of the fruit-frugivore mutualism. Questions addressed here deal with spatial aspects of foraging behavior, proximate factors governing fruit choice, geographic variation in the disperser/predator roles played by frugivores, and the response of frugivores to forest fragmentation.

Loiselle and Blake examine the correspondence between the spatial distributions of frugivorous understory birds and their food plants in a Costa Rican wet tropical forest. Although they document the expected congruence, correlations among and between bird and plant species were not particularly high. They discuss the seed dispersal implications of their results and suggest that the spatial distribution of fruiting plants may be more important than crop size in determining fruit removal rates by birds.

The diets of vertebrate and certain invertebrate frugivore/seed eaters sometimes overlap and potentially place opposing selection pressures on fruiting plants. Traveset has studied bird-wasp-*Pistacia terebinthus* interactions in Spain and finds that the influence of avian and insect frugivore guilds on each other is low because birds and wasps feed on different ripeness classes of fruits. Wasp predation can suppress maturation in a highly variable fraction of fruits, but even so, birds disperse over one-third of the fruits set by this plant. Traveset concludes that the ability of wasps to suppress fruit maturation is not the result of an evolutionary interaction with frugivorous birds.

Three papers deal with proximate factors that influence fruit choice in birds. Martinez del Rio and Restrepo review the digestive physiology of frugivorous birds and point out that, owing to differences in the ways different birds process fruit in their guts, fruit profitability cannot be calculated solely from fruit traits. They conclude that the diversity and complexity of digestive characteristics of frugivores probably results in much richter patterns of interactions between birds and fruits than previously hypothesized. Murray *et al.* report on the results of fruit choice experiments using two species of temperate thrushes in which seed size and fruit color were manipulated. Robins preferred larg-seeded fruits but had an even stronger preference for blue fruits regardless of seed size. These results indicate that fruit choice in these birds is complex and probably is influenced by several factors simultaneously. Finally, E. W, Stiles examines the relationship between lipid content in fruits and food choice in temperate birds. He notes that highly frugivorous birds feed heavily on high lipid fruits in the field and in laboratory experiments and indicates that digestive physiology probably influences food choice in these birds.

Gautier-Hion *et al.* have studied the diets of two species of *Cercopithecus* monkeys in Gabon and neighboring Zaire and report that the species are mainly fruit pulp-eaters in the former country and seed-eaters in the latter. Geographic differences in the abundance and diversity of fleshy fruits, ultimately related to differences in soil fertility, seem to dictate whether these monkeys are seed dispersers or seed predators in different plant communities.

The fruit-frugivore mutualism appears to be vulnerable to disruption resulting from the fragmentation of tropical forests. Estrada *et al.* have documented the response of frugivorous birds and mammals to forest fragmentation in Veracruz, Mexico. Their results indicate that frugivores can be common in forest fragments of < 100 ha; certain species are also common in agricultural 'islands' such as coffee, cacao, and citrus groves. Forest and agricultural islands provide food and temporary shelter for many frugivores and 'attract' a wide diversity of seeds from forest interior plants. They conclude that agricultural island provide important connections among fragmented animal populations.

Vegetatio **107/108**: 177–189, 1993.
T. H. Fleming and A. Estrada (eds). Frugivory and Seed Dispersal: Ecological and Evolutionary Aspects.

Spatial distribution of understory fruit-eating birds and fruiting plants in a neotropical lowland wet forest

Bette A. Loiselle & John G. Blake
Dept. Biology, 8001 Natural Bridge Road, University of Missouri-St. Louis, St. Louis, MO 63121, USA

Keywords: Costa Rica, Diet, Fruit-eating birds, Fruiting plants, Seed dispersal, Spatial distribution

Abstract

Spatial distribution of fruit-eating birds and fruiting shrubs of the *Melastomataceae* and *Rubiaceae* were examined on a 10 ha plot in tropical lowland wet forest of Costa Rica. Many plant species and most birds exhibited considerable spatial variation in their occurrence on the plot, as indicated by the distribution patterns of shrubs with ripe fruits and captures in mist nets, respectively. In many cases, captures of fruit-eating birds were correlated with abundance of fruiting plants, particularly for species that rely heavily on fruits. In general, fruit-eating birds concentrated their use of the plot to areas rich in fruiting shrubs. This differential use of certain areas likely results in differential visitation to fruiting plants located in these areas and in a heterogeneous dissemination of seeds into the habitat.

Introduction

Patterns of habitat use influence the distribution and abundance patterns of animals and thereby have important consequences for the extent and types of interactions that occur both within and among species (Cody 1985). The processes by which animals select and use habitats can influence distribution and abundance of prey, relative amounts of outcrossing for plants that depend on animals for pollination, and patterns of seed dissemination by dispersers. Understanding patterns and processes of habitat selection is a major goal of ecology (Rosenzweig 1985).

Habitat selection often depends to some degree on the abundance or availability of resources, particularly if those resources are limiting. Resources that are patchily distributed in space or time may produce similarly patchy distributions among consumers of that resource (Wheelwright 1983; Ford & Paton 1985; Herrera 1985; Terborgh 1985). In tropical forests, fruit is an important resource for many species of birds and other animals (Terborgh 1983; Karr *et al.* 1990). Abundance of fruit and distribution of plant species influence the distribution and abundance of birds (Herrera 1985; Levey 1988; Loiselle & Blake 1991; Blake & Loiselle 1991), thereby affecting the composition of bird communities.

Frugivores respond at both spatial (e.g., movement among habitats or microhabitats) and temporal (e.g., changes in abundance over time) levels to changes in abundance of fruits (e.g., Leighton & Leighton 1983; Terborgh 1983; Wheelwright 1983; Fleming 1988; Blake & Loiselle 1991; Loiselle & Blake 1991). For example, capture rates of avian frugivores are greatest in lowland wet forest in Costa Rica during periods of fruit abundance; increased capture rates largely reflect immigration into lowland habitats of frugivores from higher elevation forests (Loiselle & Blake 1991). Similarly, frugivores may switch habitats at the same elevation (e.g., forest to second-growth) in response to periods of fruit scarcity in mature forests (Wheelwright 1983; Blake & Loiselle 1991; see also Leighton & Leighton 1983; Terborgh 1983). On a smaller scale, differences in the abundance of flowers and fruits between treefall gaps and forest undergrowth habitats may affect capture rates of birds, particularly nectarivores and frugivores (Stiles, in Herrera 1985; Blake & Hoppes 1986; Levey 1988; but see Schemske & Brokaw 1981).

Here we examine within-habitat spatial distributions of fruiting shrubs and frugivorous birds in a tropical lowland wet forest. We ask whether or not distribution of fruiting shrubs and birds is random or clumped and whether or not distribution patterns of fruit can explain the spatial distribution of birds. We examine these relationships at several spatial scales. We discuss the implications of observed patterns of habitat use by birds in this forest for the seed dispersal of wet forest plants.

Methods

Study area

We conducted our study at Estación Biológica La Selva, located in the lowlands of northeastern Costa Rica, near Puerto Viejo de Sarapiqui, Provincia Heredia (10° 25′ N, 84° 01′ W). La Selva encompasses approximately 1,510 ha, of which about 64% is primary forest (Clark 1990). Height of primary forest canopy reaches 30 m to 40 m, with some emergents reaching 50 m (Hartshorn & Peralta 1988). La Selva lies across a transition between tropical premontane and tropical wet forest life zones (Holdridge 1967; Hartshorn 1983) and ranges in elevation from approximately 40 m to 200 m. The southern boundary adjoins Parque Nacional Braulio Carrillo (approximately 44,900 ha) so that continuous forest exists from approximately 40 m to 2,900 m asl (Pringle *et al.* 1984).

Rainfall at La Selva averages 3,900 to 4,000 mm/yr (Organization for Tropical Studies, unpubl. data). The dry season typically lasts from January or February to April or May, with a second, less pronounced dry season in September and October. We sampled birds and fruit during the main dry season (December-April). During our study, rainfall was below 'normal' for the December-April period (1958–1988 average = 1184 mm) during 1986 (743 mm) and 1987 (645 mm) and was close to normal during the remaining years (1988: 1129 mm; 1989: 1233 mm; 1990: 1048 mm).

The flora and fauna of La Selva are diverse (McDade *et al.* in press). The current checklist includes approximately 1,740 species of vascular plants; 1,450 of these are considered native 'forest species' (Hammel 1990). A total of 412 species of birds have been recorded at La Selva (Blake *et al.* 1990, pers. obs.), with approximately 360 occurring on a more or less regular basis. Breeding occurs throughout the year, with a peak in May and June (Levey & Stiles, in press). The breeding avifauna is seasonally enriched by additions of latitudinal (long-distance) and altitudinal migrants (Blake *et al.* 1990; Blake & Loiselle 1991, 1992; Loiselle & Blake 1991).

Birds

We established a study grid of approximately 5 ha in primary forest at La Selva in January 1985; the grid was doubled in size during September 1985. We used mist nets (12.5 m × 2.8 m, 36 mm mesh) to sample birds occurring in the undergrowth. Ground level mist nets effectively sample most birds that occur in forest understory (see discussions in Karr 1981; Levey 1988; Blake & Loiselle 1991), and the following discussion refers to birds of that stratum only. Capture of a fruit-eating bird in the understory is taken as an indication of activity at that level of the forest and implies the possibility of consumption and dispersal of fruit in the immediate area.

Nets were arranged in a 10 × 6 grid with grid points 40 m apart; 60 nets covered approximately 10 ha. Nets were centered on equidistant points and, thus, were placed irrespective of the surrounding vegetation (i.e., nets were not deliberately located adjacent to fruiting shrubs or tree fall gaps; nor were such areas deliberately avoided). We arranged our nets in this fashion in order to sample the entire plot in a systematic fashion.

We operated nets from approximately mid- to late December until April in 1986–1990; results from 1985 were excluded because only a portion of the plot was sampled. Each net was open on two days, from dawn until approximately 7 hrs later, during each sampling period (Table 1). We checked nets every hour. All captured birds were banded with numbered aluminum leg bands; checked for molt, presence of fat, and breeding condition; weighed; held (5 to 10 min) for collection of fecal samples; and released at the net-site. Fecal samples were later examined for presence or absence of fruit pulp, seeds, uric acid, and insect parts. All seeds were identified using a reference collection (Loiselle & Blake 1990).

Results presented here are based only on those birds classified as arboreal frugivores or arboreal frugivore-insectivores (see Blake & Loiselle 1991; Loiselle & Blake 1991), the most common frugivores captured in undergrowth of lowland forest at La Selva (refs. above; see also Levey 1988, Blake *et al.* 1990). We refer the reader to earlier papers for more detailed descriptions of our sampling protocols (Loiselle & Blake 1990, 1991; Blake & Loiselle 1991). English and scientific names of birds follow AOU (1983) and supplements.

Table 1. Number of arboreal frugivores captured and sampling effort in lowland wet forest at La Selva during 1986–1990. 1 net open 1 hour = 1 mist-net-hour.

Year	Number of captures	Number of sampling periods	Number of mist-net-hours
1986	232	3	2169
1987	241	4	3306
1988	192	3	2274
1989	140	2	1610
1990	64	1	669

Fruiting plants

We sampled fruiting plants in two transects (12.5 m × 2 m each) placed parallel to and 2 m away from each side of each mist net during the same periods that birds were sampled. We combined transects sampled at each net-site (50 m^2/net-site) for data analyses. We counted and identified all fruiting plants that displayed fruits below 10 m. Plant identifications were based on comparison with voucher specimens in the La Selva herbarium. Scientific names follow the unpublished checklist of La Selva plants.

Fruit transects covered a relatively small area relative to that commonly traversed by foraging birds. Our purpose, however, was to characterize the fruit resource immediately surrounding a capture site. Fruit samples from nearby net-sites were statistically independent (see autocorrelation discussion below) but we can not be certain that our 50 m^2 transects adequately sampled fruit resources at the capture site. This is an implicit assumption for this study. We discuss the validity of this assumption in the discussion.

We restrict our analyses to those plants known to be visited by birds for their fruit resources (pers. obs., Croat 1978, O. Vargas, pers. comm.). We further restrict our analyses to species of the *Melastomataceae* and *Rubiaceae* because of the importance of these families in forest undergrowth at La Selva (Opler *et al.* 1980; Loiselle 1987; Levey 1988) and in diets of fruit-eating birds (Loiselle & Blake 1990; F. G. Stiles & L. Roselli, pers. comm.). Plant species from other families important in mature forest undergrowth that produced fruit during our study are visited relatively rarely by fruit-eating birds (e.g., Piperaceae, Arecaceae) or have very different growth forms (e.g., Araceae) and, thus, are difficult to equate with dicot shrubs or treelets.

Data analysis

We combined data from all years to describe spatial variation in abundance of fruit and activity of fruit-eating birds; we have reported on temporal variation elsewhere (Blake & Loiselle 1991; Loiselle & Blake 1991; Blake *et al.* 1990). We used the number of captures at each net as our measure of bird activity. We used the maximum number of individuals with ripe fruits during any one sampling period as our measure of fruit abundance. Number of individuals producing ripe fruits was remarkably stable during the course of the study (not so for magnitude of fruiting), indicating that phenology of these plants varied little among years. These analyses do not consider variation in crop sizes of ripe fruits or in size and nutrient content of fruits. We lessened the variation among these fruit characters, however, by restricting our analyses to two plant families. We assumed that number of individuals with ripe fruit was an adequate representation of fruit abundance at these net-sites. Previous studies have shown that fruit-eating birds at this site respond most strongly to temporal changes in the abundance of ripe fruits or individuals with ripe fruits (e.g., Loiselle & Blake 1991).

We examined spatial pattern of bird activity and ripe fruit abundance using the variance to mean ratio (V/M) (Pielou 1977). This measure describes the degree to which distribution patterns are clumped. A value near 1.0 indicates a pattern consistent with random distribution, whereas a value greater than 1.0 indicates some degree of aggregation. We used Monte Carlo simulations to determine if observed V/M values were significantly different from values expected based on a random distribution of birds or plants among net sites. Birds (captures) or plants for each species were randomly assigned to nets (total numbers assigned to all nets equal to total numbers of birds or plants) and a V/M ratio was calculated. We repeated this simulation 1,000 times for each species. Observed V/M values were compared to mean values produced during the simulation using a *t*-test.

To determine if activity levels (capture rates) of different bird species (or abundances of fruiting plants) were correlated, we used product-moment or Spearman rank (r_s) correlations. The latter was used when data were not distributed normally (Wilks-Shapiro test, Sokal & Rohlf 1981). In addition, we checked for spatial autocorrelation among net-sites for all bird and fruit variables using SERIES programs in SYSTAT (Wilkinson 1988); results indicated that all variables were spatially independent.

We arranged the fruit data into a matrix consisting of the number of individuals with ripe fruits for all *Melastomataceae* and *Rubiaceae* species by net-site; this matrix was then used to ordinate net-sites. We first simplified the data matrix by combining rare species (i.e., present at only 1–2 net-sites) into groups (usually genera, e.g., *Miconia* sp.; see Appendix 1). The new matrix was defined by 60 net-sites and 13 species or species groups. Nets were ordinated using abundance and composition of fruiting plants with Bray-Curtis ordination and variance-regression criterion for end-point selection (Beals 1984). Fruiting plants (number/net-site) were then correlated with ordination axes.

We examined the relationship between birds and fruits at four levels. If birds respond to variation in the composition and abundance of fruiting shrubs at a spatial scale equivalent to net-sites, then the activity of birds (i.e., capture rates) at nets might be expected to reflect similarities and differences in abundances of fruits. Therefore, bird abundances (captures) were compared directly with fruit abundance at each net-site. Because birds may respond to abundance of more than one fruit, we then used multiple regression to relate bird abundance (dependent variable) with factor scores for net-sites along the first three ordination axes (see above). Larger spatial scales of analyses were generated first by combining fruiting plant abundance in the immediate area of the net-site (i.e., net-sites within a 40 m radius) to create a new variable for each net-site that reflected fruit abundance in a larger area. Second, we combined net-sites that clustered together in ordinations. Finally, areas with abundant fruit ('fruit-rich') were identified by inspection of the

Table 2. Number of captures, observed variance/mean ratios (V/M), and mean (standard deviation) of variance/mean ratios generated from Monte Carlo simulations (see text) for seven common fruit-eating bird species and all frugivores captured in undergrowth of lowland forest during 1986–1990. Results and significance of *t*-tests comparing observed and computer simulated variance/mean ratios also are given.

Species	No. of captures	Observed V/M	Mean V/M	*t*	P<
Mionectes oleagineus	169	2.57	1.003 (0.175)	-8.95	0.001
Pipra mentalis	311	3.04	0.995 (0.178)	11.48	0.001
Corapipo leucorrhoa	109	1.18	1.003 (0.182)	-0.97	NS
Hylocichla mustelina	116	1.33	1.000 (0.183)	-3.77	0.001
Turdus obsoletus	23	1.69	1.003 (0.177)	-3.88	0.001
Chlorothraupis carmioli	27	2.13	0.995 (0.174)	-6.52	0.001
Tachyphonus delatrii	31	2.67	0.994 (0.183)	-9.15	0.001
Total frugivores	869	3.01	1.002 (0.184)	-10.91	0.001

spatial pattern of fruit abundance on the study plot. Bird and fruit variables were totalled for these fruit-rich net-sites (19 of 60 net-sites) and for the remaining 'fruit-poor' net-sites (41 net-sites). Captures of fruit-eating birds were compared with these new fruit variables using correlation analyses or chi-square tests.

Results

Birds

We captured 869 arboreal fruit-eating birds of 20 species (Table 1). *Pipra mentalis* was the most frequently captured species and, together with six other species, accounted for 90% of all fruit-eating birds captured (Table 2). Further analyses of bird species are restricted to these seven species. Fruit occurred in >95% of all fecal samples collected from five of the seven species; *Mionectes oleagineus* and *Hylocichla mustelina* relied less heavily on fruit (Table 3). Several species frequently consumed insects; over 90% of samples from tanagers and *Hylocichla mustelina* contained at least a trace of insect parts or uric acid (Table 3). Even the highly frugivorous manakins occasionally fed on insects. Values based on percent of samples with insect parts may, however, exaggerate the importance of insects in the diet. A more accurate indication of the importance of insects in the diet would be given by percent volume of insects in diet samples (F. G. Stiles & L. Roselli, pers. comm.).

Table 3. Summary of occurrence of fruit in the diets of seven common fruit-eating birds in undergrowth of lowland forests at La Selva. Data are from dry season periods 1985–1989.

Species	No. of samples	Samples with fruit	Samples with insect parts
Mionectes oleagineus	130	110	82
Pipra mentalis	249	247	39
Corapipo leucorrhoa	76	76	20
Hylocichla mustelina	106	66	101
Turdus obsoletus	21	21	8
Chlorothraupis carmioli	33	33	31
Tachyphonus delatrii	23	22	21

Captures of all arboreal frugivores combined and most of the common fruit-eating species ex-

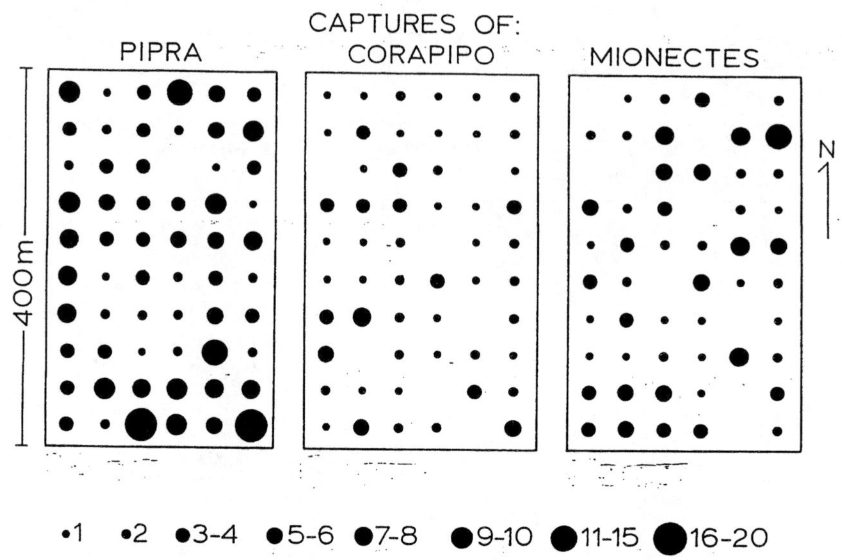

Fig. 1. Distribution of captures of *Pipra mentalis*, *Corapipo leucorrhoa*, and *Mionectes oleagineus* among 60 net-sites in lowland wet forest. Each net is approximately 40 m apart. Size of circles represents cumulative number of captures over the period from December to April 1986–1990.

amined separately showed considerable spatial variation (Figs. 1, 2). For example, the most common frugivore, *Pipra mentalis*, was captured most frequently in the southeast corner of the plot; other centers of activity included the southwest corner and the western edge of the plot (Fig. 1). Variance/mean (V/M) ratios ranged from 1.18 (*Corapipo leucorrhoa*) to over 3.00 (*Pipra mentalis* and all frugivores) (Table 2). Except for *Corapipo*, observed V/M values were significantly greater than expected if distribution of captures among nets was random (Table 2).

Captures of frugivores generally were independent. Significant (but weak) correlations in the number of captures among net-sites occurred for only three species pairs (one significant correlation would be expected from 21 comparisons based on chance alone). Captures of *Turdus obsoletus* were positively correlated with captures of both *Hylocichla mustelina* ($r_s = 0.26$, $P < 0.05$) and *Tachyphonus delatrii* ($r_s = 0.26$, $P < 0.05$); captures of *Mionectes oleagineus* and *Pipra mentalis* were significantly correlated among net-sites as well ($r = 0.28$, $P < 0.05$).

Fruiting plants

We recorded a total of 17 species of *Melastomataceae* and 11 of Rubiaceae with ripe fruits during this study. A maximum of 181 individuals (see methods) with ripe fruits occurred over all net-sites. Distribution patterns of fruiting individuals showed considerable spatial variation (Fig. 3). In particular, the eastern edge of the plot supported few plants with ripe fruits. Fruit species commonly eaten by birds (Loiselle & Blake 1990) showed different degrees of patchiness on the plot; melastome shrubs tended to be more aggregated than shrubs in the Rubiaceae (Table 4). Distributions of these shrubs on the plot were not correlated, with the exception of *Psychotria officinalis* and *Cephaelis elata*, whose abundances were positively correlated ($r_s = 0.32$, $P < 0.05$).

Individuals of *Clidemia densiflora* and *Miconia simplex* were more common on upland areas of the plot, particularly on forest ridges. *Ossaea macrophylla*, in contrast, was restricted more to forest swale areas. *Henriettea tuberculosa* and *Psychotria officinalis* were located primarily in upland

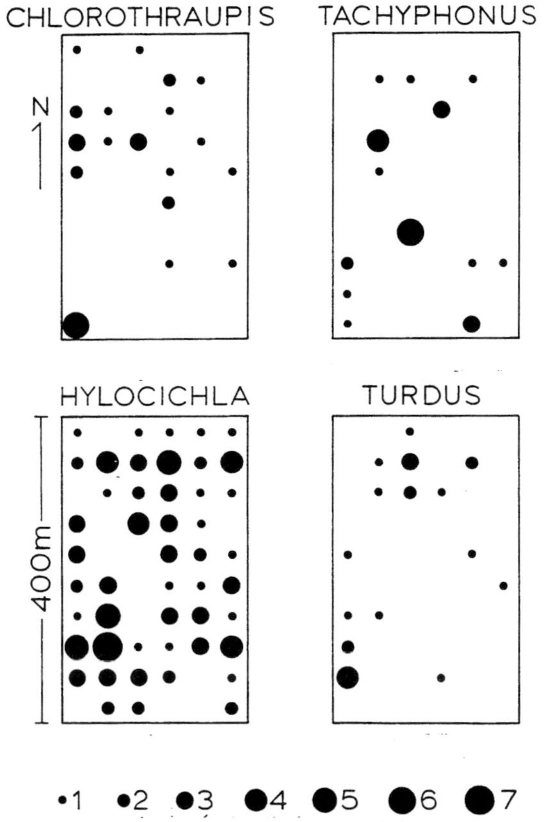

Fig. 2. Distribution of captures of *Chlorothraupis carmioli*, *Tachyphonus delatrii*, *Turdus obsoletus*, and *Hylocichla mustelina* among 60 net-sites in lowland wet forest. Each net is approximately 40 m apart. Size of circles represents cumulative number of captures over the period from December to April 1986–1990.

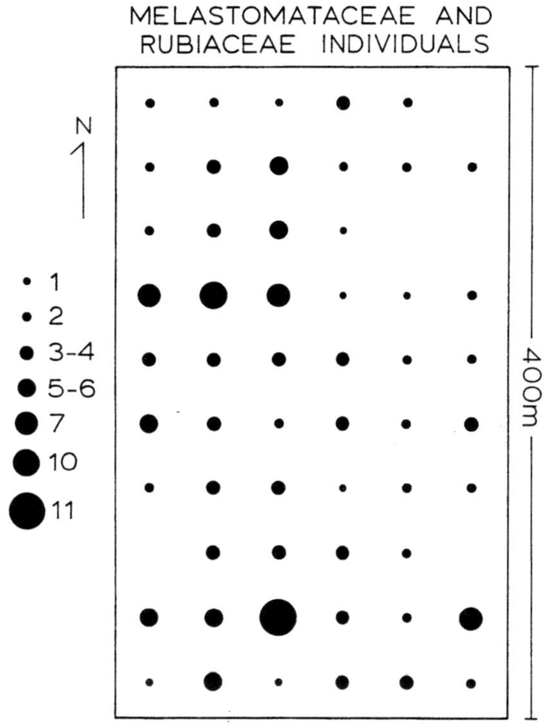

Fig. 3. Distribution of fruiting shrubs of Melastomataceae and Rubiaceae among net-sites in lowland wet forest. Numbers refer to the maximum number of fruiting shrubs displaying ripe fruits during any one sample period from December to April 1986–1990 (see text). Size of circle is indexed according to plant abundance.

areas, but not necessarily concentrated on ridges. It should be noted, however, that these spatial patterns are based solely on mature fruiting individuals and, thus, may not be representative of the population as a whole.

Ordination of nets in fruiting plant space separated net-sites into 10 groups (Fig. 4). The first three axes explained 58% of the variance in the ordination. Interpretation of these groups was based on the correlation of fruit variables with major ordination axes. Abundances of four species (*Clidemia densiflora*, *Henriettea tuberculosa*, *Leandra dichotoma*, *Coussarea hondurensis*) and the absence of fruiting shrubs at net-sites, were most important in separating net-sites along major axes, as indicated by correlation of these fruit variables with ordination axes (Table 5).

Plant-bird interactions

Relatively few significant correlations existed between abundance of birds and fruits at net-sites (Table 6). The distribution of all arboreal frugivores was significantly related to total abundance of fruiting shrubs ($P < 0.01$, Table 6). On an individual species level, captures of *Pipra mentalis* and of *Chlorothraupis carmioli* were positively correlated, respectively, with abundance of *Psychotria officinalis* and *Ossaea macrophylla*. *Psychotria* was a common component in the diet of *Pipra mentalis*; *Chlorothraupis* regularly consumed *Ossaea* fruits (Loiselle & Blake 1990). Captures of the manakin

Table 4. Maximum number of shrubs with ripe fruits at each net-site (50 sq. m) totalled over all net-sites during January to April 1986–1990. Observed variance/mean ratio (V/M), mean (standard deviation) of variance/mean ratios generated from Monte Carlo simulations (see text), and results of *t*-test comparing observed and computer simulated variance/mean ratios are given for each plant species.

Species	No. of plants	Observed V/M	Mean V/M	*t*	P<
Psychotria officinalis	12	0.94	1.002 (0.182)	−0.34	NS
Cephaelis elata	5	0.93	0.995 (0.153)	0.43	NS
Henriettea tuberculosa	26	1.61	0.997 (0.178)	3.44	0.001
Ossaea macrophylla	13	3.76	0.997 (0.178)	15.51	0.001
Miconia simplex	7	1.48	1.008 (0.177)	2.67	0.01
Clidemia densiflora	48	1.86	1.005 (0.176)	4.86	0.001
Total shrubs	181	1.66	0.996 (0.181)	3.67	0.001

Corapipo leucorrhoa and the two tanagers *Chlorothraupis* and *Tachyphonus* were negatively correlated with the abundance of *Psychotria officinalis*; these frugivores rarely, if ever, fed on fruits of *Psychotria officinalis* (no occurrences in diet samples collected during 1985–1989).

When captures of frugivores were compared to the abundance of fruit in an area encompassing 5 nets (i.e., examining resource abundance at a larger spatial scale, see methods; Table 5), results generally were similar to those at the scale of net-sites, but some differences were apparent. *Corapipo leucorrhoa*, for example, correlated with abundance of *Henriettea tuberculosa* at the larger spatial scale but not at the level of one net; this

Table 5. Correlation of plant species (Pearson's r) with major axes generated by Bray-Curtis ordination of net-sites in lowland wet forest. Only significant correlations ($P<0.05$) are shown.

Variable	Axis 1	Axis 2	Axis 3
Henriettea tuberculosa	0.440		
Clidemia densiflora	−0.642		
Leandra dichotoma	0.325		−0.258
Coussarea hondurensis		−0.421	0.594
No fruit [a]		0.848	0.244
% Variance explained	24.0	18.0	16.4

[a] Represents absence of any fruiting shrubs in Melastomataceae or Rubiaceae.

Table 6. Spearman rank correlations between bird and fruit variables at two levels: level of net-sites and fruit area (focal net plus four adjacent nets; see text). Only significant relations ($P<0.05$) are shown. Po = *Psychotria officinalis*, Om = *Ossaea macrophylla*, Ht = *Henriettea tuberculosa*, and Tot = total number of fruiting shrubs.

Bird species	Plant species			
	Po	Om	Ht	Tot
Net-sites				
Mionectes oleagineus				0.25
Pipra mentalis	0.26			0.25
Corapipo leucorrhoa	−0.25			
Tachyphonis delatrii	−0.26			
Chlorothraupis carmioli		0.27		
Total frugivores				0.36
Fruit area				
Pipra mentalis	0.27			
Corapipo leucorrhoa			0.32	
Tachyphonus delatrii		0.27		
Chlorothraupis carmioli		0.35		

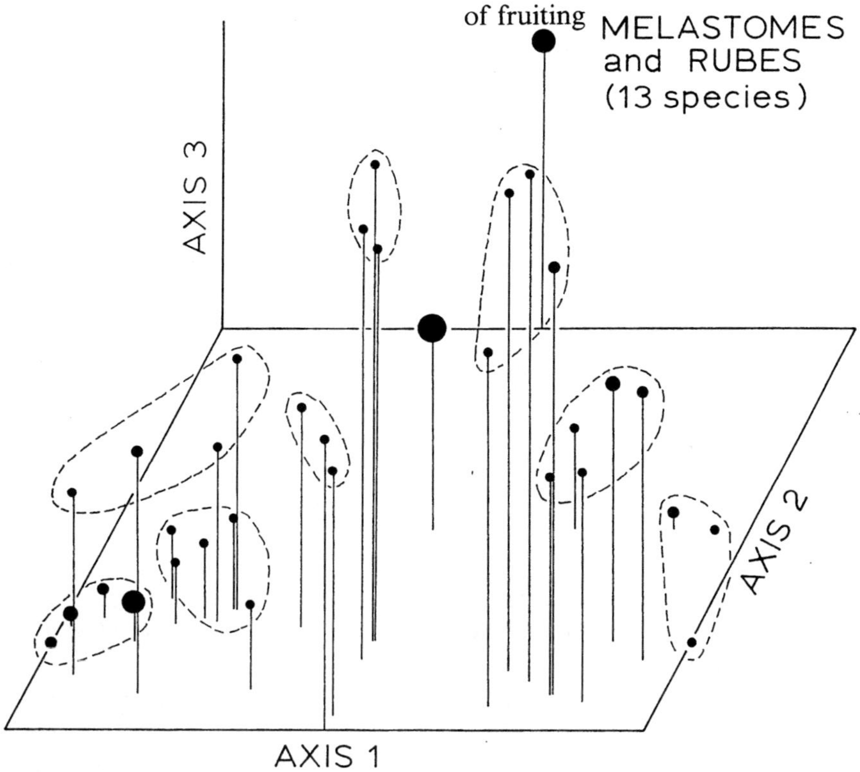

Fig. 4. Ordination of net-sites using abundance and composition of fruiting shrubs. Net-sites are plotted according to their factor scores along the first three ordination axes. Groups were drawn by visual inspection of ordination. Size of circle indicates the number of net-sites represented at that location.

fruit is common in the diet of *Corapipo*. That birds may respond to abundance or distribution of some combination of fruits, rather than to abundance of individual species of fruits, was indicated by significant correlations between bird abundance and net-site factor scores (from Bray-Curtis ordination of fruiting plants) for both *Pipra mentalis* ($F_{3,56} = 3.0$, $r^2 = 0.14$, $P < 0.03$) and all frugivores ($F_{3,56} = 3.4$, $r^2 = 0.15$, $P < 0.02$).

Captures of all frugivores differed significantly among net-groups defined by the ordinations ($X^2 = 35.8$, $P < 0.001$, df = 9). Captures of *Pipra mentalis* ($X^2 = 23.9$, $P < 0.01$, df = 9) and *Mionectes oleagineus* ($X^2 = 22.8$, $P < 0.01$, df = 9) also differed among groups. In all cases where significant differences in captures among groups were noted, fewer birds than expected were captured at sites where little or no fruit was present. Finally, when nets were grouped into fruit-rich and fruit-poor groups, based on the distribution

Table 7. Distribution of captures among high and low fruit activity areas based on total fruiting shrubs at net-sites (Fig. 3). Expected values, based on captures expected on sampling effort alone, are given in parentheses. High fruit areas represent a composite of 19 net-sites; low fruit areas a composite of 41 net-sites.

Bird species	Low fruit	High fruit	X^2	P
Pipra mentalis	205	106	0.9	NS
	(212)	(98)		
Corapipo leucorrhoa	70	39	1.0	NS
	(74)	(34)		
Mionectes oleagineus	103	66	3.6	<0.10
	(113)	(54)		
Hylocichla mustelina	71	45	2.5	NS
	(79)	(37)		
Turdus obsoletus	11	12	5.1	<0.05
	(16)	(7)		
Chlorothr./Tachyphonus	37	21	0.7	NS
	(40)	(18)		
Total frugivores	548	321	11.3	<0.001

shrubs (Fig. 3), more birds than expected were captured in nets where fruit was abundant than in nets with little fruit. Differences, however, were significant only for *Turdus obsoletus* and for all frugivores combined (Table 7).

Discussion

Distribution patterns of plants and animals within tropical forests reflect the interactions of a complex set of biotic and abiotic factors. Most fruiting shrubs and fruit-eating birds were distributed non-randomly on our 10 ha plot. Despite the fact that fruit-eating birds depend on a superficially similar resource (i.e., fruits), the species examined here showed little similarity in distribution. Different bird species have distinct preferences among fruits (Moermond & Denslow 1985; Loiselle & Blake 1990); such dietary differences may influence habitat use, thereby affecting distributions. Distribution of fruiting shrubs likely reflects past differences in seedling recruitment as influenced by variation in seed dispersal as well as microenvironmental conditions necessary for seedling growth and establishment (including both biotic – competition and predation – and abiotic – moisture, light – factors).

This study has provided additional evidence that spatial variation in resource abundance (at the scale of net-sites) can influence the distribution patterns of birds within a habitat. In most cases, however, fruit abundance accounted for relatively small (but still significant) amounts of variation in captures of birds, indicating that many other factors also help determine bird movement patterns. Why did we not find better correlations between birds and fruits at the scale of net-sites? Although Murray (1987) found that undergrowth plants with large fruit crops had more fruits removed than plants with small fruit crops, he argued that even large fruit crop sizes of forest undergrowth shrubs may not be large enough to promote sedentary behavior by birds. Moreover, frugivorous animals may need to visit many species of fruiting plants to balance nutritional needs (Foster 1977; White & Stiles 1983; Fleming 1988). Several studies have demonstrated that nutritional considerations influence selection of fruits by frugivores (Howe & Vande-Kerckhove 1981; Sorenson 1981; Herrera 1984; Fleming 1988), but distance between fruiting plants also affects fruit choice (Levey *et al.* 1984; also Schoener 1971; Charnov 1976). Indeed, Janson *et al.* (1986) found that rates of fruit removal by *Cebus* monkeys were related more to tree distribution than to nutritional rewards of fruits. The greater abundance of birds in fruit-rich areas seen in this study may reflect these points. Birds may be captured at net-sites with little fruit because these areas offer additional resources or needs (e.g., insect prey, nesting areas, etc.), because birds in such areas are more actively searching for fruits, or because fruit-poor areas occur in flyways between fruit-rich areas. Further, species that include substantial portions of insects in their diet (e.g., *Chlorothraupis*, *Tachyphonus*, and *Hylocichla*) may select foraging sites based on their ability to find insect prey and may simply feed on fruits as they are encountered.

One of our initial assumptions was that birds captured at a net-site were using the immediate area for foraging and were not simply flying through the area. If birds were captured at net-sites simply because they were flying from one site to another, then we would expect net-sites located between fruit-rich areas to have larger number of captures than net-sites located away from (and not between) fruit-rich areas. We tested this by comparing captures among net-sites located between and away from fruit-rich areas. In all such comparisons, we found that net-sites located between fruit-rich areas did not act as flyway corridors; birds were captured in similar numbers between fruit-rich areas as well as away from them (chi-square tests, $P > 0.05$).

Irrespective of the cause, the nonrandom distribution of frugivore captures on our plot (e.g., Fig. 1) likely results in a heterogeneous dissemination of seeds by these bird species. Thus, as suggested by Herrera (1985), birds, by virtue of their feeding and movement patterns, may influence the structure of the habitat in which they live. Given the probability of capture at a given site for a particular bird species and the occurrence of a fruit species in the diet of that bird species, and assuming that number of captures reflects activity in the area, one can estimate the relative probability of seeds being disseminated near net-sites. Such an expected distribution of seeds could be compared with actual distribution of seedlings to test theories concerning seedling recruitment and resource limitation. Other important factors to consider in examining these expected distributions of seeds would be density of seeds (M. Kaspari, unpubl. data) and the species composition of seeds in droppings (Stiles & White 1986; Loiselle 1990).

Differential use of an area by fruit-eating birds also may result in increased reproductive success of adult plants that are located near fruit-rich areas. For example, single fruiting individuals of *Henriettea tuberculosa* located in areas with little fruit are less likely to be visited by fruit-eating birds than are single fruiting individuals located in areas with abundant fruit. Thus, the spatial distribution of fruiting plants may, because of its influence on frugivore movements (Janson *et al.* 1986), be more important than other factors (e.g., crop size) in determining the proportion of fruit crops removed. Whether or not this inequality exists for the lifetime of the plant will depend on the dynamics of both bird and plant populations.

Long-lived plants may go through periodic bottlenecks in dispersal success only to experience relatively higher levels of fruit removal during other years. Except for mortality due to branchfalls or treefalls, few adult shrubs on our plot have died since the study began in 1985. This is in sharp contrast with results from concurrent studies in younger successional areas where dramatic turnovers in plant species composition have been observed during this same period. These differences reflect the fact that the importance of selective pressures operating on mature forest and successional shrubs likely differ; successional shrubs may be selected to produce large fruit crops that maximize absolute numbers of fruit removed, a pattern similar to that observed for shrubs reproducing only in forest gaps (Murray 1988). The structure of the forest, however, is dynamic; we have witnessed the creation of many new treefall gaps on the plot since January 1985 (pers. obs.) and fruiting shrubs likely experience different levels of frugivore activity over their lifetimes (assuming they survive gap events) (Levey 1988).

In summary, fruit-eating birds and many fruiting shrubs were distributed non-randomly in lowland wet forest. Many fruit-eating birds occur more frequently in areas with abundant fruit. Although clearly not the only factor influencing habitat selection by fruit-eating birds, fruit abundance was an important influence on spatial use of the forest, particularly for those birds that rely almost entirely on fruit for food (e.g., *Pipra mentalis*). This differential use of space by seed dispersers likely results in differential dissemination probabilities for fruiting plants. Because birds differ in their use of fruits, the species composition, as well as abundance, of fruits present in an area may affect which bird species is primarily responsible for dispersing seeds from individual plants. The relative lifetime importance of different bird dispersers to a particular plant will likely change because of the dynamic nature of this forest (Sanford *et al.* 1986).

Acknowledgements

We thank Victoria Sork, Ted Fleming, Loreta Roselli, & F. Gary Stiles for many helpful discussions and comments on this manuscript. We greatly appreciate the help and support of David & Deborah Clark & Ronald Suarez for facilitating our research at Estación Biológica La Selva. We thank the Organization for Tropical Studies, the Department of Zoology-University of Wisconsin (Guyer Fellowship), the National Geographic Society, and the Douroucouli Foundation for their financial support.

Appendix 1. Plant species (from Melastomataceae and Rubiaceae only) found with ripe fruits in sampling areas located in lowland wet forest during December to April 1986–1990.

Melastomataceae	
Clidemia densiflora	*Miconia nervosa*
Clidemia discolor	*Miconia gracilis*
Clidemia epiphytica	*Miconia aff. smaragdina*
Clidemia septuplinervia	*Miconia simplex*
Conostegia setosa	*Miconia* sp.
Henriettea tuberculosa	*Ossaea laxivenula*
Miconia appendiculata	*Ossaea macrophylla*
Miconia brenesii	*Ossaea robusta*
Miconia centrodesma	
Rubiaceae	
Bertiera guianensis	*Psychotria grandistipula*
Cephaelis elata	*Psychotria officinalis*
Coussarea hondurensis	*Psychotria polyphlebia*
Faramea talamancarum	*Psychotria suerrensis*
Psychotria acuminata	*Psychotria* sp.
Psychotria brachybotrya	

References

American Ornithologists' Union. 1983. Check-list of North American birds, Sixth edition. American Ornithologists' Union, Washington, D.C., USA.

Beals, E. W. 1984. Bray-Curtis ordination: an effective strategy for analysis of multivariate ecological data. Adv. Ecol. Res. 14: 1–55.

Blake, J. G. & Hoppes, W. G. 1986. Resource abundance and microhabitat use by birds in an isolated east-central Illinois woodlot. Auk 103: 328–340.

Blake, J. G. & Loiselle B. A. 1991. Variation in resource abundance affects capture rates of birds in three lowland habitats in Costa Rica. Auk 108: 114–130.

Blake, J. G. & Loiselle B. A. 1992. Habitat use by neotropical migrants at La Selva Biological Station and Braulio Carrillo National Park, Costa Rica. In: J. Hagen & D. W. Johnston (eds.) Ecology and conservation of neotropical landbird migrants. Smithsonian Institution Press, Washington, D. C., USA.

Blake, J. G., Stiles, F. G. & Loiselle, B. A. 1990. Birds of La Selva: habitat use, trophic composition, and migrants. In: A. Gentry (ed.) Four neotropical rainforests. Yale University Press, New Haven, CT, USA. pp. 161–182

Charnov, E. L. 1976. Optimal foraging: the marginal value theorem. Theor. Pop. Biol. 9: 129–136.

Clark, D. B. 1990. La Selva Biological Station: a blueprint for stimulating tropical research. . In: A. Gentry (ed.) Four neotropical rainforests. Yale University Press, New Haven, CT, USA. pp. 9–27

Cody, M. L. (ed.) 1985. Habitat selection in birds. Academic Press, Orlando, FL, USA.

Croat, T. B. 1978. Flora of Barro Colorado Island, Panama. Stanford University Press, Stanford, CA, USA.

Fleming, T. H. 1988. The short-tailed fruit bat. University of Chicago Press, Chicago, IL, USA.

Ford, H. A. & Paton D. C. 1985. Habitat selection in Australian honeyeaters, with special reference to nectar productivity. In: M. L. Cody (ed.) Habitat selection in birds. Academic Press, Orlando, FL, USA, pp. 368–388.

Foster, M. S. 1977. Ecological and nutritional effects of food scarcity on a tropical frugivorous bird and its fruit source. Ecology 58: 73–85.

Hammel, B. 1990. The distribution of diversity among families, genera, and habitat types in the La Selva flora. In: A. Gentry (ed.) Four neotropical rainforests. Yale University Press, New Haven, CT, USA, pp. 75–84

Hartshorn, G. S. 1983. Plants: introduction. In: D. H. Janzen, (ed.) Costa Rican natural history. University of Chicago Press, Chicago, IL, pp. 118–157.

Hartshorn, G. S. & Peralta R. 1988. Preliminary description of primary forests along the La Selva-Volcan Barva altitudinal transect, Costa Rica. In: F. Almeda and C. M. Pringle (eds.) Tropical rainforests: diversity and conservation. California Academy of Sciences, San Francisco, CA, USA, pp. 281–296.

Herrera, C. M. 1984. Adaptation to frugivory of Mediterranean avian seed dispersers. Ecology 65: 609–617.

Herrera, C. M. 1985. Habitat-consumer interactions in frugivorous birds. In: M. L. Cody (ed.) Habitat selection in birds. Academic Press, New York, NY, USA, pp. 341–365.

Holdridge, L. R. 1967. Life zone ecology: revised edition. Tropical Science Center, San Jose, Costa Rica.

Howe, H. F., & G. A. Vande Kerckhove. 1981. Removal of wild nutmeg (*Virola surinamensis*) crops by birds. Ecology 62: 1093–1106.

Janson, C. H., Stiles E. W. & White D. W. 1986. Selection on plant fruiting traits by brown capuchin monkeys: a multivariate approach. In: A. Estrada & T. H. Fleming (eds.), Frugivores and seed dispersal. Dr. W. Junk Publishers, Dordrecht, The Netherlands, pp. 83–92.

Karr, J. R. 1981. Surveying birds with mist nets. Stud. Avian Biol. 6: 62–67.

Karr, J. R., Robinson S., Blake J. G. & Bierregaard R. O. Jr. 1990. Birds of four neotropical forests. In: A. H. Gentry, (ed.) Four neotropical rainforests. Yale University Press, New Haven, CT, USA, pp. 237–269.

Leighton, M. & Leighton D. R. 1983. Vertebrate responses to fruiting seasonality within a Bornean rain forest. In: S. L. Sutton, T. C. Whitmore & A. C. Chadwick (eds.), Tropical rain forest: ecology and management. Blackwell Scientific, Oxford, England, UK, pp. 181–186.

Levey, D. J. 1988. Spatial and temporal variation in Costa Rican fruit and fruit-eating bird abundance. Ecol. Monogr. 58: 251–269.

Levey, D. J., Moermond T. C. & Denslow J. S. 1984. Fruit choice in neotropical birds: the effect of distance between fruits on preference patterns. Ecology 65: 844–850.

Levey, D. J. & Stiles F. G. In press. Birds of La Selva. In L. A. McDade, K. S. Bawa, H. A. Hespenheide & G. S. Hartshorn (eds.) La Selva: ecology and natural history of a neotropical rainforest. University of Chicago Press, Chicago, IL, USA.

Loiselle, B. A. 1987. Birds and plants in a neotropical rain forest: seasonality and interactions. Unpubl. Ph.D. dissertation. University of Wisconsin, Madison, WI, USA.

Loiselle, B. A. 1990. Seeds in droppings of tropical fruit-eating birds: importance of considering seed composition. Oecologia 82: 494–500.

Loiselle, B. A. & Blake J. G. 1990. Diets of understory fruit-eating birds in Costa Rica: seasonality and resource abundance. Stud. Avian Biol. 13: 91–103.

Loiselle, B. A. & Blake J. G. 1991. Temporal variation in birds and fruits along an elevational gradient in Costa Rica. Ecology 72: 180–193.

McDade, L. A., Bawa K. S., Hespenheide H. A. & Hartshorn G. S. (eds.) In press. La Selva: ecology and natural history of a neotropical rainforest. University of Chicago Press, Chicago, IL, USA.

Moermond, T. C. & Denslow J. S. 1985. Neotropical frugivores: patterns of behavior, morphology, and nutrition with consequences for fruit selection. Ornithol. Monogr. 36: 865–897.

Murray, K. G. 1987. Selection for optimal fruit-crop size in bird-dispersed plants. Amer. Nat. 129: 18–31.

Murray, K. G. 1988. Avian seed dispersal of three neotropical gap-dependent plants. Ecol. Monogr. 58: 271–298.

Opler, P. A., Frankie G. W. & Baker H. G. 1980. Comparative phenological studies of treelet and shrub species in tropical wet and dry forests in the lowlands of Costa Rica. J. Ecology 68: 167–188.

Pielou, E. 1977. Mathematical ecology. J. S. Wiley Publ., New York, NY, USA.

Pringle, C., Chacon I., Grayum M., Greene H., Hartshorn G., Schatz G., Stiles G., Gomez C. & Rodriguez M. 1984. Natural history observations and ecological evaluation of the La Selva Protection Zone, Costa Rica. Brenesia 22: 189–206.

Rosenzweig, M. L. 1985. Some theoretical aspects of habitat selection. in M. L. Cody (ed.), Habitat selection in birds. Academic Press, Orlando, FL, USA, pp. 517–540.

Sanford, R. L. Jr., Braker H. E. & Hartshorn G. S. 1986. Canopy openings in a primary neotropical lowland forest. J. Tropical Ecology 2: 277–282.

Schemske, D. W. & Brokaw N. V. L. 1981. Treefalls and the distribution of understory birds in a tropical forest. Ecology 62: 938–945.

Schoener, T. W. 1971. The theory of foraging strategies. Ann. Rev. Ecol. Syst. 2: 369–404.

Sokal, R. R. & Rohlf F. J. 1981. Biometry. Second edition. W. H. Freeman, San Francisco, CA, USA.

Sorenson, A. 1981. Interactions between birds and fruits in a temperate woodland. Oecologia 50: 242–249.

Stiles, E. W. & White D. W. 1986. Seed desposition patterns: influence of season, nutrients, and vegetation structure. In: A. Estrada & T. H. Fleming (eds.) Frugivores and seed dispersal. Dr. W. Junk Publishers, Dordrecht, The Netherlands, pp. 45–54

Terborgh, J. 1983. Five new world primates. Princeton University Press, Princeton, NJ, USA.

Terborgh, J. 1985. Habitat selection in Amazonian birds. In: M. L. Cody (ed.) Habitat selection in birds. Academic Press, Orlando, FL, USA. pp. 311–338.

Wheelwright, N. T. 1983. Fruits and the ecology of the Resplendent Quetzal. Auk 100: 286–301.

White, D. W. & Stiles E. W. 1983. The adaptive significance of nutritional differences in North American bird-dispersed fruits. Bull. Ecol. Soc. Amer. 64: 166–167.

Wilkinson, L. 1988. SYSTAT: the system for statistics. SYSTAT, Inc., Evanston, IL, USA.

Vegetatio **107/108**: 191–203, 1993.
T. H. Fleming and A. Estrada (eds). Frugivory and Seed Dispersal: Ecological and Evolutionary Aspects.

Weak interactions between avian and insect frugivores: the case of *Pistacia terebinthus* L. (Anacardiaceae)

A. Traveset
Estación Biológica de Doñana, CSIC, Apdo. 1056, E-41080 Sevilla, Spain; Present address: Institut d'Estudis Avançats de les Illes Balears, Crtra. de Valldemossa, Km 7.5, 07071 Palma de Mallorca, Spain

Keywords: Avian fruit removal, Insect seed predation, Bird-insect interactions, Chalcidoid wasps, *Pistacia*

Abstract

Vertebrate frugivores often feed on fruits upon or within which insects also feed, yet little information exists on the potential magnitude of interactions between these consumers. The Mediterranean shrub *Pistacia terebinthus*, the birds that consume its fruits, and the wasps that feed upon its seeds are examined in this study.

P. terebinthus produces a highly variable fraction of final-sized red fruits that never become mature (green-colored). Red fruits can be immature, parthenocarpic, aborted, or attacked by wasps, and their pulp is much less nutritious than that of mature fruits. A total of 20 bird species consumed the fruits in the study area. Legitimate dispersers accounted for 39% of the total fruit removal, while pulp eaters and seed predators accounted for the remainder. Birds strongly preferred the mature fruits (only 4% of the fruits consumed were red). The incidence of wasps in the seeds ranged from 0 to 42% of the crop in 1989 and from 0 to 24% in 1990. The influence of avian and insect frugivore guilds on each other appears to be quite low because of the narrow overlap in resource utilization by birds and wasps, and an overall low intensity of wasp seed predation. From an evolutionary perspective, the possible ability of wasps to preclude fruit maturation appears not to be attributable to the present interaction with avian frugivores.

Introduction

Vertebrate frugivores and insects often feed on the same fruits, either on the pulp or the seeds. The possible interaction between these two guilds of consumers has been explored by relatively few studies compared to those on both the plant-seed disperser mutualism and the plant-insect parasitic relationship. Such an interaction may affect insect population dynamics (Bigler & Delucchi 1981; Janzen 1982; Herrera 1984a, 1989; Drew 1987), and/or the vertebrate's choice of fruits (Manzur & Courtney 1984; Redford *et al.* 1984; Bauer 1986; Jordano 1987; Burger 1987; Borowicz 1988; Semel & Andersen 1988; Krischik *et al.* 1989; Valburg 1992a, b). In either case, there should be important consequences for seed dispersal (see Manzur & Courtney 1984; Jordano 1987).

Besides decreasing the resources available to the insects, vertebrates may kill any developing insect inside the ingested fruit (cf. Lamprey *et al.* 1974; Halevy 1974). It is generally assumed that insect and vertebrate frugivores compete in a 'race' (Janzen 1976) that results in selection against those insects that oviposit late enough in fruit maturation that larval development overlaps with vertebrate fruit consumption. The studies by Bigler & Delucchi (1981) and Drew (1987) suggest that vertebrates are the major natural enemies of tephritid flies that feed on fruit pulp. The fact that larvae leave the fruits to pupate in the soil may be an adaptation to escape from being eaten by vertebrates (Drew 1987). In the case of insect seed predators, however, there is no substantial evidence that vertebrate frugivory represents a major insect mortality factor (but see Lamprey *et al.* 1974; Halevy 1974; Herrera 1989a).

The goal of the present study is to examine the interaction between avian frugivores and the insect seed predators of the Mediterranean shrub *Pistacia terebinthus* (Anacardiaceae). The bird species that consume the fruits in the study area and the insect species that prey upon the seeds of this plant were identified. Next, the variability in both seed predation by insects and avian frugivory was examined, in relation to plant traits, in order to assess the overlap in fruit utilization. In this way it should also be possible to determine whether plant traits selected by the avian frugivores coincide with those selected against by the insect seed predators.

Study site

The study was carried out at Parque Natural de Cazorla, Segura y Las Villas, eastern Andalucía (southeastern Spain), during the fall of 1989 and 1990. Descriptions of the Sierra de Cazorla, a mountain chain 40 km long with a maximum elevation of 2107 m, can be found in Fernández Galiano & Heywood (1960), Polunin & Smythies (1973: 83–89) and Otero *et al.* (1978). One of the study sites (Agracea) is located between 1100 and 1200 m in the Guadahornillos Valley. This site is occupied by dense Mediterranean forest constituted mainly by *Quercus, Phillyrea, Arbutus* and *Juniperus*. Here *Pistacia terebinthus* usually grows in small gaps or along trails, and its density is rather low (1–15 individuals/ha). The other study site (Las Navillas) is located about 5 km of the small town of Vadillo-Castril, at an altitude of approximately 1000 m. This site is a large secondary scrub, which until recently was intensively used for sheep and goat grazing. The terrain is calcareous and very rocky, and *P. terebinthus*, *Quercus rotundifolia*, and *Pinus pinaster* are the commonest vegetation. Additional observations were made on *P. terebinthus* shrubs located on the road from Puente de las Herrerías to the source of the Guadalquivir River.

Mean annual precipitation at Vadillo-Castril, the nearest weather station, is 1201 mm (the range for 26 years is 669–1970 mm; Herrera 1991). In the two years of the study, rainfall was 916.5 mm and 553.0 mm., respectively. The summer (from June to mid-September) is dry. Snowfalls are frequent from November to March, although at the altitudes where *P. terebinthus* lives snow does not remain longer than a week or two. Mean monthly temperatures of the coldest (January) and hottest (August) months are 4.0 °C and 21.1 °C (Herrera 1991).

Study organisms

P. terebinthus is a dioecious woody shrub or small tree that can reach a height of 10 m. It is distributed around the Mediterranean from Portugal to Turkey and from Morocco to Cyrenaica, as far as the southern Alps (Zohary 1952). Flowering occurs towards the end of April, just preceding leaf growth, and lasts until the beginning of June. Inflorescences of 3- ca. 50 anemophilous flowers are always grouped in panicles. A panicle may contain up to 20 inflorescences and up to 800 flowers, although the usual is 100–400 flowers. Single branches may contain up to 10 panicles, usually quite close to each other and at the distal end of the branch. Males tend to start flowering somewhat earlier than females.

After anthesis, the ovary – but not the ovule – grows continuously until reaching its maximum size about a month later. One or two weeks later, the endocarp becomes woody. Along with the endocarp, the pericarp also develops before and independently of the ovule (or embryo if it has been fertilized). As in other species of *Pistacia* (Grundwag 1976), there is probably a delay in syngamy (Grundwag & Fahn 1969), the fertilized ovule reaching its final size several months after anthesis.

The fruit is a drupe measuring 6.6 ± 0.07 (s.d.) mm long and 5.9 ± 0.7 mm wide (N = 575). A viable seed fills the entire fruit locule with a green endosperm. Fruits lacking viable seeds are either parthenocarpic, i.e. come from unpollinated flowers (showing vestiges of funiculus and ovule), or are the result of embryo abortions, which can occur at different stages of development. In late abortion, the small seed may contain some endosperm, but this never occupies the entire fruit locule. Detailed information on the causes of these types of seeds can be found in Grundwag & Fahn (1969), Grundwag (1975, 1976) and Crane (1975). Mature, green fruits have an average weight of 0.14 ± 0.03 g (N = 147), which is significantly higher (r = 0.89, DF = 278, F = 75.62, P = 0.0001; 10 plants) than that of inviable red fruits (0.09 ± 0.02 g; N = 150). Viable seeds weigh 0.05 ± 0.01 g (N = 147).

Final-sized fruits with inviable seeds are retained on the plant for several months. Whole infructescences usually start falling by mid-end October, before fruits have become desiccated. Immature fruits are red and become green or bluish green after maturation. Fruits with inviable seeds are indistinguishable from immature viable fruits. As found in *P. lentiscus* (Jordano 1989), fruit ripening is strongly associated with seed viability. Of a total of 3700 fruits dissected in 1989, only a small proportion (1.5%) had an apparently viable seed but did not become mature. Except on six occasions in which seeds had been damaged by larvae, all the dissected green fruits (N = 705) contained viable seeds.

Chemical analyses of the pulp of both ripe and red fruits indicated that they differed considerably in lipid content (35.83 ± 8.65% vs. 3.92 ± 1.61%, respectively). Quantities of all other compounds except calcium were similar in the two types of fruits. The amount of calcium appeared to be significantly higher in red (0.30 ± 0.04) than in green fruits (0.16 ± 0.05%) ($P < 0.05$; Mann-Whitney test). The procedures of the chemical analyses are described in detail in Herrera (1987).

Four species of chalcidoid wasps emerge from the seeds of *P. terebinthus*. The most common species are *Sycophila* sp. and *Megastigmus pistaceae* Walker 1871. The two other species, *Cyrtoptix* sp. and *Eurytoma* sp. (in the *pistaciae* group) are rarely found and might be parasitoids of *M. pistaciae* (Davatchi 1956). *M. pistaciae* and *Sycophila* sp. overwinter as larvae and emerge as first-generation adults in late June or early July, when the fruits of *P. terebinthus* are already full-sized and have a woody endocarp. Adult females probably lay only one egg within each fruit because seeds bear a single larva, although on one occasion two larvae were found in the same fruit. Attacked fruits remain red and resemble intact ones. Pupation occurs within the fruits, and adults emerge beginning in mid-August. Thus, wasp developmental time takes about 40–50 days, as found by Davatchi (1956) in *P. vera*. The adult females of this second generation oviposit on the fruits still on shrubs and their offspring spend the winter as larvae. The other species, *Sycophila* sp., emerges at the same time as *M. pistacea* and has a similar developmental time. However, *Sycophila* sp. probably has three generations per year, since adults emerge as late as mid October, and females have been seen ovipositing on the fruits at this time.

The fruits are consumed by a variety of birds, some of which are migratory. Ants have sometimes been seen feeding on the ripe pulp of fruits on the shrub, and they also take fallen fruits to their nests. Carnivores, mainly the red fox (*Vulpes vulpes)*, may also consume fallen fruits; they act as seed predators since seeds appear broken in their faeces (Herrera 1989b). The field mouse, *Apodemus sylvaticus*, also eats fruits, as evidenced by broken seeds often found below fruiting shrubs.

Methods

The fruit ripening phenology of 158 marked female plants (120 from Las Navillas and 38 from

Agracea) was determined in 1990 by visiting the plants every 3–5 days and recording the proportion of mature fruits in the entire plant. From twenty-eight of these plants, a sample of 50 fruits per plant was dissected in order to determine the proportion of viable seeds and the incidence of wasp seed predation. Aborted fruits were distinguished from parthenocarpic ones by the presence of a more or less enlarged embryo. Fruit crop size, plant height, canopy diameter, and distance to the nearest two neighbors were measured for each plant. Fruit crop size was estimated by complete count of the final-sized fruits. Average fruit diameter was obtained by measuring 25 fruits from each plant (red and green fruits are of similar size). Preliminary data on the incidence of insect-damaged fruits were gathered in 1989 from a sample of 50 fruits of each of 74 plants (20 from Agracea, 25 from Las Navillas, and 29 from along the road to the Guadalquivir source).

In order to determine fruit choice by birds and their intensity of fruit removal, five fruit panicles per plant from a sample of 18 plants were randomly chosen and marked during the ripening period of 1990. The total number of final-sized fruits was counted for each panicle before fruit removal by birds began. Fruits were again counted every 3–5 days until all mature fruits disappeared and most immature fruits began falling. A set of four fruit traps ($40 \times 40 \times 6$ cm) were placed beneath each canopy to collect the falling mature fruits. These traps were under about 2/3 of the total number of mature fruits produced. Thus, the total number of mature fruits dropped was estimated as one and a half times the number of fruits in the traps and substracted from the total number of mature fruits that disappeared. Red fruits are much more strongly attached to the branches than mature fruits and do not easily fall unless they are completely dry. Usually, the whole infructescence containing red fruits dries out and falls (most mature fruits have been removed by this time). The number of fruits that began drying out was recorded in each census, and was not considered as removed by birds in the following census. Mature fruits can remain on the plant for several weeks before drying.

Direct observations on bird visitation and fruit removal were made on another 18 plants. Plants were observed for 13 days from October 16 to November 4 1990, between 8:30 am to 6:30 pm. All 18 plants were observed at randomly selected times each day for 30 minutes. Previous observations on some of those and other plants were made after September 21, mainly to learn the species that visited the plants. Total time spent in the direct observations was 163.25 h. For each bird visit to a plant I recorded the species, whether it fed on fruits, the color of the fruits removed, and the total number of fruits removed.

In order to determine whether the number of bird visits and the proportion of fruits removed was associated to some plant trait, I measured the following variables for each plant: fruit crop size, proportion of fruits maturing, plant height, canopy diameter, number of fruiting conspecific individuals within a 10 m radius, and foliage density (high, low, or intermediate). In addition, the distances to the crowns of all trees or shrubs of any species along four perpendicular transects 10 m in length were averaged. The first transect was always taken to the nearest tree or shrub. This value, here called the degree of isolation of the plant, expresses the average minimum amount of open country a bird had to cross to reach it from another perch.

Fruit preferences were tested also on caged individuals of *Sylvia atricapilla* (N = 4), *Parus caeruleus* (N = 3), *Parus ater* (N = 1) and *Parus major* (N = 1), the main fruit consumers of *P. terebinthus*. Additional trials were run for two more species (*Parus cristatus* and *Erithacus rubecula)*, but these failed because the birds refused to eat in captivity. Three different trials were run (when possible) for each individual bird and involved presenting infructescences with different proportions of mature fruits to the bird in random order: a) 30 red and 30 green, b) 40 red and 20 green, and c) 60 red and 0 green. During each trial, which lasted 20 min, I recorded the bird's choice and the number of fruits consumed.

All statistical analyses were performed using SAS (1985). Percentages and proportions in the ANOVAs and multiple regressions were arcsine transformed. Both Spearman and Pearson correlations were used and are distinguished in the results section as (r_s) and (r), respectively. Standard deviations are given with the means throughout the paper.

Results

Fruiting and ripening patterns

The average fruit crop in the plants observed in 1989 was 1,542 ± 1,401 fruits (range 150–6,230; N = 45), and the proportion of the fruit crop that ripened averaged 13.1 ± 14.9% (range 0–54%; N = 45). In plants that yielded mature fruits this average was 17.82 ± 14.73% (N = 33). Of the 120 female plants marked and monitored in Las Navillas area in 1990, 30 (25%) produced no mature fruits at all. The other 90 plants ripened from 1 to about 75% of their fruit crop, the average being 16.1 ± 19.2% (Fig. 1). Mean fruit crop size was 2,073 ± 3,579 (range 3–20,000; N = 120), whereas number of ripe fruits averaged 686 ± 1,732 (range 0–11,500; N = 120). In the Agracea area, with similar fruit crop sizes, only 3 of 38 plants (8%) yielded some mature fruits.

Figure 2 shows the fruit ripening phenology in a haphazardly chosen sample of 18 individuals observed in 1990. Data on fruit ripening phenology were not available for 1989. The fruit ripening period in a single individual lasted an average of 45 ± 8 days, ranging from 30 to 57 days. By mid September, almost all plants had ripened the fruits, but these remained on the branches for several weeks before they were removed by birds.

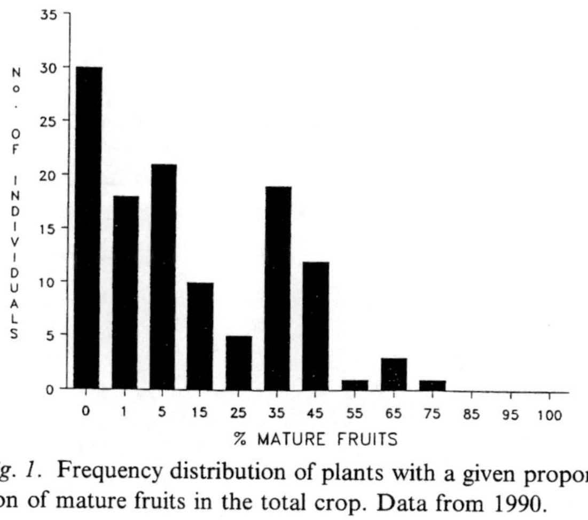

Fig. 1. Frequency distribution of plants with a given proportion of mature fruits in the total crop. Data from 1990.

Most inviable seeds of *P. terebinthus* were the result of embryo abortions (37.8 ± 16.3%; N = 74 plants in 1989; 70.1 ± 13.1%, N = 28 plants in 1990), although parthenocarpy was also important (30.7 ± 15.8%, N = 74 in 1989; 17.7 ± 13.9%, N = 28 in 1990). The proportion of embryo abortion and parthenocarpy was similar at all study sites (Table 1). A detailed account of the intra- and inter-population variation in the production of aborted and parthenocarpic fruits will be published elsewhere (Traveset, in prep.). Greater pollen limitation in 1989 (mainly because of rains during the flowering period) was apparently the cause of the higher parthenocarpy that year.

Seed predation by chalcidoid wasps

The average percentage of the fruit crop attacked by wasps in 1989 was 11.0 ± 9.9 (range 0–42%; N = 74 plants). In 1990, that percentage averaged 4.6 ± 6.3 (range 0–24%; N = 28 plants) (Fig. 3). In 16.2% of the plants from which fruits were collected in 1989 no fruits were found to be damaged by wasps. That value doubled in 1990; 9 of the 28 plants (32.1%) showed no evidence of insect seed predation. There might be some consistency between years in the incidence of wasps in an individual plant since at least four plants of the seven studied in both years had a similar proportion of insect damaged seeds. However, more data are needed to test this possibility.

In 1989, the proportion of attacked fruits per shrub varied significantly among areas (r = 0.43, ANOVA, DF = 73, F = 7.94, $P < 0.001$), the highest occurring in Las Navillas, where *P. terebinthus* occurs at high density, and the lowest in the Agracea, where the species is less common (Table 1). The incidence of wasp seed predation was neither correlated with crop size ($r_s = 0.03$, $P > 0.05$) nor to fruit dimensions (length or diameter) ($r_s = 0.18$ and $r_s = 0.16$, $P > 0.05$, respectively).

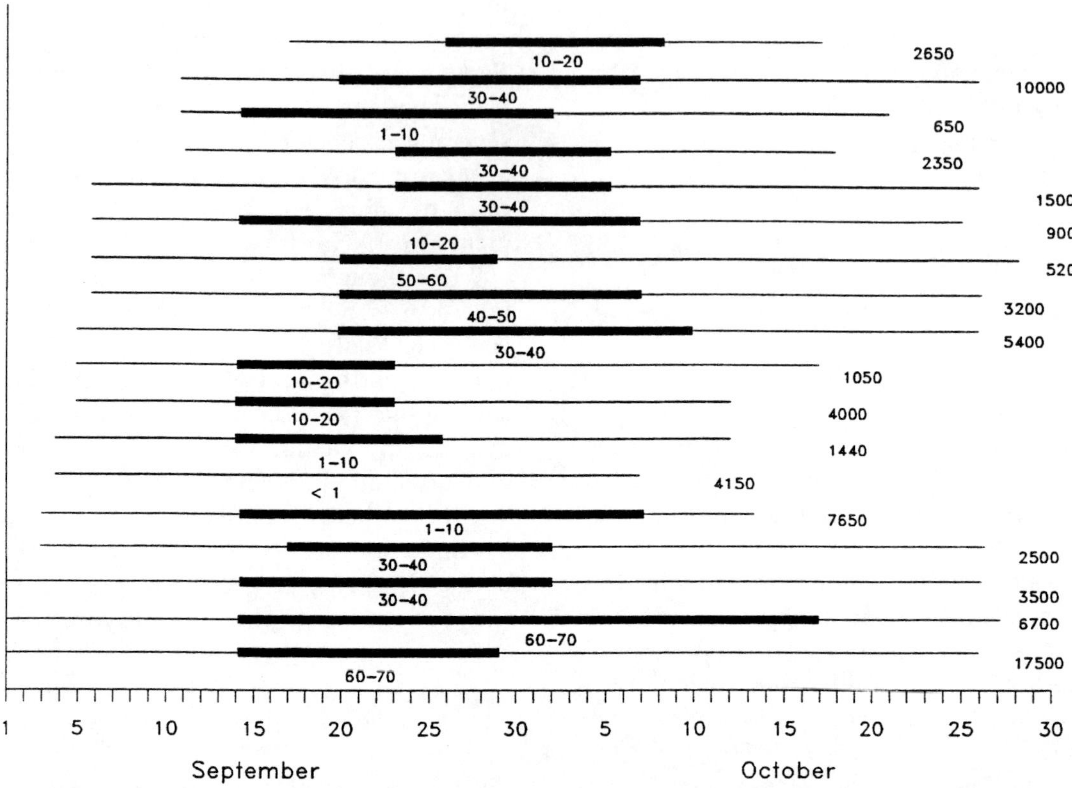

Fig. 2. Fruit ripening phenology of 18 selected individual plants in 1990. Each line represents the entire period of ripe fruit availability. The filled bars indicate the period during which the maximum percentage of mature fruits was available. This percentage was estimated visually a 10% intervals and is indicated under the bars. The numbers to the right of each line stand for total crop size.

In contrast to the results obtained in 1989, the proportion of damaged seeds at Las Navillas was not significantly different ($r = 0.12$, $DF = 27$, $F = 0.39$, $P > 0.05$) from that at the Agracea area (Table 1). These results suggest that plant density does not consistently determine the incidence of wasp seed predation each year.

The proportion of seeds attacked by wasps appeared to be significantly correlated with only fruit crop size ($r = 0.53$, $P < 0.01$) and not with any of

Table 1. Percentages of parthenocarpic, aborted, wasp-attacked, and mature fruits from the plants studied in 1989 and in 1990 in the different areas. Figures are means and standard deviations.

Area	N	Parthenocarpic	Aborted	Wasp-attacked	Mature
1989					
Navillas	20	33.4 ± 11.0	30.8 ± 11.3	15.4 ± 10.0	18.6 ± 15.0
Agracea	25	36.3 ± 20.0	50.0 ± 14.3	5.1 ± 5.7	6.2 ± 11.8
Guadalq.	29	24.5 ± 14.5	35.4 ± 16.9	11.2 ± 10.1	28.3 ± 16.5
1990					
Navillas	15	16.4 ± 15.4	67.3 ± 16.4	3.9 ± 6.1	12.3 ± 12.5
Agracea	13	19.2 ± 12.2	73.2 ± 7.2	5.4 ± 6.7	2.2 ± 4.2

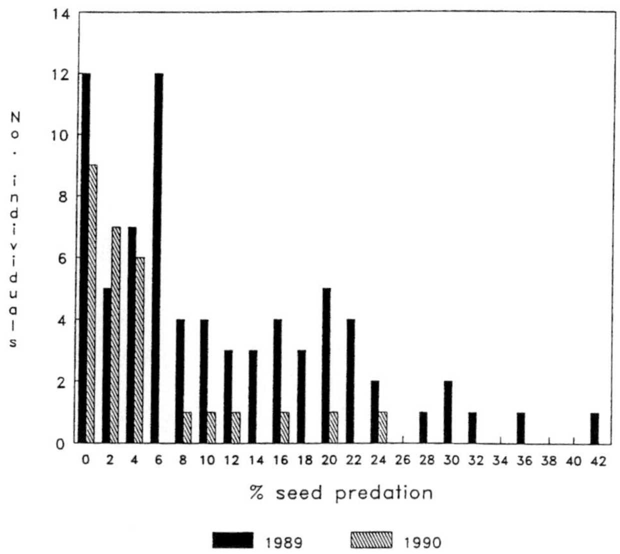

Fig. 3. Frequency distribution of individual plants with a given percentage of their total seed crop attacked by chalcidoid wasps in the two years of study. Notice the large number of plants lacking fruit damage by insects. A total of 50 seeds were examined per plant.

the other variables measured for each plant in 1990 (plant height, canopy diameter, fruit diameter, average distance to the two nearest fruiting neighbors, and number of fruiting conspecifics in a 10 m radius). The proportion of attacked seeds was not correlated to that of parthenocarpic, aborted or mature fruits, either. Wasp larvae are able to survive in seeds that are not full-sized yet. Fruit dissections showed that the seeds from which adults had emerged usually filled less than 50% of the fruit locule. At least the first generation of both *Megastigmus pistaciae* and *Sycophila* sp. had to develop in seeds that were not full-sized, since all adults have emerged by late August, and at this time seed development is not complete yet. The second generation of *Sycophila* emerges by late September, when seeds are already full-sized. However, even at this time it is not uncommon to find larvae living in undeveloped seeds, suggesting that (1) larvae have enough endosperm to feed on in late-aborted seeds, and/or (2) larvae prevent full seed development once they start feeding.

During September and October, *Sycophila* was observed only on red fruits, even on those plants with a high proportion of green fruits on them. So far, it is unknown whether wasps can discriminate between red fruits with empty seeds from those with some endosperm inside. In order to increase their offspring survival during winter, it seems that adults should be able to choose for oviposition those fruits containing seeds with larger amounts of food.

Fruit consumption by frugivorous birds

a) Indirect observations of fruit removal

Percent removal of green fruits by birds averaged 98.8 ± 2.9% (range 88.2–100%). Birds did not leave a single ripe fruit in 14 of the 18 plants monitored in Las Navillas. These values contrast strongly with the 2.7 ± 3.4% removal of red (inviable) fruits (range 0–13.2%). Fig. 4 shows the

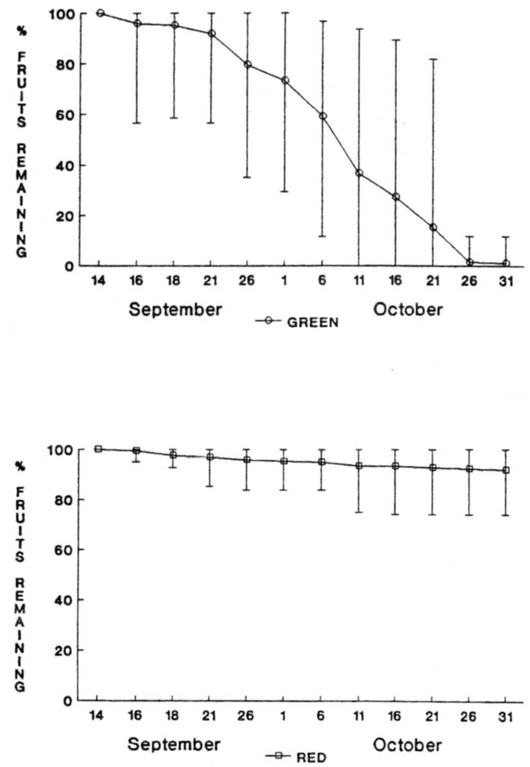

Fig. 4. Removal of green (mature) and red (immature) fruits by birds from 18 selected plants. Vertical bars indicate standard deviations around means. Notice the decline in green fruits coinciding with the arrival of migrant frugivores.

average removal rates of red and green fruits from the beginning to the end of the ripening period. Fruit removal rate increased towards the end of September, and coincided with the arrival of migratory birds. On October 21, 77.8% of the plants still had some green fruit on their branches; by October 26 this figure had dropped to 33.4%. The time during which ripe fruits were consumed varied between 25 and 47 days, with an average of 39 ± 6 days. In a correlation analysis, the duration of this period was not associated to any of the variables measured from each plant.

Red fruits were removed simultaneously with the ripe ones, not after the latter had been depleted. In only two plants did the highest proportion of red fruit removal occur after all green fruits had been consumed. After the ripe fruits of most plants in an area were exhausted, birds appeared to leave the area, and soon thereafter entire infrutescences of red fruits began to fall off.

Percent removal of red fruits (those that may contain wasp larvae inside) was significantly correlated with fruit crop size ($r_s = 0.68$, $P < 0.01$) and with the average distance to the two nearest fruiting neighbors ($r_s = 0.60$, $P < 0.01$). In a multiple regression analysis, these two variables explained 59% of the variation in percent removal of red fruits, although the effect of the second was not significant ($P = 0.07$). Fruit crop size, thus, appears to be a good predictor (explaining 52% of the variance) of the proportion of red fruits removed by birds.

Neither the proportion nor the absolute number of red fruits removed was related to the proportion or the absolute number of fruits attacked by wasps in a correlation analysis. Damaged fruits apparently look like the undamaged ones, and it seems unlikely that birds can distinguish between them. There was also no correlation between the proportion of either red or green fruits removed by birds and the other plant variables measured – height, canopy diameter, or average fruit diameter –.

b) Direct observations of fruit removal

A total of 20 bird species was recorded as fruit consumers of *Pistacia terebinthus* in the study areas (Table 2). Ten of these species are legitimate seed dispersers and represented 37.5% of the total number of visits to the fruiting trees. Five species of pulp consumers and five of seed predators accounted for the remainder (47.6% and 14.9%, respectively).

Not all plants were equally visited by all species (Fig. 5). Some attracted mostly pulp consumers while others were visited mainly by legitimate dispersers. The proportions of visits by these two types of frugivores were negatively correlated ($r_s = -0.83$, $P = 0.0001$). Seed predators, on the other hand, were seen in most (72%) plants and their visits were not correlated with those by other frugivores. Plant isolation appeared to determine, in part, the type of visiting frugivore: the more isolated (with little vegetation cover nearby) the plant, the higher the probability of being vis-

Table 2. Avian frugivores that consumed *Pistacia terebinthus* fruits and total number of visits to the fruiting plants during observations in 1990. *Garrulus glandarius* was observed visiting shrubs twice but it is not included in the table because its feeding type on this plant is unknown. Most observations come from Las Navillas.

Feeding type	Bird species	Number of visits
Pulp consumers	*Parus ater*	191
	Parus major	99
	Parus caeruleus	184
	Parus cristatus	30
	Aegithalos caudatus	55
Seed dispersers	*Sylvia atricapilla*	211
	Sylvia melanocephala	15
	Sylvia communis	1
	Phoenicurus phoenicurus	32
	Phoenicurus ochruros	93
	Erithacus rubecula	84
	Turdus philomelos	1
	Turdus merula	2
	Ficedula hypoleuca	1
	Corvus corone	1
Seed predators	*Sitta europaea*	29
	Fringilla coelebs	114
	Dendrocopos major	7
	Carduelis chloris	13
	Coccothraustes coccothraustes	12

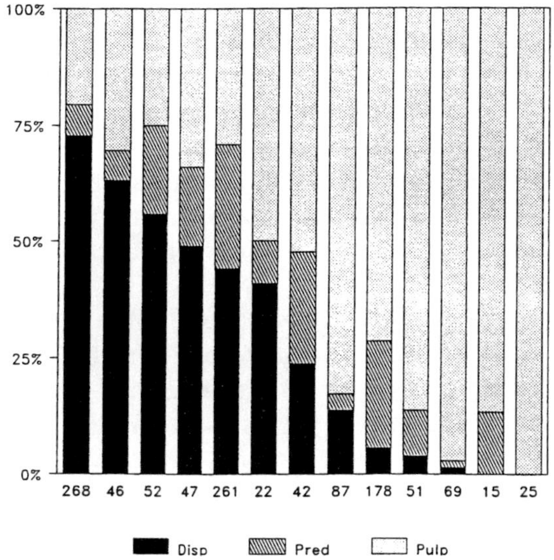

Fig. 5. Proportion of visits by legitimate dispersers (Disp), seed predators (Pred) and pulp eaters (Pulp) in 13 plants. Not included are five plants that received less than three visits. Numbers at the bottom indicate the total number of visits to each plant. Notice the negative correlation between visits by dispersers and by pulp eaters, reflecting differences in the habitat preferences of the two avian guilds.

ited by a pulp consumer ($r_s = 0.78$, $P = 0.0015$; N = 13 plants). In contrast, plants with more available perches and nearby cover were more frequently visited by either legitimate dispersers or seed predators ($r_s = -0.68$, $P = 0.01$ and $r_s = -0.70$, $P = 0.08$, respectively; N = 13 plants). Proportion of visits by a particular type of frugivore was not correlated with plant size, crop size, absolute number of ripe fruits, foliage density, distance to the two nearest fruiting conspecifics, or number of fruiting conspecifics in a 10 m radius.

Considering only feeding visits, 96.1% of the fruits removed (N = 877 fruits) were green whereas only 3.9% were red (Table 3). Legitimate dispersers accounted for 39.1% of fruit removal (0.5 and 38.7% of red and green fruits, respectively); the other 60.9% of the fruits were taken by either pulp consumers or seed predators.

Legitimate dispersers, in particular *S. atricapilla* and *E. rubecula*, accounted for only four of the 34 red fruits consumed. Six red fruits were taken by pulp consumers (*P. ater*, *P. major* and *A. caudatus*) while the remaining 24 were removed by seed predators, particularly, by *C. chloris*. In two occasions, *C. coccothraustes* was observed cutting infructescences with red fruits and rejecting them to reach the green ones.

Table 3. Consumption of fruits by birds during feeding observations in 18 plants. Numbers in parentheses represent the percentages of total fruit consumption.

Type of frugivore	Red fruits	Green fruits
Legitimate dispersers	4 (0.5)	339 (38.7)
Pulp eaters	6 (0.7)	372 (42.4)
Seed predators	24 (2.7)	132 (15.1)

The proportion of ripe fruits consumed by legitimate dispersers was negatively correlated with that consumed by pulp or seed eaters ($r_s = -0.99$, $P = 0.0001$; N = 11). That is, some plants had their fruits eaten mostly by legitimate dispersers while in other plants it was mostly other birds. It was also negatively correlated with the degree of plant isolation, although not significantly ($r_s = -0.20$, $P > 0.05$). In contrast, the proportion of ripe fruits consumed by non-legitimate dispersers was positively associated with plant isolation ($r_s = 0.21$, $p < 0.005$). There were no significant correlations between the proportion of red or green fruits consumed by either kind of birds and the variables measured for each plant (including proportion of fruits that ripened).

The nine caged individuals of the four species tested did not show any interest in red fruits. When offered the two different mixtures of green and red fruits they consumed only green ones; when offered only red fruits they did not eat. Three individuals (two males and one female) of *Sylvia atricapilla* were forced to eat red fruits with wasp larvae inside. The fruits (N = 3) were defecated within 30 and 75 minutes; they still contained much pulp, and the larvae appeared intact.

Discussion

Pistacia terebinthus is one of the Mediterranean plants whose fruits represent a rich food resource for a large coterie of frugivorous birds. At least 20 species consume fruits of this species in a forest in southeastern Spain; half of these acted as legitimate seed dispersers (seven species being migratory) but accounted for only 39% of the fruit crop removed. The remaining 61% was removed mainly by five species of pulp consumers and five species of seed predators. At the same time, the seeds of *P. terebinthus* are attacked by four species of chalcidoid wasps that can damage a large proportion (up to 42%) of a fruit crop, although much lower figures are usual.

The potential for bird-wasp interactions apparently is high in this plant since the insects are still present in the fruits when birds begin consuming them. However, a closer look at the system reveals that there is almost no interaction between them, and the few cases in which they encounter each other do not necessarily have negative consequences for either. The possibility exists that wasps are precluding fruit maturation in order to avoid being consumed by birds, as it has been suggested by Krischik *et al.* (1989) in the American holly, *Ilex opaca,* in which attacked fruits never ripen (in this case, a fungus associated with a cecidomyiid fly appears to prevent the fruit from changing color from green to red). However, that possibility seems unlikely mainly for two reasons:

(1) Wasps may develop in late-aborted seeds which contain some green endosperm, thus acting as 'seed commensalists. The observation that many individuals, especially in the Agracea area, have a high proportion of aborted fruits but a low proportion of wasp-attacked ones further supports the possibility of seed commensalism. Wasps, however, may also kill some of the seeds by feeding upon the developing embryo, interfering with fruit physiology, and preventing it to change color – but for no adaptive reason –. In those rare occasions in which green fruits contain a larva, the larva may have entered the seed when it was nearly or completely developed and it was too late to preclude maturation.

(2) Larvae within the seeds may not be affected when being ingested by the birds. Evidence is growing, at least in the case of avian frugivory, which shows that insects can survive the passage through the bird's digestive system (Chung & Waller 1986; Jordano 1989; C. A. Domínguez observed it in *Erythroxylum havanense,* and H. F. Howe in *Casearia corymbosa*, pers. comm.; present study). Chung & Waller (1986), for example, suggest that chalcid wasps living in the seeds of *Rhus glabra* may benefit from being transported between stands inside vertebrate guts.

In the case of *P. terebinthus*, if there is a selection on the wasps to prevent fruit maturation, it is exerted only by seed predators and pulp consumers; the former eat and kill any larva along with the endosperm, and the latter, by dropping most seeds under the parent plant, may increase the probability that the infested fruits are found by rodents, ants, carnivores, or even by other birds (such as chaffinches) that search for seeds below shrubs. Preference for insect-attacked fruits has been reported at least in four studies (Redford *et al.* 1984, Drew 1988, and Valburg 1992a, b). Nevertheless, both the wasp-pulp eater and the wasp-seed predator interactions appear to take place too infrequently to result in strong selection on wasps that avoid avian frugivores by precluding fruit ripeness. Furthermore, such selection would only act on the second and third wasp generations, since the first one develops before fruits are removed by birds.

Larger fruit crops, with higher absolute numbers of red fruits, harbored a greater proportion of wasp-infested seeds than smaller fruit crops. Density dependence in insect infestation of fruits or seeds has also been found in other studies (e.g. Roitberg *et al.* 1982; Courtney & Manzur 1985; Jordano 1987). It is possible that wasps locate larger crops more easily than smaller ones. Wasps may therefore be selecting for crops with a low absolute number of red fruits. However, variation in seed filling is probably determined more by pollination and fertilization success and by early growth of fruits – as Grundwag (1975) found in other *Pistacia* species – than by wasp seed predation. Plant traits other than fruit crop size do not appear to be relevant in explaining individual variation in seed predation by wasps.

Although removal of red fruits by birds also

appeared to vary with fruit crop size, the proportion removed was so small that it is improbable that the number of bird-wasp encounters is relevant to the wasp population. It appears to be even less important when considering just those bird species that can interfere with larval development. Removal of red fruits was not associated with incidence of wasps in any plant. It seems unlikely that birds can discriminate between attacked and intact fruits. In the few occasions in which red fruits were pecked and later rejected, the birds were pulp consumers (usually *P. ater)* and dissections of such fruits revealed the absence of larvae. Greenfinches, however, are able to detect unfilled seeds and reject them by bill-weighing in *Pistacia lentiscus* (Jordano 1990).

Some trees were visited mostly by pulp consumers while others mostly by legitimate dispersers. This seems to correspond, at least in part, to the degree of vegetation cover surrounding a particular fruiting tree. Pulp consumers (species in the Paridae) are not as elusive as legitimate dispersers such as *Sylvia* or *Turdus*; they feed upon fruits in open areas from quite visible perches. This may have direct consequences for plant fitness, but apparently it does not affect the wasps in any way. The five species of avian seed predators visited most plants with a similar frequency. In contrast to what has been found by other authors (Greig-Smith 1986; Foster 1990), neither the (presumed) extra conspicuousness of the bicolored fruits nor a larger fruit crop size attracted one type of frugivore more than another.

As in the case of *P. lentiscus* (Jordano 1989), birds prefer mature fruits over red fruits because pulp of the former contains more than 10 times the amount of lipids than the latter. Herrera (1987) gives even higher values for the lipid content of ripe fruits of this species (55.6%), placing it second in lipid richness among the vertebrate-dispersed plants of the Iberian Peninsula. In addition to having a much lower nutritive value, red fruits have an astringent taste (at least to humans). Seed predators also prefer green fruits to red ones because of their greater amount of endosperm (again, as in *P. lentiscus*; Jordano 1989). It is possible, though not evidenced, that they are also searching for red fruits with larvae inside.

From an evolutionary viewpoint, the possible ability of the wasps to preclude fruit maturation and hence avian predation cannot be attributed to the present avian consumers of the fruits. As is the case with virtually any plant species, *P. terebinthus* has had different dispersers throughout its evolutionary history. The subtropical genus *Pistacia* belongs to an old Tertiary flora present in the Mediterranean scrubland (see Suc 1984 and references therein). Seed dispersal of this species in ancient times was probably achieved by a tropical African avifauna (species in the families Musophagidae, Coliidae, Trogonidae among others; Cracraft 1973) present in Europe during the mid-Tertiary period. The co-occurrence of the present Mediterranean avifauna and the genus *Pistacia* appears, thus, to be much younger than the fruiting biology of *P. terebinthus* (sec Herrera 1984b and references therein). As is the case with *P. lentiscus* (see Jordano 1989), the widely different biogeographies of the interacting species further indicate that the interaction is very recent. The insects, on the other hand, have probably been associated with *P. terebinthus* for a long (but unknown) period. Hence, it is not reasonable to ascribe the observed life history traits of the insect seed predators to the current feeding behavior of avian frugivores, and vice versa. The interaction between these two guilds probably has a much lower intensity than other factors affecting them, such as weather, predators, and diseases. The traits of both insects and birds related to fruit consumption must have evolved before their present ecological interaction. It appears then that the evolutionary ecology of fruit consumption in this system, and probably in many others, would be better addressed with a synthesis of ecology, phylogeny, and biogeography.

In short, the interaction between avian and insect frugivores in the system studied here seems to be quite weak, resulting in an irrelevant influence on each other. The implications of such interaction to the plant are apparently negligible mainly because wasps do not seem to represent an important detrimental effect on the plant fitness. This appears to be more determined by the fertilization success, early fruit growth, and by the type of frugivore that disperses the seeds. In other systems, however, the vertebrate-insect-fruit interaction may come out to be quite important, mainly from the plant viewpoint. This may be more likely to occur when the incidence of insects in the fruits is relatively high and/or when the vertebrate frugivores can discriminate between attacked and intact fruits, either preferring or rejecting the former ones. Such ecological interactions certainly deserve further attention, and detailed studies will allow to determine whether they translate into evolutionary forces that act on each one of the components of the triad.

Acknowledgements

I especially want to thank J. A. Amat, C. M. Herrera, and P. Jordano for teaching me a lot about the birds of Cazorla. I am also very grateful to Quico González and Mari Carmen Mazuelos (Instituto de Recursos Naturales, C.S.I.C., Sevilla) for carrying out the chemical analyses of fruit pulp, and to María Jesús Verdú (Universitat de Valencia) for kindly identifying the wasps. The Agencia del Medio Ambiente allowed me to work in Cazorla and provided valuable logistic support. P. Jordano and C. M. Herrera were always ready for suggestions and discussions, and I greatly benefited from that. Other people at the Estación Biológica de Doñana helped in different ways during the study. P. Jordano, M. F. Willson, T. H. Fleming and P. Charles-Dominique made constructive comments on the manuscript, and I deeply thank them all.

References

Bauer, G. 1986. Life-history strategy of *Rhagoletis alternata* (Diptera: Trypetidae), a fruit fly operating in a 'non-interactive' system. J. Anim. Ecol. 55: 785–794.

Bigler, F. & Delucchi, V. 1981. Wichtigste Mortalitatsfaktoren wahrend der prapupalen Entwicklung der Olivenfliege, *Dacus oleae* Gmel. (Dipt., Tephritidae) auf Oleastern und kultivierten Oliven in Westkreta, Griechenland. Z. ang. Ent. 92: 343–363.

Borowicz, V. A. 1988. Do vertebrates reject decaying fruits? An experimental test with *Cornus amomum* fruits. Oikos 53: 74–78.

Burger, A. E. 1987. Fruiting and frugivory of *Cornus canadensis* in boreal forest in Newfoundland. Oikos 49: 3–10.

Chung, J. C. & Waller, D. M. 1986. Patterns of insect predation on seeds of smooth sumac (*Rhus glabra* L.). Amer. Midl. Nat. 116: 315–322.

Courtney, S. P. & Manzur, M. I. 1985. Fruiting and fitness in *Crataegus monogyna:* the effects of frugivores and seed predators. Oikos 44: 398–406.

Cracraft, J. 1973. Continental drift, paleoclimatology, and the evolution and biogeography of birds. J. Zool. Lond. 169: 455–545.

Crane, J. C. 1975. The role of seed abortion and parthenocarpy in the production of blank pistachio nuts as affected by rootstock. J. Am. Soc. Hort. Sci. 100: 267–269.

Davatchi, A. 1956. Sur quelques insectes nuisibles au pistachier en Iran. Rev. Path. Veg. Entomol. Agr. France 35: 17–26.

Drew, R. A. I. 1987. Reduction in fruit fly (Tephritidae: Dacinae) populations in their endemic rainforest habitat by frugivorous vertebrates. Aust. J. Zool. 35: 283–288.

Drew, R. A. I. 1988. Amino acid increases in fruit infested by fruit flies of the family Tephritidae. Zool. J. Linn. Soc. 93: 107–112.

Fernández-Galiano, E. & Heywood, V. H. 1960. Catálogo de plantas de la provincia de Jaén (mitad oriental). Instituto de Estudios Gienenses, Jaén, Spain.

Foster, M. S. 1990. Factors influencing bird foraging preferences among conspecific fruit trees. Condor 92: 844–854.

Greig-Smith, P. W. 1986. Bicolored fruit displays and frugivorous birds: the importance of fruit quality to dispersers and seed predators. Am. Nat. 127: 246–251.

Grundwag, M. 1975. Seed set in some *Pistacia* L. (Anacardiaceae) species after inter- and intraspecific pollination. Israel J. Bot. 24: 205–211.

Grundwag, M. 1976. Embryology and fruit development in four species of *Pistacia* L. (Anacardiaceae). Biol. J. Linn. Soc. 73: 355–370.

Grundwag, M. & Fahn, A. 1969. The relation of embryology to the low seed set in *Pistacia vera* (Anacardiaceae). Phytomorphology 19: 225–235.

Halevy, G. 1974. Effects of gazelles and seed beetles (Bruchidae) on germination and establishment of *Acacia* species. Israel J. Bot. 23: 120–126.

Herrera, C. M. 1984a. Avian interference of insect frugivory: an exploration into the plant-bird-fruit pest evolutionary triad. Oikos 42: 203–210.

Herrera, C. M. 1984b. A study of avian frugivores, bird-dispersed plants, and their interaction in Mediterranean scrublands. Ecol. Monogr. 54: 1–23.

Herrera, C. M. 1987. Vertebrate-dispersed plants of the Iberian Peninsula: a study of fruit characteristics. Ecol. Monogr. 57: 305–331.

Herrera, C. M. 1989a. Vertebrate frugivores and their interaction with invertebrate fruit predators: supporting evidence from a Costa Rican dry forest. Oikos 54: 185–188.

Herrera, C. M. 1989b. Frugivory and seed dispersal by carnivorous mammals, and associated fruit characteristics, in undisturbed Mediterranean habitats. Oikos 55: 250–262.

Herrera, C. M. 1991. Dissecting factors responsible for individual variation in plant fecundity. Ecology 72: 1436–1448.

Janzen, D. H. 1976. Two patterns of pre-dispersal seed predation by insects on Central American deciduous forest trees. In: Burley, J. & Styles, B. T. (eds), Tropical trees. Variation, breeding and conservation. Linn. Soc. Symp. Ser. #2: 179–188.

Janzen, D. H. 1982. Simulation of *Andira* fruit pulp removal by bats reduces seed predation by *Cleogonus* weevils. Brenesia 20: 165–170.

Jordano, P. 1987. Avian fruit removal: effects of fruit variation, crop size, and insect damage. Ecology 68: 1711–1723.

Jordano, P. 1989. Pre-dispersal biology of *Pistacia lentiscus* (Anacardiaceae): cumulative effects on seed removal by birds. Oikos 55: 375–386.

Jordano, P. 1990. Utilización de los frutos de *Pistacia lentiscus* (Anacardiaceae) por el verderón común (*Carduelis chloris*). Principios en Etología. I Congreso Nacional de Etología, Cordoba, Spain: 145–153.

Krischik, V., McCloud, E. S. & Davidson, J. A. 1989. Selective avoidance by vertebrate frugivores of green holly berries infested with a cecidomyiid fly (Diptera: Cecidomyiidae). Amer. Mid. Nat. 121: 350–354.

Lamprey, H. F. G., Halevy, S. & Makacha, S. 1974. Interactions between *Acacia*, bruchid seed beetles and large herbivores. E. Afr. Wild. J. 12: 81–85.

Manzur, M. I. & Courtney, S. P. 1984. Influence of insect damage in fruits of hawthorn on bird foraging and seed dispersal. Oikos 43: 265–270.

Otero, C., Castién, E., Senosiain, R. & Portillo, F. 1978. Fauna de Cazorla. Vertebrados. Ministerio de Agricultura, Madrid, Spain.

Polunin, O. & Smythies, B. E. 1973. Flowers of south-west Europe, Oxford University Press, London, England.

Redford, K. H., Da Fonseca, G. A. B. & Lacher, T. E. Jr. 1984. The relationships between frugivory and insectivory in primates. Primates 25: 433–440.

Roitberg, B. D., Lenteren, J. C., Alphen, J. J. M., Galis, F. & Prokopy, R. J. 1982. Foraging behaviour of *Rhagoletis pomonella*, a parasite of hawthorn (*Crataegus viridis)*, in nature. J. Anim. Ecol. 51: 307–325.

SAS Institute Inc. SAS User's Guide: Statistics, Version 5 Edition. Cary, NC: SAS Institute Inc., 1985. 956 pp.

Semel, B. & Andersen, D. C. 1988. Vulnerability of acorn weevils (Coleoptera: Curculionidae) and attractiveness of weevils and infested *Quercus alba* acorns to *Peromyscus leucopus* and *Blarina brevicauda*. Amer. Midl. Nat. 119: 385–393.

Suc, J. P. 1984. Origin and evolution of the Mediterranean vegetation and climate in Europe. Nature 307: 429–432.

Valburg, l. K. 1992a. Ealing infested fruits: interactions in a plant-disperser-pest Triad. Oikos 65: 25–28.

Valburg, L. K. 1992b. Feeding preferences of common bush-tanagers for insect-infested fruits: avoidance or attraction? Oikos 65: 29–33.

Zohary, M. 1952. A monographical study of the genus *Pistacia*. Palestine J. Bot. 5: 187–238.

Vegetatio **107/108**: 205–216, 1993.
T. H. Fleming and A. Estrada (eds). Frugivory and Seed Dispersal: Ecological and Evolutionary Aspects.

Ecological and behavioral consequences of digestion in frugivorous animals

Carlos Martínez del Rio [1] & Carla Restrepo [2]
[1] *Department of Ecology and Evolutionary Biology, Princeton University, Princeton, NJ 08544–1003, USA;*
[2] *Department of Zoology, University of Florida, Gainesville, FL 32611, USA*

Keywords: Digestion, Nutrients, Profitability, Feeding choices, Lipids, Sugars

'Ecology is nice, behavior is good,
but first you have to know the anatomy and physiology'
Horst Schwassman

Abstract

The postingestional effect of seed size and mass and nutrient composition on fruit profitability are reviewed. It is emphasized that profitability results from the interaction between fruit characteristics and the physiological and morphological traits of frugivores. The processes by which frugivores regurgitate or defecate seeds and the differential processing of the nutrient rich pulp and ballast in fruit are strongly dependent on the interaction between frugivore gut morphology and seed size. Euphonias that lack a functional gizzard defecate seeds, whereas tanagers that have a delete a functional gizzards regurgitate seeds. Some frugivores separate seeds from pulp and exocarp in the gizzard. It is hypothesized that the gizzard plays an important role in determining the postingestional fate of different pulp components. Although fruit nutrient content is often invoked as a determinant of frugivore feeding choices, studies that rely on proximal nutrient analysis, have often failed to find clear nutrient composition-preference correlations. It is argued that a partial reason for this failure is that proximal nutrient analysis ignores the complexities of fruit digestion. The ecological and physiological correlates of lipid and sugar assimilation are used to identify the limitations of traditional proximal nutrient analyses in fruit-frugivore studies. We suggest that recognizing the intricacies of the digestive characteristics of frugivores may reveal a much richer patterning in the interaction of frugivores with plants than has been previously hypothesized.

Introduction

After ingesting a fruit, frugivores have to digest and absorb the nutrients contained in it, get rid of seeds (undigestible ballast), and eventually may have to metabolize assimilated secondary compounds (Izhaki & Safriel 1989). The physiological events occurring after a frugivore handles (or mandibulates) and swallows a fruit are not immediately obvious to casual observers sitting under a fruiting tree but may be important to the frugivore. Seed processing and disposal and nutrient digestion and absorption are important components of fruit profitability and probably influence frugivore feeding choices.

Fruit profitability is determined by the net rewards provided in pulp relative to the time and effort required to acquire and process it (Schoener 1971; Martin 1985). Fruit profitability can be ranked using many variables (Martin 1985). Here we concentrate on those variables that include the interaction of fruit characteristics and frugivore traits (profitability *sensu-stricto*), and ignore the effect of the activity of frugivore predators and competitors on profitability (profitability *sensu-lato*, Martin 1985). Fruit profitability sensu-stricto can be subdivided into pre- and postingestional components. The preingestional components (fruit size, search and handling times) have received considerable attention (Moermond & Denslow 1983; Wheelwright 1985 and references there), whereas the postingestional components (digestibility, nutritional quality, processing rate) have been relatively neglected (Worthington 1989).

Although nutrition and digestion are frequently invoked as potential correlates of fruit profitability and frugivore fruit choice (Moermond and Denslow 1985), few studies have directly examined the digestive and nutritional physiology of frugivores (Walsberg 1975; Karasov & Levey 1990). Here we concentrate on those fruit characteristics that can influence the postingestional components of fruit profitability: seed size, seed mass, and fruit nutrient composition.

We first review how frugivores deal with the undigestible ballast contained in fruit (seeds and the largely undigestible exocarp). We argue that postingestional seed disposal is an important component of fruit profitability and that postingestional seed handling strongly depends on frugivore gut morphology. We then assess the contribution of fruit nutrient composition studies to current views of frugivore-plant interactions. By focusing on the ecological and physiological correlates of digestion of two kinds of nutrients commonly present in fruit, lipids and carbohydrates, we identify the limitations of traditional proximal nutrient analyses (Van Soest 1983) as a tool in fruit-frugivore studies. Relatively subtle differences in the chemical structure of nutrients can have profound implications for frugivores. We contend that the major weakness of proximal nutrient analysis is that it quantifies broad nutrient classes but ignores the identity of specific nutrients. We claim that by shifting our attention from relatively crude proximal nutrient analyses of fruit to the study of specific nutrients, we may find significant patterns of correspondence between groups of animals and groups of plants. We support this claim by providing a partial explanation for the contrasting distribution of sugars in fruits dispersed by different kinds of frugivores.

Seed processing

Many wild fruits can be characterized by a bulky, nutrient dilute pulp that is accompanied by a relatively large seed mass (30–40% of total dry fruit mass, Moermond & Denslow 1985) that is non-assimilable. The indigestible seed mass reduces gut capacity and increases the energy expenditure of flight (Norberg 1990). Liddy (1982), for example, remarked that the weight of an actively feeding mistletoebird (*Dicaeum hirundinaceum*) is increased by about 20% by the undigestible bulk of mistletoe seeds in the gut. Although birds that feed primarily on mistletoe fruits (*Euphonia, Dicaeum*) may be atypical in that they swallow fruits and defecate seeds that are large relative to their size (Richardson & Wooler 1988), they illustrate the dilemma of fruit-feeding vertebrates: how to minimize the cost of ingesting undigestible seed bulk while maximizing the rate of energy and nutrient uptake. The means used by frugivores to get rid of seeds may have consequences for plants because they may influence patterns of seed deposition (Levey 1987).

Regurgitation and seed discarding

Some fruit-eating animals discard seeds before ingesting pulp and thus avoid filling the digestive organs with undigestible ballast. Many tanagers and finches masticate fruit and drop the seeds (they are fruit 'mashers', Levey 1986). This is also the method used by stenodermatine and pteropodid fruit bats when feeding on large-seeded fruits (Thomas 1982; Fleming 1986). They consume fruit in small bites pressing out and swallowing juice and spitting out the rind (Morrison 1980). Other frugivores regurgitate seeds (Levey 1987), presumably before they are transported to the intestine where digestion and absorption of nutrients occurs (Alpers 1987). Sorensen (1984) demonstrated that European blackbirds (*Turdus merula*) increased their rate of energy intake by consuming fruits whose seeds were regurgitated. *T. merula* preferred fruits with regurgitated seeds over those with seeds that were defecated and that presumably took up intestine space for longer time periods (see also Johnson *et al.* 1985).

Internal separation of pulp and seeds

Seed discarding and regurgitating are only possible when seeds are relatively large (Levey 1987). Even stenodermatine bats, such as *Artibeus jamaicensis*, which are the quintessential mashers (*sensu* Levey 1986) swallow about 80% of the minute seeds of figs (*Ficus spp*). How do frugivores minimize the lost opportunity of digestion imposed by a gut full of ballast? Levey & Grajal (1991) and Levey & Duke (1991) used retention time measurements and radiographic techniques to show that Cedar Waxwings (*Bombycilla cedrorum*) separate seeds from pulp and process them separately in the gut. While pulp is processed in the intestine, seeds collect in the gizzard. After the pulp is processed, seeds are rapidly shunted through the intestine (Levey & Duke 1991). The gizzard to rectum transit time of seeds is extremely short (about 3 minutes). The gizzard to rectum transit time of pulp is much longer, presumably to permit nutrient assimilation (Levey & Duke 1991). The intestine of waxwings appears to function intermittently, alternating periods of assimilation with short bursts of seed disposal. Levey & Grajal (1991) have suggested that separation of seeds and pulp in cedar waxwings depends on seed size and functions more efficiently for large than for small seeds. Consequently, cedar waxwings show higher rates of pulp ingestion when feeding on large-seeded artificial fruits than when feeding on small-seeded artificial fruits (Levey & Grajal 1991).

Processing mistletoe fruits

The morphology of some fruits precludes separation of pulp and seed. The nutrients contained in the pseudoberry of many mistletoes, for example, are embedded in an undigestible viscine layer (Bhatnagar & Lori 1983). Many birds discard the epicarp before ingesting only pulp and seed. Afterwards, the seeds surrounded by the undigested viscin layer are either defecated or regurgitated (Liddy 1982). Other birds, such as phainopeplas (*Phainopepla nitens*), ingest whole mistletoe fruits including the epicarp (Cowles 1936). Phainopeplas use the gizzard to extrude the pseudoberry's seed and pulp into the intestine, separating them from the epicarps which are temporarily retained in the gizzard and then extruded in a single package (Walsberg 1975). Internal fruit processing by phainopeplas resembles seed-pulp separation in cedar waxwings, and similar mechanisms may be involved in both cases. In both cases nutrient containing portions (in one case pulp, in the other seeds surrounded by nutrients) are separated from undigestible portions (seeds and epicarps) and processed separately. The gizzard (muscular stomach, McLelland 1979) of many frugivorous birds may be an important organ implicated in the internal manipulation of fruit (Walsberg & Thompson 1990).

Why do some birds discard and regurgitate seeds whereas others defecate them?

Seed handling depends on both fruit characteristics and on the traits of birds. The same fruits are treated in different ways by different groups of birds (Levey 1987 and references there). For example, the fruits of *Phoradendron spp* in Colombia are consumed by tanagers (*Tangara spp*) and euphonias (*Euphonia spp* and *Chlorophonia spp*, Restrepo 1987). The larger tanagers regurgitate seeds whereas the smaller euphonias defecate them. This difference can probably be explained by differences in digestive tract morphology. In euphonias the gizzard is reduced to a narrow zone of the tubular tract between the proventriculus and the duodenum (Wetmore 1914); tanagers have well developed gizzards that may impede seed passage to the intestine (Forbes 1880; Desselberger 1931). Information on the functional morphology of the digestive tract of frugivores is essential to understand the mechanisms underlying 'internal' fruit processing. With the exception of the detailed study by Chivers & Hladik (1980) on primate digestive tract morphology, our information on the gut morphology of most frugivorous taxa is limited.

Seed processing and fruit design characteristics

The influence of seed packaging (pulp/seed ratios and seed size) on frugivore preferences has been demonstrated in the field (Howe & Vande Kerckhove 1980), and in the laboratory (Sorensen 1984; Levey & Grajal 1991; Murray *et al.* this volume). However, only one study has convincingly shown evidence of the influence of frugivore selectivity on the evolution of seed packaging. Herrera (1981a) found that *Smilax aspera* had lower pulp/seed ratios in areas where fruits were scarce relative to seed dispersers, and therefore where frugivores had little choice, than in areas where dispersers were scarce relative to fruits.

Seed size and morphology are probably the result of a plethora of evolutionary forces. Howe (1986) suggested that the evolution of seed size in *Virola surinamensis* is governed by two opposing forces: frugivore selectivity favoring the production of small seeded fruits and seedling vigor favoring large seeded fruits. The evolutionary consequences of frugivore selectivity on seed packaging are probably constrained by forces acting on seedling survival.

Seeds, the raison d'être of fruit, are surrounded by pulp. The nutrients contained in pulp are the reason why frugivores ingest fruit and eventually disperse fruits. The next section discusses the nutritional value of fruit pulp.

Fruit nutrient content

Fruit pulp nutrient composition varies enormously (Moermond & Denslow 1985). Although a classification of fruits according to nutrient composition can be developed using a variety of criteria, earlier classifications have emphasized a dichotomy between 'high investment fruits' and 'low investment fruits' (McKey 1975; Howe & Estabrook 1977). In this dichotomy, high investment fruits are lipid-rich fruits which are considered highly nutritious. Low investment fruits are low in lipids and high in sugars and are deemed to be nutritionally poor (Howe & Smallwood 1982). This peculiar lipid chauvinism which pervades the literature on fruit nutrient composition has been perpetuated by the wide use of Herrera's overall profitability measure (OP, Herrera 1981b),

$$OP = (RY) \cdot (d)$$

where RY equals the amount of dry pulp (seedless material) in a fruit and d is the relative richness (on a weight to weight basis) of dry pulp in crude lipid and protein. Why carbohydrates are considered of such low quality as energy sources as to be disregarded in an estimation of fruit nutritional yield is unclear from a biochemical perspective.

McKey (1975) and Howe & Estabrook (1977) established a hypothetical relationship between fruit nutritional quality (based on a lipid content scale) and the characteristics of plants. They hypothesized that low investment fruits should be produced superabundantly by plants with peaked fruiting phenologies and visited by diverse assemblages of opportunist frugivores. High investment fruits, in contrast, should have extended fruiting phenologies and be consumed by a limited set of specialized frugivores. The limited comparative data available does not strongly support this hypothesis. Although a few species that produce oily arils do show low diversity assemblages of large frugivores (*Virola spp*, *Casearea corymbosa*, Howe & Estabrook 1982), many species that produce oily fruits are visited by very large assemblages (*Stemmadenia donnell-smithii*, *Ocotea tonduzii*, *Sapium oligoneurum*, McDiarmid *et al.*

1977; Wheelwright *et al.* 1984). Many Melastomataceae produce juicy, sugar-rich fruits, and seem to be visited by low diversity assemblages (Wheelwright *et al.* 1984).

McKey's hypothesis (McKey 1975 and Howe & Smallwood 1977) stimulated a large number of studies on fruit nutrient composition (Foster & McDiarmid 1983; Johnson et al. 1985; Herrera 1987; Borowicz & Stephenson 1985, among others), and a flurry of experimental and field studies that attempted to correlate the nutrient composition of fruit with frugivore preferences (Sorensen 1984; Johnson *et al.* 1985; Borowicz 1988; Foster 1990). With few exceptions (Herrera 1982), these studies failed to uncover strong patterns. Herrera's conclusion after surveying the composition of 110 Spanish fruit species reveals a widespread and rather pessimistic point of view: 'The study failed to reveal any overwhelming influence of the disperser/dispersal environments on the characteristics of Iberian fruits taken as a whole' (Herrera 1987, Page 332).

The outcomes of preference experiments with frugivores using natural fruits were not more encouraging. Most documented a conspicuous lack of correlation between nutrient composition and preferences (Sorensen 1984; Johnson *et al.* 1985; Borowicz 1988; Foster 1990, among others).

The factors governing diet choice, fruit nutrient composition, and their influence on seed dispersal have proved to be more complicated and elusive than originally believed (Wheelwright *et al.* 1984). The lack of clear patterns in nutrient composition and frugivore preferences has two possible explanations: frugivores may not use nutrient composition as a factor in fruit selection, or we may be measuring nutrient composition in ways that are irrelevant to frugivores. Commenting on the zoocentric view of most seed dispersal studies, Janzen (1983) recommended that we think like plants. Our failure to correlate preferences of frugivores for fruits with different nutritional contents may reflect our inability to think like frugivores rather than a true absence of patterns.

Fruit composition studies usually rely on proximate nutrient analyses which emphasize the dichotomy between relatively undigestible cell wall components and readily degradable cell contents (Robbins 1983, Herbst 1988). Although this approach has been relatively successful for studying the nutrition of herbivores (folivores and grazers, Van Soest 1983; Grajal *et al.* 1989), it may be inadequate for studying the nutrition of fruit eating animals (with the possible exception of horses and gomphotheres, Janzen *et al.* 1985).

Limitations of fruit proximal nutrient analyses

In contrast with most herbivores, frugivores appear to depend very little on microorganisms for nutrient assimilation (Klite 1965; Karasov & Levey 1990). The 'nutrient averaging' that turns most nutrients into volatile fatty acids and microorganism biomass in the gut of foregut fermenting herbivores does not occur in most frugivores (Stevens 1988), which in consequence must rely on very specific enzymatic and transport pathways to digest and metabolize particular nutrients (Vonk & Western 1984). Because frugivores assimilate or fail to assimilate specific carbohydrates, fats, and proteins, the synthetic information contained in proximate analyses (lipid ether extractions, total Kjehldal nitrogen, and nonstructural carbohydrates) fails to convey the nutritional complexity of fruit; its widespread use may have greatly impoverished our perspective on fruit-frugivore interactions. Lipids are a good example of the loss of nutritional information in proximal nutrient analyses.

Lipids in fruit

Proximal nutrient analysis uses non-polar solvent extraction (ether or chloroform, methanol and water, Herbst 1988) to estimate fruit lipid contents. These extractions – summarized as dry weight percent of lipids – are constituted by complex mixtures of simple (acylglycerols and waxes), compound (galactosylglycerides, phospholipids) and derived lipids (volatile oils, pigments and resins, Nawar 1985). The identity of the components of the mixture is relevant because the processes by which different lipids are assimilated by vertebrates are quite specific (Carey *et al.* 1983).

Vertebrates digest and absorb lipids by a complex process that involves bile secretion, lipid emulsification, hydrolysis of fatty acid esters, and the absorption of the hydrolyzed products in the enterocyte (Castro 1985). Several steps in this process are substrate-specific and potentially limiting (e.g. different kinds of lipids are hydrolyzed by relatively specific lipases and stearases, Carey *et al.* 1983). Although we know extremely little about how fruit eating animals assimilate fats it is safe to predict that different groups of frugivores will show contrasting abilities to digest lipids and will probably exhibit differences in the ability to assimilate different kinds of lipids. For example, the presence of a gall bladder is very variable among fruit-eating birds. The gall bladder stores and concentrates bile which functions mainly to emulsify fats prior to digestion (Ziswiler & Farner 1972). The absence of a gall bladder may cause the inability to concentrate bile which may prevent the efficient assimilation of a lipid rich diet. The waxwing (*Bombycilla garrulus*) has no gall bladder (Kuhl 1820); among fruit pigeons some species have gall bladders (e.g *Ptilonopus spp*, *Lopholaemus antarcticus*, Garrod 1874) whereas others do not (e.g. *Treron spp* and *Goura cristata*, Garrod 1874); toucans and aracaris (Ramphastidae) have extremely well developed gall bladders (Forbes 1882 in McLelland 1979). We predict that the absence of a gall bladder will be found in birds, such as waxwings, that feed primarily on juicy fruits and berries with high sugar and low lipid contents (Pulliainen *et al.* 1981).

Two puzzling examples illustrate the potential importance of the variation in ability to assimilate lipids as an influence on feeding assemblage composition. Greenberg (1981) reported that the small waxy fruits of *Lindackeria laurina* (Flacourtiaceae) in Panama are consumed mainly by migrant warblers (mainly *Dendroica castanea*). Place and Stiles (1992) have shown that yellow-rumped warblers (*Dendroica coronata*) can assimilate the waxes present in bayberries (*Myrica spp*). The waxes in bayberries consist primarily of mono and diglycerides of myristic, palmitic and stearic fatty acids (Place & Stiles 1992). Chickens assimilate these saturated fatty acids poorly (Renner & Hill 1960). Perhaps the high wax content of *Lindackeria* fruits makes them relatively undigestible for frugivores other than warblers.

Restrepo (1987) has shown that in Colombia the fruits of the two mistletoe families Viscaceae and Loranthaceae are consumed and dispersed by largely non-overlapping assemblages of frugivores, even when plants of the two families are fruiting together in the same host tree. Species of Viscaceae produce carbohydrate-rich fruits that are consumed by euphonias and tanagers, whereas species of Loranthaceae produce lipid-rich fruits that are consumed mainly by frugivorous flycatchers (Tyrannidae, see also Davidar 1983). We predict that these two contrasting assemblages of frugivores will show contrasting fat assimilation abilities: Euphonias and tanagers (*Euphonia spp* and *Tangara spp*) should be poor lipid assimilators whereas many frugivorous flycatchers (*Mionectes spp*, *Zimmerius spp*, Fitzpatrick 1980) should be efficient lipid assimilators.

Sugar composition of fruit pulp

Total 'available' carbohydrate content in fruit is routinely estimated as the material remaining after ash, fiber, lipids, and total nitrogen have been estimated (Herbst 1988). This fraction includes simple sugars (mono- and disaccharides), starch, and amylopectin (Southgate 1969). Because humans, rats, and large herbivores assimilate these chemically similar substances with more or less the same very high efficiency (Alpers 1985; Stevens 1988), it is usually assumed that the total available carbohydrate fraction of fruit is readily assimilable by frugivores and that therefore its precise composition is irrelevant (Janson *et al.* 1986; White 1974). A remarkable data set collected by Baker *et al.* (1992) challenges this assumption.

Baker *et al.* (1992) analyzed the content of the three sugars most commonly found in fruit in a large sample of species (more than 200 species).

They discovered that the pulp of most bird-dispersed fruits is rich in glucose and fructose but contains only very small amounts of sucrose. Cultivated fruits used for human consumption and mammal dispersed fruits contain significantly higher sucrose proportions (Baker *et al.* 1992; Freeman & Worthington 1989). This broad pattern in the composition of fruit pulp appears to hold within restricted phylogenetic groups. In the Rosaceae, for example, mammal-dispersed fruits have significantly higher sucrose proportions (mean $\pm$ SD $= 0.13 \pm 0.01$, $n = 8$) than bird-dispersed fruits (0.041 ± 0.03, $n = 14$; $t = 15.3$, $p < 0.05$, data kindly provided by Herbert Baker).

Why does the pulp of fruits dispersed by different groups of animals contain different kinds of sugars? Conventional ecological wisdom suggests that the sugar composition of the rewards that plants offer should be the result of selection by pollinators and seed dispersers that vary in their preferences for simple sugars (Baker & Baker 1983; Martínez del Rio & Stevens 1989). Although the data available are limited, the known preferences of frugivores seem to nicely mirror the composition of fruit pulp: many frugivorous primates strongly prefer sucrose (Dennys & Hladik 1988; Sunderland & Sclafani 1988; Ramirez 1990), whereas frugivorous birds appear to prefer glucose and fructose over sucrose (Martínez del Rio *et al.* 1988, 1989). The sugar preferences of fruit-eating bats are unknown.

Sugar digestion and the preferences of fruit-eating birds

The variation in sugar preferences among frugivores is puzzling. Why should animals show marked preferences among readily assimilable sugars which have very similar energetic content? Neither standard foraging theory (Stephens and Krebs 1986), which ignores the complexities of digestion, nor standard 'proximal nutrient' nutritional ecology, which ignores the details of nutrient assimilation, can provide an explanation. The physiological details of sugar assimilation provide a hypothesis. Sucrose is a disaccharide that has to first be hydrolyzed into its monosaccharide constituents, glucose and fructose, by intestinal enzymes to be assimilated (Norén *et al.* 1986). Glucose and fructose are absorbed intact in the intestine (Semenza & Corcelli 1986). Martínez del Rio & Stevens (1989) hypothesized that differences among frugivores in the ability to digest sucrose can explain their contrasting preferences.

The digestive correlates of sucrose preference in primates are unknown. There is enough data, however, to at least outline some of the reasons why avian frugivores prefer glucose and fructose over sucrose. Briefly, fruit-eating birds appear to be relatively inefficient at assimilating sucrose, but to be very efficient at assimilating glucose and fructose (Martínez del Rio *et al.* 1988, 1989). Two physiological mechanisms seem to account for the inefficient sucrose digestion in avian frugivores: lack of functional intestinal sucrase activity and extremely fast food passage rates (Martínez del Rio *et al.* 1989; Martínez del Rio 1990).

Some fruit-eating birds lack intestinal activity of sucrase, the enzyme responsible for the hydrolysis of sucrose into glucose and fructose (Martínez del Rio & Stevens 1989). For these birds sucrose is useless as an energy source and can cause osmotic diarrhea (Brugger & Nelms 1991). Birds lacking intestinal sucrase activity reject concentrated sucrose solutions and develop a conditioned aversion to even relatively dilute sucrose solutions (Brugger 1991; Martínez del Rio *et al.* 1988). Lack of functional sucrase activity has been reported in five species of passerine birds in three very closely related frugivorous families (Muscicapidae, Sturnidae, and Mimidae, Martínez del Rio 1990). Martínez del Rio (1990) has hypothesized that lack of functional sucrase in the sturnid-muscicapid lineage (sensu Sibley & Ahlquist 1990) is an example of a phylogenetic constraint with profound ecological consequences. Because the sturnid-muscicapid lineage is widely distributed and includes a large number of seed dispersers, its lack of intestinal sucrase activity probably contributes to the maintenance of low sucrose contents in bird-dispersed fruits.

Lack of sucrase is a sufficient but not necessary

condition for birds to prefer hexoses over sucrose. The extremely fast passage rates of some frugivores (Karasov & Levey 1990) may preclude the efficient hydrolysis and absorption of sucrose relative to the hexoses glucose and fructose, even when these animals have significant amounts of intestinal sucrase. Cedar waxwings, for example, show intestinal sucrase activity, but prefer glucose and fructose over sucrose (Martínez del Rio *et al.* 1989). Cedar waxwings absorb 92% and 88%, respectively, of the glucose and fructose they consume, but only 60% of the sucrose (Martínez del Rio *et al.* 1989). Time at first appearance of marked food in cedar waxwings is about ten minutes, and mean retention times of fluid markers equal 30 minutes (Levy & Grajal 1991). These extremely fast passage rates probably do not provide the processing time required to assimilate complex substrates such as sucrose that must be hydrolyzed to be absorbed (Martínez del Rio & Karasov 1990; Levey & Grajal 1990). In some frugivorous birds, preference for the monosaccharides glucose and fructose over sucrose may be a consequence of the constraint imposed by high food passage rates.

A weakness of the sugar preference and assimilation information available is that all the species that have been well-studied belong to a single lineage, Sibley & Ahlquist's (1990) superfamily Muscicapoidea. Much more data on the sugar preferences and digestive abilities of birds in other frugivorous lineages are needed to assess how general the proposed patterns are. Preliminary data on two fruit-eating emberizids (*Cacicus melanicterus* and *Icteria virens*) support the hypothesis that fruit-eating birds prefer glucose and fructose over sucrose and are relatively inefficient at digesting sucrose (Fig. 1).

The sugar composition of fruit pulp and the sugar preferences and digestive abilities of birds clearly disclose the most important weakness of proximal nutrient analysis approaches to fruit-frugivore interactions. Frugivores assimilate specific nutrients by relatively specific physiological pathways; therefore subtle differences in the chemical composition of the nutrients in fruit can have profound consequences. The amount of carbohydrates (or lipids) in a fruit is an important parameter. Its usefulness is limited however, without information about a) the kinds of carbohydrates present, and b) the digestive abilities of the frugivores in question.

Nutritional ecology and plant-frugivore interactions

Nutritional physiology can contribute to understand several aspects of plant-frugivore interactions. Knowledge about digestive tract morphology and physiology is clearly an essential, albeit neglected, component in comprehensive explanations of the feeding behavior of frugivores. We have suggested that internal seed and fruit-pulp processing depend on the interaction between fruit and frugivore characteristics and that they are important components of fruit profitability. In order to better understand why animals process fruit in contrasting ways, however, we need to acquire adequate morphological and physiological data. We perceive the lack of information on basic aspects of the internal morphology and physiology of frugivores as one of the crucial shortcomings of current approaches to frugivore-plant interactions. Students of frugivory cannot continue to ignore classical functional morphology and modern physiology.

Nutritional ecology can also contribute to the identification of patterns of correspondence between groups of fruits with similar pulp chemistry and groups of frugivores (chemical 'syndromes', Howe 1986). Furthermore, nutritional ecology can contribute to the elucidation of the physiological mechanisms and behavioral processes responsible for maintaining these patterns. However, if nutritional ecology is to succeed in becoming a useful component of plant-frugivore studies it has to recognize the intricacies of digestion and metabolism. We suspect that earlier nutritional approaches were hindered by an inadequate nutritional paradigm (proximal nutrient analysis) that relied on an extremely simplistic view of the nutritional physiology of frugivores. Recognizing the diversity and complexity of the

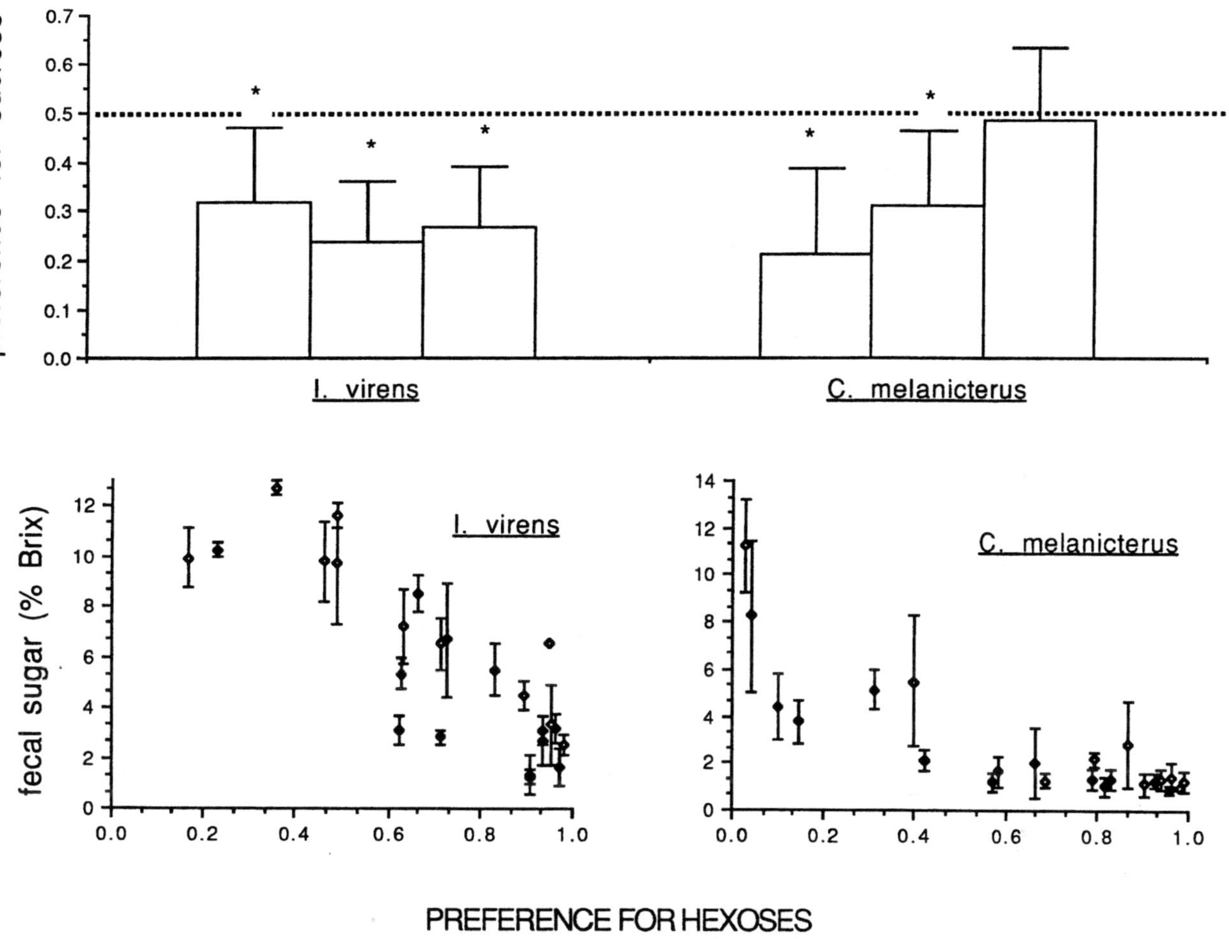

Fig. 1. Sugar preferences and fecal sugar as a function of sugar preferences in *Icteria virens* and *Cacicus melanicterus*. Birds were offered a 15% (wt/wt) sucrose solution paired with a 15% mixture (1:1) of glucose and fructose. Preference for sucrose equals the proportion of sucrose solution ingested. Bars are mean preference values for individual birds (10 replicates/bird), and error bars are 95% confidence intervals. Individuals that exhibited significant preference for the hexose mixture are starred. The dotted line is the line of equipreference, indicating equal consumption of sucrose and hexoses. Lower graphs plot fecal sugar measured with a refractometer as a function of preference for the hexose mixture. Values are mean measurements ± SD for a single trial ($n = 5$ excreta samples). The negative correlations indicate lower sugar assimilation (i.e. more sugar in excreta) when sucrose was preferentially ingested.

digestive characteristics of frugivores may reveal much richer patterns in the interaction of frugivores with plants than previously hypothesized.

Acknowledgements

We thank T. H. Fleming, D. J. Levey, K. G. Murray, and E. W. Stiles for many helpful suggestions.

References

Alpers, D. H. 1987. Digestion and absorption of carbohydrates and proteins. In: Johnson L. R. (ed.). Physiology of the gastrointestinal tract, Vol 2, Raven Press, N. Y. pp. 1469–1486.

Baker, H. G. & Baker, I. 1983. Floral nectar sugar constituents in relation to pollinator type. In: Jones, C. E. & Little, R. J. (eds.), Handbook of experimental pollination ecology. Scientific and Academic Editions, N. Y. pp. 131–171.

Baker, I., Baker, H. G. & Hodges, S. A. 1992. Patterns in the sugar composition of nectar and fruit juices taken by Microchiroptera and Megachiroptera. Biotropica, in press.

Bhatnagar, S. P. & Johri, B. M. 1983. Embryology of Loranthaceae. In: Calder, M. & Bernhardt, P. (eds.) The biology of mistletoes. Academic Press, London, pp. 47–66.

Borowicz, V. A. 1988. Fruit consumption by birds in relation to fat content of pulp. Amer. Midl. Nat. 119: 121–127.

Borowicz, V. A. & Stephenson, A. G. 1985. Fruit composition and patterns of fruit dispersal of two *Cornus* spp. Oecologia 67: 435–441.

Brugger, K. E. 1992. Sucrose as a fruit repellent to American robins (*Turdus migratorius*). J. Wildlife Manag., in press.

Brugger, K. E. & Nelms, C. O. 1991. Sucrose avoidance by American robins (*Turdus migratorius*): implications to control of bird damage in fruit crops. *Crop Protection*, 10: 455–460.

Carey, M. C., Small, D. M. & Bliss, C. M. 1983. Lipid digestion and absorption. Ann. Rev. Physiol. 45: 651–677.

Castro, G. A. 1984. Digestion and absorption. In: Johnson, L. R. (ed.). Gastrointestinal physiology. C. V. Mosby Co., St Louis Missouri, pp. 105–128.

Chivers, D. J. & Hladik, C. M. 1980. Morphology of the gastrointestinal tract in primates: comparison with other mammals in relation to diet. J. Morphol. 166: 337–386.

Cowles, R. B. 1937. The relation of birds to seed dispersal of the desert mistletoe. Madroño 3: 352–356.

Davidar, P. 1983. Birds and neotropical mistletoes: effects on seedling recruitment. Oecologia 60:271–273.

Dennys, V. & Hladik, C. M. 1988. Le seuil préférence du saccharose chez les Prosimiens de Madagascar et son implication dans le comportement alimentaire. Biol. Behav. 13: 43–48.

Desselberger, H. 1931. Der verdauungskanal der dicaediden nach Gestalt un funktion. Jour. f. Ornithol. 79: 353–370.

Fitzpatrick, J. W. 1980. Foraging behavior of neotropical tyrant flycatchers. Condor 82: 43–57.

Fleming, T. H. 1986. Opportunism versus specialization: the evolution of feeding strategies in frugivorous bats. In: Estrada A. & Fleming T. H. (eds.). Frugivores and seed dispersal. Dr. W. Junk Publishers Dordrecht, The Netherlands, pp. 105–118.

Forbes, W. A. 1880. On the structure of the stomach in certain genera of tanagers. Proc. Zool. Soc. London: 143–147.

Foster, M. S. 1990. Factors influencing bird foraging preferences among conspecific fruit trees. Condor 92: 844–854.

Foster, M. S. & McDiarmid, R. W. 1983. Nutritional value of the aril of *Trichilia cuneata*, a bird dispersed fruit. Biotropica 15: 26–31.

Freeman, C. E. & Worthington, R. D. 1988. Is there a difference in the sugar composition of cultivated sweet fruits of tropical/subtropical and temperate origins? Biotropica 21: 219–222.

Garrod, A. H. 1874. On some points in the anatomy of the Columbae. Proc. Zool. Soc. Lond. 249.

Grajal, A., Strahl, S. D., Parra, R., Dominguez, M. G. & Neher, A. 1989. Foregut fermentation in the Hoatzin, a neotropical leaf-eating bird. Science 245: 1236–1238.

Greenberg, R. 1981. Frugivory in some migrant tropical forest wood warblers. Biotropica 13: 215–223.

Herbst, L. H. 1988. Methods of nutritional ecology of plant-visiting bats. In: Kunz, T. H. (ed.). Ecological and behavioral methods for the study of bats. Smithsonian Institution Press, Washington D. C., pp. 233–246.

Herrera, C. M. 1981a. Fruit variation and competition for dispersers in natural populations of *Smilax aspera*. Oikos 36: 51–58.

Herrera, C. M. 1981b. Are tropical fruits more rewarding than temperate ones? Am. Nat. 118: 896–907.

Herrera, C. M. 1982. Seasonal variation in the quality of fruits and diffuse coevolution between plants and avian dispersers. Ecology 63: 773–785.

Herrera, C. M. 1987. Vertebrate-dispersed plants of the Iberian peninsula: a study of fruit characteristics. Ecol. Monographs 57: 305–331.

Howe, H. F. 1986. Seed dispersal by fruit-eating birds and mammals. In: Murray, D. R. (ed.). Seed dispersal. Academic Press, London, pp. 123–189.

Howe, H. F. & Estabrook, G. F. 1977. On intraspecific competition for avian dispersers in tropical trees. Am. Nat. 111: 817–832.

Howe, H. F. & Vande Kerckchove, G. A. 1980. Nutmeg dispersal by tropical birds. Science 210: 925–927.

Howe, H. F. & Smallwood, J. 1982. Ecology of seed dispersal. Ann. Rev. Ecol. Syst. 13: 201–208.

Izhaki, I. & Safriel, U. N. 1989. Why are there so few exclusively frugivorous birds? Experiments on fruit digestibility. Oikos 54: 23–32.

Janzen, D. H. 1983. Seed and pollen dispersal by animals: convergence in the ecology of contamination and sloppy harvest. Biol. J. Linn. Soc. 20: 103–113.

Janzen, D. H., Demmet, E. M. & Robertson, J. B. 1985. How fast and why do germinating guanacaste seeds (*Enterolobium cyclocarpum*) die inside cows and horses. Biotropica 17: 322–325.

Janzon, C. H., Stiles, E. W. & White, D. J. 1986. Selection on plant fruiting traits by brown capuchin monkeys: a multivariate approach. In: Estrada A. and Fleming T. H. (eds.). Frugivores and seed dispersal. Dr. W. Junk Publishers Dordrecht, The Netherlands. pp. 83–92.

Johnson, R. A., Willson, M. F., Thompson J. N. & Bertin, R. I. 1985. Nutritional values of wild fruits and consumption by migrant frugivorous birds. Ecology 66: 819–827.

Karasov, W. H. & Levey, D. J. 1990. Digestive system tradeoffs and adaptations of frugivorous passerine birds. Phys. Zool. 63: 1248–1270.

Klite, P. D. 1965. Intestinal bacterial flora and transit time of three neotropical bat species. J. Bact. 90: 375–379.

Kuhl, H. 1820. Betrage zur Zoologie und vergleichenden Anatomie. Abt. Urban and Schwarzenberg, Frankfurt-am-Main.

Levey, D. J. 1986. Methods of seed processing by birds and seed deposition patterns. In: Estrada, A. & Fleming, T. H. (eds.). Frugivores and seed dispersal. Dr. W. Junk Publishers Dordrecht, The Netherlands, pp. 147–158.

Levey, D. J. 1987. Seed size and fruit-handling techniques of avian frugivores. Am. Nat. 129: 471-485.

Levey, D. J. & Duke, G. E. 1991. How do frugivores process fruit? radiographic studies of gastrointestinal transit in cedar waxwings (*Bombycilla cedrorum*). Auk, in press.

Levey, D. J. & Grajal, A. 1991. Evolutionary implications of fruit processing by cedar waxwings. Am. Nat., 138: 171–189.

Liddy, J. 1982. Food of the mistletoebird near Pumicetown Passage, south-eastern Queensland. Corella 6: 11–15.

Martin, T. E. 1985. Resource selection by tropical frugivorous birds: integrating multiple interactions. Oecologia 66: 563–573.

Martínez del Rio, C. 1990. Dietary and phylogenetic correlates of intestinal sucrase and maltase activity in birds. Physiol. Zool. 63: 987–1011.

Martínez del Rio, C. & Karasov, W. H. 1990. Digestion strategies in nectar and fruit eating birds and the sugar composition of plant rewards. Am. Nat. 136: 618–637.

Martínez del Rio, C. & Stevens, B. R. 1989. Physiological constraint on feeding behavior: intestinal membrane disaccharidases of the starling. Science 243: 794–796.

Martínez del Rio, C., Stevens B. R., Daneke, D. & Andreadis, P. T. 1988. Physiological correlates of preference and aversion for sugars in three species of birds. Physiol. Zool. 61: 222–229.

Martínez del Rio, C., Levey, D. J. & Karasov, W. H. 1989. Physiological basis and ecological consequences of sugar preferences in cedar waxwings. Auk 106: 64–71.

McDiarmid, R. W., Ricklefs, R. E. & Foster, M. S. 1977. Dispersal of *Stemadenia donnell-smithii* (Apocynaceae) by birds. Biotropica 9: 9–25.

McLeeland, J. 1979. The digestive system. In: King, A. S. & McLelland, J. (eds.) Form and function in birds, Vol. 1. Academic Press, N. Y, pp. 1–181.

McKey, D. 1975. The ecology of coevolved dispersal systems. In: Gilbert, L. E. & Raven, P. H. (eds.). Coevolution of animals and plants. University of Texas Press, Austin, pp. 159–191.

Moermond, T. C. & Denslow, J. S. 1983. Fruit choice in neotropical birds: effects of fruit type and accessibility on selectivity. J. Anim. Ecol. 52: 407–420.

Moermond, T. C. & Denslow, J. S. 1985. Neotropical avian frugivores: patterns of behavior, morphology, and nutrition with consequences for food selection. In: Buckley, P. A., Foster, M. S., Morton, E. S., Ridgely, R. S. & Smith, N. G. (eds.). Neotropical Ornithology. Ornithological Monographs 36, pp. 865–897.

Morrison, D. W. 1980. Efficiency of food utilization by fruit bats. Oecologia 45: 270–273.

Nawar, W. W. 1985. Lipids. In: Fennema, O. R. (ed.). Food chemistry. Marcel Dekker Inc. Basel, pp. 139–244.

Norberg, U. M. 1990. Vertebrate flight. Springer Verlag, Berlin.

Norén, O., Sjöström, H., Danielsen E. M., Cowell G. H. & Skovbjerg H. 1986. The enzymes of the enterocyte plasma membrane. In Desnuelle P., Sjöström H., and Norén A. (eds.). Mollecular and cellular basis of digestion. Elsevier, N. Y. pp. 335–365.

Place, A. R. & Stiles, E. W. 1991. Bayberries and warblers – living off the wax of the land. Auk, 109: 334–345.

Ramirez, I. 1990. Why do sugars taste good? Neuroscience & Biobehav. Rev. 14: 125–134.

Renner, R. & Hill, W. F. 1960. The utilization of corn oil, lard, and tallow by chickens of various ages. Pult. Sci. 39: 849.

Restrepo, C. 1987. Aspectos ecologicos de la diseminacion de cinco especies de muerdagos por aves. Humboldtia 1: 65–116.

Richardson, K. C. & Wooller, R. D. 1988. The alimentary tract of a specialist frugivore, the mistletoebird, *Dicaeum hirundinaceum*, in relation to its diet. Aust. J. Zool. 36: 373–382.

Robbins, C. T. 1983. Wildlife feeding and nutrition. Academic Press, New York.

Semenza, G. & Corcelli, A. 1986. The absorption of sugars and amino acids across the small intestine. In: Desnuelle P., Sjöström H., and Norén A. (eds.). Molecular and cellular basis of digestion. Elsevier, N. Y. pp. 381–412.

Stevens, C. E. 1988. Comparative physiology of the vertebrate digestive system. Cambridge Univ. Press, Cambridge.

Schoener, T. W. 1971. Theory of feeding strategies. Ann. Rev. Ecol. Syst. 2: 369–404.

Sibley, C. G. & Ahlquist, J. E. 1990. Phylogeny and classification of birds: a study in molecular evolution. Yale University Press, New Haven.

Sorensen, A. E. 1984. Nutrition, energy and passage time: experiments with fruit preference in European Blackbirds (*Turdus merula*). J. Anim. Ecol. 53: 545–557.

Southgate, D. A. T. 1969. Determination of carbohydrates in foods I. – Available carbohydrates. J. SCi. Fd. Agic. 20: 331-335.

Stephens, D. W. & Krebs, J. R. 1986. Foraging theory. Princeton Univ. Press, Princeton, New Jersey.

Stevens, C. E. (1988). Comparative physiology of the vertebrate digestive system. Cambridge University Press, Cambridge.

Sunderland, G. & Sclafani, A. 1988. Taste preferences of squirrel monkeys and bonnet macaques for polycose, maltose and sucrose. Phys. Behav. 43: 685–690.

Thomas, D. 1982. The ecology of an African savanna fruit bat community: resource partitioning and role in seed dispersal. Ph. D. Dissertation, University of Aberdeen.

Van Soest, P. J. 1983. Nutritional ecology of the ruminant. O & B Books Inc., Corvallis, Oregon.

Vonk, H. J. & Western, J. R. H. 1984. Comparative biochemistry and physiology of enzymatic digestion. Academic Press, New York.

Walsberg, G. E. 1975. Digestive adaptations of *Phainopepla nitens* associated with the eating of mistletoe berries. Condor 77: 169–174.

Walsberg, G. E. & Thompson, C. W. 1990. Annual changes in gizzard size and function in a frugivorous bird. Condor 92: 794–795.

Wetmore, A. 1914. The development of the stomach in the Euphonias. Auk 31: 458–461.

Wheelwright, N. T. 1985. Fruit size, gape width, and the diets of fruit-eating birds. Ecology 66: 808–818.

Wheelwright, N. T., Haber, W. A., Murray, K. G. & Guindon, C. 1984. Tropical fruit-eating birds and their food plants: a survey of a Costa Rican lower montane forest. Biotropica 16: 173–192.

White, S. C. 1974. Ecological aspects of growth and nutrition in tropical-fruit eating birds. Ph. D. Dissertation, University of Pennsylvania.

Worthington, A. H. 1989. Adaptation for avian frugivory: assimilation efficiency and gut transit time of *Manacus vitellinus* and *Pipra mentalis*. Oecologia 80: 381–389.

Ziswiler, V. & Farner, D. S. 1972. Digestion and the digestive system. In: Farner, D. S. & King, J. R. (eds.). Avian Biology. Vol II. Academic Press, N. Y. pp. 343–430.

Vegetatio **107/108**: 217–226, 1993.
T. H. Fleming and A. Estrada (eds). Frugivory and Seed Dispersal: Ecological and Evolutionary Aspects.

The influence of seed packaging and fruit color on feeding preferences of American robins

K. G. Murray, K. Winnett-Murray, E. A. Cromie, M. Minor & E. Meyers
Department of Biology, Hope College, Holland, MI 49423, USA

Keywords: Fruit consumption rate, Fruit-eating birds, Pulp:seed ratio, Seed size, Seed retention time, *Turdus migratorius*

Abstract

We investigated the role of seed packaging (division of total seed volume among individual seeds) and fruit color in determining feeding preferences of American Robins (*Turdus migratorius*). Experiments were conducted using artificial fruits with either 8 small plastic beads or a single large one with equivalent volume. Other fruit characters were held constant. As predicted, large seeds were voided rapidly by regurgitation, resulting in higher pulp consumption rates for large-seeded fruits than for small-seeded ones, whose seeds were passed through the gut. Most birds apparently used this difference in profitability as a choice criterion: four of seven preferred large-seeded fruits. That three individuals did not do so suggests that birds may differ in their ability to perceive minor differences in fruit profitability, or to use them as choice criteria. Pulp color was also important: blue fruits were preferred by all seven birds. This preference was surprising, since Robins commonly feed on red fruits in the field.

Introduction

A central theme in the study of plant-frugivore interactions has been the concept of fruit 'quality', its influence on the feeding behavior of fruit-eating animals, and the concommitant consequences for the dispersal services that animals provide to parent plants (McKey 1975; Howe & Estabrook 1977; Howe & Smallwood 1982; Howe 1986). In general, a fruit's 'quality' is roughly equivalent to its 'profitability' in optimal foraging models, i.e., the nutritional reward gained by a consumer minus the energetic costs associated with obtaining, handling, and digesting it. Most studies of fruit quality and frugivore feeding preferences focus on the concentrations of major nutrients (Howe & Vande Kerckhove 1980; Sorensen 1981; Johnson *et al.* 1985; Denslow 1987; Levey 1987a; McPherson 1987; Martinez del Rio *et al.* 1989), caloric content (Sorensen 1984; Johnson *et al.* 1985; McPherson 1987), pulp:seed ratios (Howe & VandeKerckhove 1980; Herrera 1981; McPherson 1987), or seed passage rates (Sorensen 1984; Levey & Grajal, 1991) as determinants of frugivore preferences. The upshot of such studies, conducted with a diverse array of bird and plant species from a variety of locations, is that birds in many cases indeed make rational foraging decisions – fruits with higher nutritional content in the pulp or higher pulp:seed ratios are generally preferred over those recognized by researchers as less profitable.

In many cases, however, other factors exert significant confounding effects, such that feeding preferences are not accurately predicted by the nutritional quality of fruits. Fruit color (Turĉek

1963; Morden-Moore & Willson 1982; Wheelwright & Janson 1985; McPherson 1987, 1988; Willson *et al.* 1990), secondary chemicals (Sherburne 1972; Sorensen 1983), and microbial infection (Buchholz & Levey 1990) have all been shown to be important in some species. In addition, several community-level studies demonstrate the importance of factors extraneous to the fruits themselves, e.g., the relative abundances (Baird 1980; McPherson 1987) and accessibilities (Moermond & Denslow 1983) of fruits of different species. Presumably, this complexity in the determination of feeding preferences is a reflection of the complexity of nutritional demands of fruit-eating animals, which are poorly understood at best. All but a few studies rely on comparisons among naturally occurring fruits, which vary not only in the characteristic of interest (e.g., sugar or lipid content, pulp:seed ratio, etc.), but also in secondary chemistry, flavor, color, abundance, etc. Because natural fruits vary in diverse ways, well-controlled manipulative experiments to determine the bases for feeding preferences of frugivores are exceedingly rare (but see Levey & Grajal, 1991).

We report here on a study designed to evaluate the ability of American Robins (*Turdus migratorius)* to distinguish fine differences in fruit profitability due to different 'packaging' of the total seed load. By using artificial fruits of identical size, pulp composition, and pulp:seed ratio, we manipulated a single characteristic – the packaging of total seed volume into a single large seed or several small ones. Because large seeds are usually reguritated while small ones are passed through the gut, we predicted that pulp consumption rates, hence net rates of energy gain, would be higher for fruits with large seeds. As a consequence, we predicted that Robins would prefer the more profitable, large-seeded fruits.

Methods

The seven Robins used in this study were captured on 7 October 1989 during their southward migration at the Kalamazoo Nature Center, Kalamazoo, MI, USA. All were hatching-year birds. When not used in feeding experiments, birds were housed in large walk-in aviaries on the Hope College campus and maintained on a synthetic diet comprised primarily of bananas and soy protein (Denslow *et al.* 1987). All birds maintained or gained weight until 28 March 1990, when they were released. All birds readily ate both the synthetic diet and experimental fruits throughout the period of captivity, despite their strong reliance on fruits in the field only during the fall and winter (Wheelwright 1986).

All feeding experiments were conducted in small (50 × 30 × 46 cm) cages with one-way mirrors. Artificial fruits were presented in a small dish attached to a perch in each cage. Water was available at all times, as was maintenance food except when experimental fruits were present.

Artificial fruits were prepared by boiling 7.5 g glucose, 7.5 g fructose, and 2.0 g agar in 100 ml distilled water colored with either red or blue food coloring. The hot mixture was then injected into a fishing sinker mold, the individual wells (9.3 mm diameter) of which contained either a single large (6 mm diameter) plastic bead (a seed 'mimic', we use the terms interchangeably hereafter) or eight small (3 mm) ones. Both fruits thus had equivalent pulp (0.308 ml) and total bead (0.113 ml) volume, and differed only in the way the total bead volume was packaged. Beads used in a given fruit were the same color as the pulp; as a result, the bead content of a fruit was probably predictable by birds only by association with pulp color.

Birds were placed in the small cages one day prior to the first feeding trial in each series. To measure consumption rates, we allowed birds to feed freely on large- or small-seeded fruits for a period of one hour. So as not to bias birds' responses to fruit color in the subsequent preference trials, we colored the pulp in these fruits dark purple (equal concentrations of red and blue food coloring). We performed three replicates with each bead size on each bird.

Seed retention times were determined according to methods outlined in Murray (1988). Birds were allowed to feed freely on either large-seeded or small-seeded fruits (again, with purple pulp) for a period of five minutes, after which remaining experimental fruits were removed and maintenance food replaced. Food was thus present at all times. Regurgitated and defecated seeds were then collected at five-minute intervals until all ingested seeds had been voided.

Preference trials were conducted by presenting birds with 10 small-seeded and 10 large-seeded fruits, each with a different pulp color (e.g., large-seeded red and small-seeded blue). We recorded the number of each type eaten (without replacement) until a total of 10 fruits had been consumed. After 15 such trials (one per day) had been conducted on each bird with a given color/seed size combination, we returned all birds to the aviary for one week. Birds were then returned to the small cages, and 15 trials were run on each with the reciprocal color/seed size combination. To avoid systematically biasing birds' color preferences, we randomly assigned three birds to large-beaded red and small-beaded blue fruits, the other four birds to large-beaded blue and small-beaded red fruits for the first set of trials.

To determine whether Robins can use taste or seed:pulp ratio to assess profitability directly, we also conducted preference trials like those above with two types of equicaloric fruits that differed only in the presence/absence of a single large (6 mm) bead and in the sugar concentration of the pulp. Fruits with beads were prepared as described above, with total sugar concentration of 15%. Beadless fruits were prepared with a total sugar concentration of 10.97% (equal concentrations of glucose and fructose); total sugar/fruit was thus equivalent in both fruit types (0.0462 g). For these trials, all fruits were colored blue.

Results

With rare exceptions, the Robins used in this study always ingested fruits whole. Occasionally, an individual mashed experimental fruits in its bill such that seeds were popped out before the pulp was swallowed. In such cases, we disregarded the fruits from which the seeds were removed, and replaced them with intact ones. Data used in the analyses below include only cases in which fruits were ingested whole.

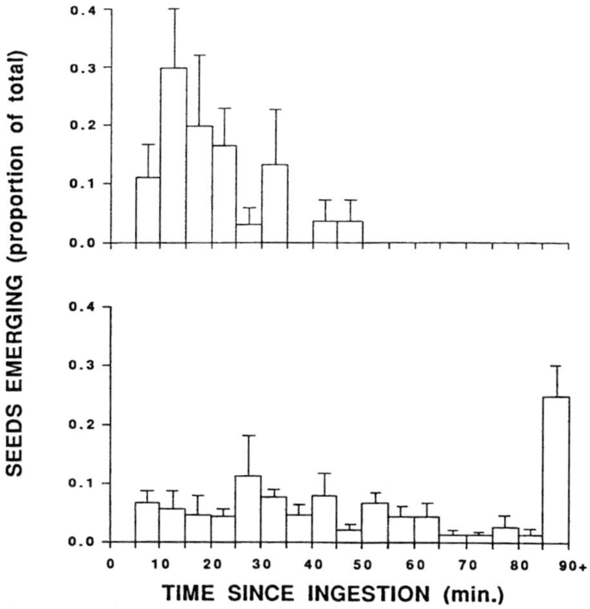

Fig. 1. Retention times for large and small beads. Bars represent the mean proportion voided by 7 Robins in each time interval; error bars are ± 1 standard error.

Seed retention times and fruit consumption rates

Large seeds were invariably regurgitated, while the great majority of small seeds passed through the entire gut and were voided by defecation. A

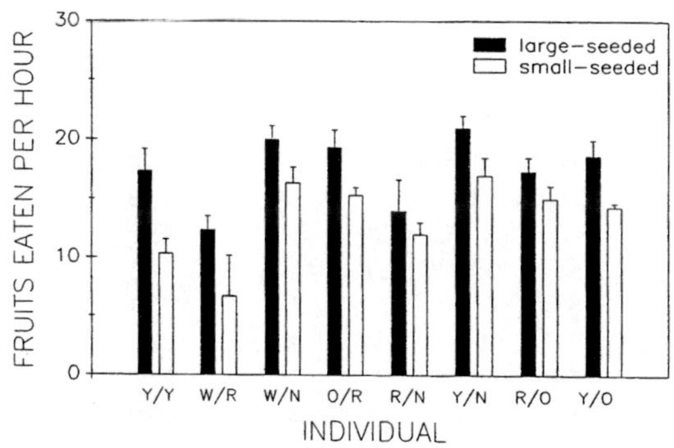

Fig. 2. Consumption rates for large- and small-beaded fruits. Bars for each bird represent the mean of 3 consumption rate determinations; error bars are ± 1 standard error.

Table 1. Results of randomized-blocks ANOVA on consumption rates of large- and small-beaded fruits, with individual birds as blocks. Data used were means of three consumption rate determinations on each bird with each fruit type.

Source	SS	d.f.	MS	F
Bead size	48.29	11	48.29	63.00***
Blocks (birds)	132.40	6	22.07	28.79***
Error	4.60	6	0.77	
Total	185.29	13		

*** $P<0.001$.

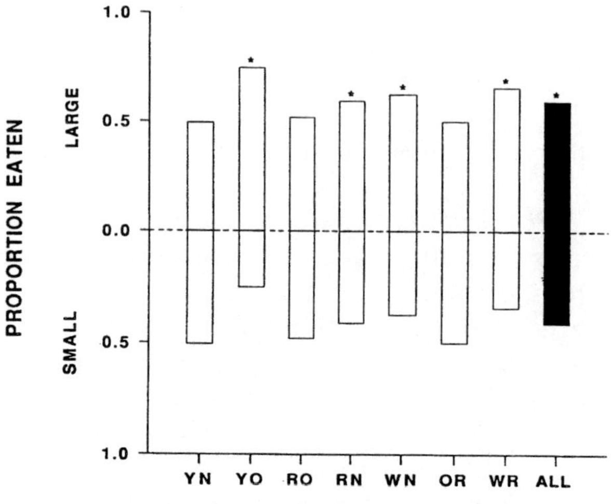

Fig. 3. Robin preferences for large- and small-beaded fruits. Data from all trials and both pulp colors were combined for each individual; the last bar combines data from all 7 individuals. * Denotes statistically significant preference $P<0.005$, χ^2 one-sample test).

few small seeds (9% of total) were regurgitated, however. Correspondingly, retention times were much greater for small seeds than large ones (Fig. 1). Median retention times for large and small beads were 16.1 and 44.8 min, respectively, and the retention time distributions for the two seed sizes were very different (Kolmogorov-Smirnoff $D_{max} = 0.616$, $n = 19$, $P \ll 0.01$).

As we predicted, the more rapid processing of large seeds translated into higher consumption rates for large-seeded fruits (Fig. 2): overall means ($\pm$ s.e.) for large- and small-seeded fruits were 17.5 ($\pm$ 1.22) and 13.8 ($\pm$ 1.33) fruits/hour, respectively. To evaluate the differences between these means, we used the average of three consumption rate measurements for each bird/fruit type combination in a randomized blocks ANOVA, with individual birds as blocks. Although consumption rates varied among birds, the difference between mean consumption rates

Table 2. Partial associations among variables in hierarchical loglinear analysis of fruit preference data. The categorical variables BIRD, TRIAL, SEEDS, and COLOR have 7, 15, 2, and 2 levels, respectively. See text for explanation.

Effect	d.f.	χ^2	P
BIRD * TRIAL * SEEDS	84	34.8	1.0
BIRD * TRIAL * COLOR	84	133.3	0.0005
BIRD * SEEDS * COLOR	6	23.5	0.0006
TRIAL * SEEDS * COLOR	14	14.3	0.4287
TRIAL * SEEDS	14	6.1	0.9626
TRIAL * COLOR	14	17.8	0.2158
BIRD * SEEDS	6	33.5	<0.0001
BIRD * COLOR	6	210.0	<0.0001
SEEDS * COLOR	1	43.4	<0.0001
SEEDS	1	67.7	<0.0001
COLOR	1	1160.3	<0.0001

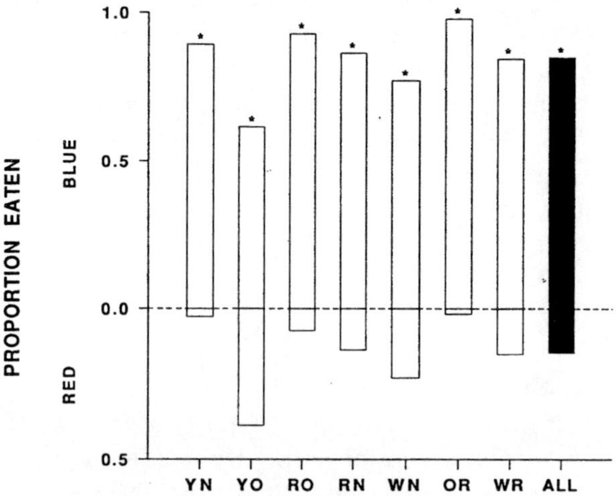

Fig. 4. Robin preferences for blue and red fruit pulp. Data from all trials and both bead sizes were combined for each individual; the last bar combines data from all 7 individuals. * Denotes statistically significant preference ($P<0.005$, χ^2 one-sample test).

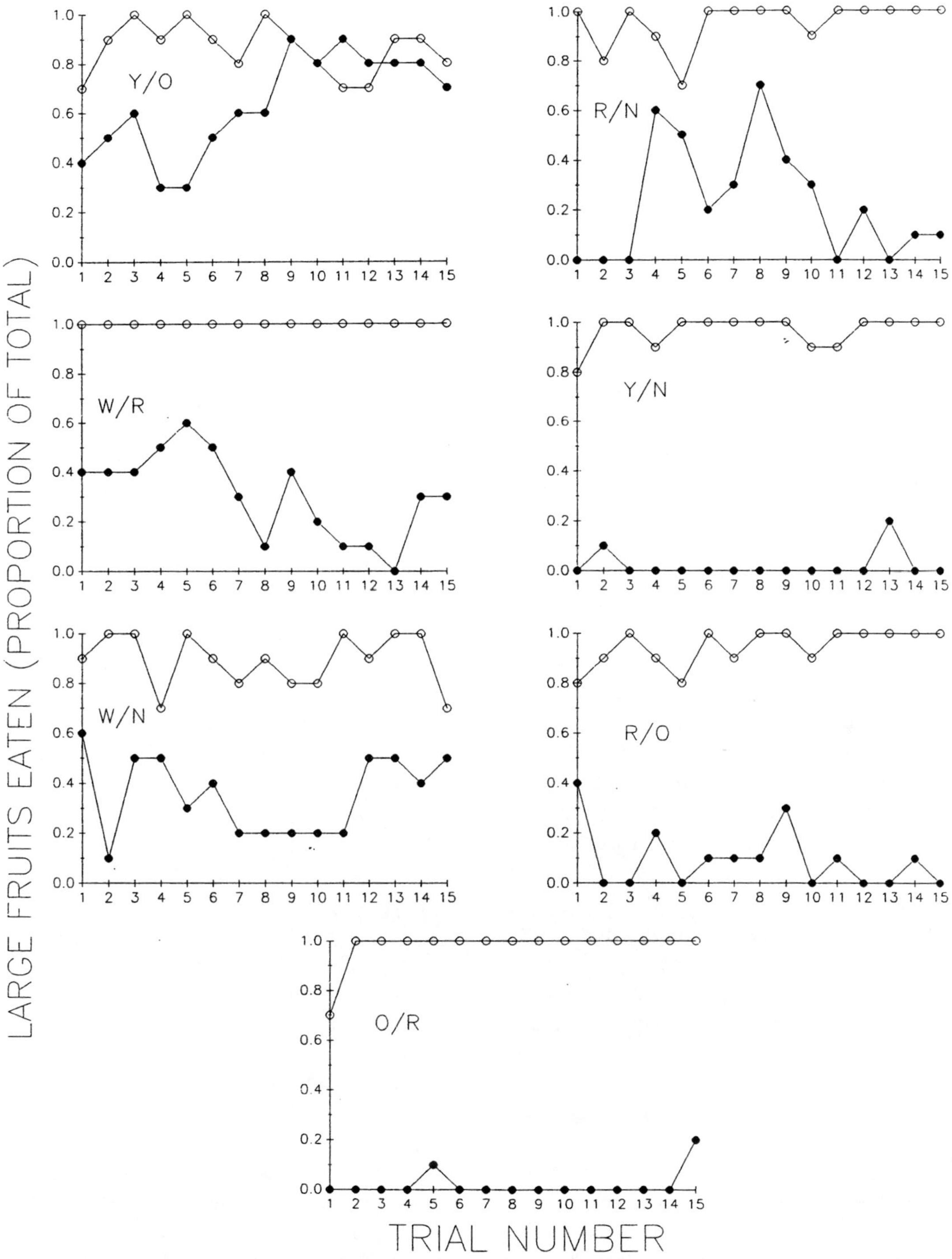

Fig. 5. Results of fruit preference trials on seven captive Robins. Open circles represent trials in which the large-beaded (= more profitable) fruit had blue pulp, filled circles represent trials in which large-beaded fruits had red pulp.

for large- and small-beaded fruits was highly significant (Table 1).

Fruit preference experiments

To distinguish between effects of seed size, fruit color, and experience (i.e., preference shifts resulting from learning of relative fruit profitabilities over the 15 trials on each bird with each seed size/pulp color combination), we fit the data from preference experiments to a hierarchical loglinear model (Bishop *et al.* 1975). This analysis is essentially a multiway contingency table, with significance tests based on the marginal totals for each factor. Partial associations between number of fruits eaten and each of the three factors (seed size, pulp color, and trial number; Table 2), along with Figures 3 and 4 (see below), suggest that Robins indeed preferred large-seeded fruits overall (effect SEEDS), that they preferred blue fruits over red (effect COLOR), but that the strengths of these preferences did not change systematically over the courses of the two 15-trial experiments (effects TRIAL * SEEDS and TRIAL * COLOR). However, the significant interaction terms (BIRD * SEEDS and BIRD * COLOR) in Table 2 indicate that individual birds differed in the degree of color- and seed size-based fruit preference. Moreover, the significant 3-way interactions suggest that the effects of seed size and pulp color on fruit preference varied widely among individual birds. Therefore, we examined the effects of pulp color, seed size, and trial number separately for each bird. Four of the seven Robins displayed significant preferences for large-seeded fruits (Fig. 3), and all seven strongly preferred blue to red (Fig. 4). Figure 5 shows that bird preference for the more profitable, large-seeded fruits did not increase monotonically with experience. When the more profitable fruit also had the preferred color, most birds overwhelmingly chose 'correctly'. However, when the more profitable fruit had the non-preferred color, bird preferences were still dominated by color, and the proportion of large-seeded fruits did not increase consistently with experience.

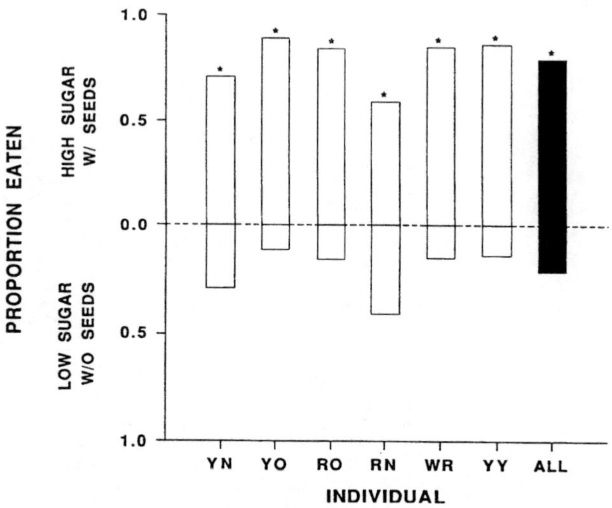

Fig. 6. Results of preference trials with equicaloric fruits differing in sugar concentration and 'seed':pulp ratio. Data from 15 trials on each individual were combined; the last bar combines data from all 6 individuals. * Denotes statistically significant preference ($P < 0.005$, χ^2 one-sample test).

Results of preference tests using equicaloric fruits that differed in bead presence and sugar concentration were more clear-cut. All six birds tested clearly preferred fruits with higher sugar concentration, even though these also contained beads (Fig. 6). Apparently, Robins use sugar concentration as an important indicator of profitability.

Discussion

Data presented here show clearly that differences in the way that total seed volume is 'packaged' within fruits have important consequences for fruit profitability, and that these consequences can be used by American Robins as a basis for fruit preferences. As expected, 6 mm-diameter beads were regurgitated while 3 mm-diameter ones were passed through the gut. Consequently, retention times were significantly longer for the smaller beads (Fig. 1). The same sort of relationship is well-known in other frugivorous birds; probability of regurgitation (vs. defecation) increases with seed size and decreases (for seeds of a given size) with increasing bird size (e.g., Levey 1986, 1987b; Murray & Winnett-Murray unpublished data).

Our data show clearly that differences in passage rates of large and small beads resulted in higher rates of pulp (and, consequently, sugar) consumption for birds feeding on large-beaded fruits. Based on values of 0.308 ml pulp per fruit and 0.15 g total sugar per ml pulp, Robins consumed, on average, 17.5 fruits/h × 0.308 ml pulp/fruit × 0.15 g sugar/ml pulp = 0.8085 g sugar per hour when eating large-beaded fruits but only 13.8 fruits/h × 0.308 ml pulp/fruit × 0.15 g sugar/ml = 0.6376 g/h when eating small-beaded fruits. Although we have no data on extraction efficiencies for glucose and fructose in the seven birds used in our experiments, complete oxidation of glucose or fructose to CO_2 and H_2O yields approximately 15.94 kJ/g (Guyton 1986). Assuming complete oxidation, the above consumption rates suggest that feeding on large-beaded fruits yielded (0.8085 g/h–0.6376 g/h) × 15.94 kJ/g = 2.72 kJ/h (a 27% increase) more energy than did feeding on small-beaded fruits, for the average Robin. Thus, in spite of the fact that our experimental fruits were virtually identical in all other respects, the way in which total 'seed' volume was packaged had an important impact on the rate of net energy gain by individual birds.

An impact of this magnitude represents a significant proportion of a Robin's energy budget. The standard metabolic rate of a 76.5 g passerine (the mean weight of our Robins), as predicted by Lasiewski & Dawson's (1967) equation, is 3.51 kJ/h. A 2.72 kJ/h increase in energy gain is thus potentially very important, especially to free-living birds with higher energetic demands than our captives.

Presumably, differences in profitability were responsible for the preferences of four of the seven birds for large-beaded fruits. Given the considerable differences in net energy gain, however, it remains puzzling why only four of the birds significantly preferred large-beaded fruits. Moreover, why wasn't the degree of preference stronger among the birds that did display them? In a similar study with Cedar Waxwings (*Bombycilla cedrorum*), Levey & Grajal (1991) found that only three of seven birds preferred the more profitable fruits, and one even preferred the less profitable one (in their study, profitability was also a function of seed passage rate, but among seeds that were all defecated). They suggested that their use of an unusual visual cue (visibility of beads through translucent fruit pulp) for distinguishing fruits of different profitability may have delayed birds from learning to associate profitability with such a subtle visual cue. Our design, however, which paired bead content with a clearly recognizable color cue, should have been free from such a bias, yet three of the birds failed to develop preferences for large-beaded fruits. We can offer three possible explanations.

First, perhaps the difference in profitability of our fruits, though significant, was too slight to be perceived by all of the birds. Second, perhaps the birds' strong color preferences (discussed below) prevented them from learning preferences for the more profitable fruits in the 15 choice trials in which large beads were paired with the less-preferred, red pulp. Color preference, whether learned or innate, may have simply overridden any preference for large-beaded fruits. Indeed, all seven Robins continued to use color as their primary choice criterion, even when the preferred color was paired with the less profitable fruit (Fig. 5). Third, the ability of birds to taste our experimental fruits, and thus directly assess profitability on the basis of sugar concentration and size, may have prevented them from learning relative profitabilities through post-ingestional effects. Data presented here (Fig. 6) show clearly that Robins preferred fruits with sweeter pulp, even when they contained seeds. This was an unexpected result, since Robins typically ingest fruits whole (i.e., they are 'gulpers', *sensu* Moermond & Denslow 1985) and thus rarely have the opportunity to taste fruit pulp before ingestion. Indeed, Levey (1987a) found that two species of tropical gulpers (manakins) were unable to discriminate between fruits that differed slightly in sugar concentration, but that three species of 'mashers' (tanagers that normally crush fruits in the bill, thereby releasing juices onto their tongues) readily detected differences in sugar concentration as small as 2%. Cedar Waxwings, which are gulpers, however, readily use taste as a criterion in selecting artificial fruits (Martinez Del Rio *et al.* 1989). If Robins use sweetness as a primary choice criterion, a very long period of experience may be necessary to learn preferences for more profitable fruits through post-ingestional effects. Interestingly, in identical experiments in our lab, Swainson's Thrushes (*Catharus ustulatus*) clearly preferred seedless fruits over sweeter, seeded fruits with equal caloric content (i.e., they *avoided* fruits with seeds; D. Quint, unpublished data). Thus, even closely related gulpers can differ widely in their use of fruit characteristics as preference criteria.

In contrast to the inconsistent effect of seed size and its concommitant effect on profitability, pulp color was far more important as a determinant of fruit choice in these experiments. This was surprising, since color had no effect on profitability and because Robins commonly feed on both red and blue fruits in the field (Wheelwright 1986). Other experiments in our lab (Murray & Winnett-Murray, unpublished data) have produced similar results with Swainson's Thrushes, in that virtually all birds displayed strong color preferences. However, the thrush color preferences were highly idiosyncratic; approximately equal numbers of birds (out of 15) preferred red and blue fruits. Willson and her colleagues (Willson *et al.* 1990; Willson & Whelan 1991) have recently addressed color preferences in North American fruit-eating birds (*C. ustulatus* among them), and their findings parallel ours for *C. ustulatus*: most individuals display strong color preferences, but different birds prefer different colors. The Robins in our study may or may not differ from this pattern; we used only seven birds, and this cohort may have included only individuals that preferred blue by chance.

Despite the fact that red and black predominate among fruits adapted for avian dispersal at many locations (Turĉek 1963; Janson 1983; Wheelwright *et al.* 1984; Wheelwright & Janson 1985; Willson 1986; Willson *et al.* 1989, 1990), few data support the notion that these patterns are the result of evolutionary responses to avian color preferences (Willson & Whelan 1991). Despite the fact that many frugivorous birds have strong color preferences (Willson *et al.* 1990, this study), the preferred color varies widely both within and among species, such that the coterie of dispersers in any particular location is unlikely to exert consistent selection for any particular color.

Conclusions

After nearly a decade of investigation by a number of researchers, the factors underlying fruit choice by birds remain obscure. Field studies demonstrate that numerous factors (e.g., abundance, how fruits are displayed, pulp:seed ratio, seed size, caloric content) may be important, but suggest strongly that such factors interact in complex fashion. Laboratory studies that successfully isolate one or a few factors suggest that many birds can discriminate among fruits that differ in their profitability, whether it results from different nutrient concentration, pulp:seed ratio, or seed passage rate. Nevertheless, even these carefully controlled studies suggest that fruit choice is a complex process; the criteria used differ both inter- and intraspecifically. Clearly, many frugivores are capable of using sweetness, pulp:seed ratio, and/or seed passage rate to assess fruit profitability, yet we lack detailed information on the importance that such factors have in determining the feeding choices of individuals. Detailed laboratory studies are needed to determine how such factors are used by individuals. Are they used in a distinct hierarchical fashion by individual birds, or are there important interactions among cues? If such a hierarchy of criteria does exist, is there significant intraspecific and/or interspecific variation in it?

Acknowledgements

Numerous individuals have contributed to our studies of avian fruit preferences and we gratefully acknowledge their assistance. R. Adams (Kalamazoo Nature Center) procurred the Robins, D. Levey shared unpublished manuscripts and advice on preparing artificial fruits, and several other undergraduate students (T. Avolio, K. Buege, M. Camp, S. Clearwater, S. Dawson, J. Dumm, D. Quint, K. Tilman, and W. Wooley) helped work out the logistics of maintaining fruit-eating birds and performing choice experiments. Hope College generously provided space and financial support, and NSF REU grant (DIR 8804082-01, to Hope College) supported some of the earlier work on Swainson's Thrushes.

References

Baird, J. W. 1980. The selection and use of fruit by birds in an eastern forest. Wilson Bull. 92: 63–73.

Bishop, Y. M. M., Fienberg, S. E. & Holland, P. W. 1975. Discrete multivariate analysis. MIT Press, Cambridge, MA.

Buchholz, R. & Levey, D. J. 1990. The evolutionary triad of microbes, fruits, and seed dispersers: an experiment in fruit choice by cedar waxwings, *Bombycilla cedrorum*. Oikos 59: 200–204.

Denslow, J. S. 1987. Fruit removal rates from aggregated and isolated bushes of the red elderberry, *Sambucus pubens*. Can. J. Bot. 65: 1229–1235.

Denslow, J. S., Levey, D. J., Moermond, T. C. & Wentworth, B. C. 1987. A synthetic diet for fruit-eating birds. Wilson Bull. 99: 131–134.

Guyton, A. C. 1986. Textbook of medical physiology, 7th ed. W. B. Saunders, Philadelphia, PA.

Herrera, C. M. 1984. A study of avian frugivores, bird dispersed plants, and their interaction in Mediterranean scrublands. Ecol. Monogr. 54: 1–23.

Herrera, C. M. & Jordano, P. 1981. *Prunus mahaleb* and birds: the high efficiency seed dispersal system of a temperate fruiting tree. Ecol. Monogr. 51: 203–218.

Howe, H. F. 1986. Seed dispersal by fruit-eating birds and mammals. In: Murray, D. R. (ed), Seed Dispersal. pp. 123–189. Academic Press.

Howe, H. F. & Estabrook, G. F. 1977. On intraspecific competition for avian dispersers in tropical trees. Am. Nat. 111: 817–832.

Howe, H. F. & Smallwood, J. 1982. Ecology of seed dispersal. Ann. Rev. Ecol. Syst. 13: 201–228.

Howe, H. F. & Vande Kerckhove, G. A. 1980. Nutmeg dispersal by tropical birds. Science 210: 925–927.

Janson, C. H. 1983. Adaptation of fruit morphology to dispersal agents in a neotropical forest. Science 219: 187–189.

Johnson, R. A., Willson, M. F, Thompson, J. N. & Bertin, R. I. 1985. Nutritional values of wild fruits and consumption by migrant frugivorous birds. Ecology 66: 819–827.

Lasiewski, R. C. & Dawson, W. R. 1967. A reexamination of the relation between standard metabolic rate and body weight in birds. Condor 69: 13–23.

Levey, D. J. 1986. The role of seed size and bird species on seed deposition patterns. In: Estrada, A. & Fleming, T. H. (eds), Frugivores and seed dispersal. pp. 147–158. Junk, Dordrecht, The Netherlands.

Levey, D. J. 1987a. Sugar-tasting ability and fruit selection in tropical fruit-eating birds. Auk 104: 173–179.

Levey, D. J. 1987b. Seed size and fruit-handling techniques of avian frugivores. Am. Nat. 129: 471–485.

Levey, D. J. & Grajal, A. 1991. Evolutionary implications of fruit processing limitations in cedar waxwings. Am. Nat., 138: 171–189.

Martinez del Rio, C., Karasov, W. H & Levey, D. J. 1989. Physiological basis and ecological consequences of sugar preferences in cedar waxwings. Condor 106: 64–71.

McKey, D. 1975. The ecology of coevolved seed dispersal systems. In: Gilbert, L. E. & Raven, P. H. (eds), Coevolution of animals and plants. pp. 159–191. Univ. of Texas Press, Austin, TX.

McPherson, J. M. 1987. A field study of winter fruit preferences of cedar waxwings. Condor 89: 293–306.

McPherson, J. M. 1988. Preferences of cedar waxwings in the laboratory for fruit species, colour and size: a comparison with field observations. Anim. Behav. 36: 961–969.

Moermond, T. C. & Denslow, J. S. 1983. Fruit choice in neotropical birds: effects of fruit type and accessibility on selectivity. J. Anim. Ecol. 52: 407–420.

Moermond, T. C. & Denslow, J. S. 1985. Neotropical frugivores: patterns of behavior, morphology and nutrition with consequences for fruit selection. In: Buckley, P. A., Foster, M. S., Morton, E. S., Ridgely, R. S. & Buckley, N. G. (eds), Neotropical Ornithology. Ornithol. Monogr. No. 36. pp. 865–897. American Ornithologists' Union. Washington, D.C.

Morden-Moore, A. L. & Willson, M. F. 1982. On the ecological significance of fruit color in *Prunus serotina* and *Rubus occidentalis:* field experiments. Can. J. Bot. 60: 1554–1560.

Murray, K. G. 1988. Avian seed dispersal of three neotropical gap-dependent plants. Ecol. Monogr. 271–298.

Sherburne, J. A. 1972. Effects of seasonal changes in the abundance and chemistry of the fleshy fruits of northeastern woody shrubs and patterns of exploitation by frugivorous birds. Ph.D. dissertation. Cornell University, Ithaca, New York.

Sorensen, A. E. 1981. Interactions between birds and fruit in a temperate woodland. Oecologia 50: 242–249.

Sorensen, A. E. 1983. Taste aversion and frugivore preference. Oecologia 56: 117–120.

Sorensen, A. E. 1984. Nutrition, energy and passage time: experiments with fruit preference in european blackbirds (Turdus Merula). J. Anim. Ecol. 53: 545–557.

Turĉek, F. J. 1963. Color preference in fruit- and seed-eating birds. In: Sibley, C. G. (ed), Proc. XIII Int. Ornithol. Congr., vol. 1. pp. 285–292. American Ornithologists' Union, Baton Rouge, Louisiana.

Wheelwright, N. T. 1986. The diet of American Robins: an

analysis of U.S. Biological Survey records. Auk 103: 710–725.

Wheelwright, N. T., Haber, W. A., Murray, K. G. & Guindon, C. 1984. Tropical fruit-eating birds and their food plants: a survey of a Costa Rican Lower Montane Forest. Biotropica 16: 173–192.

Wheelwright, N. T. & Janson, C. H. 1985. Colors of fruit displays of bird-dispersed plants in two tropical forests. Arn. Nat. 126: 777–799.

Willson, M. F. 1986. Avian frugivory and seed dispersal in eastern North America. In: Johnston, R. F. (ed), Current Ornithology, vol. 3. pp. 223–279. Plenum Press, New York.

Willson, M. F., Graff, D. A. & Whelan, C. J. 1990. Color preferences of frugivorous birds in relation to the colors of fleshy fruits. Condor 92: 545–555.

Willson, M. F., Irvine, A. K. & Walsh, N. G. 1989. Vertebrate dispersal syndromes in some Australian and New Zealand plant communities. Biotropica 21: 133–147.

Willson, M. F. & Whelan, C. J. 1991. The evolution of fruit color in fleshy-fruited plants. Am. Nat. 136: 790–809.

Vegetatio **107/108**: 227–235, 1993.
T. H. Fleming and A. Estrada (eds). Frugivory and Seed Dispersal: Ecological and Evolutionary Aspects.

The influence of pulp lipids on fruit preference by birds

E. W. Stiles
Department of Biological Sciences;, Rutgers University, Piscataway, NJ 08855, USA

Keywords: Fleshy fruit, Frugivore, Lipids

Abstract

Use patterns of lipid-rich, fleshy fruits were examined for frugivorous birds. Approximately 20% of fleshy-fruited species within plant communities bear fruit with > 10% dry weight lipids. These fruits are fed upon by the most heavily frugivorous bird species in those communities studied and are the focus of highly specialized relationships among birds and fruits. In aviary tests some temperate frugivores show preference for higher lipid native fruits and prefer higher-lipid artificial fruits with as little as 3% wet weight difference in lipid content. The importance of lipids in fruit preference by birds, and the possible effects of differing digestive physiologies in birds are discussed.

Introduction

The selection of fleshy fruits by birds is influenced by a diverse array of fruit and fruiting characteristics. Some of these factors include the size of the fruit or fruiting display (Johnson *et al.* 1985; Pratt & Stiles 1985; Wheelwright 1985; McPherson 1987; Debussche & Isenmann 1989), crop size (Snow 1971; Howe & Estabrook 1977; Howe & De Steven 1979; Howe & Vande Kerckhove 1979; Jordano 1984), color (Turcek 1963; Stiles 1982; Willson & Thompson 1982; Wheelwright & Janson 1985; Knight & Siegfried 1983; Willson *et al.* 1989; Willson & Whelan 1990), presentation (Denslow & Moermond 1982; Moermond & Denslow 1983, 1985), availability (Stiles 1980; Jordano 1984, 1988; Wheelwright *et al.* 1984; McPherson 1987), and habitat (Moore & Willson 1982; Manasesse & Howe 1983; Denslow *et al.* 1986; Piper 1986). In addition to these factors the nutritional composition of fruit pulp also influences selection of fruits by birds (Snow 1971; Stiles 1980; Herrera 1982; Snow & Snow 1988; Martinez del Rio *et al.* 1988, but see Johnson *et al.* 1985).

It is becoming apparent that differences in fruit selection are related to many of these factors acting in concert and no single character will completely explain a bird/fruit preference. It is also increasingly clear that one must be very careful in formulating generalizations covering bird/fruit preferences because different bird species react quite differently to the same stimulus. For example *Manacus candei* hovers to pick fruit whereas *Tangara larvata* perches when feeding, but *Tangara* can reach below its perch a distance 50% greater than can *Manacus*. Avian morphology determines the energetics of fruit access by different species of birds. This illustrates the care that must be employed in generalizing the preferences exhibited by birds. The recent illustration of differences in sucrase activity within the same order of birds (Martinez del Rio *et al.* 1988; Martinez del Rio *et al.* 1989) provides ample evidence that preference for high carbohydrate fruits cannot be generalized, even within the Passeriformes.

Birds eat fruits to gain the nutritional rewards in fruit pulp. The preferences exhibited by different bird species must, in part, be influenced by the nutritional components assimilated. Fruits vary widely in the nutritional content with less than 10% water in some sumacs (*Rhus*) and poison ivy (*Toxicodendron*) to over 90% water in blackberries (*Rubus*) and strawberries (*Fragaria*). Carbohydrates vary from less than 5% in bayberry and wax myrtle (*Myrica*) and sumacs to more than 90% in some blueberries (*Vaccinium*) and strawberries measured as percentage dry weight of pulp (Stiles 1989). As estimated from nitrogen content, proteins of fruit pulp vary from less than 2% in blueberries, strawberries, some cherries (*Prunus*), and mountain ash (*Sorbus*) to over 20% in some members of the Lauraceae.

In this paper, I examine the importance of high-lipid fruits in bird/fruit interactions. The importance is reflected in the extensive use of high-lipid fruits by highly frugivorous species and by the preferences expressed for high-lipid fruits. I will show that fruits bearing larger quantities of lipids are dominant and/or preferred food items in the diets of the most highly frugivorous species of birds. Also for a variety of fruit characteristics, lipids are most strongly correlated with bird preference.

Lipids are one of the four major groups of compounds found in living tissues. They are defined by their general solubility in organic solvents and insolubility in water. Lipids include not only compounds with fatty acid moieties, but also many compounds derived from isoprene including sterols, carotenoids and chlorophylls (Goodwin & Mercer 1983). Lipids are high-energy compounds and yield approximately twice the energy on catabolism as either carbohydrates or proteins (Paine 1971), so for birds, high-lipid fruits provide potentially high energy food packets generating twice the energy for a given ingested pulp mass. Among 76 temperate fruits White (1989) found strong positive correlations ($P<0.001$) among water, carbohydrates and fruit mass, but strong inverse relations ($P<0.001$) of water carbohydrate and fruit mass with lipids. High-lipid fruits are hydrophobic and smaller with low carbohydrate load but high energy.

High-lipid fruits are found in many plant families. For over 500 species of fruits that have been analyzed for lipid content from diverse areas of the globe, approximately one quarter are composed of greater than 10% dry weight lipids (Stiles 1980; Snow 1981; Stiles & White 1982; Herrera 1984; Dinnerstein 1985; Stiles unpublished). These include a variety of genera that are so far represented by three or more species with this level of lipid (Table 1). Over half of the high-lipid

Table 1. Genera and families of plants bearing fleshy fruits with large numbers of species having greater than 10% dry weight lipid in fruit pulp. Compiled from a sample of 500 species from North America, Europe, Africa, South America, and Papua New Guinea.

Families with large numbers of high-lipid species		Genera with large numbers of high-lipid species		
Family	Percentage of 131 known species with >10% lipid	Genus	Number of species	Family
Lauraceae	21	*Ocotea*	11	Lauraceae
Anacardiaceae	9	*Cornus*	9	Cornaceae
Cornaceae	7	*Juniperus*	6	Cupressaceae
Cupressaceae	5	*Rhus*	6	Anacardiaceae
Magnoliaceae	5	*Magnolia*	5	Magnoliaceae
Palmae	5	*Nectandra*	4	Lauraceae
		Virola	4	Myristicaceae
		Euronymous	4	Celastraceae
Total % from six families	52			

species found in this sample are members of only six plant families (Table 1). One must treat these data with some caution because they may not represent a random sample of fleshy fruited species; they may represent slightly more tropical high-lipid species than if fruits were taken at random. From two fairly complete temperate samples, 16 of 75 species (21%) from eastern North America (White 1989) and 15 of 92 species (16%) from Spain (Herrera 1984) contain about one fifth high-lipid species.

Evidence for preference of high-lipid species

Birds appear to select high-lipid fruits first, when choices of fruits are available. In temperate North America summer ripening fruits are nearly all low-lipid (Stiles 1980; White 1989). Species bearing high-lipid fruits begin to ripen in late August and early September in New Jersey as large numbers of neotropical migrants begin to move to wintering areas in the Caribbean, Central and South America. High-lipid fruits disappear first as birds remove them. Baird (1980) found for woody species that the two most rapidly disappearing species were spicebush (*Lindera benzoin*, 34.6% lipid) and flowering dogwood (*Cornus florida*, 24.0% lipid), the two highest lipid species in the study of eight species. A similar preference for *Lindera* was found among woody species in Illinois (Thompson & Willson 1979), and for *Lindera* when compared with a low-lipid species, *Viburnum acerifolium* in New Jersey (Stiles 1980). White (1989) monitored disappearance of fruits of 12 woody species over 2 years in New Jersey and determined that although fleshy, lipid-rich fruits were eaten rapidly when frugivores were present, some lipid-poor species were removed as fast as lipid-rich species. For example, *Cornus amomum* (6% lipid) was removed as fast as *Cornus racemosa* (33% lipid) (White 1989), a result also found by Borowicz and Stephenson (1985). Also, rate of field removal was not always linearly related to the percentage lipid as in the use of *Lindera benzoin* and *Cornus florida* where fruit use suggested preference for the lower lipid *Cornus*. Individual fruit profitability, however, is greater in *Cornus* due to larger, drier fruits, and this may influence choice.

White's (1989) study of 76 species noted an important difference in high-lipid fruits. Two groups of relatively imperishable fruits are included in the high-lipid fruits. These are the waxy fruits of bayberries (*Myrica*) and poison ivies (*Toxicodendron*) and the oily, tannin-rich fruits of sumacs (Rhus). These dry fruits were not taken rapidly upon ripening even though percentage dry weight lipids were high (e.g. *Myrica pensylvanica*, 43.5% lipid; *Rhus copallina*, 15.8% lipid). These species differed from not only the other high-lipid species but also from all other fleshy fruits studied by having significantly lower fruit mass, water, and percentage dry weight carbohydrate as well as higher seed load (seed mass/fruit mass). No significant differences existed between waxy and oily fruits except that waxy fruits are 2.9 times higher in lipid content.

Herrera (1981a, 1981b), working with blackbirds (*Turdus merula*)and robins (*Erithacus rubecula*), and Jordano and Herrera (1981), working with blackcap (*Sylvia atricapilla*), demonstrated that overwintering frugivores in southern Spanish mediterranean scrub rely most heavily on the lipid-rich species that ripen in winter in this area, preferring these over concurrently available low-lipid species. Snow found similar patterns in the United Kingdom, and Skeate (1985), working in San Felasco Hammock in Florida, found higher numbers of wintering frugivores than in New Jersey, and also greater numbers of winter-ripening high-lipid fruits that were the focus of the feeding of these birds.

Bird reliance on high-lipid fruits: special cases

Large tropical frugivores

McKey (1975) identified the special association of certain birds in the tropics with high-lipid fruits based primarily on the work of David and Barbara Snow. David Snow (1962), working in Trinidad, showed that fruits that oilbirds (*Steatornis caripensis*) fed to their nestlings included 7 species of palms, 8 of Lauraceae, 2 of Burseraceae and 1 of Araliaceae. Those he could measure were high in fat and protein. Barbara Snow (1970) showed that the bearded bellbird (*Procnias averano*) also fed nestlings high-lipid, high-protein fruits primarily from the Lauraceae (*Ocotea oblonga*, 19% lipid; *Cinnamomum elongatum*, 44% lipid; these two species formed 74% of the diet). Wheelwright (1983) found that resplendent quetzals (*Pharomachrus mocinno*) ate at least 41 different fruit species annually at Monteverde, Costa Rica, including low-lipid species, but they depend mostly on the large, lipid-rich drupes of 18 species of Lauraceae. The fruiting phenology of these species seem to dictate the timing and direction of seasonal movements by quetzals. The fruit diet of the cock-of-the-rock (*Rupicola rupicola*) included 40% *Didymopanax morototoni* (Araliaceae, 33.5% lipid), 22% Burseraceae, 12% Meliaceae, 5% Palmae, and 4% Lauraceae (Snow 1971). Nutritional analyses are not available for the species eaten, but they are members of plant families containing high-lipid genera. In Papua New Guinea the highly frugivorous fruit pigeons, *Ptilinopus rivoli*, *P. superbus*, and *Gymnophaps albertisii* feed on large fruits in the Icacinaceae (1 sp.), Rosaceae (1 sp.), Lauraceae (3 spp.), Elaeocarpaceae (1 sp.), and Meliaceae (1 sp.). The latter three families are known to have mostly high-lipid species.

Waxy fruits and special digestion

Plant waxes are complex mixtures of highly non-polar lipids, and are found in fruit pulp of some species. The ability to digest waxes is rare in vertebrates (Roby *et al.* 1986). Waxy fruits provide another example of high-lipid fruits and special use by birds. Specialized relationships among birds and fruits are not only found in the tropics as conceived by McKey, nor are they necessarily focused on the large-seeded, lipid-rich drupes found in the palms and laurels. Yellow-rumped warblers (*Dendroica coronata*) and tree swallows (*Iridoprocne bicolor*) are the species of their respective taxonomic groups that winter the farthest north. Both species are known to consume and be associated with bayberry or wax myrtle (*Myrica*) during winter, and yellow-rumped warblers also feed on the waxy pulp of poison ivy (*Toxicodendron radicans*). Although woodpeckers also feed on poison ivy and bayberry, most frugivorous species rarely eat these fruits (White 1989). This special set of associations is facilitated by the special physiological capacity of these birds to metabolize the high molecular weight lipids found in waxy fruits (Place & Stiles 1992). The capacity to digest wax is also found in honeyguides that digest beeswax (Diamond & Place 1985) and in several pelagic seabirds that consume marine plankton high in wax esters (Obst 1986; Roby *et al.* 1986; Place *et al.* 1986; Jackson & Place 1990), but not in the yellow warbler (*Dendroica petechia*), a congener of the yellow-rumped warbler. The capacity to assimilate waxes may exist in some other species. Greenberg (1981) found that bay-breasted warblers (*Dendroica castanea*) and chestnut-sided warblers (*D. pensylvanica*) feed selectively on fruits of *Lindackeria laurina* Presl. (Flacourtiaceae) that have 'a distinctly waxy texture and odor'. These waxy fruits are morphologically and chemically distinct from other fleshy fruits (White 1989).

Mistletoes and frugivores

Some of the most highly specialized fruit/frugivore relationships are found among mistletoes and the birds that consume their fruits. Docters van Leeuwen (1954) described the highly specialized behavior of two species of flowerpecker (*Dicaeum*) feeding on mistletoe fruits. Waslberg (1975) studied the phainopepla (*Phainopepla nitens*) feeding on desert mistletoe (*Phoradendron californicum*, 15% lipid) in California. He showed that the phainopepla's stomach is highly adapted for processing mistletoe berries, much more so than a secondary consumer of the fruits, the house finch (*Carpodacus mexicanus*).

Euphonias (neotropical tanagers) are also well known as mistletoe consumers (Wetmore 1914). Davidar (1983) and Sargent (unpublished) have shown the specialized behavior of euphonias and chlorophonias feeding on *Phoradendron*. Lipid content of these species are not known, but Susan White (1974) measured a *Phoradendron* sp. in Panama with 52.7% lipid. Skeate (1985) found that 80.7% of visits to *Phoradendron serotinum* (6.4% lipid) were from cedar waxwings (*Bombycilla cedrorum*) and that waxwings made 70.5% of their feeding visits to *Phoradendron*.

Snow and Snow (1988) present evidence on the use of mistletoe (*Viscum album*, Viscaceae, 8.61% lipid) by the mistle thrush (*Turdus viscivorous*). The thrush feeds on a variety of other fruits but 251 of 262 visits to mistletoe were by the mistle thrush. Snow feels this specialized feeding (recognized in the common and scientific names) is the result of defense of the plants by the mistle thrush and not the inability of other species to utilize the fruits. Further experimental work is needed here. The white-cheeked cotinga (*Zaratornis stresemanni*) was found to feed solely on the fruits of the mistletoe *Tristerix chodatianum* (Viscaceae) in Peru. Five to ten minutes after consuming up to 5 fruits in succession, *Zaratornis* regurgitated seeds one by one and wiped them onto the surface of a limb. No seeds were ever seen to fall to the ground (Snow 1982). No data are available on the lipid content of this fruit.

Thrushes and junipers

Junipers (*Juniperus*, Cupressaceae) have also been identified as fruits with special relations with birds. Townsend's solitaire (*Myadestes townsendi*) feeds heavily on cones of *Juniperus monosperma* in northwest Arizona in winter (Salomonson & Balda 1977; Salomonson 1978), and cones of *Juniperus occidentalis* (17.6% lipid) are used as the sole winter food of *Myadestes* in northeast California (Lederer 1977a, 1977b, 1980; Poddar & Lederer 1982). Jordano (this volume) has also recognized the importance of the relationship among southern Spanish junipers and the thrushes (*Turdus*) wintering in this region. Four of the five species of *Juniperus* in Spain have $> 10\%$ dry weight lipid in their fruit pulp (Herrera 1984).

The relationship between lipids and proteins in fruits

McKey (1975) described the fruits eaten by specialized frugivores as having 'firm dense flesh, rich in fats and proteins'. While a positive correlation between high lipids and high proteins may be true for some species, it is far from universal, even in the tropics. An evaluation of mean protein content in fruit pulp from Europe, North America and the Neotropics yielded mean values of $5.5 \pm 0.4\%$ protein for 52 species of 35 genera in Europe, $5.2 \pm 0.4\%$ protein for 101 species of 51 genera in North America, and $8.9 \pm 0.6\%$ protein for 122 species in 77 genera in the Neotropics (White 1989). European and North American means were not significantly different but the Neotropical fruits had higher protein for those species evaluated.

Fruits eaten by the more heavily frugivorous birds described above do not lend support to an hypothesis that these fruits generally use protein as an attractant. Considering first the fruits eaten by large frugivores considered specialized by McKey (1975), mean protein for 9 species of *Ocotea* (Lauraceae) was above the mean at 11.14%, but mean value for 4 species of *Nectandra* (Lauraceae) was at the mean (8.75%), and the mean for 4 species of *Virola* (Myristicaceae) (5.38%) was below the overall mean.

High lipid mistletoe (*Phoradendron*) from Costa Rica had only 5.8% protein while two species of temperate *Phoradendron* had 7.5% and 11.5% protein and two temperate *Myrica* had 6.1% and 7.3% protein, both above the mean. Some other high-lipid temperate species, however, had low protein content. The mean of 4 Spanish junipers was 4.5% protein and North American *Juniperus occidentalis* had 3.3% protein. Also two species of *Toxicodendron* had only 1.9% protein. Although protein in fruits may be important in some fruit/frugivore interactions (see Herbst 1986, Dinerstein 1986 for information on bats), existing evidence does not yield the same strong relationships as those interactions with lipids.

Laboratory evidence for lipid preference

Selection of fruits is influenced by the myriad of factors mentioned in the Introduction to this chapter. Selectivity, however, differs from preference in that preference indicates the likelihood that a fruit will be chosen if offered on an equal basis with another fruit (Johnson 1980).

Fruit taste and fruit mass consumption

Sorensen (1983) used pellets of dough, 'dough fruits', that she flavored with the juice of wild fruits to test bird preference based on odor. Her experiments reduced the numbers of interacting variables that might affect selectivity by offering equal dough fruits that differed only in the fruit flavors. She found birds showed a distinct order of preference with only 3 of 32 paired preference tests for 11 species of fruits being out of place in a linear hierarchy. Ivy (*Hedera helix*) was the most preferred species. Sorensen reported the lipid content of *Hedera* as 4.7% dry weight pulp but this may have been an error in analysis. Herrera reports *Hedera* in Spain as having 31.9% lipid and my analysis of a sample from England shows *Hedera* to have 20% lipid. In Sorensen's study dough fruits with ivy juice were preferred over other species that have between 2.2% and 5.4% lipid (Herrera 1984).

Johnson *et al.* (1985) evaluated fruit consumption using 18 species of fruits and 11 migrant frugivore species. Different species of fruits were given in sequence to birds. Johnson *et al.* determined that fruit mass consumed by the birds was correlated best with the dry pulp mass per fruit. In these experiments the investigators found that high-lipid fruits are the most rewarding for the birds when other fruit traits are equal, but choices were not given to the birds and preferences were not tested.

Preference experiments with real and artificial fruits

In recent experiments with gray catbirds (*Dumetella carolinensis*) and American robins (*Turdus migratorius*) Stiles (unpublished) evaluated the preferences of these frugivores for different species of native fruits. Most aspects of availability were held constant to assess preference based solely on the characteristics of the fruits. Wild-caught birds were given pairwise choices of fruits placed in petri dishes 5 cm apart in the same tray.

For gray catbirds I used 11 species of fruits and 42 different pairwise test combinations. Percentage dry weight lipids in these fruits ranged from 34.6% to 0.0%. In 36 of 42 trials, gray catbirds selected the higher lipid member of the fruit pair offered and in 27 trials this was significant. For only 6 pairwise comparisons did the gray catbirds choose the lower lipid fruit and in only 2 was this result significant.

I evaluated the percentage lipid, percentage carbohydrate, lipid per fruit, carbohydrate per fruit, percentage water, energy index 1 (= twice percentage lipid plus percentage carbohydrate plus percentage protein), and energy index 2 (= twice the lipid per fruit plus carbohydrate per fruit plus protein per fruit). Lipids were doubled to estimate energy per fruit since lipids yield about twice the energy on catabolism as carbohydrates or proteins (Paine 1971). For individual regressions, percentage lipid, percentage carbohydrate, lipids per fruit, and energy index 1 were all significant, but in multiple regression of these traits only percentage lipid was significant.

Similar experiments were performed with American robins (Stiles unpublished) using 12 species of fruits that ranged from 34.6% to 0.0% dry weight lipid. In 42 pairwise comparisons, American robins selected the higher lipid fruit 28 times (19 comparisons were significant, and in 14 comparisons they selected the low lipid member of the pair (only 4 comparisons were significant. For most situations in which the low lipid member of the pair was chosen, the difference in lipid values were small.

Individual regressions of the mean difference in the number of fruits selected in the pairwise comparisons against the differences in percentage lipids, percentage carbohydrates, lipids per fruit, carbohydrates per fruit, energy index 1 and energy index 2 showed only percentage lipid and lipid per fruit to be significant. When all characters were considered in a multiple regression, only the difference in lipid per fruit was significant.

Both these experiments suggest that lipids in fruits are either strong determinants of preference or that they are correlated with some character that is preferred by birds.

I also ran preference experiments with 3 to 10 American robins using three species of imperishable fruit species (1 *Myrica* and 2 *Rhus*). These species have high percentage dry weight lipids (43.5% to 13.1% dry weight lipid) but have significantly lower water content, lower carbohydrate content, smaller fruit size, and larger seed load (seed mass/fruit mass) than the other fruits (White 1989). In each of 23 choice experiments pairing these three species with eight other species (24.0% to 0.0% dry weight lipid), the non-imperishable member of the pair was selected. The selection was significant in 12 of the comparisons. Although American robins appear to make choices of fruits based in part on lipid content, the specific nature of the lipids also influences their decisions.

To explore fruit preference based on lipid content more fully, I (Stiles unpublished) offered fruits on yet a more equal basis (Johnson 1980) by manufacturing agar-based artificial fruits that contained carbohydrates, proteins and fiber equal to the mean value for 75 species of local fruits (White 1989), but that differed in lipid content (soy oil). Lipids were added to the artificial fruits to equal 0%, 3%, 6%, and 12% wet weight lipid (to span the range on naturally occurring lipid values). Birds preferred 12%, 6%, and 3% over 0% in each comparison. Birds also preferred 12% lipid fruit over 3%, but comparisons of 12% to 6% were less clear cut.

Conclusions

The lipid content of bird-dispersed fleshy fruits ranges from 0% to over 60% dry weight. Even with many other factors interacting to influence fruit selection in the field, birds tend to remove higher-lipid species before lower-lipid species. A linear hierarchy should not be expected under these conditions nor does it exist. But a clear relationship is present.

High-lipid fruits, often the species with the highest percentage dry weight lipid in any particular community of plants, have been identified by many studies of a variety of taxa as being important in relationships where birds exhibit high dependency on fruits. Many large-seeded, high-lipid species of the Lauraceae, Palmae, Myristicaceae, and Meliaceae are primary foods for large tropical birds that depend on fruits. Mistletoes have special relationships with silky flycatchers, waxwings and flowerpeckers; junipers are linked on two continents with thrushes, and waxy fruits require a digestive physiology that has been found in only a limited number of birds.

In addition to these more specific relationships, American robins and gray catbirds, which consume a wide variety of fruits (Martin *et al.* 1951), also show strong preferences for higher-lipid fruits in laboratory tests. American robins clearly differentiate and select artificial fruits that differ only by having higher soy oil content.

The evidence presented here suggests that for a variety of bird species, lipids are important factors in their selection of fruits. Lipid contents greater than 10% dry weight are found in only about one quarter of fleshy fruits, but these fruits are involved in many of the close associations between birds and fruits that have been identified in the literature. Single factor studies seldom provide complete explanations for interactions as complex as bird/fruit interactions. Furthermore, lipids are a diverse array of chemicals and birds differ in their ability to assimilate them as in the case of waxy fruits. As discussed by Martinez del Rio and Restrepo in this volume, we must look beyond proximate analysis to explore the relationships among more specific fruit biochemistry and bird physiology to understand more fully bird/fruit interactions. The evidence presented here suggests that a detailed examination of lipids is a good place to start.

Acknowledgements

I would like to thank M. Cipollini and J. Zurovchak for valuable help and discussions. The study was supported by a grant from the Morris County Park Commission.

References

Baird, J. W. 1980. The selection and use of fruit by birds in an eastern forest. Wilson Bulletin 92: 63–73.

Borowicz, V. A. & Stephenson, A. G. 1985. Fruit composition and patterns of fruit dispersal of two *Cornus* spp. Oecologia 67: 435–441.

Davidar, P. 1983. Birds and neotropical mistletoes: effects on seedling recruitment. Oecologia 60: 271–273.

Debussche, M. & Isenmann, P. 1989. Fleshy fruit characters and the choices of bird and mammal seed dispersers in a Mediterranean region. Oikos 56: 327–338.

Denslow, J. S. & Moermond, T. C. 1982. The effect of accessibility on rates of fruit removal from tropical shrubs: an experimental study. Oecologia 54: 170–176.

Denslow, J. S., Moermond, T.C. & Levey, D. J. 1986. Spatial component of fruit display in understory trees and shrubs. Pages 37–44 in A. Estrada and T. H. Fleming, eds. Frugivores and seed dispersal. Junk, Dordrecht.

Diamond, A. W. & Place, A. R. 1988. Wax digestion in black-throated honeyguides, *Indicator indicator*. Ibis 130: 558–561.

Dinerstein, E. 1986. Reproductive ecology of fruit bats and the seasonality of fruit production in a Costa Rican cloud forest. Biotropica 18: 307–318.

Docters van Leeuwen, W. M. 1954. On the biology of some Loranthaceae and the role birds play in their life-history. Beaufortia 4: 105–208.

Goodwin, T. W. & Mercer, E. I. 1983. Introduction to plant biochemistry, 3rd edition. Pergamon Press, Oxford.

Greenberg, J. W. 1981. Frugivory in some migrant tropical forest wood warblers. Biotropica 13: 215–223.

Herbst, L. H. 1986. The role of nitrogen from fruit pulp in the nutrition of the frugivorous bat *Carollia perspicillata*. Biotropica 18: 39–44.

Herrera, C. M. 1981a. Datos sobre la dieta frugivora del Mirlo (*Turdus merula*) en dos localidades del sur de espana. Donana Acta Vertebrata 8: 306–310.

Herrera, C. M. 1981b. Fruit food of robins wintering in southern Spanish mediterranean scrubland. Bird Study 28: 115–122.

Herrera, C. M. 1982. Seasonal variation in the quality of fruits and diffuse coevolution between plants and avian dispersers. Ecology 63: 773–785.

Herrera, C. M. 1984. A study of avian frugivores, bird-dispersed plants, and their interaction in Mediterranean scrublands. Ecological Monographs 54: 1–23.

Howe, H. F. & DeSteven, D. 1979. Fruit production, migrant bird visitation, and seed dispersal in *Guarea glabra* in Panama. Oecologia 39: 185–196.

Howe, H. F. & Estabrook, G. F. 1977. On intraspecific competition for avian dispersers in tropical trees. American Naturalist 111: 817–832.

Howe, H. F. & Smallwood, J. 1982. Ecology of seed dispersal. Annual Review of Ecology and Systematics 13: 201–228.

Howe, H. F. & Vande Kerckhove, G. A. 1980. Nutmeg disperal by tropical birds. Science 210: 925–927.

Jackson, S. & Place, A. R. 1990. Gastrointestinal transit and lipid assimilation efficiencies in three species of high latitude seabird. J. Exper. Zool. 255: 141–154.

Johnson, D. H. 1980. The comparison of usage and availability measurements for evaluating resource preference. Ecology 61: 65–71.

Johnson, R. A., Willson, M. F., Thompson, J. N. & Bertin, R. I. 1985. Nutritional values of wild fruits and consumption by migrant frugivorous birds. Ecology 66: 819–827.

Jordano, P. 1984. Seed weight variation and differential avian dispersal in blackberries (*Rubus ulmifolius*). Oikos 43: 149–153.

Jordano, P. 1988. Diet, fruit choice and variation in body condition of frugivorous warblers in Mediterranean scrublands. Ardea 76: 193–209.

Knight, R. S. & Siegfried, W. R. 1983. Inter-relationships between type, size and colour of fruits and dispersal in southern African trees. Oecologia 56: 405–412.

Lederer, R. J. 1977a. Winter territoriality and foraging behavior of the Townsend's solitaire. American Midland Naturalist 97: 101–109.

Lederer, R. J. 1977b. Winter feeding territories in the Townsend's solitaire. Bird-banding 48: 11–18.

Lederer, R. J. 1980. Facultative territoriality in the Townsend's solitaire (*Myadestes townsendi*). Southwest Naturalist 25: 129–136.

Levey, D. J., Moermond, T. C. & Denslow, J. S. 1984. Fruit choice in neotropical birds: the effect of distance between fruits on preference patterns. Ecology 65: 844–850.

Manasse, R. S. & Howe, H. F. 1983. Competition for dispersal agents among tropical trees: influence of neighbors. Oecologia 59: 185–190.

Martin, A. C., Zim, H. S. & Nelson, A. L. 1951. American wildlife and plants. McGraw-Hill, New York.

Martinez del Rio, C. & Stevens, B. R. 1989. Physiological constraint on feeding behavior: intestinal membrane disaccharides of the starling. Science 243: 794–796.

Martinez del Rio, C., Stevens, B. R., Daneke, D. E. & Andreadis, P. T. 1988. Physiological correlates of preference and aversion for sugars in three species of birds. Physiol. Zool. 61: 222–229.

McKey, D. 1975. The ecology of coevolved seed dispersal systems. Pages 159–191 in L. E. Gilbert and P. H. Raven eds. Coevolution of animals and plants. University of Texas Press, Austin.

McPherson, J. M. 1987. A field study of winter fruit preferences of cedar waxwings. Condor 89: 293–306.

Moore, L. A. & Willson, M. F. 1982. The effect of microhabitat, spatial distribution, and display size on dispersal of *Lindera benzoin* by avian frugivores. Canadian Journal of Botany 60: 557–560.

Moermond. T. C. & Denslow, J. S. 1983. Fruit choice in neotropical birds: effects of fruit type and accessibility on selectivity. Journal of Animal Ecology 52: 407–420.

Moermond, T. C. & Denslow, J. S. 1985. Neotropical avian frugivores: patterns of behavior, morphology, and nutrition with consequences for fruit selection. Pages 865–897 in P. A. Buckley, M. S. Foster, E. S. Morton, R. S. Ridgely, and F. Buckley, eds. Neotropical ornithology. Ornithological Monograph 36.

Obst, B. S. 1986. Wax digestion in Wilson's storm-petrel. Wilson Bull. 98: 189–195.

Paine, R. T. 1971. The measurement and applications of the calorie to ecological problems. Annual Review of Ecology and Systematics 2: 145–164.

Piper, J. K. 1986. Seasonality of fruit characters and seed removal by birds. Oikos 46: 303–310.

Place, A. R. & Stiles, E. W. 1992. Living off the wax of the land: bayberries and yellow-rumped warblers. Auk 109: 334–345.

Place, A. R., Stoyan, N., Butler, R. G. & Ricklefs, R. R. 1989. The physiological basis of stomach oil formation in Leach's storm-petrel, *Oceanodroma leucorhoa*. Auk 106: 687–699.

Poddar, S. & Lederer, R. J. 1982. Juniper berries as an exclusive winter forage for Townsend's solitaires. American Midland Naturalist 108: 34–40.

Pratt, T. K. & Stiles, E. W. 1985. The influence of fruit size and structure on composition of frugivore assemblages in New Guinea. Biotropica 17: 314–321.

Roby, D., Place, A. R. & Ricklefs, R. R. 1986. Assimilation and deposition of wax esters in planktivorous seabirds. J. Exp. Zool. 239: 29–41.

Salomonson, M. J. 1978. Adaptation for animal dispersal of one seed juniper seeds. Oecologia 32: 333–339.

Salomonson, M. J. & Balda, R. P. 1977. Winter territoriality of Townsend's solitaires (*Myadestes townsendi*) in a pinon juniper-ponderosa pine ecotone. Condor 79: 148–161.

Skeate, S. T. 1985. Mutualistic interactions between birds and fruits in a northern Florida hammock community. PhD Thesis. University of Florida, Gainesville, Florida.

Snow, B. K. 1970. A field study of the bearded bellbird in Trinidad. Ibis 112: 299–329.

Snow B. & Snow, D. 1988. Birds and berries. Poyser, Calton.

Snow, D. W. 1962. The natural history of the oilbird, *Steatornis caripensis*, in Trinidad, W.I. Part II. Population, breeding ecology and food. Zoologica 47: 199–221.

Snow, D. W. 1971. Evolutionary aspects of fruit-eating by birds.

Snow, D. W. 1981. Tropical frugivorous birds and their food plants: a world survey. Biotropica 13: 1–14.

Snow, D. W. 1982. The cotingas. British Museum (Natural History).

Sorensen, A. E. 1983. Taste aversion and frugivore preference. Oecologia 56: 117–120.

Stiles, E. W. 1980. Patterns of fruit presentation and seed dispersal in bird-disseminated woody plants in the eastern deciduous forest. American Naturalist 116: 670–688.

Stiles, E. W. 1982. Fruit flags: two hypotheses. American Naturalist 120: 500–509.

Stiles, E. W. 1989. Fruits, seeds, and dispersal agents. Pages 87–122 in W. G. Abrahamson ed. Plant-animal interactions, McGraw-Hill, New York.

Stiles E. W. & White, D. W. 1982. Additional information on birddisseminated fruits: response to Herrera's comments. American Naturalist 120: 823–827.

Thompson, J. N. & Willson, M. F. 1979. Evolution of temperate fruit/bird interactions: phenological strategies. Evolution 33: 973–982.

Turcek, F. J. 1963. Color preference in fruit and seed-eating birds. Proceedings of the International Ornithological Congress 13: 285–292.

Walsberg, G. E. 1975. Digestive adaptations of *Phainopepla nitens* associated with the eating of mistletoe berries. Condor 77: 169–174.

Wetmore, A. 1914. The development of the stomach in the euphonias. Auk 31: 458–461.

Wheelwright, N. T. 1983. Fruits and the ecology of the resplendent quetzal. Auk 100: 286–301.

Wheelwright N. T. 1985. Fruit size, gape width, and the diets of fruit-eating birds. Ecology 66: 808–818.

Wheelwright, N. T., Haber, W. A., Murray, K. G. & Guindon, C. 1984. Tropical fruit-eating birds and their food plants: a survey of a Costa Rican lower montane forest. Biotropica 16: 173–192.

Wheelwright, N. T. & Janson, C. H. 1985. Colors of fruit displays of bird-dispersed plants in two tropical forests. American Naturalist 126: 777–799.

White D. W. 1989. North American Bird-dispersed fruits: ecological and adaptive significance of nutritional and structural traits. PhD Thesis. Rutgers University, New Brunswick, New Jersey.

White, S. C. 1974. Ecological aspects of growth and nutrition in tropical fruit-eating birds. PhD thesis. University of Pennsylvania, Philadelphia, Pennsylvania.

Willson, M. F., Irvine, A. K. & Walsh, N. G. 1989. Vertebrate dispersal syndromes in some Australian and New Zealand plant communities, with geographic comparisons. Biotropica 21: 133–147.

Willson, M. F. & Thompson, J. N. 1982. Phenology and ecology of color in bird-dispersed fruits, or why some fruits are red when they are 'green'. Canadian Journal of Botany 60: 701–713.

Willson, M. F. & Whelan, C. J. 1990. The evolution of fruit color in fleshy-fruited plants. American Naturalist 136: 790–809.

Vegetatio **107/108**: 237–244, 1993.
T. H. Fleming and A. Estrada (eds). Frugivory and Seed Dispersal: Ecological and Evolutionary Aspects.

Seed dispersal versus seed predation: an inter-site comparison of two related African monkeys

Annie Gautier-Hion, Jean-Pierre Gautier & Fiona Maisels
CNRS URA 373 & University of Rennes, Station Biologique, 35380 Paimpont, France

Keywords: C.(mona) pogonias, C.(mona) wolfi, Diets, Fruit availability, Gabon, Zaire

Abstract

C. pogonias and *C. wolfi* plant diets were studied in two sites, in Gabon and Zaire and compared with fruit availabilities. Monkeys in Gabon were found to be mainly fruit pulp-eaters while monkeys in Zaire were alternately seed-eaters, aril-eaters or leaf-eaters. These differences were related to differences in the availability of fruit categories: fleshy fruits were found to be much more abundant in Gabon than in Zaire forests. As a result, monkeys in Gabon were found to be mainly seed-dispersers while monkeys in Zaire were found, to a large extent, to be seed-predators. Results are discussed in terms of phenotypic flexibility in monkey feeding behavior, diversity of plant-monkey interactions, geographic variability of keystone plant resources, and their implications for forest management practices. The low availability of fleshy fruit species in Zaire is hypothezized to result from the poor soil conditions.

Introduction

The importance of frugivory in tropical rain forests and the role of large vertebrate frugivores in the dynamics of forest regeneration is now well established (e.g. Emmons *et al.* 1983; Leighton & Leighton 1983; Terborgh 1983, 1986a, b; 1990; Dubost 1984; Howe 1984; Estrada & Fleming 1986). To date, mammalian frugivory has been comprehensively studied at only a few sites (e.g. Charles-Dominique *et al.* 1981; Janson 1983, Howe 1990 for the neotropics; Gautier-Hion *et al.* 1985a, 1985b, for Africa). However, we do not know to what extent what happens in one rain forest can be generalized to another or even to other parts of the same forest block.

This study compares our quantitative observations on the plant feeding ecology of two related monkeys inhabiting rain forests, in Gabon and Zaire. In both study areas, monkey plant diets were analysed and compared with fruit availabilities and the effect of consumers on seeds was studied. The observations from Zaire are still in progress whereas the other data come from our previous work (Gautier-Hion 1980, 1990; Gautier-Hion *et al.* 1985a, b; Gautier-Hion & Michaloud 1989).

Methods

Study sites

The study sites are close to the equator. Makokou in N-E Gabon, lies at 0° 4′ N, and Botsima, (Salonga National Park) in the Central Zaire Basin, at 1° 15′ S. The two sites are ca. 1050 km apart (respectively 12° 46′ E and 22° E). Whereas the tropical rain forest at Makokou has been intensively studied for three decades, forests of the

Central Zaire Basin remain rather poorly known.

The Makokou site is mainly covered with seasonal evergreen rain forest. In this forest, the Leguminosae is most numerous in tree species (Reitsma 1988) and in individuals (Caballé 1978). The Botsima site is a mosaic of riverine forests on alluvial levees and mature forests on hydromorphic soils. As in Gabon, Leguminosae is the most species-rich group in the tree stratum.

The mean annual rainfall is comparable at both sites: 1755 mm at Makokou (Gautier-Hion *et al.* 1985a); 1756 mm at Ikela, a locality situated 120 km from Botsima (Ergo & Halleux 1979). Both sites undergo two rainy seasons and two dry seasons (a small dry season occurring around January, February and a main dry season in June and July).

The monkeys and their diets

At each site, one arboreal *Cercopithecus* species was studied: *C. pogonias* in Gabon and *C. wolfi* in Zaire; the two species are representatives of the *mona* superspecies.

In this study, only the quantitative composition of diets in terms of food plant categories is considered. In Gabon, quantitative data come from the analysis of stomach contents sampled every month but April (a total of 52 stomachs; Gautier-Hion 1980). In Zaire, all data come from direct observations (a total of 3293 feeding records over the annual cycle). They were taken according to the frequency method which gives the relative importance of each food category in the diet (Struhsaker 1975). The two methods are liable to systematic errors; they are nevertheless reliable for examining gross differences in feeding behaviour (Clutton-Brock 1977).

Fruit availability

Fruit availability was assessed as follows (Gautier-Hion *et al.* 1985a). A 5 km-trail in Gabon and in Zaire was traveled every second week and numbers of fruiting species and of fruiting individuals were noted. Results are expressed as the number of fruiting species per month and as the mean monthly number of fruiting sites per 100 m of trail. The availability of the following mature food types was considered in this study: fleshy fruit (either drupes or berries), fruit with arillate seeds, and pods of legumes.

Consumer effect on seeds

Monkeys are regarded as seed dispersers when they eat only the pulp or the aril of seeds and disperse the seeds by endo- or synzoochory. They are considered as predators when they destroy the seeds by eating them (in the case of seeds in pods or capsules) or by eating both the pulp and the seed (in the case of immature fruits). They are regarded as 'neutral' when they simply discard the intact seed under the parent tree.

Results

Mean annual composition of monkey diets

The diets of the two monkeys differed greatly in their overall composition (Table 1). The pulp of mature fleshy fruit formed the main part of the diet of *C. pogonias* in Gabon (an annual mean of 77%) followed by arils (18.4%). Whatever the period, flower and leaf consumption as well as the consumption of seeds were only occasionally observed (an average of 4.5%).

Table 1. Relative intake (in %; ± standard deviation) of the different food categories during the annual cycle by *C. pogonias* (Gabon) and *C wolfi* (Zaire).

Sites	Fleshy fruits	Arils	Legume seeds	Other seeds	Leaves	Flowers
Gabon	77.1 ± 19.87	18.4 ± 16.70	2.4 ± 4.57		2.1 ± 2.18	
Zaire	4.2 ± 3.38	27.4 ± 20.12	20.0 ± 25.38	7.33 ± 4.69	29.5 ± 24.30	11.41 ± 19.0

In Zaire, leaves (29.5%), seeds (27.3%), and arils (27.4%) constituted the bulk of the annual diet of *C. wolfi* while the pulp of fleshy fruit only accounted for 4.2%. Most of the seeds were taken from pods of legumes.

Seasonal variation in diets

Whatever the month, the diet of *C. pogonias* was dominated by fleshy fruits (range: 44% to 98%; Fig. 1). The decrease in fruit consumption observed around the main dry season was correlated with an increase in the consumption of both arils and leaves (Spearman rank correlation, $p<0.05$ and $p<0.01$ respectively).

In Zaire, the diet of *C. wolfi* varied greatly throughout the year (Fig. 1). Diet was alternately dominated by leaves (a maximum of 76% in January), leaves and flowers (32% and 51% respectively in March), arils (70% in May, 43% in July), or legume pods (60% to 57.5% from August to November). Whatever the period, the consumption of mature fleshy fruit was low, with monthly percentages reaching a maximum of 10% in July. The consumption of seeds of immature fleshy fruit took place year round (maximum = 15% in December).

Leaf consumption was negatively correlated with the consumption of legume seeds ($p<0.01$) and of arils ($p<0.05$). The increase in consump-

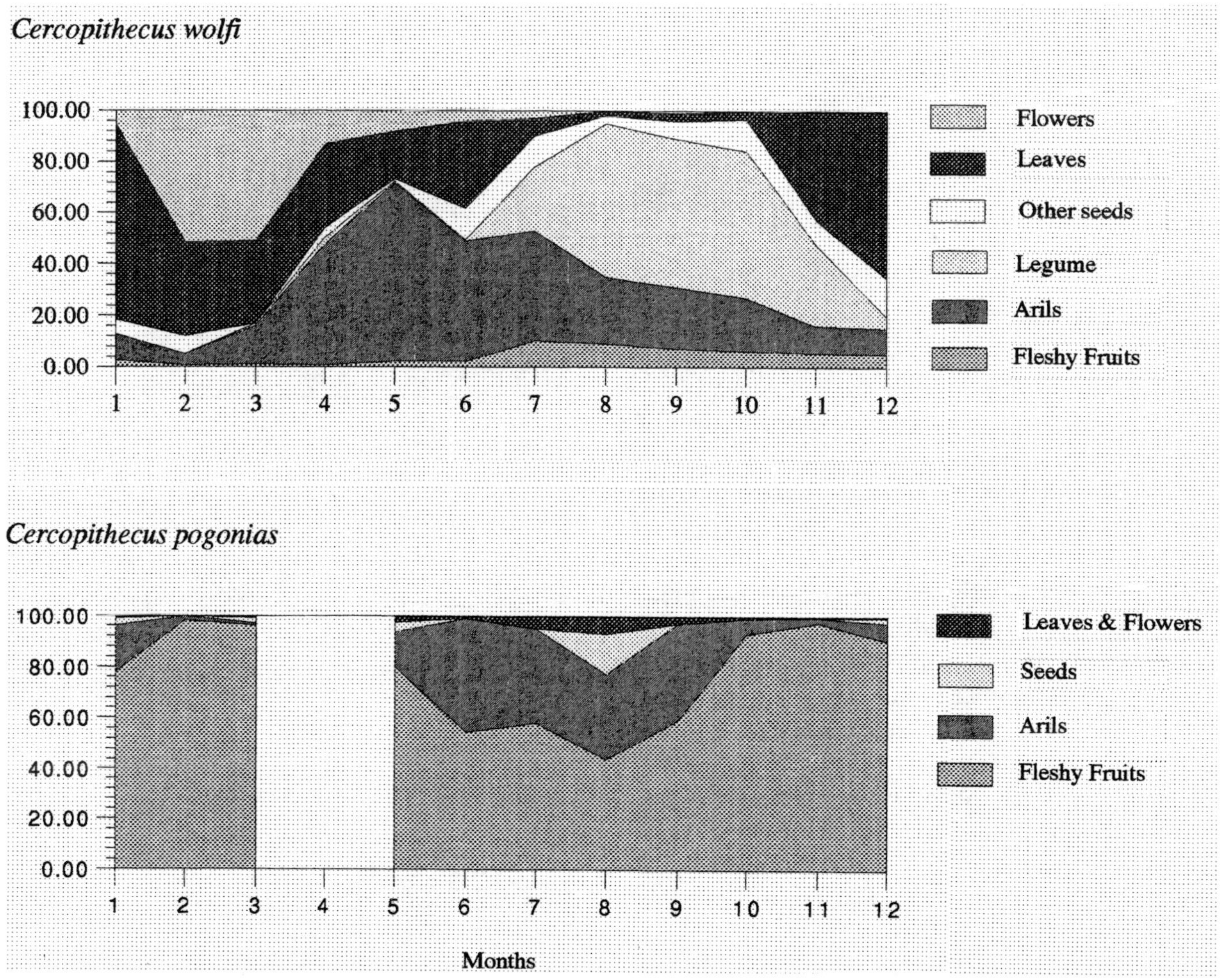

Fig. 1. Relative intake (in %) of plant food categories throughout the year for the two monkey species.

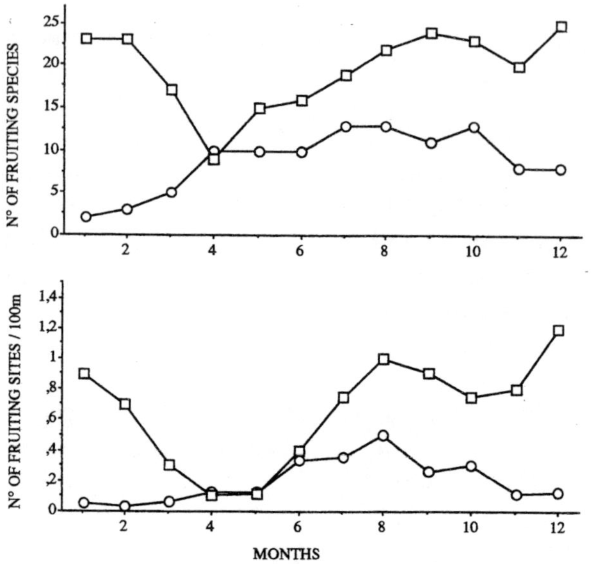

Fig. 2. Patterns of fleshy fruit production in terms of the number of fruiting species and of the mean number of fruiting sites/100m. Zaire: circles; Gabon: squares.

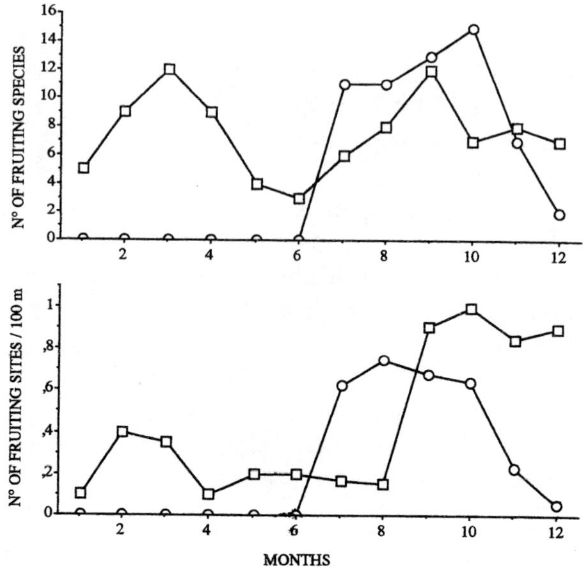

Fig. 3. Patterns of legume pod production in terms of the number of fruiting species and of the mean number of fruiting sites/100m. Zaire: circles; Gabon: squares.

tion of legume seeds was correlated with an increase in fleshy fruit intake ($p<0.01$).

Fruit availability

The two study sites differed in fruit availability. For fleshy fruits, both the monthly number of fruiting species and the mean number of fruiting sites were significantly higher in Gabon than in Zaire ($p>0.001$, Wilcoxon matched pair test; Fig. 2). A maximum of 24 species simultaneously produced mature fleshy fruit in Gabon vs a maximum of 13 in Zaire, with the minimum being 9 and 2 species respectively. The number of fruiting sites per 100 m per month ranged from 0.1 to 1.2 in Gabon versus 0.03 to 0.5 in Zaire. For the pods of Leguminosae (Fig. 3), results were similar at the two sites, both for the maximum numbers of fruiting species and of fruiting sites ($p>0.05$ in both cases). Finally the number of species producing arillate seeds per month (Fig. 4) was greater in Gabon than in Zaire ($p<0.05$), while the number of fruiting sites was higher in Zaire than in Gabon ($p<0.05$).

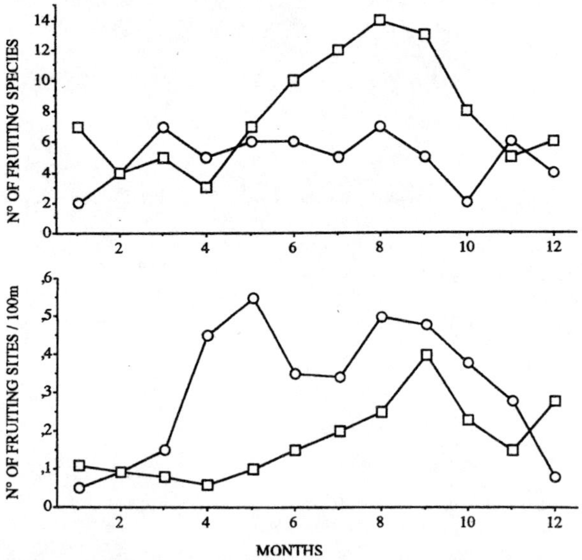

Fig. 4. Patterns of arillate seed production in terms of the number of fruiting species and of the mean number of fruiting sites/100m. Zaire: circles; Gabon: squares.

Fruit availability and monkey diets

In Gabon, the high annual consumption of fleshy fruits was related to their overall availability.

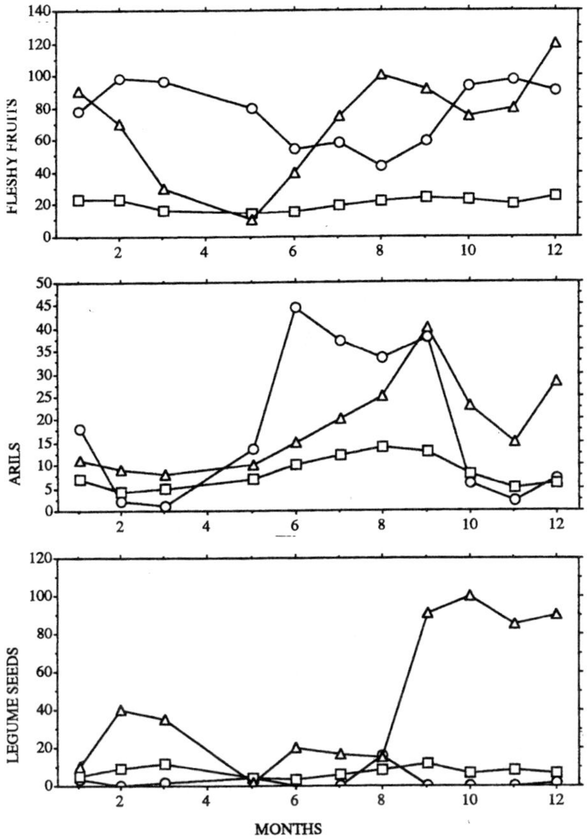

Fig. 5. Seasonal variation in plant intake by *C. pogonias* and availability of three food categories (circles: percent of intake; squares: number of fruiting species; triangles: mean number of fruiting sites/100m × 100).

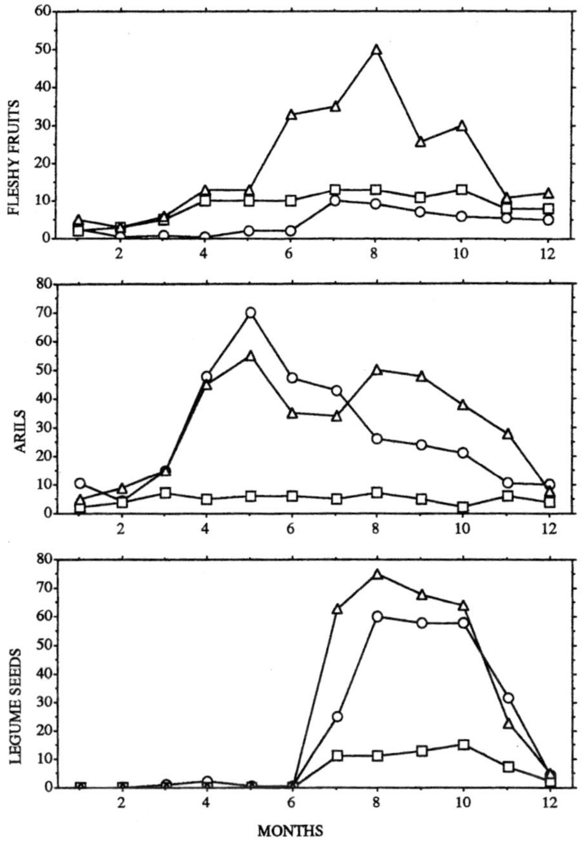

Fig. 6. Seasonal variations in plant intake by *C. wolfi* and availability of three food categories (circles: percent of intake; squares: number of fruiting species; triangles: mean number of fruiting sites/100m × 100).

However, their monthly rate of consumption was not correlated with their monthly availability either in terms of the number of fruiting species or fruiting sites (Fig. 5; Spearman rank correlation, $p > 0.05$ in both cases). Similarly, the consumption of arils was not correlated with the number of fruiting sites ($p > 0.05$). It was however correlated with the number of fruiting species ($p < 0.05$). The low consumption of legume seeds obviously diet not relate to their availability. In Zaire, both the availability of fleshy fruits and their rate of consumption were low (Fig. 6). Furthermore, the monthly rate of fleshy fruit consumption was positively correlated with their monthly availability both in number of species and number of sites ($p < 0.05$ in both cases). Similarly, the consumption of legume seeds was positively correlated with the number of fruiting species and fruiting sites ($p < 0.01$ in both cases). In the case of arils, the consumption rate was correlated with their number of fruiting sites ($p < 0.01$).

Consumer effects on seeds

In Gabon, monkeys served as seed dispersers for 82% of the species whose fruit they ate and were seed-predators for 4% of their food species. No positive or negative effects on the seeds were recorded for the remaining 14% (Gautier-Hion *et al.* 1985a).

In Zaire, monkeys are seed dispersers for 58% of the fruit species they ate and destroyed the seeds of at least 40% of the species. This included all of the tree species whose fruits were only eaten when immature and all the Leguminosae. For the latter, seed predation was important in five months of the year (Fig. 1).

Discussion and conclusions

Striking differences occur between Gabon and Zaire in the diets of this pair of related species. In Gabon, whatever the season, the pulp of fleshy fruits forms the bulk of the diet of *C. pogonias* which can therefore be considered basically a pulp-eater. Arils complement this diet during the main dry season. In Zaire, *C. wolfi* diet is characterized by a seasonal procession of different major food items; depending on the season, monkeys may successively be considered either seed-eaters, leaf-eaters or aril-eaters. Their consumption of fleshy fruits remains low throughout the year.

These differences between diets of closely-related monkeys and between seasonal diets within the same population emphasize the plasticity of their feeding behavior. They lead to major differences in the nature of interrelations between monkeys and plants. In Gabon, pulp-eating monkeys play an important role in seed dispersal. In Zaire, seed predation by monkeys is important especially for legumes.

Differences recorded in the relative availabilities of fruit types at the two sites mainly concern fleshy fruits which were found to be more abundant in number of species and density of fruiting trees in the Makokou forest than in the Botsima forest. They account for the huge differences in fleshy fruit intake between the two monkey species. They may also explain why dietary differences between the two sites are not correlated to the same extent with the availability of plant foods. Indeed, monkeys in Zaire seasonally adjust their diet by alternately taking the most common food category. In contrast, monkeys in Gabon maintain a high intake of fleshy fruits year round; they increase their consumption of arils when available but ignore legume seeds and leaves.

Leaf-eating or seed eating could be a last resort for basically frugivorous primates. While monkeys in Gabon suffer from fruit shortage only during a limited period, monkeys in Zaire are forced to consume seeds or leaves in the absence of fleshy fruits. Gautier-Hion (1983) previously pointed out that monkeys as a whole are more folivorous in East African forests than in Gabon, apparently as a result of a lack of fleshy fruits. This view was supported by studies of the lowland gorillas of the Lope Reserve (Gabon); these apes were found to largely rely on fruit and seeds, contrary to the folivorous mountain gorillas (Williamson *et al.* 1990). This led Rogers *et al.* (1990) to conclude that '*G. g. gorilla* was not strictly folivorous, except in habitats which allow it no alternative.'

Seed consumption by *Cercopithecus* monkeys as extensive as that observed for *C. wolfi* has not yet been documented elsewhere, but is reminiscent of the black colobus, *Colobus satanas*, in the Douala-Edea Reserve in Cameroun (McKey 1978). Seed-eating by these colobines was postulated to result from the vegetation growing on the impoverished sandy soils of the Douala-Edea Reserve containing a greater concentration of toxic secondary compounds that equivalent foods growing on richer soils. As seeds are more digestible, richer in fats, and generally poorer in tannins than leaves, the colobus, which are mainly leaf-eaters in the Eastern African forests, avoid leaves and rely on seeds at the Douala-Edea Reserve.

As at Douala-Edea, forests in the Salonga National Park in Zaire also grow on acid sandy soils, containing very small amounts of silt and clay. These areno-ferralsols and/or ferralsols are very poor in nutrients; the humus layer is very shallow and is generally incompletely decomposed (Evrard 1968). At Botsima, poverty of the soil is indicated by the exceptionally low rates of regeneration after clearing. It remains to be seen whether plants growing on such impoverished soils develop more toxic secondary compounds in their foliage and seeds than those found on richer soils, as suggested by Janzen (1974) and/or whether the low fleshy fruit availability in this forest results from this soil poverty.

Considering the amount of dietary flexibility in closely-related monkeys, what can be inferred about diets from gut specializations (cf. Martinez del Rio & Restrepo, this volume)? More generally, is 'the high level of correspondence between food type, plant chemistry, and digestive anatomy,' previously described by Chivers & Hladik (in Chivers & Hladik 1984), the expression of actual constraints? Could such a tight correspondence have been overestimated because of our poor understanding of phenotypic variability due to the limited number of comparative studies conducted on related species, populations of the same species, and/or groups of the same population, as suggested by Hauser (1990)?

Because they stress the plasticity of the monkeys' feeding behavior, these results highlight the geographic variability of keystone plant resources for animal populations (Terborgh 1986b; Gautier-Hion & Michaloud 1989). By highlighting the diversity of interactions between African primates and their food plants, they warn anyone concerned with management practices aimed at conserving rain forest biodiversity against accepting broad generalizations.

Acknowledgements

We are particularly indebted to Dr Mankoto ma Mbaelele, President of the Institut Zaïrois pour la Conservation de la Nature, who encouraged our research in the Salonga National Park. We wish to thank Nathalie Schildwachter and the staff of Botsima Station for their help in collecting phenological data. The field work was supported by grants from the Ministère de l'Environnement and the CNRS (France). Dr F. Maisels was also supported by grants from the Flora and Fauna Preservation Society, the Peoples Trust for endangered species, the Percy Sladen Memorial Fund, the Leakey Trust, Edinburgh Zoo, and the Boise Fund. Finally, we are grateful for logistic help from the Commission des Communautés Européennes.

References

Caballé, G. 1978. Essai sur la géographie forestière du Gabon. Adansonia. 17: 425–440.

Charles-Dominique, P., Atramentowitcz, M., Charles-Dominique, M., Gerard, H., Hladik, C.M. & Prevost, M.F. 1981. Les mammifères frugivores arboricoles nocturnes d'une forêt guyanaise: interrelations plantes-animaux. Rev. Ecol. (Terre Vie). 31: 341–436.

Chivers, D.J. & Hladik, C.M. 1984. Food acquisition and processing in primates: concluding discussion. In: Chivers, D.J., Wood, B.A & Bilsborough, A. (eds.). Food acquisition and processing in Primates. Plenum Press, London.

Clutton-Brock, 1977. Methodology and measurement. In: Clutton-Brock T.H. (ed). Primate Ecology: studies of feeding and ranging behaviour in lemurs, monkeys and apes. Acad.Press, London.

Dubost, G. 1984. Comparisons of the diets of frugivorous forest ruminants of Gabon. J. Mammal. 65: 298–316.

Emmons, L., Gautier-Hion, A. & Dubost, G. 1983. Community structure of the frugivorous-folivorous forest mammals of Gabon. J. of Zool. Lond. 199: 209–222.

Ergo, A.B. & Halleux, B. 1979. Catalogue mondial des données climatiques moyennes. Vol.II, Fasc.I, C.I.D.A.T., Tervuren.

Estrada, A. & Fleming, T.H. (eds.) 1986. Frugivores and seed dispersal. W. Junk Pub. The Hague.

Evrard, C. 1968. Recherches écologiques sur le peuplement forestier des sols hydromorphes de la cuvette centrale congolaise. ONRD, INEAC.

Gautier-Hion, A. 1980. Seasonal variations of diet related to species and sex in a community of *Cercopithecus* monkeys. J. An. Ecol. 49: 237–269.

Gautier-Hion, A. 1983. Leaf consumption by monkeys in Eastern and Western Africa: a comparison. Afr. J. Ecol. 21: 107–113.

Gautier-Hion, A. 1990. Interactions among fruit and vertebrate fruit-eaters in an African tropical rain forest. In: Bawa, K.S. & Hadley, M. (eds). Reproductive ecology of tropical forest plants. Mab Series, Vol.7, Parthenon Pub. Carnforth.

Gautier-Hion, A., Duplantier, J-M., Quris, R., Feer, F., Sourd, C., Decoux, J-P., Dubost, G., Emmons, L., Erard, C., Hecketsweiler, P., Roussilhon, C. & Thiollay, J-M., 1985a. Fruit characters as a basis of fruit choice and seed dipersal in a tropical forest vertebrate community. Oecologia 65: 324–337.

Gautier-Hion, A., Duplantier; J-M., Emmons, L., Feer, F., Heckestweiler, P., Moungazi, A., Quris, R. & Sourd, C. 1985b. Coadaptation entre rythmes de fructification et frugivorie en forêt tropicale humide du Gaon: mythe ou réalité. Rev. Ecol. (Terre Vie) 40: 405–434.

Gautier-Hion, A. & Michaloud, G. 1989. Are figs always keystone resources for tropical frugivorous vertebrates? A test in Gabon. Ecology 70: 1826–1833.

Hauser, M. 1990. Speciation and adaptation in the African guenons: what we know and have yet to understand? Am. J Primatol. 20: 299–302.

Howe, H.F. 1984. Implications of seed dispersal by animals for the management of tropical reserves. Biol. Conserv. 30: 261–281.

Howe, H.F. 1990. Seed dispersal by birds and mammals: implications for seedling demography. In: Bawa, K.S. & Hadley, M. (eds). Reproductive ecology of tropical forest plants. Mab Series, Vol.7. Parthenon Pub. Carnforth.

Janson, C.H. 1983. Adaptation of fruit morphology to dispersal agents in a Neotropical forest. Science 219: 187–198.

Janzen, D.H. 1974. Tropical blackwater rivers, animals, and mast fruiting by the Dipterocarpaceae. Biotropica 6: 69–103.

Leighton, M., Leighton, D.R. 1983. Vertebrate responses to fruiting seasonality within a Bornean rain forest. In: Sutton, S.L., Whitmore, T.C. & Chadwick, A.C. (eds.) Tropical rain forest: ecology and management. Blackwell Scientific, Oxford.

McKey, D.B. 1978. Phenolic content of vegetation in two-African rain forests: ecological implications. Science 202: 61–64.

Rogers, M.E., Maisels, F., Williamson, E.A., Fernandez, M. & Tutin, C.E.G. 1990. Gorilla diet in the Lope Reserve, Gabon: a nutritional analysis. Oecologia 84: 326–339.

Reitsma, J.M. 1988. Forest vegetation of Gabon. Tropenbos series I. Drukkerij Veenman B.V., Wageningen.

Struhsaker, T.T. 1975. The Red Colobus. Univ. of Chicago Press, Chicago.

Terborgh, J. 1983. Five New-World Primates: a study of comparative ecology. Princeton Univ., Princeton.

Terborgh, J. 1986a. Community aspects of frugivory in tropical forests. In: Estrada A. & Fleming T.H. (eds.) Frugivores and seed dispersal. W. Junk, Pub. The Hague.

Terborgh, J. 1986b. Keystone plant resources in the tropical forest. In: Soule, M.E. (ed.) Conservation biology: the science of scarcity and diversity. Sinauer Ass., Sunderland.

Terborgh, J; 1990. Seed and fruit dispersal. Commentary. In: Kawa, B.S. & Hadley, M. (eds.) Reproductive ecology of tropical forest plants. Mab Series, Vol.7. Parthenon pub. Carnforth.

Williamson, E.A., Tutin, C.E.G., Rogers, M.E. & Fernandez, M. 1990. Composition of the diet of lowland gorillas in Gabon. Am.J Primatol. 21: 265–278.

Vegetatio **107/108**: 245–257, 1993.
T. H. Fleming and A. Estrada (eds). *Frugivory and Seed Dispersal: Ecological and Evolutionary Aspects.*

Patterns of frugivore species richness and abundance in forest islands and in agricultural habitats at Los Tuxtlas, Mexico

A. Estrada [1], R. Coates–Estrada [1], D. Meritt, Jr. [2], S. Montiel [3] & D. Curiel [1]
[1] *Estación de Biología 'Los Tuxtlas', Instituto de Biología–UNAM, Apartado Postal 176, San Andrés Tuxtla, Veracruz, Mexico;* [2] *Lincoln Park Zoological Gardens, 2200 North Cannon Drive, Chicago, Ill, USA;* [3] *Universidad Veracruzana, Cordoba, Veracruz, México*

Keywords: Tropical rain forest, Los Tuxtlas, Mexico, Conservation, Birds, Mammals, Bats

Abstract

Destruction and fragmentation of tropical rain forest result in a loss of species and of generating capacity of the ecosystem via animal vectors such as seed dispersal agents. To gather quantitative data regarding this ecological problem, birds and mammals were censused in 30 forest fragments, 15 agricultural islands representing five types of vegetation (coffee, cacao, citrus, pepper and mixed–crops) and in three pastures in Los Tuxtlas, southern Veracruz, Mexico. More than 6000 animals of 257 species were detected thus suggesting the existence of a rich species pool in the fragmented landscape. Frugivores accounted for 60% of species, for 72% of individuals censured and for 85% of the total animal biomass recorded. Clusters of small forest fragments (< 100 ha) were richer in species and individuals than clusters of large area (> 100 ha) forest islands. Pastures were especially poor in forest birds and mammals. While the agricultural islands studied contributed to only 1% of the total area of vegetation sampled, they contained 58% of all species detected and 34% of all individual birds and mammals censured. Recaptures indicated inter–island movements of forest birds and mammals. Forty percent of the species were detected in forest habitats only, the rest were detected in forest and in agricultural habitats. Seeds of forest interior plants dispersed by birds and bats were detected in the agricultural habitats. The value of agricultural islands as landscape features providing some degree of biotic connectivity among fragmented animal populations is discussed.

Introduction

Tropical rain forest are the most diverse and complex terrestrial ecosystems on this planet. They harbor about 50–70% of all existing species of organisms and are a source of climatic stability and of many resources used today by man, ranging from food crops to pharmaceutical products (Myers 1988). In Latin America the conversion of tropical rain forest to pastures to raise cattle has meant the destruction and fragmentation of the natural habitats of a great number of tropical vertebrates whose ecology, behavior, and sensitivity to fragmentation is poorly known. This process has also meant the severe and sometimes irreversible weakening of the regenerating and self sustaining capacity of the tropical ecosystem. Further, the disappearance of these resources represent an irreversible loss to the world community and to the countries in which they occur.

As natural habitats are destroyed by man, they become fragmented and their component populations are subdivided and reduced, with local and regional extinction of species a common result (Gilbert 1988; Karr 1982a, 1982b, 1990; Lynch & Whitcomb 1978; Lovejoy *et al.* 1984). When animals are capable of migrating between fragments the effects of small population size may be partly or greatly mitigated. The demographic and genetic contributions of immigrants can provide a buffer against extinction (Brown & Kodric–Brown 1977; Burkey 1989; Laurance 1991; Lomolino 1986). Because food and other resources of ecological specialists are often patchy in space and time these species tend to be rare and more vulnerable to habitat destruction (Diamond 1984; Hubbel & Foster 1986; Karr 1982a, b; Terborgh 1976, 1986; Wilcox 1978). Specialists can be threatened also by successional changes in fragments (Picket & Thompson 1978), by the collapse of coevolved mutualisms or food webs (Gilbert 1980) or by edge effects (Janzen 1983; Lovejoy *et al.* 1986).

Frugivores depend on a type of food that is patchy in space and time and thus display high sensitivity to fragmentation and isolation of their habitat (Laurance & Yensen 1991; Leighton & Leighton 1983; Terborgh 1986). However, in a fragmented landscape frugivore species that are good dispersers and ecological generalists may continue to survive (Gilbert 1980; Laurance 1991). Interfragment dispersal may promote survival of forest species through the genetic and demographic contribution of immigrants (Brown & Kodric–Brown 1977; Laurance 1991). Such dispersal may be a critical prerequisite as a source of seeds of rain forest trees and other plants (Hopkins & Graham 1984) with arrival of viable seeds declining sharply as distance between fragments increases. Precise evaluation of the effects of fragmentation on the regenerating capacity of tropical rain forests is difficult because the ecology of the majority of the species is poorly known and reliable data on responses of entire guilds or assemblages to habitat fragmentation are rare (e.g. Klein 1989; Lovejoy *et al.* 1986; Zimmerman & Bierregaard 1986; Willis 1974).

Tropical rain forests in Mexico today consist of a collection of fragments of various sizes and with different histories of isolation. These fragments continue to be reduced in size at a rate of 4000 ha/yr. In the region of Los Tuxtlas, in southern Veracruz tropical rain forest reaches its northernmost distribution in the American continent and is notable for its high biological diversity (Estrada *et al.* 1985). Of the original 2,500 square kilometers of tropical rain forest that once covered the region of Los Tuxtlas, only about 10% remains today in a very fragmented condition (Estrada & Coates–Estrada 1988).

Land in this region is used primarily for cattle ranching. The cultivation of plants such as coffee, cacao, citrus and allspice, among others, is in the form of small parcels which occur sporadically in a 'sea' of pastures and represents only 3% of the total land use. The land–use pattern allowed us to experimentally assess the effects of rain forest destruction and fragmentation on the fate of two of the most sensitive animal components of this ecosystem: birds and mammals. In addition, it enabled us to assess the value of man–made islands of vegetation to frugivorous forest birds and mammals.

The specific aims of this paper are to report data on the bird and mammal species richness detected in forest fragments occurring amidst pastures in Los Tuxtlas southern Veracruz, Mexico. Further papers will report additional and complementary data both at the community and species level.

In addition to the forest fragments studied, birds and mammals were also censured in five types of agricultural vegetation: coffee (*Coffea arabica*), cacao (*Theobroma cacao*), allspice (*Pimienta dioica*), citrus (*Citrus sinensis*) and mixed crops (coffee, cacao, citrus, banana). Shade in the case of the coffee, cacao and mixed plantations was provided by tall (> 15 m) rain forest trees left by the farmers. Since pastures are now the dominant form of vegetation in Los Tuxtlas, we also conducted faunistic surveys in these habitats to determine the presence of forest birds and mammals.

Because of the important role played by frugivores as dispersal agents in the reproductive strategy of tropical rain forest plants and in the regenerating capacity of the ecosystem, special attention was paid to the occurrence of the frugivore components of these two animal communities.

Methods

The sector in the region of Los Tuxtlas in which field work was carried out encompasses 100–800 square kilometers within which the biological field station of 'Los Tuxtlas' of UNAM is located. Original vegetation cover in this area was tropical rain forest. The altitudinal gradient ranges from sea level to 1600 m above sea level and the landscape, dominated by pastures, contains forest fragments and a few agricultural islands which were the subject of our biological surveys (Fig. 1).

We studied forest fragments differing in size and in years of isolation. Area of each fragment was measured by digitizing aerial photographs and by corroboration in the field. Time since isolation was determined using aerial photographs taken at 5–10 year intervals during the last 30 years and by field surveys. The forest fragments we studied represented different stages of regeneration and disturbance. In the case of the five agricultural habitats (coffee, cacao, allspice, citrus and mixed crop), we sampled three replicates per habitat type to control for spatial variation in animal composition and distances to forest sites, but their selection depended on their

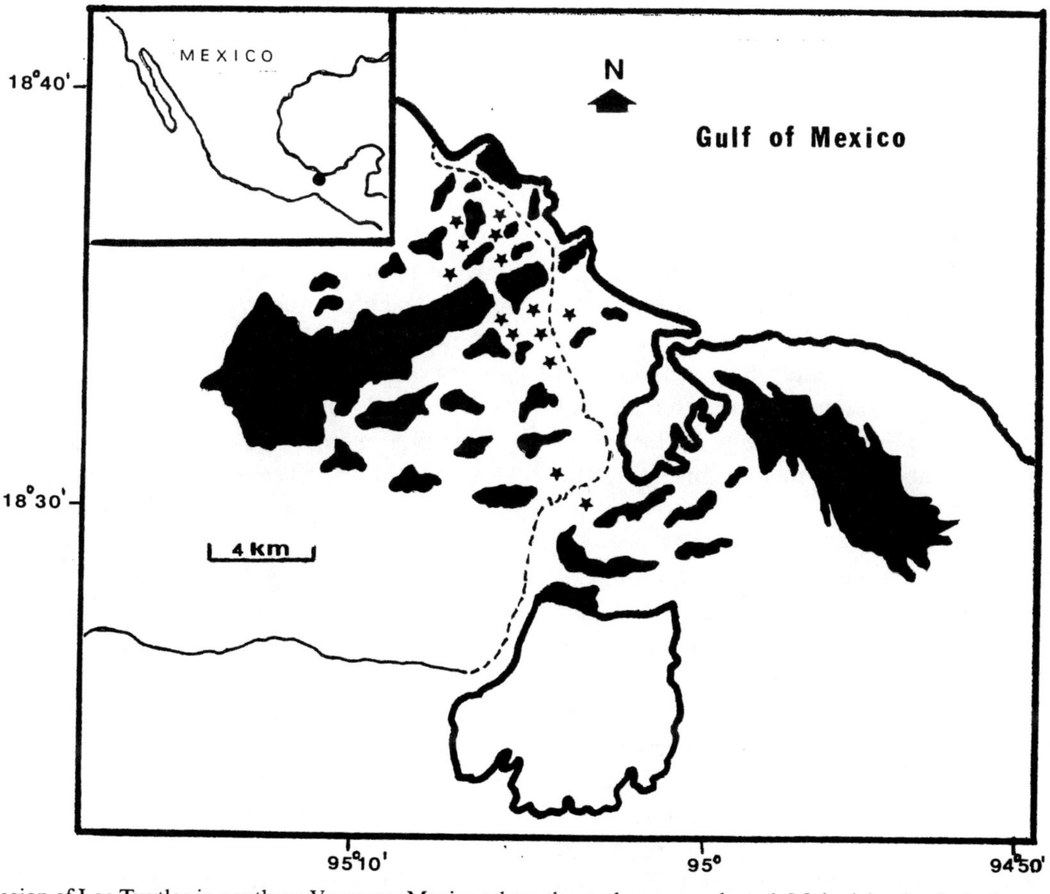

Fig. 1. Region of Los Tuxtlas in southern Veracruz, Mexico where the study was conducted. Major islands of rain forest remaining are shown in dark. Not shown are forest fragments < 100 ha in size. The approximate location of the agricultural islands studied is indicated by stars. The discontinuous line is a dirt road, Lake Catemaco is the large open area at the bottom and the paved road to Veracruz City is the continuous line bottom left.

availability because less than 3% of the land is currently used for growing these particular crops. Three areas of pasture totally devoid of the original forest vegetation were also studied; the three replicates permitted comparison with forest and agricultural islands of similar area.

Our interest in contrasting the biological richness across sites and vegetation types led to emphasize an equal or similar sampling effort in each island of vegetation. Thus, the number of nets (day and night; N = 11 in each period), net hours per day (125), net hours per night (50), number of diurnal visual and nocturnal censuses (2 in each case) as well as number of traps (46 Sherman and 15 Tomahawk), and trap nights (1), were standardized across all sites and vegetation types. The sampling quadrat or the area where netting, trapping and visual censuses were carried out in all sites (except the very small ones (1.0 ha), was 140 m in length by 80 m in width. At all sites the sampling quadrat was located at least 30–50 m from the fragment's border(s). In the case of very small sites (1.0 ha), the sampling area was adjusted in shape to have all sampling units at least 5–10 m from any border. A standard number of replicates (N = 3) per condition (e.g. island size class, vegetation type) was used in the comparisons presented below.

Sampling of birds and mammals was carried out during two days at each site. At each fragment one transect of 11 adjacent mist nets (12 m long × 3 m high) about 140 m long bisecting the site was set up. Nets were operated to capture understory birds from 0600 to 1130–1200 for two consecutive mornings and for 2–3 beginning at dusk on two consecutive nights to capture understory bats. In another parallel transect, located about 50 m away, 46 Sherman (bait = oats and vanilla) and 15 Tomahawk traps (bait = sardines) were placed on the forest floor every 5 and 10 m, respectively, and were operated for one full night (12 h). Visual censuses of birds and mammals were conducted at six sampling points at 30 m intervals along the latter transect. For each animal detected we recorded species, sex, age, height in the vegetation, time at which it was observed, and general activity.

Nocturnal censuses were carried out using flash lights and a nightvision scope (Javelin Electronics, model 221) to minimize disturbance to detected animals. Captured animals were identified to species and marked with coloured and numbered bands. Body weight, wing and/or forearm length, incubation patch and/or reproductive condition, age, molt, and the presence or absence of ectoparasites were recorded before releasing each animal. Average body weights obtained from the literature and from our own surveys were assigned to those animal records derived from visual censuses for calculations of total and partial animal biomass.

Species were classified in one of six foraging guilds: frugivore–insectivore, granivore, insectivore, insectivore–carnivore, carnivore and nectarivore–insectivore according to their foraging preferences as reported in the literature and by direct field observations by us at Los Tuxtlas over the last 15 years. For the purposes of this paper, analysis here is restricted to the frugivore (seed dispersers) and granivore (seed predators) components of these guilds which were classified as fruit–users; for some analyses these two groups were treated separately.

At each site the vegetation was censused in six random located plots measuring 10 × 10 m and separated by at least 30 m. All trees ≥ 5.0 cm in dbh and 1.5 meters in height that occurred within the plot were counted, measured (dbh, height) and identified to species. We also measured maximum canopy height and foliage density (number of foliage interceptions along a vertical pole at each of the following height classes: 1 = 0–0.50 m; 2 = 0.50–1.0 m and 3 = 1.0–3.0 m) at four random locations within each plot.

Results

Number of organisms detected

Data for this paper came from 30 forest islands, 15 agricultural islands and three pastures. The total area of vegetation accumulated by forest fragments was 7158 ha. The 15 agricultural islands encompassed 45 ha and the three pastures totaled 100 ha. A total of 8050 net hours, 184 visual censuses, 598 hours of censusing and 2806 trap nights were completed. This resulted in the censusing of 6846 animals (65% birds and 35% mammals) of 257 species, of which 189 were birds and 68 were mammals (34 species of bats and 34 species of non flying mammals (NFM from here on). Ninety percent of the species detected occurred in forest islands, 58% occurred in agricultural habitats and 4% occurred in pastures. Sixty five percent of all individuals were censused in forest fragments, 34% were detected in the agricultural habitats and 1% in the pastures (Table 1). Total animal biomass recorded in all vegetation islands was 1.607 kg of which 84% came from forest fragments and 16% from agricultural islands.

Ten of the 189 bird species detected (*Cyanocorax morio*, *Habia fuscicauda*, *Henicorhina leucosticta*, *Amazona autumnalis*, *Phaethornis superciliosus*, *Turdus grayi*, *Dendroica magnolia*, *Wilsonia citrina*, *Hylocichla mustelina* and *Oporornis formosus*) accounted for 37% of individual records and the remaining species (95%) accounted for 63% of the bird records. The community of 34 bat species detected was strongly dominated by seven species (*Carollia brevicauda*, *Sturnira lilium*, *Pteronotus parnelli*, *Vampyrops major*, *Artibeus phaeotis*, *A. toltecus*, *A. jamaicensis*) which accounted for 79% of all bat records. The remaining 27 species (79%) accounted for 22% of the remaining records. One species, *C. brevicauda*, accounted for 24% of all bats captured. Of the 34 NFM species recorded, two marsupials (*Didelphis marsupialis* and *Philander opossum*) and a rodent (*Peromyscus mexicanus*) accounted for 46% of the individuals censused. The remaining species (91%) accounted for 54% of the NFM records.

Fruit–eating birds and mammals

We detected the presence of 153 fruit–eating species representing 60% of all species detected (N = 257) and for 72% of all animal records. Within this category frugivores accounted for 110 (72%) of these species and for 83% of all fruit–eating animals censused (N = 4896); granivores represented 28% of fruit–eating species and 17% of all fruit–eating animals censused (Table 2).

In forest habitats we detected 86% of fruit–eating species and 65% of fruit–eating individuals. Fruit–eating species recorded in agricultural habitats represented 62% of all fruit–eating species detected and 35% of fruit–eating individuals censused at these sites (Table 2). Fruit–eating birds and mammals accounted for 85% of the total animal biomass recorded (1.607 kg) with frugivores accounting for 71% and granivores for 14%.

Table 1. Birds and mammals detected in 30 forest islands, 15 agricultural islands representing five types of vegetation (cacao, coffee, mixed, citrus and allspice) and 3 pastures. The number of forest islands sampled per size class and the number of sites sampled per agricultural habitat as well as the number of pastures studied are shown at the bottom of the table.

Organism	Forest		Agricultural		Total		Pasture		Total	
	Spp	N	Spp	N	Spp	N	Spp	N	Spp	N
Birds	165	2806	120	1665	189	4471	8	13	189	4484
Nonflying mammals	34	630	8	56	34	686	1	16	34	702
Bats	33	1037	22	623	34	1660	0	0	34	1660
Total	232	4473	150	2344	257	6817	9	29	257	6846

Forest size classes: 1000 ha N = 2, <800 ha N = 2, <600 ha N = 2, <400 ha N = 3, <200 ha N = 3, <100 ha N = 3, <50 ha N = 3, <20 ha N = 3, <10 ha N = 3, <5 ha N = 3, 1 ha N = 3.
Agricultural habitats: cacao 9 ha N = 3, coffee 9 ha N = 3, mixed 8 ha N = 3, citrus 9 ha N = 3, allspice 10 ha N = 3.
Pastures: 100 ha N = 3.

Table 2. Occurrence of fruit-eating birds and mammals in forest and in agricultural habitats.

	Forests		Agricultural		Total	
	Spp	N	Spp	N	Spp	N
Frugivores						
Birds	59	1383	55	921	72	2304
Bats	23	846	17	554	24	1400
Nonflying mammals	14	333	5	38	14	371
Subtotal	96	2562	77	1513	110	4075
Granivores						
Birds	23	314	15	197	30	511
Bats	0	0	0	0	0	0
Non flying mammals	13	292	3	18	13	310
Subtotal	36	606	18	215	43	821
Grand total	132	3186	95	1728	153	4896

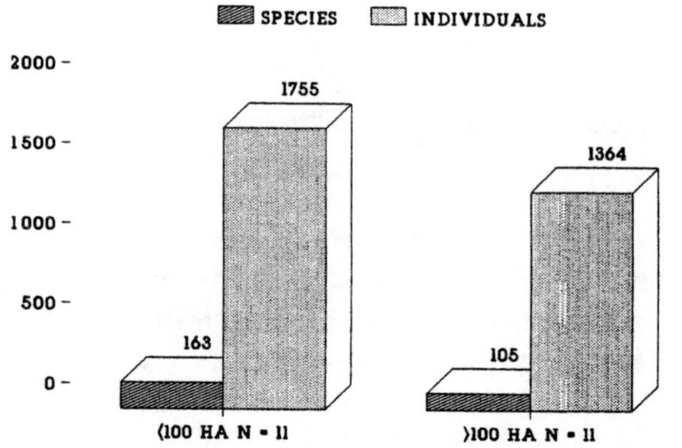

Fig. 2. Species richness and individual birds and mammals censused in forest islands < 100 ha in size and in forest islands > 100 ha in size. In this comparison, sampling effort and number of sites are equal in each condition.

Birds accounted for 67% of the species detected and for 54% of organisms censused in the fruit–eating category, mammals accounted for 33% of the species detected and for 42% of the animals censused. Bats accounted for 47% of the fruit–eating mammal species and 67% of the fruit–eating mammals censused. Birds accounted for 65% and for 70% of the frugivore and granivore species recorded and for 56% and 62% of the frugivore and granivore individuals censused. Bats were the second most numerous component among frugivores accounting for 22% of the species and for 34% of the individual records (Table 2).

Proximate effects of forest destruction and fragmentation

Replacement of forest by pasture. Our surveys in three pastures totalling an area of 100 ha resulted in no detections of forest interior birds and mammals and in the census of only nine open habitat species. These species were one rodent (*Sigmodon hispidum*) and eight bird species, four raptors (*Buteo magnirostris*, *Falco peregrinus*, *F. rufigularis*, *F. sparverius*), three opportunistic seed–eaters (*Dives dives*, *Sturnella magna*, *Quiscalus mexicanus*) and a specialized seed–eater (*Sporophila turqueola*). A comparison with a similar area of forest habitats (100 ha) represented by three forest islands, showed that forest habitats were significantly richer in forest interior birds and mammals and contained significantly higher numbers of frugivorous animals than pastures. Even a 1.0 ha forest patch was significantly richer in forest birds and mammals than the pasture habitat (Fig. 4).

Fragmentation of forest. A comparison among forest sites, varying in size, but with sampling effort kept constant, showed that, in general, larger forest fragments tend to have more bird and mammal species ($r = 0.857$, $P < 0.05$) and more frugivore species ($r = 0.746$, $P < 0.05$) than smaller forest fragments. However, broken down by type of organism (birds, bats, NFM), the trend held true only for NFM (number of species $r = 0.74$ $P < 0.05$; number of individuals $r = 0.763$ $P < 0.05$; $N = 11$ size classes).

An examination of the number of new species added when forest islands were ordered from small to large showed that an area of only about 60 ha contributed by several small forest fragments was necessary to account for 70% of the species recorded; addition of new species was very slow as additional area was added. A cluster of new species was added to the species' pool between 300–1000 ha (Fig. 3). This cluster con-

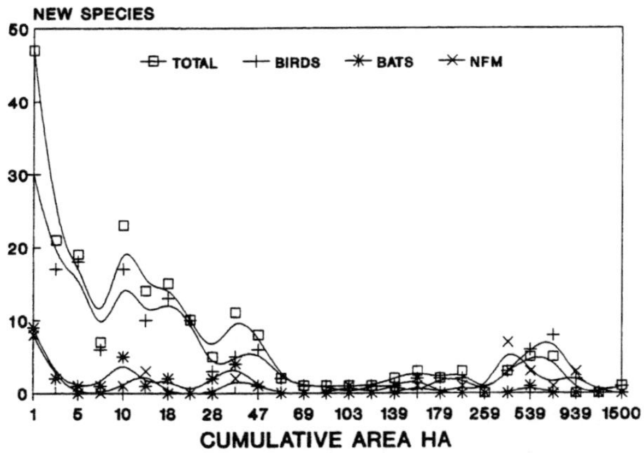

Fig. 3. Cumulative area of forest fragments and addition of new species. Note the rapid accumulation of species between 1–60 ha and the cluster of new species added with larger accumulated areas (>200 ha) of rain forest. Shown also are the patterns detected for birds, bats and nonflying mammals (NFM).

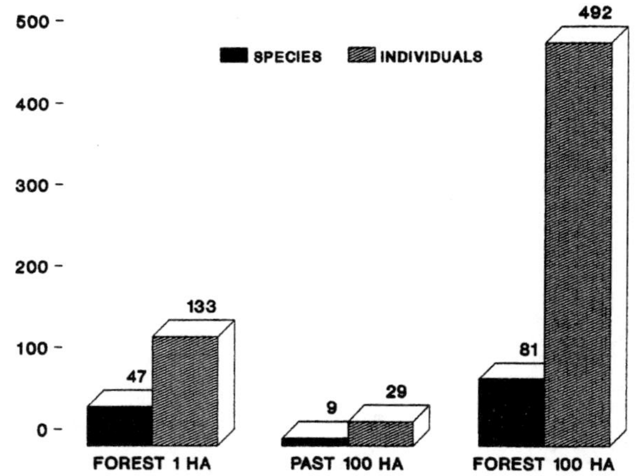

Fig. 4. Comparison of species richness (forest birds and mammals) between pastures and forest islands of equal area with sampling effort kept constant. Also shown is the species richness detected at a forest iland 1.0 ha in size. None of the species detected in pastures were forest interior birds and mammals.

sisted of frugivores such as *Ateles geoffroyi* and granivores such as *Tinamus major*, *Crax rubra* and *Penelope purpurascens*. Other species recorded in censuses at large sites were carnivores such as *Felis concolor*, *F. pardalis*, and *Spizaetus ornatus*.

An area effect was consistently found when we compared large (>100 ha; range 200–1000 ha N = 11) versus small (<100 ha; range 1–150 ha N = 11) forest islands in species richness. Small forest islands had significantly more species (small = 163; large = 105; $p < 0.001$) and more individuals (small = 1755; large = 1364; $p < 0.001$) than large forest islands (Fig. 2).

Table 3. Number of birds and mammals detected at each of the agricultural habitats studied (3 replicates per habitat).

	Birds		Bats		Nonflying animal Mammals		Nonflying animal Biomass
	N	Spp	N	Spp	N	Spp	kg
Cacao	463	75	277	15	17	7	60
Coffee	186	64	53	12	9	5	37
Mixed	330	63	89	15	25	5	107
Citrus	458	53	78	16	4	2	50
Allspice	228	39	126	14	1	1	28
Total	1665	120	623	22	56	8	282

Forest birds and mammals in agricultural habitats

The five types of agricultural vegetation studied contained a total of 120 different species of birds and 30 different species of mammals (22 bats and 8 NFM species). Total animal biomass at these sites was 282 kg (range 28.0 kg (allspice) to 107 kg (mixed)) with frugivore biomass accounting for 73% of the total (Table 3). While agricultural sites contributed only 1% of the total area of vegetation sampled, they accounted for 58% of all species and 62% of all fruit–eating species (N = 153) censused. Fruit–eating species accounted for 79% (77% frugivores, 18% granivores) of all species and for 73% of all birds and mammals detected at these habitats (N = 2344).

The number of bird species detected at each type of agricultural vegetation (N = 5; 3 replicates per type) ranged from 39 (allspice) to 75 (cacao). Mammal species recorded varied from 2 (allspice) to 30 (mixed) (mean ± sd = 15.2 ± 4.0). Mean (± sd) number of bat species detected was 14.6 (± 1.7; range 12–16) and mean number of NFM species was 4.0 (± 1.1; range 1–7) (Table 3).

More bird species (63–75 spp) were detected

at shaded (cacao, coffee, mixed) than at unshaded (citrus and allspice) habitats (39–53 spp). NFM species were significantly more numerous at shaded than at unshaded habitats (mean ± sd = 5.6 ± 1.1 vs 1.5 ± 0.71 spp; $P < 0.05$). In the case of bats, no difference existed in the mean (± sd) number of species detected at shaded ($\bar{X} = 14.0 \pm 1.7$) and unshaded (15.5 ± 0.71) agricultural habitats (Table 3).

Comparison between agricultural and forest islands in frugivore species richness

A comparison of agricultural islands and forest islands of similar total area (35 vs 36 ha) and equal sampling effort (10 sites in each case; 2 replicates per type of agricultural vegetation type), showed that forest and agricultural islands contained similar numbers of fruit–eating species (85 vs 95) and frugivore species (77 vs 77), but significantly more granivore species (8 vs 18, $P < 0.05$) were present in the man–made habitats. Frugivore biomass accounted for 80-88% of the total biomass accounted for at the 10 sites in each case, but forest sites had a significantly higher frugivore biomass than the man–made islands (Fig. 5).

Fruit-eating North American American migrant (NAM) birds

NAM birds accounted for 24% of all bird species (N = 4471) in both forest and agricultural islands. Seven NAM bird species (*Dendroica magnolia, Wilsonia citrina, W. pusilla, Hylocichla mustelina, Vireo griseus, Setophaga ruticilla*, and *Oporornis formosus*) were among the top 20 ranking species in the total sample of birds censused at all sites. No difference in number of NAM fruit–eating bird species was evident between agricultural and forest sites of equal area and sampling effort but significantly more NAM birds were censused at the agricultural habitats than at the forest sites $P < 0.001$; Fig. 6).

Number of vegetation types occupied by species

The distribution of species censused was not homogeneous across six of the vegetation types studied (tropical rain forest, cacao, coffee, mixed, citrus and pepper). Only 8% of the 257 species occurred in all six types of vegetation and 42% of all species occurred in forest islands only. Only 9% of the species occurred in agricultural habitats, and 41% of the species occurred in 2–5 types

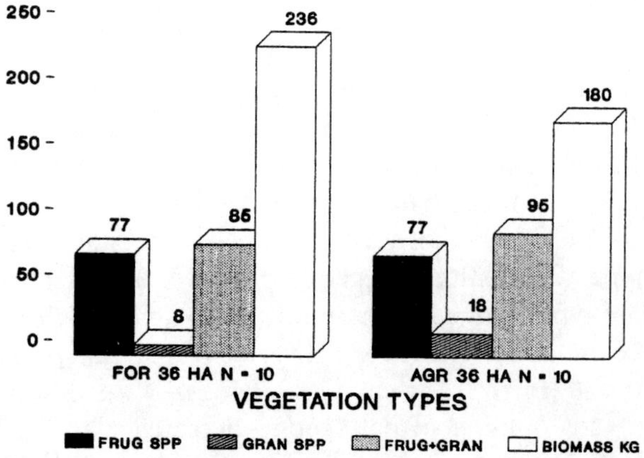

Fig. 5. Comparison of frugivore species richness and frugivore biomass in forest islands and in agricultural islands, of equal accumulated area and equal sampling effort.

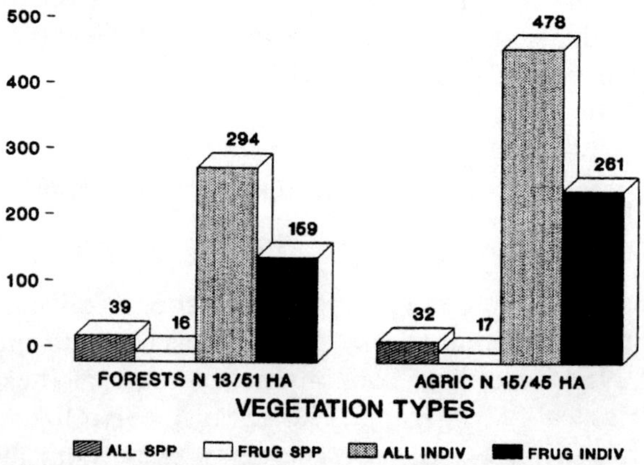

Fig. 6. Comparison of forest and agricultural habitats of similar accumulated area (sampling effort kept constant) in number of North American migrant birds censused. Also shown are the number of fruit–eating species and individuals in this subcommunity.

of vegetation which included forest and agricultural habitats (Fig. 7).

A similar proportion of bat (35%) and bird (37%) species were detected only in forest habitats, but 76% of the NFM species were detected only in this habitat. Forty seven percent of bird species were detected in agricultural habitats only and only 7% occurred in the six vegetation types studied. In the case of bats, 21% of the species were present in all six habitats, and 42% occurred in 2–5 types encompassing both forest and agricultural habitats (Fig. 7). One bat species (*Vampyrum spectrum*) was captured only in an agricultural habitat (mixed plantation). Non flying mammals showed, as indicated above, a much more restricted occurrence in the vegetation types. None of them occurred in more than four vegetation types, 24% were detected in 2–4 habitats and only one species (*Galictis vittata*; Mustelidae) was detected in one of the cacao plantations (Fig. 7).

The number of vegetation types in which a species occurred was not correlated with its mean body weight (rs birds = −0.07, rs bats = 0.04, rs NFM = −0.01, ns). The rank order of individuals censused per species was strongly correlated with the number of vegetation types used (rs birds = 0.78 $P<0.05$, rs bats = 0.92 $P<0.05$, rs NFM = 0.79 $P<0.05$).

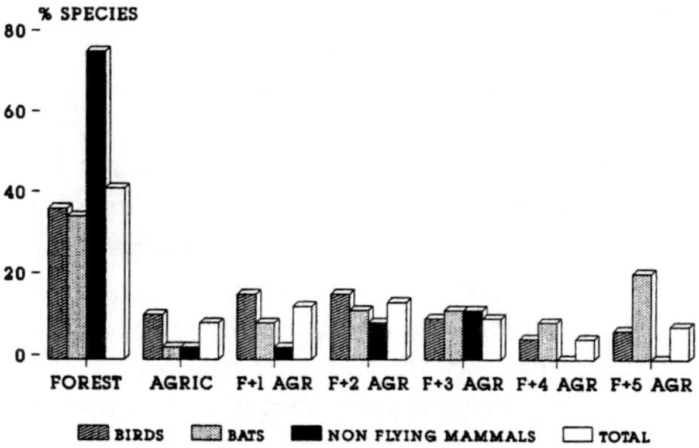

Fig. 7. Proportion of bird and mammal species detected in forest habitats only, in agricultural habitats only and in forest and agricultural habitats. F + 1 Agr indicates species occurring in forest and in one agricultural habitat. Rest of codes follow the same explanation.

Recaptures from other sites

We recorded 61 recaptures of animals banded at other vegetation islands. Of these, 44 were recaptures of birds and 17 of bats. Distances from the original banding site ranged from 0.5 to 2.0 km in the case of birds, and from 0.5 to 8.0 km in the case of bats. Such recaptures provide tangible evidence for the inter–island movement of forest interior birds with strong frugivorous habits such as *Pipra mentalis*, *Icterus virens*, *Turdus gravi*, *Momotus momota* and of forest interior bats such as *Carollia brevicauda*, *Artibeus jamaicensis*, *A. phaeotis*, *Vampyrops major*, *Glossophaga soricina*, *Vampyrops helleri*, *Sturnira lilium* and *Choeronisens godmani*, among others. In fifteen cases, birds (N = 6) and bats (N = 9) which were originally banded at forest sites were recaptured at agricultural sites.

Discussion

Although much further analysis of our data is called for, especially with respect to the relative influence of time since isolation and of isolation distance on species richness and their interaction with area effects, some general and preliminary conclusions are possible. It is obvious that at Los Tuxtlas man's management of the tropical land is having a very important impact on the survival of forest interior birds and mammals.

Our study showed that (a) disappearance of the forest totally eliminates the forest's faunistic components and that pastures, the now dominant form of vegetation, are unsuitable habitats for forest interior birds and bats, (b) a great number of species still occur in the forest fragments surveyed, but the majority of the species detected occurred in very low numbers, (c) the component most affected by forest fragmentation in the animal communities was the non flying mammals, and (d) the agricultural habitats studied were intensively used by forest birds and mammals.

The surveys also showed that fruit–eating birds and mammals constitute an important component of the animals still existing in the area. The different types of agricultural habitats merit further investigation in terms of the opportunities they may offer to forest birds and mammals. For example, shaded habitats (mixed plantation, cacao and coffee) were significantly richer in species of birds and mammals than unshaded habitats (citrus and allspice). This may possibly be a result of the greater vertical and horizontal structural complexity of the vegetation. Arboreal mammals such as *Alouatta palliata*, *Bassariscus sumichrasti*, and *Potos flavus* which require tall and structurally complex vegetation were only detected in shaded agricultural habitats.

Our study also showed that birds and flying mammals moved between forest patches and between forests and agricultural islands in the landscape at Los Tuxtlas, but this seemingly applies only to 60% of all species detected. The remaining species are less resilient. Bird species such as *Amazona autumnalis*, *Amazilia candida*, *Dendroica magnolia*, *Melanerpes aurifrons*, *Mniotilla varia*, *Tityra semifasciata*, and *Ramphastos sulfuratus* among others and bat species such as *C. brevicauda*, *A. phaeotis*, *A. jamaicensis*, *S. lilium* and *V. major* were among the very few species present in forest and in the five agricultural habitats studied. Bird species such as *Cotinga amabilis*, *Ciccaba virgata*, *Amblycercus holoseri*, *Catharus minimus* and *Onychorhynchus mexicanus*, and bat species such as *Centurio senex*, *Diclidurus virgo*, and *Macrotus waterhousi* among others were never seen and/or captured outside the forest.

While disappearance of the forest totally eliminates the forest's faunistic components, a species' capacity to use the resulting environmental matrix may have important consequences for their persistence in time and space. The ability of a species to disperse over the matrix from areas of native vegetation may depend on physical and biotic factors such as the distance at which other opportunities are located, the ability to move such distances and the availability of nutrients and the existence of competitive interactions (Saunders *et al.* 1991). The persistence of species in the matrix has been suggested to depend on enough suitable habitat to maintain sufficient numbers to avoid extinction (Soule 1987; Ewens *et al.* 1987; Neward 1991; Blake 1991). Our study showed that a species' ability to use new opportunities offered by the matrix may allow such persistence to occur.

Our study showed that about 60% of all species detected were making use of the man–made islands which were located at distances from a few hundred to several thousand meters from forest fragments. Such islands of man–made vegetation provide these species with stop–over points when moving from forest fragment to forest fragment, with energy resources which animals were observed to harvest, and with shelter from variations in climatic conditions and from predators existing in open areas. Such benefits may have the net result of reducing time and energy expenditure and exposure in addition to providing some degree of connectivity among the species' fragmented populations.

Removal of the native vegetation and the isolation of remnants, which become the only area of suitable habitat remaining for biota displaced by clearing, may lead to concentration of mobile elements of the biota in the remnants (Lovejoy *et al.* 1986). This crowding effect may be rapid and result in saturation of the remnant by some species. Crowding can alter intra and interspecific interactions, and competition and predation can increase resulting in changes in fecundity and the potential collapse of the population (Karr 1990). Resource availability may also be affected by overexploitation (Walker 1981). As a result of their value as stepping stones to birds and mammals, agricultural islands of the sort reported here may ease some of the above pressures. Species which use them, as well as some of those remaining in the forest remnants, may be buffered from the more drastic short and medium term effects of forest fragmentation and isolation.

Large forest remnants usually contain greater habitat diversity than smaller ones, but a collection of smaller forest patches may, however, contain a greater array of habitats than a single large one, simply because a single large reserve will not contain all of the habitats likely to occur in an area (Sanders *et al.* 1991). Al Los Tuxtlas, collections of small forest fragments may provide those bird and mammal species capable of reaching them with a greater variety of habitats in which to find resources and meet survival requirements. Such diversity of opportunities will increase significantly if a species can also make use of the man–made islands of vegetation. Agricultural islands are habitats that were never present before the destruction and fragmentation of the forest. The majority of animals detected there are not, as is the case in some of the forest fragments studied, part of populations that were fragmented and isolated. These animals come from nearby forest islands to specifically make use of such new opportunities. What is salient about this scenario is that (a) 40% of the species detected apparently cannot take advantage of such opportunities by being restricted to the forest habitat and thus may be in serious danger of extinction and (b) 60% of the species detected were also censused in the man–made habitats. Of these, about 93% were forest interior species.

In Los Tuxtlas, species that have very low population numbers, limited mobility and specialized habits are at risk of extinction because they cannot take advantage of the opportunities present outside of the forest habitats they occupy. In this position are the majority of the forest's non flying mammals, species that occurred in forest sites only and species with large area requirements. Some species with these characteristics such as *Eira barbara*, *Tamandua mexicana*, *Felis onca*, *Diclidurus virgo* and fruit–eaters such as *Ateles geoffrovi*, *Dasyprocta mexicana*, *Agouti paca* and *Crax rubra* among others have become extinct in several localities in Los Tuxtlas. To preserve these species, large areas of continuous rain forest are needed (Newmark 1991).

The fact that many species of birds and mammals disperse to and obtain resources at other forest fragments and at the man–made islands of vegetation means that the rate of seed rain may be relatively high in the landscape matrix which could mean the retention, via animal vectors, of some degree of genetic connectivity in plant populations. Many bats and birds captured in agricultural habitats regurgitated, defecated or were carrying seeds of forest canopy trees and plants such as *Nectandra ambigens* (Lauraceae), *Brosimum alicastrum*, *Pseudolemdia oxyphyllaria*, *Poulsenia armata* (Moraceae), *Turpinia occidentalis* (Staphylaceae), *Quararibea funebris* (Bombacaceae), *Allophylus campostachys* (Sapindaceae), *Dendropanax arboreus* (Araliaceae), and *Trichilia martiana* (Meliaceae) as well as seeds of pioneer species such as *Spondias mombin* (Anacardiaceae), *Ficus* sp. and *Cecropia obtusifolia* (Moraceae), and *Piper* sp. (Piperaceae).

Our study of seed rain in agricultural islands has shown the presence, not only of pioneer species such as *C. obtusifolia* and *Piper* sp., but also of forest species such as *N. ambigens*, *Cordia stellifera*, *Guarea grandifola*, *Sapindus saponaria* and *Cynometra retusa* among others, under the plants of allspice, cacao, citrus and coffee (AE, unpublished data). It is not uncommon to observe seedlings of many of these species in untended agricultural plots. The shade of domestic plants provides suitable conditions of humidity and light not present in pastures for seed germination and establishment. Although humans remove these seedlings as part of a regular program of crop–care, they nevertheless constitute a collection of genes removed from forest fragments by animals dispersing to other vegetation islands in the matrix. Such collections could be harvested and planted elsewhere or could be left alone to act as foci of forest regeneration if the crop–raising programs are abandoned. The conservation of collections of forest fragments not far from one another and the use of agricultural islands as stepping stones for forest animals carrying seeds and as centers of seed deposition may be a potentially effective strategy to promote forest regeneration.

Tropical rain forest destruction and reduction of area of native habitat available to birds and mammals leads in time to total extinction and to an irreversible weakening of the regenerating capacity of the ecosystem via the disappearance of animals acting as seed dispersal vectors. However, our discovery of the value of agricultural islands to birds and mammals in the remaining forest fragments suggests that these man–made islands compensate partly for the fragmentation and isolation of the forest. This may permit, if suitable breeding habitats are preserved as well, the persistence in time and space of many frugivore species and seed dispersal agents, especially those capable of taking advantage of the opportunities represented by agricultural habitats. The net impact of such a role is a potential reversal of species' extinction trends resulting from loss of area and habitat isolation.

Agricultural islands apparently allow enhanced biotic movement, provide extra foraging areas of vegetation among other benefits to forest birds and mammals. As a result of the inter–island movement (forest island–forest and forest island–agricultural island–forest island) of seed dispersal agents, agricultural islands may serve as units of connectivity among fragmented animal and plant populations. Further investigations are needed to determine the roles different agricultural habitats may play in conjunction with forest remnants in sustaining animal populations and species. Turning more pasture area into agricultural parcels of the sort studied by us rather than turning more forest into pasture may result in (1) the persistence in time and space of forest birds and mammals, (2) the retention of some degree of regenerating capacity of the ecosystem, and (3) direct economic benefits to human populations.

Conclusions

1) At Los Tuxtlas, loss of area of native habitat leads to the extinction of some forest birds and mammals.
2) The forest islands we studied contained rich assemblages of forest birds and mammals including many species of frugivores.
3) Clusters of small forest islands had greater numbers of species and more individuals than clusters of larger forest islands.
4) A significant number of species (40%) were restricted to forest habitat, but 60% were present in other types of vegetation in addition to the forest habitat.
5) Man–made islands of vegetation are temporary habitats for many forest interior birds and mammals and many provide a degree of biotic connectivity among isolated forest islands.

Acknowledgements

We thank the Lincoln Park Zoological Society and its Scott Fund for Neotropical Research for support to launch this research program and for its continued support. The Center for Field Research and Earthwatch provided additional support permitting an expansion of the project. We thank 96 Earthwatch volunteers for their enthusiastic and tireless help in the field during 1990 and 1991. We thank the Explorers Club of New York and the US Fish & Wildlife Service for support in the development of this research. Finally we thank Ted Fleming and an anonymous reviewer for constructive comments on earlier drafts of this paper.

References

Blake, J. G. 1991. Nested subsets and the distribution of birds on isolated woodlots. Cons. Biol. 5: 58–66.

Brown, J. H. & Kodric–Brown 1977. Turnover rates in insular biogeography: effect of immigration on extinction. Ecology 58: 445–449.

Burkey, T. V. 1989. Stochastic extinction in nature reserves: the effect of fragmentation and the importance of migration between reserve fragments. Oikos 55: 75–81.

Estrada, A., Coates–Estrada, R. & Martinez, M. 1985. La Estacion de Biologia 'Los Tuxtlas': un recurso para el estudio y conservacion de la selvas del tropico humedo en Mexico. In: del Amo S. & Gomez–Pompa, A. (eds) Regeneration de Selvas II. Instituto Nacional de Investigaciones sobre Recursos Bioticos. Editorial Alhambra Mexicana, S. A. de C. V. Mexico. Pp. 379–393.

Estrada, A. & Coates–Estrada, R. 1988. Tropical rain forest conver-sion and perspectives in the conservation of wild primates (*Alouatta* and *Ateles*) in Mexico. Am. J. Prim. 14: 315–327.

Ewens, W. J., Brockwell, J., Gari, M. & Resnick, S. I. 1987. Minimum viable population size in the presence of catastrophes. In: Soule, M. E. (ed) Viable populations for conservation. Cambridge University Press, Cambridge, England. Pp. 59–68.

Gilbert, L. E. 1980. Food web organization and conservation of neotropical diversity. In: Soule, M. E. & Wilcox, B. A. (eds) Conservation biology: an evolutionary – ecological perspective. Sinauer Assoc., Sunderland, Mass. Pp. 11–33.

Hopkins, M. S. & Graham, A. W. 1984. Viable soil seed banks in disturbed lowland tropical rainforest sites in north Queensland. Austr. J. Ecol. 9: 71–79.

Hubbel, S. P. & Foster, R. B. 1986. Commoness and rarity in a neotropical forest: implications for tropical tree conservation. In: Soule, M. E. (ed) Conservation biology: the science of scarcity and diversity. Sinauer Assoc., Sunderland, Mass. Pp. 205–231.

Janzen, D. H. 1983. No park is an island: increase in interference from outside as park size decreases. Oikos 41: 402–410.

Karr, J. 1982a. Avian extinction on Barro Colorado Island, Panama; a reassessment. Am. Nat. 119: 220–239.

Karr, J. 1982b. Population variability and extinction in the avifauna of a tropical land–bridge island. Ecology 63: 1975–1978.

Karr, J. 1991. Biological integrity and the goal of environmental legislation: lessons for conservation biology. Cons. Biol. 4: 244–250.

Klein, B. C. 1989. Effects of forest fragmentation on dung and carrion beetle communities in central Amazonia. Ecology 70: 1715–1725.

Laurance, W. F. 1991. Ecological correlates of extinction proneness in Australian tropical rain forest mammals. Cons. Biol. 5: 79–89.

Laurance, W. F. & Yensen, E. 1991. Predicting the impacts of edge effects in fragmented habitats. Biol. Cons. 55: 77–92.

Leighton, M. & Leighton, D. R. 1983. Vertebrate responses to fruiting seasonality within a Bornean rain forest. Pp. 181–196. In: Sutton, S. L., Whitmore, T. C. & Chadwich, A. C. (eds) Tropical rain forest: ecology and management. Blackwell Scientific Publications, Oxford, England.

Lomolino, M. V. 1986. Mammalian community structure on islands: the importance of immigration, extinction and interactive effects. Biol. J. Linn. Soc. 28: 1–21.

Lovejoy, T. E., Bierregaard, R. O., Brown, K. S., Emmons, L. H. & Van der Voort, M. E. 1984. Ecosystem decay of Amazon forest fragments. In: Niteki, M. H. (ed) Extinctions. University of Chicago Press, Illinois. Pp. 295–325.

Lovejoy, T. E., Bierregaard, R. O. Jr., Rylands, A. B., *et al.* 1986. Edge and other effects of isolation on Amazon forest fragments. In: Soule, M. E. (ed) Conservation biology: a science of scarcity and diversity. Sinauer Assoc., Sunderland, Mass. Pp. 257–285.

Lynch, J. F. & Whitcomb, R. F. 1978. Effects of insularization of the eastern deciduous forest on avifaunal diversity and turnover. In: Marmemstein, A. (ed) Classification inventory and analysis of fish and wildlife habitat. U. S. Fish and Wildlife, Washington, D. C. Pp. 461–489.

Myers, N. 1988. Tropical forests: much more than stocks of wood. J. Trop. Ecol. 4: 209–221.

Newmark, W. D. 1991. Tropical forest fragmentation and the local extinction of understory birds in the eastern Usambasa Mountains, Tanzania. Cons. Biol. 5: 67–78.

Pickett, S. T. A. & Thompson, J. N. 1978. Patch dynamics and the design of nature reserves. Biol. Cons. 13: 27–37.

Saunders, D. A., Hobbs, R. J. & Margules, C. R. 1991. Biological consequences of ecosystem fragmentation: a review. Cons. Biol. 5: 18–32.

Soule, M. E. 1987. Viable Populations for Conservation. Cambridge University, Press, New York.

Terborgh, J. 1976. Island biogeography and conservation: strategy and limitations. Science 193: 1029–1030.

Terborgh, J. 1986. Keystone plant resources in the tropical forest. In: Soule, M. E. (ed) Conservation biology: the science of scarcity and diversity. Sinauer Assoc., Sunderland, Mass. Pp. 330–334.

Walker, B. H. 1981. Is succession a viable concept in African savannah ecosystems? In: West, D. C., Shugart, H. H. & Botkin, D. B. (eds) Forest Succession: Concepts and application. Springer, New York. Pp. 431–447.

Wilcox, B. A. 1978. Supersaturated island faunas: a species–age relationship for lizards on post–Pleistocene land–bridge islands. Science 199: 996–998.

Willis, E. O. 1974. Populations and local extinctions of birds on Barro Colorado Island, Panama. Ecol. Mon. 44: 153–169.

Zimmerman, B. L. & Bierregaard, R. O. 1986. Relevance of the equilibrium theory of island biogeography and species–area relations to conservation with a case from Amazonia. J. Biogeo. 13: 137–143.

Part 4

Consequences of seed dispersal

Introduction

The fate of seeds once they are taken or released from parent plants continues to be a subject of intense research interest. Papers in this section examine the results of seed dispersal from a variety of different perspectives. Willson, for example, reviews existing data on the seed shadows of plants with different modes of dispersal. Her results indicate that the seeds of many plants lack specialized morphology for dispersal and achieve only short-distance dispersal. Wind-dispersed trees and shrubs tend to achieve shorter dispersal distances than vertebrate-dispersed trees and shrubs. She concludes that colonization patterns in new habitats are influenced by a greater array of factors than simply seed dispersal mechanism.

The genetic consequences of different dispersal mechanisms in three species of tropical trees are addressed by Hamrick *et al.* Analysis by size class reveals that family structure exists in small plants of all three species but that this structure is lost as plants mature and densities are reduced. Genetic structure is coarsest in a wind-dispersed species and is least in an animal-dispersed species. The largest size classes of all three species were biased towards heterozygotes, indicating that heterozygous individuals may have a selective advantage at this study site.

The influence of local vs. 'distant' seed dispersal as a source of new seedlings during the regeneration of tropical forests is examined by Gorchov *et al.* and Martinez-Ramos and Soto-Castro. The former authors document seed rain falling into a logged opening in a Peruvian primary forest for one year and find that wind- and bat-dispersed species occur in greater numbers than bird-dispersed species. Pioneer species of low timber value dominate the seedlings produced by dispersal during the first year whereas commercially valuable tree taxa predominate among the stump sprouts and survivors of the logging operation. Martinez-Ramos and Soto-Castro monitored seed rain in intact forest at Los Tuxtlas and find that about 50% of the seed species come from 'distant' (i.e. outside the study plot) sources. Nevertheless, large-seeded local species have higher establishment rates in closed-canopy forest, although some immigrant species also become established. They suggest that the influence of seeddispersal on advanced forest regeneration depends on the disturbance history of the sites where seeds land.

Guevara and Laborde report on the diversity of frugivores that land in isolated fig trees in pastures in Veracruz, Mexico and on the diversity of seeds dropped under or near these trees. In one year 73 species of birds perched in the trees and deposited seeds of 56 species of plants (81% of the total seeds). Their study demonstrates that frugivorous birds provide substantial mobility for the seeds of a wide variety of trees and that isolated fruiting trees potentially serve as recruitment foci for tropical plants.

The mobility provided for seeds of tropical plants dispersed by ants and birds should differ substantially. Horvitz and LeCorff examine the consequences of ant-vs. bird-dispersal in a group of tropical understory members of the genus *Calathea*. They find that spatial scale, defined as the average distance between conspecifics, is positively correlated with dispersal distance whereas dispersion pattern is independent of dispersal distance (and disperser type). Byrne and Levey demonstrate that ants are attracted to seeds in frugivore (bird) defecations and that they can provide appreciable levels of secondary dispersal for these seeds. Owing to the highly patchy distribution of ants in tropical forests, however, the ultimate effect of ants on bird-dispersed seeds in difficult to determine.

The effects of distance from parent plant and vulnerability to insects and mammals on the survival of five species of Peruvian canopy trees was experimentally investigated by Terborgh *et al.* Their results indicate that seed predation by mammals is less distance-responsive than is that by insects. Small mammals have the greatest impact as predators on seeds of *Astrocaryum*, *Bertholletia*, and *Dipteryx* whereas large mammals are the major predators of seeds of *Hymenaea*. One species (*Calatola*) suffered no seed predation in a one-year period. Since there was no correlation between the intensity of seed predation and the abundance of adults of the five species, the demography of these species is likely to be affected by important post-germination bottlenecks, even in plants suffering heavy seed predation.

Vegetatio **107/108**: 261–280, 1993.
T. H. Fleming and A. Estrada (eds). Frugivory and Seed Dispersal: Ecological and Evolutionary Aspects.

Dispersal mode, seed shadows, and colonization patterns

M. F. Willson
Forestry Sciences Laboratory, 2770 Sherwood Lane, Juneau AK 99801, USA

Keywords: Seed dispersal, Seed shadow, Dispersal mode, Colonization

Abstract

This review assesses the state of our knowledge about comparative seed shadows. Using data presently available in the literature, I compare the slopes (on a log-linear scale) of seed shadows for plants with different morphologically characterized modes of dispersal. The seeds of many species have no evident morphological adaptation for dispersal and seem to achieve only short-distance dispersal. Seed shadows for herbaceous species with devices for wind have flatter slopes and more distant modes and maxima than those of ballists, which in turn exceed those with no special devices. Seed shadows for wind-dispersed trees and shrubs had similar or steeper slopes than those for vertebrate-dispersed species in this sample. Species with poor mechanisms for dispersal in space only sometimes had the capacity for better dispersal in time (dormancy). Although some species exhibited seed shadows sufficiently steep to be predicted to colonize new habitat in a 'front' or phalanx pattern, actual colonization patterns must reflect many other factors.

Introduction

Many plant species have special contrivances that are assumed to enhance dispersal of the seeds, including wings and plumes for wind-dispersal, fleshy fruits and elaiosomes for dispersal by animal consumers, hooks and barbs for attachment on animal exteriors, and ballistic mechanisms (van der Pijl 1982). Inferences about selection on dispersal devices are best made from comparisons between closely related entities. Several studies have shown that seeds with well-developed dispersal devices can achieve greater dispersal distances than conspecific seeds without such devices (e.g., Cremer 1971; Lanner 1985; Venable & Levin 1985; Sacchi 1987). Therefore, the assumption that these morphological devices enhance dispersal receives support. These comparisons are limited, however, to species that produces some seeds that have well-developed wings or plumes and others whose morphology is less well-suited to dispersal by wind. To make comparisons for other dispersal modes, ideally one would examine polymorphic species or closely related species exhibiting a variety of dispersal modes (e.g., *Acacia*, *Calathea* for ant-vs. bird-dispersal; Liliaceae for many dispersal modes). In the absences of published comparisons of the other modes in closely related species, we may make exploratory comparisons among unrelated species (e.g., Augspurger 1986; Matlack 1987 for different types of wind dispersal). Although such comparisons necessarily sweep together species that have many ecologically important differences, such as seed size, they provide some indication of the variation in seed shadows associated with different modes of seed dispersal. By summarizing the available published information on the length of seed shadows in relation to dispersal mode, I hope to set the stage for more refined comparisons that examine selection on dispersal modes via the resulting seed shadow. I emphasize that this review only documents what the existing literature tells us about comparative seed shadows; persistant lacunae in available information (especially for dispersal by vertebrates) preclude complete and definitive comparisons. For purposes of this paper, comparisons are confined to 'local' seed shadows (see Howe & Smallwood 1982) and do not include the serendipitous events leading to ultra-long distance dispersal of the type discussed by Carlquist (1967).

In addition to documenting comparative seed shadows, I use published data on seed shadows to examine several other questions: a) To what extent do seed shadows conform to the commonly assumed negative exponential shape? b) Are there tradeoffs between dispersal in space (as related to dispersal mode) and dispersal in time (dormancy)? Dormancy potentially can increase total reproductive success many fold (Murray 1988). Species that are relatively poor dispersers in space might be predicted to be better dispersers in time (e.g., Venable & Levin 1983; Olivieri & Berger 1985; Venable & Brown 1988). c) Are seed shadows and dispersal capacity related to patterns of local colonization? Harper (1977), extending a model for the spread of plant pathogens (van der Plank 1949, 1960) to the dispersal of seeds, offered the generalization that when the seed-shadow slope is steeper than that of the inverse square law (-2, on a log-log scale), colonization should occur on an advancing front rather than by isolated outposts of establishment. In this view, the shape of the seed shadow may have direct consequences for spatial patterns of population recruitment in previously unoccupied sites, with concomitant implications for natural regeneration (e.g., Hanson *et al.* 1990) and avoidance of predators and pathogens (Harper 1977). I briefly examine a set of transformed seed shadows, asking if the steepness of the slope is associated with the mode of dispersal.

Methods

Measures of seed shadows

There are several ways, of differing utility, to index dispersal capacity and seed-distribution patterns. I have used three simple ones for the comparisons in this paper (Tables 1 and 2). 1) The slope of the linear regression of seed number (ln-transformed) vs distance measures the rate of decline of seed fall with increasing distance; the flatter the slope, the greater the proportion of seeds reaching distances farther from the source. The probability that at least one propagule will travel a given distance from a source is proportional to the number of propagules released (van der Plank 1960), and therefore seed crop size can have a major effect on the length of the seed shadow in nature (see Dirzo & Dominguez 1986). The number of propagules dispersed is reflected by the intercepts of the regression equations when slopes are equal. However, I did not examine intercepts in this survey, because so few samples were based on total numbers in actual seed crops. 2) The modal dispersal distance is the distance reached by the greatest number of dispersing seeds. Thus, it indicates the most common 'expectation' for any particular seed in a seed crop. 3) The maximum dispersal distance achieved reflects conditions of seed release and numbers of seeds released; it estimates the limits of dispersal capacity, for each sample, under those conditions. Central tendencies and maxima of seed shadows can vary independently (e.g., Greene & Johnson 1989) and may differ in their fitness consequences (e.g., Antonovics & Ellstrand 1985; Portnoy & Willson ms.). Maximum distances were seldom reported for the samples of tree seeds, so this measure is used only for herbaceous species.

Treatment of data

A review of the literature yielded a number of studies of seed distribution patterns, assessed under a wide range of circumstances. For each set of data, the seed-distribution pattern was repre-

Table 1. Comparison of seed shadows of herbaceous species with different modes of seed dispersal (data from Appendix 1). All distances in meters. Significant differences between adjacent entries for slope and maximum are marked with a * (Mann-Whitney U, $p<0.017$, adjusted for multiple comparisons[b]). A * in the last column marks significant differences between wind and ballistic dispersal. Only species with significant regressions are included in comparisons of slope. No summary data are presented unless at least two species are represented in each category. Differences in the frequency of fit to the semilog regression were tested by G-test, $p<0.05$; significant differences marked with **.

		Wind	No special mechanism	Ballistic
Circular distributions				
Slope	n	9	9	18
	x	– 1.09	– 2.03	– 1.43
	median	– 1.08*	– 1.72*	– 1.15
Mode	all means and medians < 1			
Maximum[a]	n	11	9	19
	x	~9.1	3.8	3.6
	median	> 10*	2.0*	3.8*
Frequency of fit**		9/11 = 82%	9/10 = 90%	18/23 = 78%
Linear distributions				
Slope	n	9	17	
	x	– 0.13	– 2.75	
	median	– 0.12*	– 2.37	
Mode	n	16	17	
	x	< 10	<0.4	
	median	7.5*	0.3	
Maximum	n	16	17	
	x	321	>2.3	
	median	77*	2.5	
Frequency of fit**		9/12 = 75%	17/17 = 100%	

[a] These 'maxima' are underestimates, because a number of authors reported only that the greatest distances achieved were greater than a certain distance.

[b] All Mann-Whitney tests are one-tailed, because of *a priori* predictions that dispersal devices enhance dispersal, and wind leads to greater dispersal than ballistic mechanisms.

sented by the equation $\ln y = mx + b$, where y is the number of seeds occurring at a given distance (x, in meters) from the seed source. This transformation allows comparisons of the slopes (m) of the regressions without regard for seed abundance. In all cases, the slope was calculated, not for the entire distribution, but from the mode outward. Use of this equation reflects the general expectation that seed-distribution patterns commonly follow a negative exponential form (e.g., Bullock 1976; McCanny & Cavers 1987), although a number of exceptions are found (see below, and references in Okubo & Levin 1989). Furthermore, the focus on dispersal distances means that variations in the directionality of seed-distribution patterns are neglected. Because the emphasis here is on the potential 'reach' of seeds from a seed-parent, only distances where seeds actually landed are included (empty intervals are excluded).

Table 2. Comparison of seed shadows of trees (and a few shrubs) with different modes of seed dispersal (dat from Appendix 2). All distances are in meters. Significant differences between entries are marked by * (Mann-Whitney U, $p<0.05$, for slopes and modes; G-test, $p<0.05$, for frequency of fit. Only samples with significant regressions are included in comparisons of slopes. Maxima were seldom reported and, for linear distributions, most modes were adjacent to the source. No summary data are included unless at least two species are represented in each category.[a]

		Wind	Vertebrate
Circular distributions			
Slope	n	10	5
	x	− 0.10	− 0.06
	median	− 0.05	− 0.04
Mode	n	11	6
	x	27.4	9.3
	median	10*	3.5
Frequency of fit		10/11 = 91%	5/6 = 87%
Linear distributions			
Slope	n	37	10
	x	− 0.04	− 0.37
	median	− 0.02*	− 0.39
Frequency of fit		37/44 = 84%	10/14 = 71%

[a] All available samples except Hoppes 1985 were used in comparing slopes, although some species are represented much better than others. Given the marked variation within species, this bias is probably not very serious for present purposes. Recalculation of the means using no more than three sampes per species gave results similar in direction of difference and in statistical significance.

Where seed dispersal was measured around a point source, the number of seeds at each distance was estimated (if not given) by multiplying some measure of seed density for each distance by the area at that distance interval (circular distribution). However, this correction-for-area could not be made for some samples (linear distributions), because the source was not a point source or because insufficient information was given by the authors, and the data then represent a measure of density (rather than number) of seeds as a function of distance. In general, non-point sources are expected to yield flatter slopes than point sources (Gregory & Read 1949). Other seed shadows were measured in a downwind direction only. These are treated here as linear distributions. Such measures omit seeds that might have moved in other directions in natural conditions, and therefore may underestimate the number falling near the source; thus they may yield an estimated curve less steep than one including all seeds released under all conditions. Linear and circular distributions are treated separately in the Results. Some naturally generated seed shadows may have been subject to unassessed seed predation before they were measured, and thus may not reflect the actual distribution of seeds. However, most of the data in the Appendices do not have this problem.

For statistical comparisons, only slopes associated with at least marginally significant ($p<0.10$) correlation coefficients are used. The alpha values were relaxed somewhat from the usual criterion level of 0.05 in order to increase the sample size. Some seed shadows did not produce even a marginally significant fit to a semilog regression line, in a few cases because very few sampling points were used, but more often because the seed distribution pattern showed no resemblance to a negative exponential decline with distance.

Because height of seed-release can have major effects on seed shadows (both directly, e.g., Morse & Schmitt 1985; Trapp 1988, but see Rabinowitz & Rapp 1981, and indirectly, e.g., Watkinson 1978), at least for air-borne diaspores including ballistic seeds (Beer & Swaine 1977), comparisons of seed shadows are made separately among herbaceous species and among trees as a partial control for the effect of height. This comparison assumes that the distributions of diaspore-release heights are similar for each dispersal mode within each growth-form category. Some possible violations of this assumption and their effects on the comparisons made here are discussed below.

For consideration of the inverse square law and expected patterns of colonization, both seed numbers and distances were ln-transformed (Appendices 1 and 2). Double-log transformations remove the effect of scale from both axes, which can create problems of interpretation if scales of distance measurement differ between samples (Minogue 1986), but in this survey the scale on the x axis was consistent for all samples. Many of the data for trees measure seed rain from a forest edge into an opening and thus represent seed distributions from multiple sources. Such data do not reflect seed-distribution patterns from single sources within forest, which are likely to be more restricted, but can be used to examine Harper's generalization about expected colonization patterns.

It is not entirely clear whether Harper's rule for colonization is based on seed density or total seed numbers. Harper (1977) uses both 'number' and 'density' in discussing van der Plank's model; van der Plank uses 'number' but seems to mean 'number per plant' or 'number per field' (i.e., a measure of density). Consultation with four persons having strong backgrounds in mathematics or mathematical biology concerning which is correct has yielded a split decision, with a majority favoring 'density'. Unfortunately, the sample for seed densities is inadequate for comparisons in any category, and the results are presented only for seed numbers.

Results

Fit to negative exponential equation

How frequently do seed shadows fit the commonly assumed negative exponential shape? The coefficient of determination (r^2) for the slope of seed distributions from the mode outward can be evaluated for 73 samples of herbaceous species (Appendix 1, Table 1) and 75 tree and shrub samples (Appendix 2, Table 2). It is significant or marginally so for 63 of the herbaceous species (86%) and 63 of the trees and shrubs (84%). Four of seven grass species examined by Rabinowitz and Rapp (1981) fit this expectation. Furthermore, 18 of 20 individuals of *Panicum mileaceum* (McCanny & Cavers 1987) but only four of eight individuals of *Penstemon digitalis* (P. A. Thomas, pers. comm.) fit a negative exponential. The frequency of fit for herbs was greater for species with no special dispersal devices than for those dispersed by wind or ballistic mechanisms (Table 1). For trees, no difference between dispersal modes in frequency of fit was detectable (Table 2).

For significant and marginally significant correlations, r^2 values were $>50\%$ in the great majority of cases, and commonly $>75\%$. Thus, the equation accounted for much of the observed variation in most cases that reached at least marginal statistical significance.

Comparison of seed shadows

a) Does having a special dispersal mechanism improve dispersal capacity? One would predict that slopes should be flatter and achieved distances greater for seeds dispersed by some special mechanism. This question was examined using the data for herbaceous species, many of which exhibit no evident special mechanism of dispersal. In general, the data tend to support the prediction. Wind- and ballistically-dispersed diaspores of herbaceous species in both circular and linear distributions have significantly greater maxima than those with no special dispersal mechanism, and the slopes are flatter (Table 1).

The seed shadows considered in Table 1 were generated under many different conditions, and more refined comparisons can be made among seed shadows produced under similar conditions. Two such comparisons are possible with existing data. Three prairie herbs with plumed seeds had slopes averaging -0.44, under natural conditions, whereas three species with no special dispersal devices had slopes averaging -1.16, a 2.6-fold difference (from Platt & Weis 1977; Mann-Whitney U, $p<0.05$, one-tailed). Several prairie grasses produce seeds bearing long hairs thought to be important for dispersal (e.g., Campbell 1983, but see also Peart 1979). Three such species exhibited terminal velocities of falling seeds similar to those of three other species with unhaired seeds, and the slopes of the seed shadows of hairy seeds in the field were not flatter than those for unhaired seeds, although lateral movement of hairy seeds dropped in still air was greater (Rabinowitz & Rapp 1981). Other diaspore traits of species studied by Rabinowitz and Rapp (1981) apparently counteracted the effect of hairs, and the importance of other factors should not be neglected (Johnson 1988; Greene & Johnson 1989). Thus, achievement of greater potential dispersal distances often may depend on traits other than particular dispersal devices (see also Guries & Nordheim 1984).

b) Does wind dispersal achieve greater distance than ballistic dispersal? For circular distributions of herbaceous species, the slopes are similar, but the maximum distances are significantly greater for wind dispersal (Table 1). Thus, the distribution of seeds within the seed shadow for these dispersal modes are indistinguishable from each other, but the absolute distances achieved are greater for wind dispersal.

c) Do wind and vertebrate dispersal yield detectably different seed shadows? This comparison was made for the samples from trees and shrubs, making no *a priori* predictions. There were no significant differences in the slopes of circular seed shadows, but the modes were more distant for wind-dispersed species than for vertebrate-dispersed species (Table 2). Vertebrate-generated linear seed shadows had significantly steeper slopes than wind-dispersed trees and shrubs (see also Dirzo & Dominguez 1986; Willson & Crome 1989).

Dispersal in space and in time

From a compendium of data on seed germination phenology of herbaceous plants (Baskin & Baskin 1988), I extracted data on 89 species for which I had good information on the mode of seed dispersal. All species showed peak germination within a year of maturation in greenhouse conditions. However, the species that probably have good long-distance dispersal (by wind or vertebrate exteriors or interiors; n = 33) are less likely to be able to germinate two years after planting than those with shorter-range dispersal in space

Table 3. Frequency of slopes > − 2 on ln-ln scale, for different dispersal modes, presented for three categories of seed distributions. No statistically significant differences were found (Fisher exact test, p > 0.05). Data were inadequate for compariosn of linear distributions of tree seeds.

	Dispersal mode	> − 2	< − 2
Herbs-circular	Wind	0	8
	Ballistic	5	14
Herbs-linear	Wind	6	10
	No special device	9	5
Trees-circular	Wind	5	5
	Verbebrate	0	6

(ballists, by ants, no special mechanisms; n = 56) (12% vs 30% of species; $G = 4.13$, $p < 0.05$). Several species with no special dispersal mechanisms exhibited prolonged dormancies greater than four years (in greenhouse conditions), but this was seldom true in species with other dispersal modes. Wind-dispersed species (n = 19) had less well-developed temporal dispersal than species lacking special mechanisms (n = 38) (0% vs. 26% with > 2 yr dormancy, $G = 9.14$, $p < 0.05$).

Harper's generalization

Slopes steeper than −2 on a ln-ln scale occurred with statistically similar frequencies among dispersal modes (Table 3). Very few (only 14) seed shadows had very steep slopes (greater than −3).

Discussion

Shapes and lengths of seed shadows

A negative exponential provides an adequate descriptor of the shape of seed shadows (post-modal) for roughly 85% of the samples surveyed here, and for varying proportions of species or individuals in other studies. Therefore the assumption of a negative exponential for general theoretical purposes seems to have some practical justification, but for particular cases the risk of making such an assumption may be considerable.

One could also ask how frequently a negative exponential *best* describes the shape of seed shadows, by comparing the fit to other possible shapes, such as the negative power function (Minogue 1986; Okubo & Levin 1989). For a number of samples, the ln-ln-transformed data provided a higher r^2 than the semilog data, suggesting that negative exponentials may not always be the best fit. However, I chose not to examine this question statistically, because so many of the data sets are inadequate for the purpose.

As presented here, the fit to a negative exponential was calculated only for the part of the curve distal to the mode, and it often does not apply to the entire curve. This is true in general when dealing with absolute numbers in circular distributions (rather than densities), because this automatically places the mode at some distance from the source (Clark & Clark 1984; Peart 1985; Okubo & Levin 1989; D. F. Greene, pers. comm.). A number of other studies have also shown curves that failed to fit a negative exponential (e.g., Wolfenbarger 1946; Swaine & Beer 1977; Lee & Brothers 1981; Kjellsson 1985). In addition, for some curves the mode is closer to the tail of the distribution than to the origin, such that the shape of the entire curve is virtually the reverse of that commonly pictured. This is true of a number of ballistically dispersed seed shadows (e.g., Stamp & Lucas 1983; Lee 1984 and pers. comm., D. Thiede pers. comm., MFW and J. Agren unpublished), and may occur with some animal-dispersed seeds as well. It is thus often misleading to assume a negative exponential shape for whole seed shadows.

Comparison of dispersal modes

Wind-dispersal often can, as expected, produce a considerably longer seed shadow than can be achieved without special aerial devices or by ballistic mechanisms. Some studies examined seed fall from forest on all sides of a rectangular clearcut, but no consistent differences in slope of the log-linear dispersal curve were evident in relation to direction of seed flow (relative to prevailing winds etc.) (e.g., Carkin *et al.* 1978; Hughes & Fahey 1988). Although ballistic species tend to have rather flat slopes, they also have a sharply constrained maximum dispersal distance (Levin & Kerster 1974; Stamp & Lucas 1983) determined by the mechanism itself. Furthermore, it seems to me that the average stature of the herbaceous ballists in this sample may be small compared to the other categories, and this may mean that the 'reach' of their seeds is further constrained.

Dispersal distances for ant-dispersed herbs overlap broadly both with ballistic species (see also Bulow-Olsen 1984) and species lacking special mechanisms. For a variety of species, the dispersal distances generally average <3 m (Beattie & Lyons 1975; Barrett 1978; Culver & Beattie 1978; O'Dowd & Hay 1980; Auld 1986; Horvitz & Schemske 1986; Lack & Kay 1987; Ganeshaiah & Uma Shaanker 1988; Ohara & Higashi 1990) and maxima rarely exceed 10 m (Auld 1986; Horvitz & Schemske 1986; Smith 1989).

Seeds carried externally or internally by vertebrates may achieve rather long dispersal distances (e.g., Bullock & Primack 1977; Sorensen 1986; Willson & Crome 1989), although this survey suggests that internally dispersed seeds travel less far than wind-dispersed seeds, on average. The relatively steep slopes for vertebrate-dispersed species compared to wind-dispersed species in this survey were surprising; this result may reflect the small number of vertebrate-produced seed shadows and the high proportion of seed shadows produced by nonvolant vertebrates. Moreover, the general dispersal ability for wind-dispersed species in this survey may be over-estimated, because so many of the samples were taken only in a downwind direction or only into unobstructed areas (especially linear distributions). However, none of these samples assessed the full dispersal capacity that can be achieved in some circumstances (Webber 1934; Solbreck & Andersson 1987). Hubbell (1979) reported that regression slopes for the density of juvenile neotropical trees was steeper for wind-dispersed than for bird- and bat-dispersed species. However, his data are not directly comparable, because they are densities rather than total numbers, mortality between seed and juvenile life-history stages were, perforce, neglected, and wind dispersal inside the forest should be far more constrained than in most of the studies in my survey. Across a rain forest border in Australia, wind- and vertebrate-dispersed seeds moved equivalent distances from oldfield into forest, while vertebrate-dispersed seeds moved farther than wind-dispersed seeds from forest into oldfield (Willson & Crome 1989). However, most of the wind-dispersed seeds that travelled into the field were of only one species, and this sample did not necessarily measure seed rain in a downwind direction.

Dispersal mechanism and fecundity

There are two ways for a maternal plant to increase the probability of at least some of its propagules dispersing relatively far: 1) use of a long-distance dispersal mechanism (probably often by wind or vertebrates, perhaps also water), and 2) production of many seeds, which increases the chance that at least a few seeds go far, thus raising the tail of the distribution on an absolute scale (e.g., van der Plank 1960; Dirzo & Dominguez 1986). Thus, if there is selection for long dispersal distances, one might predict that plants with no special dispersal mechanisms should produce seed more prolifically, on average, than those with long-distance mechanisms (i.e., the reverse of Salisbury's 1975, 1976 prediction). Among wind-dispersed species, there is little support either for an inverse correlation of distance achieved and number produced (Greene & Johnson 1986) or for Salisbury's prediction. Moreover, I think it most unlikely that wind-dispersed plants have lower average fecundity than plants of similar stature that have short-distance dispersal mechanisms (and see also Salisbury 1975).

Why do so many species have no special dispersal mechanism?

If long seed shadows are generally advantageous (e.g., Hamilton & May 1977; Howe & Smallwood 1982), but many species seem to have only short-range dispersal, how can this be rationalized? There are several possibilities: 1) Some species may be prevented from achieving long seed shadows by phylogenetic constraints on seed dispersal mechanisms or correlations between dispersal mode and other traits related to floral morphology (Primack 1987; Uma Shaanker *et al.* 1990). Given the existence of well-developed seed polymorphisms within species (e.g., Venable & Levin 1985; Venable *et al.* 1987, and many others), and of variation in dispersal mode among related species and genera (e.g., Salisbury 1975; Sorensen 1986, and others), phylogenetic constraints do not offer a likely general explanation. The role of constraints based on genetic and developmental correlations should be examined more closely in relation to this question.

2) Selection for increased dispersal distance is weak or absent. Two kinds of examples are discussed here:

a) Successful dispersal may be less a function of distance (or of any factors correlated with distance) than of arrival in particular kinds of sites (e.g., deposition in ant nests). However, not even all ant-dispersed species obtain advantages from deposition in ant nests, either because the principal advantage of dispersal is escape from predation or because the ants, after removing the elaiosomes, discard the seeds some distance from the nest. Most modes of dispersal seem to be less directed to special sites than ant-dispersal.

b) There may be a correlation between dispersal mode and the distribution of safe sites for offspring establishment: Safe sites for long-range dispersers might be more far-flung than those of short-range dispersers (e.g., Green 1983, Collins & Uno 1985; but see Geritz *et al.* 1984; Augspurger & Kitajima ms.). For instance, tree species dependent upon treefall gaps for establishment might be expected to have more long-range dispersal than shade-tolerant species that can establish under the forest canopy (*ceteris paribus*). Augspurger (1986) found a weak correlation between potential dispersal capacity and shade intolerance (and, thus, dependence on treefall gaps) in a sample of neotropical trees. However, for deciduous trees of North America (Fowells 1965), shade-tolerant and shade-intolerant trees have equivalent frequencies of three dispersal modes with different distance potentials (squirrel, wind, and bird) (χ^2, $p<0.05$; MFW unpublished data). The distribution of safe sites for most herbs is more difficult to approximate, because many forest herbs can establish readily throughout the understory and the conditions constituting a safe site are undetermined. For many herbs of prairie and oldfield, the same is true, except for a few species that require recent disturbances. Of the fugitive species of an Iowa prairie that are quite dependent on badger mounds, four have plumed seeds, three have no special dispersal devices but are airborne, and two are vertebrate-dispersed (Platt 1975). Safe-site requirements are broadly overlapping, but the means of reaching them differ greatly. The seed shadows of the plumed seeds are far more likely to include extant badger mounds than are those of the unplumed seeds (the vertebrate-dispersed species were not examined in this respect; Platt & Weis 1977). Thus, there is no evident strong correlation between the distribution of safe sites and dispersal mode, in general.

3) Short-distance dispersers might compensate for their poor spatial dispersal by dispersing in time. Indeed, short-distance dispersers in eastern North America seem to have the capacity for extended dormancy more often than long-range dispersers (at least as measured in the greenhouse). But many species in both categories have only short dormancies (Baskin & Baskin 1988), so dispersal in time cannot account entirely for the apparent success of species with short seed shadows (see also Murray 1988). A parallel situation was reported for intraspecific comparisons by Venable and Lawlor (1980). Furthermore, a buried-seed strategy is apparently rare in some vegetation types (e.g., Johnson 1975); to the extent that short-distance dispersers are found in such habitats, dispersal in time may not be an available option.

4) Seeds without evident dispersal devices may achieve long-distance dispersal by less evident means. Some of the plants with no obvious dispersal mechanism might be dispersed at least sometimes by vertebrates that consume seeds along with the foliage (Janzen 1984; Collins & Uno 1985; Vickery *et al.* 1986), by other means such as water (Waser *et al.* 1982; Vickery *et al* 1986), by vertebrates that eat seeds but do not digest them all (Staniforth & Cavers 1977), or by being shaken from the plant. Some seeds may produce dispersal devices only under certain circumstances (e.g., mucilage produced only when the seed is wetted – Vaughn & Whitehouse 1971; Grubert 1974; Hurka & Haase 1982 – although this trait may have several other functions as well). This possibility cannot be assessed at present, but it seems likely to pertain only to some of the seeds that seem to lack special dispersal mechanisms.

5) Secondary dispersal, by various means, could enhance the dispersal distances achieved by the initial mechanism. Some seeds (e.g., certain *Viola,* and many Australian plants; Westoby & Rice 1981) have dual dispersal mechanisms. A device that functioned initially in one capacity (e.g., sticking a seed to a substrate) may function secondarily for external dispersal (e.g., *Arceuthobium*, Punter & Gilbert 1989). Others may be moved serendipitously, after initial deposition, by divers agents (e.g., Harper 1977; Watkinson 1978; Estrada & Coates-Estrada 1986; Gonzalez-Espinosa & Quintana-Ascencio 1986; Roberts & Heithaus 1986). Some secondary movement of seeds may be considerably greater than that achieved by the initial means (e.g., Friedman & Stein 1980). However, at present, we have no evidence that secondary dispersal of species lacking evident dispersal mechanisms is different from that of species with such mechanisms.

6) The development of dispersal mechanisms might be costly – their elaboration could necessitate tradeoffs with other components of fitness. For instance, dispersal might be improved by a decrease in seed size but such decrease could impair seedling survivorship, or the cost of con-

structing a dispersal mechanism might detract from fecundity. This possibility is difficult to evaluate with present information, and there is no information to suggest why particular species can afford to pay such costs but others cannot.

7) Various combinations of the above factors may extend the effective seed shadow of species that seem (with present information) to achieve only short-range dispersal. Some species may employ more than one method, or various methods may serve as alternatives. It seems likely that the single-variable considerations detailed above are oversimplified, and that a multifactorial approach would be profitable (but not yet achieved).

Seed shadows and colonization

Although some kinds of dispersal modes might sometimes yield steeper slopes than others, considerable intraspecific variation in the steepness of the seed-distribution slope (ln-ln) suggests that both maternal traits and dispersal conditions may exert major influences on the seed shadow. Observed patterns of colonization must depend on many factors: 1) requirements for germination and establishment and the distribution of safe sites in the colonized area; 2) conditions under which dispersal is achieved – slow or fast windspeeds (e.g., Cremer 1971; Green 1983; Guries & Nordheim 1984), upwind or downwind direction, presence of updrafts (e.g., Solbreck & Andersson 1987), or squirrel vs jay dispersal (Stapanian & Smith 1984; 1986; Johnson & Adkisson 1986) – which can have very large effects on the dispersal distances achieved; 3) structure of the vegetation in the colonized area, which can affect deposition patterns (e.g., Willson & Crome 1989). In general, the van der Plank-Harper rule may apply more to particular conditions than to particular dispersal modes, but this should be examined again when appropriate samples become available.

Van der Plank (1960) suggested that slopes of relatively low magnitude 'can be expected if the movement of [propagules] is oriented with a restriction on random scattering, as would occur, for example, if the [propagules] were carried by birds migrating toward a particular destination.' Similar restrictions might be imposed on dispersal by ants, water-currents, or trade-winds, and the restriction is not simply a function of dispersal mode per se.

Problems and future studies

The problems pervading the comparisons attempted here are many. Although one of the functions of a published literature on a subject is the eventual synthesis of the information therein, such a synthesis is hampered by the vast differences in authors' methods of sampling and presentation of results. Sampling intervals vary enormously in spacing and in number, which renders some correlations insignificant (for sampling rather than biological reasons). Some studies used single seed sources, while others (such as those sampling at a forest edge) used multiple sources with overlapping individual seed shadows. Many of the studies measuring seed rain into a clearcut did not measure seedfall from particular sources within the forest, so the estimated seed shadow applies only to clearcut conditions. (Such studies were conducted chiefly by foresters, whose limited goal was to assess seed rain into clearcuts; the difficulty comes in trying to use such data for broader purposes.) Many data were collected in transects radiating from a source and sampling effort was not proportional to the area sampled. In some cases, it was possible to adjust the data to account for this (at the cost of multiplying any inherent errors), but in other cases the authors gave too little information to allow this to be done. Sometimes data were presented as percent of seeds trapped vs distance, which, in the absence of a total N, permits calculation of a slope but not of an intercept that is comparable to those calculated from numbers of seeds. Several authors presented equations for semilog regressions but did not give correlation coefficients or coefficients of determination, merely stating (and not even always that!) that statistical significance occurred. For many species, it will be forever impossible to document maximum dispersal distances, but many studies stop remarkably short of the far tail of the seed distribution. Only recently have investigators showed much interest in variation in seed shadows, either in relation to habitat (e.g., McEvoy & Cox 1987) or among individuals in a population. Many of the studies in Appendices 1 and 2 are reports concerning a single seed shadow (one seed crop or one set of experimental releases), and yet variation among conspecific individuals can be marked (e.g., Augspurger 1983; Guries & Nordheim 1984; McCanny & Cavers 1987, D. Thiede, unpubl.). For epiphytes, whose seeds must land on other plants to be successful, measurement of the seed shadow in a vertical dimension, as well as the customary horizontal one, would be necessary (R. Mesquita, pers. comm.). These numerous problems mean that the comparisons in this paper are necessarily preliminary and imprecise. Nevertheless, perhaps they will serve as a beginning that prompts studies that are designed explicitly to examine comparative seed shadows.

For purposes of comparing the shapes and lengths of seed shadows, even in the limited way attempted in this paper, data sets should be characterized by several features: clear distinction between density and total numbers, point (or single plant) sources with sampling proportional to distance from source, numerous sampling intervals (certainly > 10), inclusion of as much of the distal portion of the distribution as possible, assessment of variation in conspecific seed shadows (both environmental and parental). Good quantitative comparisons of seed shadows may then become possible.

All attempts at interspecific comparisons are dogged by our general ignorance about the ecological consequences of dispersal to different distances from the source. How does the probability of offspring survival change with increasing dispersal and how does it vary among species and circumstances? Although a few studies have addressed the first part of this question for certain species and certain circumstances, the variation in outcome of dispersal events, so critical to understanding the evolution of dispersal and other ecological matters (Thompson 1988), is still virtually unknown.

Acknowledgements

I thank J. Su, K. R. Spitze, J. Herkert, C. J. Whelan, C. Link, and J. F. Fox, E. Alvarez-Buylla, and L. Venable for helpful consultation. C. K. Augspurger, P. J. Burton, C. A. Kelly, D. Thiede, and P. A. Thomas provided constructive discussion of early drafts of the manuscript and/or access to unpublished data. D. F. Greene offered pointed and very helpful criticism of the manuscript.

Appendix 1. Seed shadows of herbaceous species with different modes of seed dispersal. Relative seed-shadow length is indicated by the slope of the regression ln (seed number) = $m \times$ (distance in meters) + b. Slopes in parentheses are associated with statistically insignificant ($p > 0.10$) correlation coefficients but are included for comparative purposes (see text). The slope of the regression on a ln-ln scale is also presented. All values constrained to two places after the decimal point. Within each dispersal category, species are listed alphabetically by genus, first those with circular distributions of sampled seeds and those with other distributions. Some species are represented by more than one sample; these are indicated by separate entries labelled a, b, c...

		Semilog (ln)		ln-ln		Mode	Max	Site conditions	References
		m	r^2	m	r^2				
A. Species with special devices for wind dispersal									
Circular distributions									
Cryptantha flava	a	− 1.39	0.98	− 1.98	0.86	< 1	3	semidesert	Casper 1987[a]
	b	− 1.89	0.42	− 0.99	0.73	< 1	> 10		
	c	(− 0.05)	0.07	(− 0.42)	0.39	< 1	> 10		
Heterotheca latifolia	a	− 0.81	–	–	–	< 1	> 20	?	Venable & Levin 1985
	b	− 1.08	–	–	–	< 1	> 16		

Appendix 1. Continued.

		Semilog (ln)		ln-ln		Mode	Max	Site conditions	References
		m	r^2	*m*	r^2				
Hieracium aurantiacum	a	−1.97	0.90	0.68	0.94	<0.1	>1.9	oldfield	Stergios 1976[b]
	b	−1.44	0.88	0.50	0.96	<0.1	>1.9		
Senecio jacobaea	a	−0.38	0.80	−1.63	0.82	1	11	oldfield	McEvoy & Cox 1987[a]
	b	−0.37	0.93	−1.51	0.77	1	11		
	c	−0.47	0.97	−1.98	0.77	2	8		
	d	(−0.41)	0.64	−1.51	0.85	1	7		
Linear and downwind distributions									
Agrostis hiemalis (tumbling)		−0.22	0.74	−1.01	0.87	0.25	11.5	prairie	Rabinowitz & Rapp 1979
Apocynum sibirica		−0.05	0.62	−2.13	0.75	21	75	prairie	Platt & Weis 1977[a e]
Asclepias syriaca		−0.09	0.69	−2.46	0.83	11	46	prairie	Platt & Weis 1977[a e]
	a	(−0.01)	0.07	−0.67	0.50	<2.5	>150	oldfield	Morse & Schmitt 1985[a e]
	b	−0.020	0.54	−1.05	0.90	<2.5	>150		
	c	(−0.02)	0.30	−0.92	0.65	<2.5	>150		
	d	(−0.01)	0.03	−0.79	0.44	<2.5	>150		
Carduus nutans	a	–	–	−1.1	0.76	<10	40	parking lot	Smith & Kok 1984
	b	–	–	−1.3	0.79	<10	60		
	c	–	–	−0.6	0.59	10	80		
	d	–	–	−1.1	0.76	<10	>100		
Cirsium undulatum		−0.17	0.95	−4.24	0.95	16.5	64	prairie	Platt & Weis 1977[a e]
Liatris aspera		−0.50	0.91	−2.20	0.90	1.6	8.5	garden	Levin & Kersten 1969[a]
Senecio jacobaea		−0.12	0.90	−2.01	1.00	5	>36	oldfield	Poole & Cairns 1940[b]
Solidago rigida		−0.33	0.70	−2.72	0.82	3.8	15	prairie	Platt & Weis 1977[a e]
Tussilago farfara		−0.01	0.92	−0.59	0.96	10	>4000	pasture	Bakker 1961
B. Species dispersed without special morphological devices or mechanisms (including those shaken out of a fruit by wind action)									
Circular distributions									
Amphicarpaea bracteata (subterranean)		−2.36	0.90	−3.79	0.77	0.75	3.25	lab	Trapp 1988[a]
Dipsacus sylvestris		−2.88	–	–	–	<0.5	1.5	oldfield	Werner 1975
Gentianella germanica		−2.51	0.82	(−1.16)	0.74	0.5	0.8	grassland	Verkaar *et al.* 1983[a]
Heterotheca latifolia	a	−1.36	–	–	–	<1	≥12	?	Venable & Levin 1985
	b	−1.36	–	–	–	<1	≥10		
Lithospermum carolinense	a	−1.47	0.69	−1.85	0.78	0.7	1.8	sand dune	Westelaken & Maun 1985[a]
	b	−1.49	0.65	−1.62	0.76	0.4	1.7		
	c	−1.72	0.53	−1.89	0.66	0.5	2.0		
Panicum mileaceum[f]		−3.1	–	–	–	–	–	crop fields	McCanny & Cavers 1987
Scabiosa columbaria		(−2.41)	0.54	(−0.69)	0.33	0.45	~1.2	grassland	Verkaar *et al.* 1983[a]
Linear and downwind distributions									
Agrostis hiemalis (single seeds)		−1.50	–	–	–	<0.1	1.5	prairie	Rabinowitz & Rapp 1979
Anthoxanthum odoratum		−2.14	0.92	−1.81	0.83	0.3	~2.0	mowed field	Antonovics & Ellstrand 1985[b]
Chaenorrhinum minus	a	−9.80	0.61	−0.51	0.31	<0.1	0.2	railroad	Arnold 1981[a]
	b	−2.60	0.91	−1.10	0.74	<0.2	1.4		
Mimulus guttatus	a	−2.3	0.98	−2.3	0.90	0.25	>2.5	riparian	Vickery *et al.* 1986
	b	−2.3	0.76	−2.6	0.93	0.25	>2.5		

Appendix 1. Continued.

		Semilog (ln)		ln-ln		Mode	Max	Site conditions	References
		m	r^2	*m*	r^2				
Mirabilis hirsuta		−2.48	0.75	−2.25	0.91	0.35	1.6	prairie	Platt 1976[a e]
Mirabilis hirsuta		−3.61	0.85	−3.17	0.95	0.35	1.6	prairie	Platt & Weis 1977[a e]
Oenothera biennis		−1.08	0.72	−3.36	0.83	1.6	5	prairie	Platt & Weis 1977[a e]
Papaver argemone		−1.61	0.84	−1.62	0.65	0.6	3.1	?	Salisbury 1942[a]
P. dubium		−1.30	0.91	−1.96	0.83	0.9	2.1	?	Salisbury 1942[a]
P. hybridum		−2.37	0.97	−3.27	0.92	0.4	2.4	?	Salisbury 1942[a]
Penstemon digitalis	a	−3.05	0.85	(−1.98)	0.62	0.25	2.0	prairie	P. A. Thomas pers. comm.
	b	−2.66	0.89	−3.71	0.94	0.25	2.5		
	c	−2.83	1.00	−4.79	1.00	1.25	2.5		
	d	−3.35	0.93	(−2.27)	0.74	0.25	2.0		
Verbena stricta		−1.74	0.75	−3.36	0.89	0.85	3.5	prairie	Platt & Weis 1977[a e]
C. Species with ballistic dispersal									
Circular distributions									
Amphicarpaea bracteata (aerial)	a	−1.13	0.75	−1.34	0.46	0.25	4.25	lab	Trapp 1988[a]
	b	−1.05	0.83	−1.17	0.56	0.25	3.75		
Cardamine hirsuta		(−1.97)	0.42	(−0.59)	0.28	0.3	0.9	?	Salisbury 1961
Cassia fasciculata	a	−1.62	1.00	−4.75	1.00	2.25	4	lab	Lee 1984, pers. comm.
	b	−0.56	0.62	−0.73	0.39	0.25	5		
Crotalaria sp.		−0.96	0.87	−1.61	0.89	0.3	4.3	field	Levin and Kerster 1974
Geranium maculatum		−1.07	0.86	−4.29	0.89	3.2	5	?	Stamp & Lucas 1983[a e]
	a	(−0.19)	0.23	−0.49	0.25	1.2	5.2	lab	MFW and J. Agren unpub.
	b	(−0.01)	0.04	(−0.24)	0.01	1.4	6.2		
	c	−1.55	0.73	−7.96	0.74	4.6	5.6		
Lepidium vulgare	a	−1.11	0.75	−0.82	0.73	<0.4	>1.9	oldfield	D. Thiede pers. comm.
	b	−0.88	0.58	−0.58	0.46	0.2	>1.9		
	c	−1.06	0.77	−1.08	0.69	0.2	>1.9		
	d	−0.83	0.93	−0.81	0.89	0.4	>1.9		
	e	−0.89	0.56	−0.63	0.50	0.2	>1.9		
Lupinus texensis		−1.30	0.91	−1.33	0.84	<0.2	3.4	?	Schaal 1980[a]
Impatiens capensis		−2.30	0.76	−1.46	0.92	0.3	1.8	?	Stamp & Lucas 1983[a e]
	a	−1.96	0.91	−1.27	0.79	0.1	2.1	lab	Schmitt *et al.* 1985[a]
	b	−3.44	0.90	−1.57	0.80	0.1	1.3		
Phlox drummondii		−2.80	0.81	−2.78	0.72	0.7	1.6	?	Stamp & Lucas 1983[a e]
P. pilosa		−1.17	0.97	−2.08	0.91	0.75	3.8	lab	Levin & Kerster 1968[a]
Viola papilionacea		(−0.34)	0.21	(−0.87)	0.23	1.5	4.5	greenhouse	Beatie & Lyons 1975
V. stricta		(−0.31)	0.37	(−0.64)	0.63	0.5	3.4	?	Stamp & Lucas 1983[a e]

[a] Estimated from graph presented by the indicated source. I could check the accuracy of this estimate for *Lonchocarpus* and *Cassia*, with data provided by authors. In both cases the slope estimated from the graphs differed by <7% from that calculated from tabulated data.

[b] Transects around a source, uncorrected for area (see Methods). The remaining data in sections A and B of the table are downwind measures only. For Poole and Cairns (1940), the slope for all seeds was the same as that for downwind-moving seeds.

[c] *Panicum*: median value; range −0.09 to −0.73 for 20 different individuals.

[d] If r^2 values for a slope are missing, the slope was given by the source without that value.

[e] A known number of seeds were released and retrieved, so the curve is based on all seeds, and the tail of the distribution was actually observed.

[f] No special dispersal device present, but data for actual dispersal in wind.

Appendix 2. Seed shadows of trees and a few shrubs with different modes of seed dispersal. See legend for Appendix 1. In these samples the mode was generally near the source, and maximum dispersal distances were seldom reported.

		Semilog (ln)		ln-ln		Site conditions	References
		m	r^2	*m*	r^2		
A. Wind-dispersed species							
Circular distributions							
Betula uber		−0.02	0.92	−0.14	0.47	forest	Ford *et al.* 1983
Fraxinus sp.		−0.03	0.98	−1.32	0.86	?	*in* Levin and Kerster 1974
Lonchocarpus pentaphyllus		−0.09	0.96	−2.66	0.82	forest	Augspurger & Hogan 1983 [a]
Melaleuca quinquenervia		−0.05	0.96	−1.24	0.79	field	Woodall 1982
Nuytsia floribunda		−0.15	0.92	−2.6	0.93	grassy field	Lamont 1985 [a]
Platypodium elegans	a	−0.04	0.87	−2.37	0.79	forest	Augspurger 1983, pers. comm.
	b	−0.04	0.98	−2.05	0.97		
	c	−0.08	0.96	−3.23	0.93		
	d	−0.06	0.93	−1.82	0.89		
Shorea contorta	a	(−0.02)	0.10	(−0.61)	0.18	plantation	Tamari & Jacalne 1984
	b	−0.50	0.44	−1.86	0.56		
Linear distributions							
Acer saccharum		−0.02	0.63	−1.47	0.74	into oldfield	P. J. Burton pers. comm.
		−0.03	0.87	(−0.41)	0.46	into clearcut	Hughes & Fahey 1988
Ateleia herbert-smithii		−0.02	0.57	−2.13	0.79	into clearing	Gladstone 1979 [a b]
Betula alleghaniensis		−0.03	0.96	(−0.33)	0.50	into clearcut	Hughes & Fahey 1988
B. ermanii	a	−0.04	0.63	−1.78	0.58	into clearcut	*in* Tamari & Jacalne 1984
	b	−0.05	0.96	−2.32	0.88		
B. papyrifera		−0.03	0.99	−0.17	0.53	into clearcut	Bjorkbom 1971 [a]
Eucalytus regnans	a	−0.04	0.88	−1.48	0.93	into clearcut	Cremer 1966 [a]
	b	−0.02	0.91	−0.76	0.88		
Fraxinus americana		−0.03	0.52	−2.01	0.69	into oldfield	P. J. Burton pers. comm.
Abies amabilis		−0.01	0.96	(−0.33)	0.23	into clearcut	Carkin *et al.* 1978
A. procera		−0.02	0.91	−0.18	0.92	into clearcut	Carkin *et al.* 1978
Abies sp.	a	−0.02	0.88	−0.37	0.89	into clearcut	Franklin and Smith 1974b
	b	−0.02	0.90	−0.47	0.96		
Picea sitchensis	a	−0.18	0.83	(−2.80)	0.73	into clearcut	Mair 1973 [a]
	b	−0.17	0.96	−5.33	0.96		
	c	−0.11	0.92	−3.02	0.86		
P. glauca		−0.01	0.72	−0.62	0.86	into clearcut	Dobbs 1976
		(−0.01)	0.21	−0.74	0.33	into clearcut	Zasada & Lovig 1983
P. engelmannii		−0.02	0.81	−0.26	0.86	into clearcut	Alexander 1969 [a]
	a	−0.01	0.81	−0.17	0.85	into clearcut	McCaughey & Schmidt 1987
	b	(−0.01)	0.50	−0.23	0.84		
	c	−0.01	0.65	−0.26	0.83		
	d	−0.02	0.92	−0.27	0.82		
	e	−0.02	0.88	−0.28	0.81		
	a	−0.02	1.00	−1.41	0.90	into clearcut	Roe 1967
	b	−0.02	0.98	−1.34	0.89		

Appendix 2. Continued.

		Semilog (ln)		ln-ln		Site conditions	References
		m	r^2	*m*	r^2		
	c	−0.02	1.00	−2.14	0.89		
	d	−0.02	1.00	−1.32	0.89		
Pinus palustris		−0.06	0.98	−0.51	0.53	into clearcut	Boyer 1958[a]
P. halapensis		−0.15	0.97	−0.41	0.55	into oldfield	Acherar *et al.* 1984[a]
Pseudotsuga menziesii		(−0.01)	0.24	−0.41	0.84	into clearcut	Pickford 1929
		−0.02	1.00	−0.93	0.98	into clearcut	Gashwiler 1969
	a	−0.01	0.61	−1.55	0.83	into clearcut	Isaac 1930
	b	(−0.03)	0.95	−2.50	0.99		
	c	(−0.01)	0.41	−0.92	0.57		
Thuja plicata		−0.01	0.42	−0.90	0.82	into clearcut	Pickford 1929
		−0.03	1.00	−1.78	0.96	into clearcut	Gasgwiler 1969
Tsuga heterophylla		(−0.01)	0.33	−0.27	0.66	into clearcut	Pickford 1929
		−0.02	0.99	−1.10	0.99	into clearcut	Gashwiler 1969
T. mertensiana	a	(−0.01)	0.15	(−0.10)	0.68	into clearcut	Franklin & Smith 1974a
	b	−0.03	0.96	−0.26	0.79		
	c	−0.02	0.83	−0.27	0.83		
	d	−0.04	0.98	(−0.30)	0.61		
B. Vertebrate-dispersed species							
Circular distributions							
Andira inermis	a	−0.02	0.45	−0.96	0.91	pasture	Janzen *et al.* 1976
(bat)	b	−0.03	0.29	−1.11	0.36		
Hymenaea courbaril	a	−0.10	0.89	−1.71	0.98	forest	Hallwachs 1986
(agouti)	b	−0.04	0.45	−1.00	0.64		
Macrozamia riedlei (possum)		−0.20	0.61	−1.62	0.92	pasture	Burbidge & Whelan 1982
Virola surinamensis (bird)		(−0.04)	0.30	−0.81	0.50	forest	Howe *et al.* 1984[a]
Various spp., artifical displays (bird)		−0.19 to −0.75	0.34 to 0.95	–	–	treefall gaps	Hoppes 1985
Linear distributions							
Juniperus virginiana		−0.01	0.64	–	–	along fenceline	Holthuijzen & Sharik 1985b
(bird)	a	−0.59	0.67	–	–	oldfield	Holthuijzen & Sharik 1985a
	b	−0.58	0.63	–	–		
	c	−0.41	0.67	–	–		
	d	(−0.03)	0.21	–	–		
	e	−0.54	0.84	–	–		
	f	−0.64	0.78	–	–		
	g	(−0.70)	0.43	–	–		
	h	(−0.17)	0.11	–	–		
Sambucus nigra (bird)		−0.10	0.98	–	–	mixed habitat	Debussche *et al.* 1985[b]

Appendix 2. Seed shadows of trees and a few shrubs with different modes of seed dispersal. See legend for Appendix 1. In these samples the mode was generally near the source, and maximum dispersal distances were seldom reported.

		Semilog (ln)		ln-ln		Site conditions	References
		m	r^2	*m*	r^2		
Trichilia martiana (bird)		− 0.26	0.88	–	–	?	*in* Dirzo & Dominguez 1986
Various spp. (primates)		(− 0.01)	0.38	(− 0.04)	0.16	forest	Garber 1986[b]
Various spp. (bird)	a	− 0.35	0.98	–	–	abandoned orchard	Debussche *et al.* 1985
	b	− 0.27	0.46	–	–		
C. Ballistically dispersed species							
Circular distribution							
Hura crepitans		− 0.02	0.14	− 0.37	0.10	field	Swaine & Beer 1977[a]
D. Ant-dispersed species							
Circular distribution							
Acacia suaveolens		− 0.30	0.64	− 1.48	0.81	woodland	Andersen 1988[b]

[a] Seed shadow estimated from graph; see footnote in Appendix 1.
[b] Seed shadow derived from the equivalent of a point source; others are from multiple sources.

References

Acherar, M., Lepart, J. & Debussche, M. 1984. La colonisation des friches par le pin d'Alep (*Pinus halapensis* Miller) en Languedoc mediterraneen. Acta Oecol./Oecol. Plant. 5: 179–189.

Alexander, R. R. 1969. Seedfall and establishment of Engelmann spruce in clearcut openings: A case history. USDA For. Serv. Res. Pap. RM-53: 1–8.

Andersen, A. N. 1988. Dispersal distance as a benefit of myrmecochory. Oecologia 75: 507–511.

Antonovics, J. & Ellstrand, N. 1985. The fitness of dispersed progeny: experimental studies with *Anthoxanthum odoratum*. In: P. Jacquard, G. Hein, & J. Antonovics (eds.). Genetic differentiation and dispersal in plants. Pp. 369–381. Springer-Verlag, Berlin.

Arnold, R. M. 1981. Population dynamics and seed dispersal of *Chaenorrhinum minus* on railroad cinder ballast. Am. Midl. Nat. 106: 80–91.

Augspurger, C. K. 1983. Seed dispersal of the tropical tree, *Platypodium elegans*, and the escape of its seedlings from fungal pathogens. J. Ecol. 71: 759–771.

Augspurger, C. K. 1986. Morphology and dispersal potential of wind-dispersed diaspores of neotropical trees. Amer. J. Bot. 73: 353–363.

Augspurger, C. K. & Kitajima, K. 1991. Experimental studies of seedling recruitment from contrasting seed distributions. (ms. submitted).

Augspurger, C. K. & Hogan, K. P. 1983. Wind dispersal of fruits with variable seed number in a tropical tree (*Lonchocarpus pentaphyllus*: Leguminosae). Amer. J. Bot. 70: 1031–1037.

Auld, T. D. 1986. Population dynamics of the shrub *Acacia suaveolens* (Sm.) Willd: Dispersal and the dynamics of the soil seed-bank. Austr. J. Ecol. 11: 235–254.

Bakker, D. 1961. A comparative life-history study of *Cirsium arvense* (L.) Scop. and *Tussilago farfara* L., the most troublesome weeds in the newly reclaimed polders of the former Zuiderzee. In: J. L. Harper (ed.). The biology of weeds. Pp. 205–222. Blackwell, Oxford.

Barrett, S. C. H. 1978. Heterostyly in a tropical weed: The reproductive biology of the *Turnera ulmifolia* complex (Turneraceae). Can. J. Bot. 56: 1713–1725.

Baskin, C. C. & Baskin, J. M. 1988. Germination ecophysiology of herbaceous plant species in a temperate region. Amer. J. Bot. 75: 286–305.

Beattie, A. J. & Lyons, N. 1975. Seed dispersal in *Viola* (Violaceae): adaptations and strategies. Amer. J. Bot. 62: 714–722.

Beer, T. & Swaine, M. D. 1977. On the theory of explosively dispersed seeds. New Phytol. 78: 681–694.

Bjorkbom, J. C. 1971. Production and germination of paper birch seed and its dispersal into a forest opening. USDA For. Serv. Res. Pap. NE-209: 1–14.

Boyer, W. D. 1958. Longleaf pine seed dispersal in south Alabama. J. Forest. 56: 265–268.

Bullock, S. H. 1976. Consequences of limited seed dispersal within simulated annual populations. Oecologia 24: 247–256.

Bullock, S. H. & Primack, R. B. 1977. Comparative experimental study of seed dispersal on animals. Ecology 58: 681–686.

Bulow-Olsen, A. 1984. Diplochory in *Viola*: A possible relation between seed dispersal and soil seed bank. Amer. Midl. Nat. 112: 251–260.

Burbidge, A. H. & Whelan, R. J. 1982. Seed dispersal in a cycad, *Macrozamia riedlei*. Aust. J. Ecol. 7: 63–67.

Campbell, C. S. 1983. Wind dispersal of some North American species of *Andropogon* (Gramineae). Rhodora 85: 65–72.

Carkin, R. E., Franklin, J. F., Booth, J. & Smith, C. E. 1978. Seeding habits of upper-slope tree species. IV. Seed flight of noble fir and pacific silver fir. USDA For. Serv. Res. Note PNW-312: 1–10.

Carlquist, S. 1967. The biota of long-distance dispersal. V. Plant dispersal to Pacific islands. Bull. Torrey Bot. Club 94:129–162.

Casper, B. B. 1987. Spatial patterns of seed dispersal and post-dispersal seed predation of *Cryptantha flava* (Boraginaceae). Amer. J. Bot. 74: 1646–1655.

Clark, D. A. & Clark, D. B. 1984. Spacing dynamics of a tropical rain forest tree: evaluation of the Janzen-Connell model. Am. Nat. 124: 769–788.

Collins, S. L. & Uno, C. E. 1985. Seed predation, seed dispersal, and disturbance in grasslands: a comment. Am. Nat. 125: 866–872.

Cremer, K. W. 1966. Dissemination of seed from *Eucalyptus regnans*. Aust. For. 30: 33–37.

Cremer, K. W. 1971. Speeds of falling and dispersal of seed of *Pinus radiata* and *P. contorta*. Austr. For. Res. 5: 29–32.

Culver, D. C. & Beattie, A. J. 1978. Myrmecochory in *Viola*: Dynamics of seed-ant interactions in some West Virginia species. J. Ecol. 66: 53–72.

Debussche, M., Lepart, J. & Molina, J. 1985. La dissemination des plantes à fruits charnus par les oiseaux: rôle de la structure de la végétation et impact sur la succession en région méditerranéenne. Acta Oecol./Oecol. Gener. 6: 65–80.

Dirzo, R. & Dominguez, C. A. 1986. Seed shadows, seed predation and the advantages of dispersal. In: A. Estrada & T. Fleming (eds.). Frugivores and seed dispersal. Pp. 237–250. Junk, Dordrecht.

Dobbs, R. C. 1976. White spruce seed dispersal in central British Columbia. For. Chron. 52: 225–228.

Estrada, A. & Coates-Estrada, R. 1986. Frugivory in howling monkeys (*Alouatta palliata*) at Los Tuxtlas, Mexico: dispersal and fate of seeds. In: A. Estrada & T. Fleming (eds.). Frugivory and seed dispersal. Pp. 93–104. Junk, Dordrecht.

Ford, R. H., Sharik, T. L. & Feret, P. P. 1983. Seed dispersal of the endangered Virginia round-leaf birch (*Betula uber*). For. Ecol. Mgt. 6: 115–128.

Fowells, H. A. 1965. Silvics of forest trees of the United States. USDA Agric. Handbook 271.

Franklin, J. F. & Smith, C. E. 1974a. Seeding habits of upper-slope tree species. II. Dispersal of a mountain hemlock seedcrop on a clearcut. USDA For. Serv. Res. Note PNW-214: 1–9.

Franklin, J. F. & Smith, C. E. 1974b. Seeding habits of upper-slope tree species. III. Dispersal of white and Shasta red fir seeds on a clearcut. USDA For. Serv. Res. Note PNW-215: 1–9.

Friedman, J. & Stein, Z. 1980. The influences of seed–dispersal mechanisms on the dispersion of *Anastatica hierochuntica* (Cruciferae) in the Negev Desert, Israel. J. Ecol. 68: 43–50.

Ganeshaiah, K. N. & Uma Shaanker, R. 1988. Evolution of a unique seed maturity pattern in *Croton bonplandianum* Baill strengthens ant-plant mutualism for seed dispersal. Oecologia 77: 130–134.

Garber, P. A. 1986. The ecology of seed dispersal in two species of callitrichid primates (*Saguinus mystax* and *Saguinus fuscicollis*). Am. J. Primatol. 10: 155–170.

Gashwiler, J. S. 1969. Seed fall of three conifers in west-central Oregon. For. Sci. 15: 290–295.

Geritz, S. A. H., de Jong, T. J., & Klinkhamer, P. G. L. 1984. The efficacy of dispersal in relation to safe site area and seed production. Oecologia 62: 219–221.

Gladstone, D. E. 1979. Description of a seed-shadow of a wind-dispersed tropical tree. Brenesia 16: 81–86.

Gonzalez-Espinosa, M. & Quintana-Ascencio, P. F. 1986. Seed predation and dispersal in a dominant desert plant: *Opuntia*, ants, birds, and mammals. In: A. Estrada & T. Fleming (eds.). Frugivory and seed dispersal. Pp. 273–284. Junk, Dordrecht.

Green, D. S. 1983. The efficacy of dispersal in relation to safe site density. Oecologia 56: 356–358.

Greene, D. F. & Johnson, E. A. 1986. Dispersal capacity and seed production in anemochorous plants. Oecologia 68: 629–631.

Greene, D. F. & Johnson, E. A. 1989. A model of wind dispersal or winged or plumed seeds. Ecology 70: 339–347.

Gregory, P. H. & Read, D. R. 1949. The spatial distribution of insect-borne plant-virus diseases. Ann. Appl. Biol. 36: 475–482.

Grubert, M. 1974. Studies on the distribution of myxospermy among seeds and fruits of Angiospermae and its ecological importance. Acta Biol. Venez. 8: 315–551.

Guries, R. P. & Nordheim, E. V. 1984. Flight characteristics and dispersal potential of maple samaras. For. Sci. 30: 434–440.

Hallwachs, W. 1986. Agoutis (*Dasyprocta punctata*), the inheritors of guapinal (*Hymenaea courbaril:* Leguminosae). In: A. Estrada & T. H. Fleming (eds.). Frugivores and seed dispersal. Pp. 285–304. Junk, Dordrecht.

Hamilton, W. D. & May, R. M. 1977. Dispersal in stable habitats. Nature 269: 578–581.

Hanson, J. S., Malanson, G. P. & Armstrong, M. P. 1990.

Landscape fragmentation and dispersal in a model of riparian forest dynamics. Ecol. Model. 49: 277–296.

Harper, J. L. 1977. Population biology of plants. Academic Press, London.

Holthuijzen, A. M. A. & Sharik, T. L. 1985a. The avian seed dispersal system of eastern red cedar (*Juniperus virginiana*). Can. J. Bot. 63: 1508–1515.

Holthuijzen, A. M. A. & Sharik, T. L. 1985b. The red juniper (*Juniperus virginiana* L.) seed shadow along a fenceline. Am. Midl. Nat. 113: 200–202.

Hoppes, W. G. 1985. Seed dispersal by fall migrant frugivorous birds in an east-central Illinois woodland. Ph.D. Thesis, University of Illinois. 155 pp.

Horvitz, C. C. & Schemske, D. W. 1986. Seed dispersal of a neotropical myrmecochore: variation in removal rates and dispersal distance. Biotropica 18: 319–323.

Howe, H. F. & Smallwood, J. 1982. Ecology of seed dispersal. Annu. Rev. Ecol. Syst. 13: 201–228.

Howe, H. F., Schupp, E. W. & Westley, L. C. 1984. Early consequences of seed dispersal for a neotropical tree (*Virola surinamensis*). Ecology 66: 781–791.

Hubbell, S. P. 1979. Tree dispersion, abundance, and diversity in a tropical dry forest. Science 203: 1299–1309.

Hughes, J. W. & Fahey, T. J. 1988. Seed dispersal and colonization in a disturbed northern hardwood forest. Bull. Torrey Bot. Club: 115: 89–99.

Hurka, H. & Haase, R. 1972. Seed ecology of *Capsella bursa-pastoris* (Cruciferae): Dispersal mechanism and the soil seed bank. Flora 172: 35–46.

Isaac, L. A. 1930. Seed flight in the Douglas fir region. J. Forest. 28: 492–499.

Janzen, D. H. 1984. Dispersal of small seeds by big herbivores: Foliage is the fruit. Amer. Nat. 123: 338–353.

Janzen, D. H., Miller, G. A., Hackforth-Jones, J., Pond, C. M., Hooper, K. & Janos, D. P. 1976. Two Costa Rican bat-generated seed shadows of *Andira inermis* (Leguminosae). Ecology 57: 1068–1075.

Johnson, E. A. 1975. Buried seed populations in the subarctic forest east of Great Slave Lake, Northwest Territories. Can. J. Bot. 53: 2933–2941.

Johnson, W. C. 1988. Estimating dispersibility of *Acer*, *Fraxinus* and *Tilia* in fragmented landscapes from patterns of seedling establishment. Landscape Ecology 1: 175–187.

Johnson, W. C. & Adkisson, C. S. 1986. Airlifting the oaks. Nat. Hist. 95(10): 40–47.

Kjellsson, G. 1985. Seed fate in a population of *Carex pilulifera* L. I. Seed dispersal and ant-seed mutualism. Oecologia 67: 416–423.

Lack, A. J. & Kay, Q. O. N. 1987. Genetic structure, gene flow and reproductive ecology in sand-dune populations of *Polygala vulgaris*. J. Ecol. 75: 259–276.

Lamont, B. 1985. Dispersal of the winged fruits of *Nuytsia floribunda* (Loranthaceae). Aust. J. Ecol. 10: 187–193.

Lanner, R. M. 1985. Effectiveness of the seed wing of *Pinus flexilis* in wind dispersal. Great Basin Nat. 45: 318–320.

Lee, T. D. 1984. Effects of seed number per fruit on seed dispersal in *Cassia fasciculata* (Caesalpiniaceae). Bot. Gaz. 145: 136–139.

Lee, M. A. B. & Brothers, T. 1981. Seed dispersal in hybrid *Salsola*. Great Basin Nat. 41: 370–376.

Levin, D. A. & Kerster, H. W. 1968. Local gene dispersal in *Phlox*. Evolution 22: 130–139.

Levin, D. A. & Kerster, H. 1969. Density-dependent gene dispersal in *Liatris*. Am. Nat. 103: 61–74.

Levin, D. A. & Kerster, H. W. 1974. Gene flow in seed plants. Evol. Biol. 7: 139–220.

Mair, A. R. 1973. Dissemination of tree seed Sitka spruce, western hemlock and Douglas fir. Scot. For. 27: 308–314.

Matlack, G. R. 1987. Diaspore size, shape, and fall behavior in wind-dispersed plant species. Amer. J. Bot. 74: 1150–1160.

McCanny, S. J. & Cavers, P. B. 1987. The escape hypothesis: a test involving a temperate, annual grass. Oikos 49: 67–76.

McCaughey, W. W. & Schmidt, W. C. 1987. Seed dispersal of Engelmann spruce in the intermountain west. Northwest Sci. 61: 1–6.

McEvoy, P. B. & Cox, C. S. 1987. Wind dispersal distances in dimorphic achenes of ragwort, *Senecio jacobaea*. Ecology 68: 2006–2015.

Minogue, K. P. 1986. Disease gradients and the spread of disease. In: K. J. Leonard & W. E. Fry (eds.). Plant Disease Epidemiology. Pp. 285–310. MacMillan, New York.

Morse, D. H. & Schmitt, J. 1985. Propagule size, dispersal ability, and seedling performances in *Asclepias syriaca*. Oecologia 67: 372–379.

Murray, K. G. 1988. Avian seed dispersal of three neotropical gap-dependent plants. Ecology 58: 271–298.

O'Dowd, D. J. & Hay, M. E. 1980. Mutualism between harvester ants and a desert ephemeral: Seed escape from rodents. Ecology 61: 531–540.

Ohara, M. & Higashi, S. 1987. Interference by ground beetles with the dispersal by ants of seeds of *Trillium* species (Liliaceae). J. Ecol. 75: 1091–1098.

Okubo, A. & Levin, S. A. 1989. A theoretical framework for data analysis of wind dispersal of seeds and pollen. Ecology 70: 329–338.

Olivieri, I. & Berger, A. 1985. Seed dimorphism for dispersal: physiological, genetic and demographical aspects. In: P. Jaquard *et al.* (eds.). Genetic Differentiation and Dispersal in Plants. Pp. 413–429. Springer-Verlag, Berlin.

Peart, D. R. 1985. The quantitative representation of seed and pollen dispersal. Ecology 66: 1081–1083.

Peart, M. H. 1979. Experiments on the biological significance of the morphology of seed-dispersal units in grasses. J. Ecol. 67: 843–863.

Pickford, A. F. 1929. Studies of seed dissemination in British Columbia. For. Chron. 5: 8–16.

Platt, W. J. 1975. The colonization and formation of equilibrium plant species associations on badger disturbances in a tall-grass prairie. Ecol. Monogr. 45: 285–305.

Platt, W. J. 1976. The natural history of a fugitive prairie plant (*Mirabilis hirsuta* (Parsh) MacM). Oecologia 22: 399–405.

Platt, W. J. & Weis, M. I. 1977. Resource partitioning and competition within a guild of fugitive prairie plants. Amer. Nat. 111: 479–513.

Poole, A. L. & Cairns, D. 1940. Botanical aspects of ragwort (*Senecio jacobaea* L.) control. DSIR Bull. 82, Wellington, NZ.

Portnoy, S. & Willson, M. F. ms. Seed dispersal curves: Behavior of the tail of the distribution.

Primack, R. B. 1987. Relationships among flowers, fruits, and seeds. Annu. Rev. Ecol. Syst. 18: 409–430.

Punter, D. & Gilbert, J. 1989. Animal vectors of *Arceuthobium americanum* seed in Manitoba. Can. J. For. Res. 19: 865–869.

Rabinowitz, D. & Rapp, J. K. 1979. Dual dispersal modes in hairgrass, *Agrostis hiemalis* (Walt.) B.S.P. (Gramineae). Bull. Torrey Bot. Club 106: 32–36.

Rabinowitz, D. & Rapp, J. K. 1981. Dispersal abilities of seven sparse and common grasses from a Missouri prairie. Amer. J. Bot. 68: 616–624.

Roe, AL. L. 1967. Seed dispersal in a bumper spruce seed year. USDA For. Serv. Res. Pap. INT-39: 1–10.

Roberts, J. T. & Heithaus, E. R. 1986. Ants rearrange the vertebrate-generated seed shadow of a neotropical fig tree. Ecology 67: 1046–1051.

Sacchi, C. F. 1987. Variability in dispersal ability of common milkweed, *Asclepias syriaca,* seeds. Oikos 49: 191–198.

Salisbury, E. J. 1942. The reproductive capacity of plants. Bell, London.

Salisbury, E. J. 1961. Weeds and Aliens. Collins, London.

Salisbury, E. J. 1975. The survival value of modes of dispersal. Proc. Roy. Soc. Lond. B 188: 183–188.

Salisbury, E. J. 1976. Seed output and the efficacy of dispersal by wind. Proc. Roy. Soc. Lond. B 192: 323–329.

Schaal, B. A. 1980. Measurement of gene flow in *Lupinus texensis*. Nature 284: 450–451.

Schmitt, J., Ehrhardt, D. & Swartz, D. 1985. Differential dispersal of self-fertilized and outcrossed progeny in jewelweed (*Impatiens capensis*). Am. Nat. 126: 570–575.

Smith, J. M. B. 1989. An example of ant-assisted plant invasion. Austr. J. Ecol. 14: 247–250.

Smith, L. M. & Kok, L. T. 1984. Dispersal of musk thistle (*Carduus nutans*) seeds. Weed Sci. 32:120–125.

Solbreck, C. & Andersson, D. 1987. Vertical distribution of fireweed, *Epilobium angustifolium,* seeds in the air. Can. J. Bot. 65: 2177–2178.

Sorensen, A. E. 1986. Seed dispersal by adhesion. Annu. Rev. Ecol. Syst. 17: 443–463.

Stamp, N. E. & Lucas, J. R. 1983. Ecological correlates of explosive seed dispersal. Oecologia 59: 272–278.

Staniforth, R. J. & Cavers, P. B. 1977. The importance of cottontail rabbits in the dispersal of *Polygonum* spp. J. Appl. Ecol. 14: 261–267.

Stapanian, M. A. & Smith, C. C. 1984. Density-dependent survival of scatterhoarded nuts: an experimental approach. Ecology 65: 1387–1396.

Stapanian, M. A. & Smith, C. C. 1986. How fox squirrels influence the invasion of prairies by nut-bearing trees. J. Mamm. 67: 326–332.

Stergios, B. G. 1976. Achene production, dispersal, seed germination, and seedling establishment of *Hieracium aurantiacum* in an abandoned field community. Can. J. Bot. 54: 1189–1197.

Swaine, M. D. & Beer, T. 1977. Explosive seed dispersal in *Hura crepitans* L. (Euphorbiaceae). New Phytol. 78: 695–708.

Tamari, C. & Jacalne, D. V. 1984. Fruit dispersal of dipterocarps. Bull. For. and For. Prod. Res. Inst. 325: 127–140.

Thompson, J. N. 1988. Variation in interspecific interactions Annu. Rev. Ecol. Syst. 19: 65–88.

Trapp, E. J. 1988. Dispersal of heteromorphic seeds in *Amphicarpaea* bracteata (Fabaceae). Amer. J. Bot. 75: 1535–1539.

Uma Shaanker, R., Ganeshaiah, K. N. & Radhamani, T. R. 1990. Associations among the modes of pollination and seed dispersal – Ecological factors and phylogenetic constraints. Evol. Trends Plants 4: 107–111.

van der Pijl, L. 1982. Principles of dispersal in higher plants. (3rd ed.) Springer-Verlag, Berlin.

van der Plank, J. E. 1949. The relation between the size of fields and the spread of plant-diseases into them. II. Diseases caused by fungi with air-borne spores, with a note on horizons of infection. Empire J. Exper. Agric. 17: 18–22.

van der Plank, J. E. 1960. Analysis of epidemics. Plant Pathology 3: 229–289.

Vaughn, J. G. & Whitehouse, J. M. 1971. Seed structures and the taxonomy of the Cruciferae. Bot. J. Linn. Soc. 64: 383–409.

Venable, D. L. & Lawlor, L. 1980. Delayed germination and dispersal in desert annuals: escape in space and time. Oecologia 46: 272–282.

Venable, D. L. & Levin, D. A. 1983. Morphological dispersal structures in relation to growth habit in the Compositae. Plant Syst. Evol. 143: 1–16.

Venable, D. L. & Levin, D. A. 1985. Ecology of achene dimorphism in *Heterotheca latifolia*. I. Achene structure, germination and dispersal. J. Ecol. 73: 133–145.

Venable, D. L., Burquez, A., Corral, G., Morales, E. & Espinosa, F. 1987. The ecology of seed hetermorphism in *Heterosperma pinnatum* in central Mexico. Ecology 68: 65–76.

Venable, D. L. & Brown, J. S. 1988. The selective interactions of dispersal, dormancy, and seed size as adaptations for reducing risk in variable environments. Am. Nat. 131: 360–384.

Verkaar, H. J., Schenkeveld, A. J. & van de Klashorst, M. P. 1983. The ecology of short-lived forbs in chalk grasslands: dispersal of seeds. New Phytol. 95: 335–344.

Vickery, R. K., Phillips, D. R. & Wonsavage, P. R. 1986. Seed dispersal in *Mimulus guttatus* by wind and deer. Amer. Midl. Nat. 116: 206–208.

Waser, N. M., Vickery, R. K. & Price, M. V. 1982. Patterns of seed dispersal and population differentiation in *Mimulus guttatus*. Evolution 36: 753–761.

Watkinson, A. R. 1978. The demography of a sand dune annual: *Vulpia fasciculata*. III. The dispersal of seeds. J. Ecol. 66: 483–498.

Webber, M. L. 1934. Fruit dispersal. Malay. Forest. 3: 18–19.

Werner, P. A. 1975. A seed trap for determining patterns of seed deposition in terrestrial plants. Can. J. Bot. 53: 810–813.

Westelaken, I. L. & Maun, M. A. 1987. Spatial pattern and seed dispersal of *Lithospermum caroliniense* in Lake Huron sand dunes. Can. J. Bot. 63: 125–132.

Westoby, M. & Rice, B. 1981. A note on combining two methods of dispersal-for-distance. Aust. J. Ecol. 6: 189–192.

Willson, M. F. & Crome, F. H. C. 1989. Patterns of seed rain at the edge of a tropical Queensland rain forest. J. Trop. Ecol. 5: 301–308.

Wolfenbarger, D. O. 1946. Dispersion of small organisms. Amer. Midl. Nat. 35: 1–152.

Woodall, S. L. 1982. Seed dispersal in *Melaleuca quinquenervia*. Florida Sci. 45: 81–93.

Zasada, J. C. & Lovig, D. 1983. Observations on primary dispersal of white spruce, *Picea glauca*, seed. Can. Field-Nat. 97: 104–106.

Vegetatio **107/108**: 281–297, 1993.
T. H. Fleming and A. Estrada (eds). Frugivory and Seed Dispersal: Ecological and Evolutionary Aspects.

The influence of seed dispersal mechanisms on the genetic structure of tropical tree populations

J. L. Hamrick, Darlyne A. Murawski[1] & John D. Nason
Departments of Botany and Genetics, University of Georgia, Athens, GA 30602, USA; [1]*Present address: Biology Department, University of Massachusetts at Boston,Boston, MS 02125, USA*

Keywords: Allozymes, Genetic diversity, Genetic relatedness, Population density, Seed shadows, Size classes

Abstract

Seed dispersal mechanisms should have a direct impact on the genetic structure of populations. Species whose seeds are dispersed near the maternal plant (e.g. gravity or wind dispersal) or species whose seeds are deposited in clumps or patches should have more fine-scale genetic structure than species whose seeds are dispersed singly by mobile animals. Furthermore, due to the overlap of seed shadows, species with high adult densities should have less genetic structure than species with lower densities. Allozyme analyses of three tropical tree species belonging to the moist tropical forest of Barro Colorado Island, Republic of Panama, were used to describe variation in the scale and intensity of genetic structure within their populations. The genetic structure of seedlings and immature trees in the low-density, wind-dispersed species (*Platypodium elegans*) was the coarsest and strongest whereas genetic structure in a population of *Swartzia simplex* var. *ochnacea* (high density, bird-dispersed) was both the finest and the weakest. The genetic structure of *Alseis blackiana*, a high-density, wind-dispersed species was intermediate in both degree and scale. In *P. elegans* and *A. blackiana*, which had 'J' shaped size distributions, the significant genetic structure seen in the smaller and intermediate diameter classes disappeared in the largest diameter class. The loss of genetic structure was not observed in *S. simplex*, a species with a more even size distribution.

Nomenclature: follows Croat, T. B. 1978. Flora of Barro Colorado Island. Stanford University Press, Stanford. Ca. 943 pp.

Introduction

Seed dispersal patterns can shape the genetic composition and structure of plant populations. Species with limited seed dispersal are likely to have considerable genetic heterogeneity among patches of new seedlings while species with more extensive seed dispersal should have less spatial genetic structure (Fleming & Heithaus 1981; Howe 1989; 1990). Genetic structure is also affected by the interaction of seed dispersal with other ecological and genetic processes. Seed deposition patterns, pollen dispersal, adult densities, microhabitat selection and several aspects of the recruitment ecology of species could have significant effects on the patterning of genetic variation within populations (Howe & Smallwood 1982; Howe *et al.* 1985; Hamrick & Loveless 1986; Loveless 1991).

Sexually reproducing plants disperse genes in

two ways (Levin 1981). Male gametes are dispersed twice; from the pollen parent to the maternal parent via pollen and from the maternal parent as part of the genetic complement of the embryo. Maternal gametes, in contrast, move only once via seeds. The potentially greater variance in the dispersal distance of pollen relative to seeds may, in part, explain why knowledge of a species' pollination system allows relatively accurate predictions concerning the distribution of genetic diversity within and among plant populations (Table 1A). An equivalent understanding of seed dispersal mechanisms provides less satisfactory predictions of the distribution of genetic diversity (Table 1B). The range of among-population genetic heterogeneity (F_{ST}) among seed dispersal categories is not as large and the mean F_{ST} values are not consistent with the species' apparent potential for seed dispersal. In particular, species with animal dispersed seeds have higher mean levels of among population genetic heterogeneity than their potential for long-distance seed dis-

Table 1. The distribution of allozyme variation at polymorphic loci within and among populations of plant species classified according to their breeding system and their seed dispersal mechanism. From Hamrick & Godt (1989).

Categories	N[a]	Mean no. populations	Mean no. loci	H_T	H_S	F_{ST}
A. Breeding system				NS	***	***
Selfing	78	20.3 (3.8)[c]	16.2 (0.7)	0.334ab[b] (0.016)	0.149c (0.016)	0.510a (0.035)
Mixed-animal	60	8.9 (2.1)	14.4 (0.8)	0.304b (0.022)	0.221b (0.017)	0.216b (0.024)
Mixed-wind	11	10.0 (3.1)	12.5 (3.6)	0.378a (0.057)	0.342a (0.054)	0.100c (0.022)
Outcrossing-animal	124	10.7 (2.1)	17.7 (0.7)	0.310b (0.010)	0.243b (0.010)	0.197b (0.017)
Outcrossing-wind	134	10.7 (1.6)	16.7 (0.9)	0.293b (0.011)	0.259b (0.011)	0.099c (0.012)
B. Seed dispersal				***	***	***
Gravity	161	10.1 (1.1)	16.9 (0.6)	0.306b (0.010)	0.207b (0.011)	0.277a (0.021)
Gravity-attached	11	29.2 (15.8)	18.6 (1.8)	0.211c (0.038)	0.171b (0.031)	0.124b (0.031)
Attached	52	20.8 (5.7)	16.5 (0.9)	0.325b (0.024)	0.236ab (0.021)	0.257ab (0.032)
Explosive	23	12.4 (2.7)	18.6 (1.4)	0.302b (0.021)	0.217b (0.023)	0.243ab (0.048)
Ingested	39	17.6 (6.5)	13.2 (1.1)	0.394a (0.020)	0.305a (0.022)	0.223ab (0.033)
Wind	121	8.7 (0.9)	16.6 (0.9)	0.292b (0.012)	0.241ab (0.011)	0.143ab (0.020)

[a] N, number of taxa; H_T, total genetic diversity; H_S, genetic diversity within populations; F_{ST}, proportion of the total diversity among populations. Levels of significance: *** $P<0.001$; NS, not significant.

[b] Means followed by the same letter in a column are not significantly different at the 5% probability level.

[c] Standard errors are in parentheses.

persal would indicate (Loveless & Hamrick 1984; Hamrick & Godt 1989; Hamrick *et al.* 1991).

The magnitude of spatial genetic heterogeneity resulting from seed dispersal depends on several factors: (1) The proportion of seeds that immigrate and contribute gametes to the next generation of an established population (i.e. gene flow) versus the proportion that colonize new habitats. Slatkin (1977), and more recently Wade & McCauley (1988) and McCauley (1991), have shown that the colonization of vacant habitats is not strictly analogous to gene flow among established populations. Thus, if most seed movement is among established populations there will be less genetic structure than if the majority of the seeds colonize newly available habitats; (2) The density of reproducing adults. If seeds from several adults colonize an open habitat, genetic structure would be less than if successful colonists had come from one or a few adults. The effect of low adult densities should be especially apparent for wind-dispersed species. Maternal plant density undoubtedly effects species with animal-dispersed seeds but may do so in less direct and predictable ways; (3) The foraging and deposition behavior of the seed dispersal agent. At one extreme are animals that forage within the canopy of a single individual and subsequently deposit several seeds (at least half-sibs) in a site suitable for seedling recruitment. High genetic heterogeneity among recruitment sites will result, but may be tempered if several independent deposition events occur. At the other extreme are seed dispersal agents that forage across several maternal individuals and deposit seeds individually across the landscape. This pattern of foraging and seed dispersal should produce less genetic structure among recruitment patches.

In this paper these predictions are tested by examining the fine-scale genetic structure of three tropical tree species with different combinations of seed dispersal mechanisms and population densities. The specific questions addressed include: (1) Do neighboring individuals have higher genetic correlations (i.e. more alleles in common) than individuals located farther apart? (2) Do species with the same seed dispersal mechanisms but with different adult densities maintain different levels of spatial genetic heterogeneity? (3) Do species with similar densities but different seed dispersal mechanisms have different levels of spatial heterogeneity? (4) Does the spatial genetic heterogeneity found in smaller size classes carry over into larger size classes?

Procedures

The study site is in the moist tropical forest of Barro Colorado Island, Republic of Panama. Barro Colorado Island (BCI) is the location of a field station managed by the Smithsonian Tropical Research Institute and of a 50 ha. Forest Dynamics Plot (FDP) established by S. P. Hubbell and R. B. Foster. Every individual on the FDP larger than 1.5 meters in height and 1 cm in DBH has been identified and its location mapped (Hubbell & Foster 1983; Hubbell & Foster 1990). Much of the research described below is located on the FDP because a 5 m × 5 m grid system allows efficient and accurate mapping of individuals. The study site for *Swartzia simplex* var. *ochnacea* was not on the FDP because the selected site had higher densities of this species.

The forest of BCI and the FDP is the location of a long-term study effort to describe the genetic structure (e.g. Hamrick & Loveless 1989) and the breeding structure (e.g. Hamrick & Murawski 1990) of several tropical tree species. We are currently studying the fine-scale genetic structure of four species. These species were chosen because they are relatively common on the FDP, their size structure is indicative of active seedling recruitment, they maintain high levels of allozyme polymorphism and they represent a variety of breeding systems, pollinator syndromes, life forms and seed dispersal mechanisms.

Alseis blackiana Hemsl

Alseis blackiana (Rubiaceae) is a canopy tree with bisexual flowers (Table 2). Little is known of its breeding biology but its small, protogynous flowers are probably pollinated by butterflies or small

Table 2. Characteristics of four tropical tree species studied for their fine-scale genetic structure.

Species	Family	Growth form	Breeding system	Pollinator	Dispersal agent
Alseis blackiana	Rubiaceae	Canopy tree	Bisexual	Bees, Butterflies	Wind
Brosimum alicastrum	Moraceae	Canopy tree	Dioecious	Wind?, Insects?	Bats, Arboreal mammals
Platypodium elegans	Fabaceae (Papil)	Canopy tree	Bisexual	Small bees	Wind
Swartzia simplex var. *ochnacea*	Fabaceae (Caesal)	Understory tree	Bisexual	Large bees	Birds

bees. *Alseis blackiana* flowers in April or May and its slender wind-dispersed seeds are released from January to March during the dry season. Augspurger (1986) estimated that the mean seed dispersal distance was 94 m with a wind speed of 1.75 m per second. *Alseis blackiana* is a common canopy tree on the FDP with more than 230 individuals greater than 32 cm DBH (Table 3). It also has a large number of smaller individuals distributed throughout the FDP.

A one-hectare study site was established on the FDP from which the basal diameter and map coordinates of approximately 1000 individuals of all size classes were obtained. Adult individuals (> 15 cm in diameter) were also sampled from a 20 m border surrounding the core study area (Fig. 1A). Leaf material was collected from each individual, freeze-dried on BCI and returned to the University of Georgia for electrophoretic analysis. Every individual was genotyped for 31 polymorphic loci.

Brosimum alicastrum (Pitt) C. C. Berg

Brosimum alicastrum (Moraceae) is a dioecious, insect-pollinated canopy tree (Table 2). It may flower from November to May but the majority of the flowering occurs from January to March. Fruits mature from May to October and are eaten by a variety of arboreal mammals. Monkeys have been observed to eat the outer layers of the fruit and to drop the seeds, often leaving a large accumulation of seeds under the crown of the maternal tree. Bats are thought to play a principal role in long-distance seed dispersal. *Brosimum alicastrum* occurs at moderate densities on the FDP (Table 3). It is somewhat unusual in that it does not show a strong 'J' shaped age distribution in the sapling stages (Table 3). Rather, individuals in the 1, 2, and 4 cm diameter classes occur at nearly equal densities.

Since *Brosimum* has relatively few individuals on the FDP we have begun to genotype all tagged individuals. Unfortunately, this work is not yet complete. We have systematically searched 60 m × 60 m areas around six known female trees to determine seedling densities. The location and size of each seedling was recorded. In order to compare levels of seedling recruitment due to localized seed dispersal with recruitment due to long-distance seed dispersal, four 40 m × 40 m plots, established at least 60 m from known female trees, were similarly searched for seedlings.

Table 3. Size distribution of four tropical tree species located on the Forest Dynamics Plot on Barro Colorado Island. Data are based on the 1985 census (Foster & Hubbell, unpubl. data).

Species	Diameter class (cm)								Total
	1	2	4	8	16	32	64	128	
Alseis blackiana	4137	2000	894	451	335	212	26	0	8055
Brosium alicastrum	193	208	257	137	45	32	25	1	898
Platypodium elegans	30	41	24	26	5	12	17	2	157
Swartzia simplex var. *ochnacea*	905	1000	688	174	20	0	0	0	2787

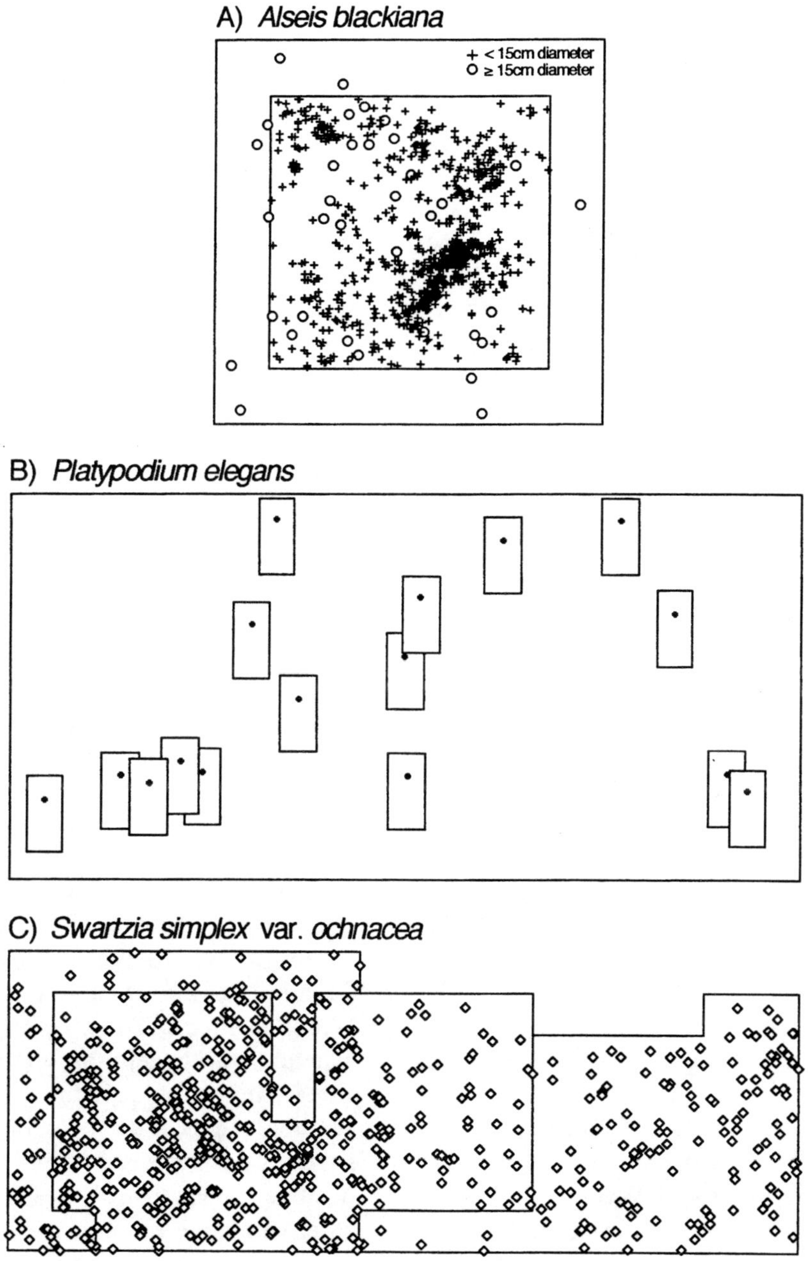

Fig. 1. The sampling areas used to study fine-scale genetic structure in three tropical tree species. A. *Aleis blackania.* The *Aleis* study plot consisted of a one hectare study site surrounded by a 20 m border. B. *Platypodium elegans.* The FDP is shown with the location of the 16 seedling shadow plots. C. *Swartzia simplex* var. *ochnacea.* The WH2 site is shown with the nested seedling plot. Seedlings and tagged plants are plotted.

Platypodium elegans J. Vogel

Platypodium elegans (Fabaceae) is a canopy tree with perfect flowers that are pollinated by small bees (Table 2). The large (mean fruit wt. = 2 g; fruit length = 10 cm) one-seeded samaras are dispersed up to 100 m by the prevailing north-east winds during the dry season (Augspurger 1983).

Mean dispersal distances of 34 m were estimated at wind speeds of 1.75 m per second (Augspurger 1986). Flowering is synchronous and may occur twice a year, with many more seeds being produced in some years than in others (Augspurger 1983; Augspurger & Kelly 1984). Flowering generally occurs from April to June and fruits are dispersed the following February to April. A large tree is capable of producing 4000–5000 seeds during each flowering period. When seed production is high, compact seed shadows are formed downwind from the seed parent (Augspurger 1983).

Every tagged *Platypodium* individual (> 1.0 cm DBH) on the FDP was collected and genotyped for 24 polymorphic loci as part of the original genetic survey of this species (Hamrick & Loveless 1989). To better analyze fine-scale genetic structure, we established a 50 m × 100 m plot around each of 16 reproducing adults on the FDP (Fig. 1B). These plots extended 25 m upwind and 75 m downwind from each maternal individual. Seedlings found within these plots were tagged, mapped, and analyzed for 20 polymorphic loci. Three 40 m × 40 m plots located at least 60 m away from mature trees were also searched to determine background levels of seedling recruitment outside of seed shadows associated with maternal trees.

Swartzia simplex var. *Ochnacea* (A. DC) Cowan

Swartzia simplex var. *ochnacea* is an understory tree or shrub. It is bisexual and self-compatible (Harcombe & Riggins 1968; Wyatt 1981) with large showy lemon-colored flowers that are pollinated by large bees (Table 2). *Swartzia* flowers in May or June and its fruits mature in December or January. There are typically two to four seeds per fruit and the seeds are bird dispersed. Flowering and fruiting is highly irregular and unpredictable. *Swartzia* is a common element in the understory of the FDP with nearly 3000 individuals above 1 cm in DBH (Table 3). Large individuals are rare but individuals 4cm in DBH or larger may flower and produce fruit.

The study site (WH2) was approximately 180 m × 70 m and was located in second-growth forest where *Swartzia* occurred at high densities (Fig. 1C). In site WH2 every *Swartzia* individual taller than 1.5 m was tagged, mapped, its basal diameter measured, and a leaf was collected for electrophoretic analysis. In a centrally located area of about 110 m × 50 m (exact shape subject to dense treefalls) every plant below 1.5 m was identified, tagged, measured, and collected. Each individual was analyzed for 19 polymorphic loci.

Data analysis

Since the four species had been previously studied, measures of genetic diversity among collection sites separated by 100–200 meters on the FDP were available (Hamrick & Loveless 1989). On the sites established to study fine-scale genetic structure, individuals representing different diameter classes were divided into subplots and, where sample sizes allowed, standard measures of genetic structure were calculated. These included Wright's (1951) F_{ST} and F_{IS} values. The F_{ST} value measures the excess of homozygotes relative to Hardy-Weinberg expectations in the population which are caused by allele frequency differences among population subdivisions (i.e. the Wahlund effect). Thus, F_{ST} measures the level of genetic heterogeneity among population subdivisions. The F_{IS} value is the deviation from Hardy-Weinberg expectations within each population subdivision averaged over all subdivisions. A positive F_{IS} value indicates an excess of homozygotes within the subdivisions under consideration. Statistical significance of F_{ST} values were examined by chi-square test: $\chi^2 = 2N\ F_{ST}\ (a - 1)$ with $df = (a - 1)\ (n - 1)$, where N is the total number of individuals in the population subdivision, a is the number of alleles per locus, and n is the number of population subdivisions (Workman and Niswander 1970). The statistical significance of F_{IS} was also examined by chi-square test: $\chi^2 = F_{IS}{}^2N(a - 1)$ with $df = (a - 1)$ (Li & Horowitz 1953). The F_{ST} and the F_{IS} values were compared among different diameter classes within species and were also compared among species.

Individuals within different diameter classes of each species were compared to determine if near-neighbors have more alleles in common than randomly chosen individuals. Three diameter classes were chosen for the three species with genetic data (*Alseis*, 0–2 cm, 2–8 cm and > 8 cm; *Platypodium*, 0–2 cm, 2–15 cm and > 15 cm; *Swartzia*, 0–2 cm 2–4 cm, and >4 cm.) For each diameter class the mean number of alleles in common per locus (NAC) between near-neighbors surrounding up to 100 different randomly chosen focal individuals was calculated by modifying the procedure developed by Surles *et al.* (1990). Where sample sizes were adequate this analysis was conducted for neighbor groups ranging from two to 60 individuals. A grand mean and variance for each near-neighbor group size was calculated across the 100 focal individuals. The mean NAC for a second set of individuals was calculated by randomly sampling from 2 to 60 individuals of the same diameter class for comparison. This procedure was replicated 100 times for each group size and mean NAC values were calculated. The NAC values generated by the random comparisons were compared to the near-neighbor NAC values. Ratios of the near-neighbor NAC value to the random NAC value were calculated for each diameter class. These ratios were necessary for interspecific comparisons since species with more genetic diversity will have lower NAC values.

Results

Alseis blackiana

A total of 999 individuals were sampled from the one hectare study site on the FDP. The 0–2 cm diameter class accounted for 794 individuals (0–1 cm = 517 individuals, 1–2 cm = 277 individuals) while the 2–8 cm and > 8 cm diameter classes had 168 and 37 individuals, respectively. Ten adults were also sampled from a 20 meter border surrounding the study plot. Distinct clumps of individuals in the smallest diameter class were found in a central location on the plot with a sparse canopy and were not close to any adults. Smaller clusters of individuals in the 0–2 cm and 2–8 cm diameter classes occurred on other parts of the study site (Fig. 1A).

Measures of genetic diversity were calculated at various spatial scales on the FDP and on the one hectare study site. The four sample sites on the FDP consisted of individuals greater than 1 cm DBH and were separated by approximately 200 m (Hamrick & Loveless 1989). The F_{ST} value found among these four sites was 0.034 (Table 4), indicating that moderate but significant heterogeneity ($\chi^2_3 = 14.28; P < 0.001$) in allele frequencies occurs at this spatial scale (Table 4). The mean F_{IS} value within each sample site (Table 5) was 0.104 ($\chi^2_1 = 2.34$; $P < 0.25$). Since these sample sites included all diameter classes greater than 1 cm and covered rather large areas, the F_{IS} value includes, in addition to inbreeding, a Wahlund effect among diameter classes and among spatial subdivisions within each sample site.

On the one hectare study site, different subdivision sizes were used for the three diameter classes to insure that the F_{ST} and F_{IS} values were not affected by small sample sizes. A minimum of 15 individuals were required for a subdivision to be included for analysis. Individuals in the 0–2 cm diameter class were assigned to 20 m × 20 m subdivisions of the one-hectare study site to maximize the number of subdivisions included in the

Table 4. Levels of genetic diversity among population subdivisions (F_{ST}) of various spatial scales. The F_{ST} values on the FDP represent collection sites separated by approximately 100 m (from Hamrick & Loveless 1989). The study sites were subdivided at different spatial scales for the different diameter classes (I, II, III). See the text for the actual spatial scales and diameter classes used for each species.

Species	FDP	Study site		
		I	II	III
Alseis blackiana	0.034	0.041	0.027	0.015
Brosium alicastrum	0.050	–	–	–
Platypodium elegans	0.051	0.091[1]	–	–
Swartzia simplex	0.037	0.021	0.019	0.031

[1] f_{ST} among seedling shadows.

Table 5. Deviations from Hardy-Weinberg expectations within population subdivisions (F_{IS}) of various spatial scales. The F_{IS} values from the FDP are for collection sites separated by approximately 100 m. The study sites were subdivided at different spatial scales for the different diameter classes (I, II, III). See text for the actual spatial scales and diameter classes used for each species.

Species	FDP	Study site		
		I	II	III
Alseis blackiana	0.104	0.048	0.054	− 0.014
Brosium alicastrum	0.120	–	–	–
Platypodium elegans	0.092	0.012[1]	–	–
Swartzia simplex	0.161	0.103	0.010	0.043

[1] F_{IS} within seedling shadows.

analyses. The F_{ST} value among the 14 subdivisions containing a minimum of 15 individuals was 0.041 indicating that highly significant heterogeneity ($\chi^2_{13} = 57.4$, $P < 0.001$) in allele frequencies occurred among these arbitrarily defined subdivisions. The F_{IS} value over all loci was 0.048 ($\chi^2_1 = 1.61$; $P < 0.25$). To obtain adequate sample sizes for the 2–8cm diameter class, the one hectare plot was divided into four 50 m × 50 m subdivisions. The F_{ST} value among these subdivisions was 0.027 ($\chi^2_3 = 4.07$; $P < 0.05$) and the F_{IS} value over all loci was 0.054 ($\chi^2_1 = 0.49$; $P < 0.50$). The study site was divided into two equal subdivisions for the analysis of the largest diameter class. The F_{ST} value among these two sites was 0.015 ($\chi^2_1 = 1.35$; $P < 0.25$) and F_{IS} averaged across all loci was 0.014 ($\chi^2_1 = 0.009$; $P < 0.99$). We also calculated F_{ST} among the three size classes. The F_{ST} value was low (0.004) but was significant ($\chi^2_2 = 7.26$; $P < 0.05$) because of the large sample sizes involved. The mean F_{IS} value within each diameter class averaged across all loci was 0.083 ($\chi^2_1 = 6.94$; $P < 0.010$). The significance of the F_{IS} values indicates that there is a significant heterozygote deficiency within the diameter classes. However, since each diameter class covered the one-hectare plot, the F_{IS} values include a significant spatial Wahlund effect. The F_{IS} value calculated for the spatial subdivisions of each age class is largely free of this confounding effect and is more representative of the true inbreeding coefficient.

In the 0–2 cm diameter class the number of alleles in common (NAC) among nearest neighbors (1.399) was significantly higher ($P < 0.05$) than the value for randomly chosen individuals (Fig. 2A). The NAC values drop sharply as group size increases from 4 to 20 individuals. In near-neighbor groups of 20 to 60 individuals, NAC values are nearly equal. Additional examination indicated that the near-neighbor NAC value converges with the random NAC value at a group size of approximately 240 individuals. The ratio of the near-neighbor NAC value to the random NAC value for group size 2 was 1.032 and decreased to 1.020 at group size 10 and to 1.018 at group size 60 (Table 6). In the 0–2 cm diameter class the mean distance of neighbors from the focal individual was approximately 1m for group size 2, increased steeply to approximately 5 m for group size 20 and then increased more slowly to 7.6 m for group size 60 (Fig. 2A).

The NAC value for near-neighbor individuals in the 2–8 cm diameter class was 1.400 for a group size two, and decreased sharply to 1.358 for group size 20 (Fig. 2B). For group size 60 the NAC value was 1.338. The random NAC values across all group sizes in this diameter class (1.335) were somewhat lower than that for the 0–2 cm class. By group size 60 the near-neighbor NAC

Table 6. The ratio of near-neighbor NAC values to random NAC values for different diameter classes and group sizes.

Species	Diameter class (cm)	Group size			
		2	10	20	60
Alseis blackiana	0–2	1.032	1.040	1.020	1.019
	2–8	1.040	1.034	1.025	1.002
	>8	1.018	0.992	0.994	0.998
Platypodium elegans	0–2	1.071	1.042	1.031	1.001
	2–15	1.064	1.036	1.017	1.000
	> 15	1.016	0.991	1.001	–
	seedlings	1.049	1.079	1.060	1.062
Swartzia simplex	0–2	1.026	1.015	1.003	0.998
	2–4	1.031	1.006	1.004	1.003
	>4	1.037	1.013	1.011	1.010

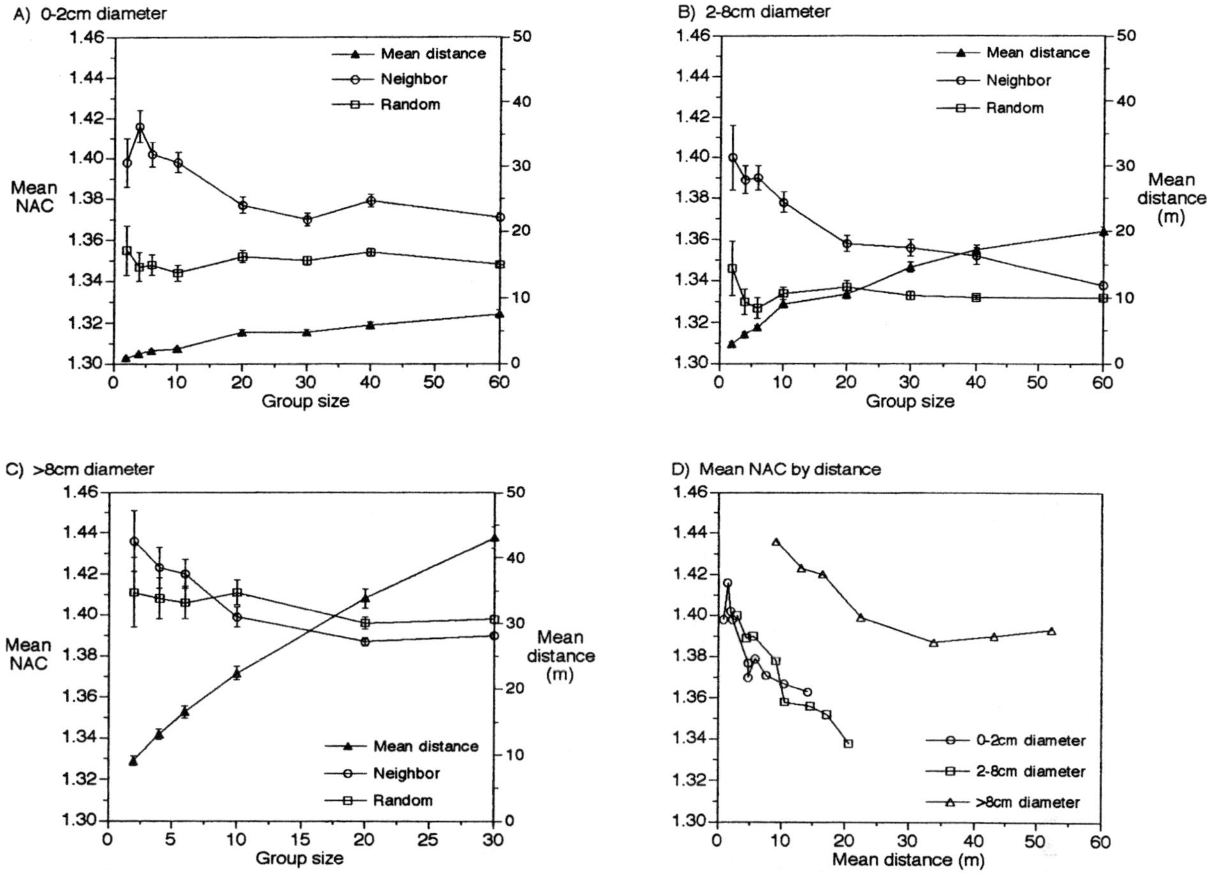

Fig. 2. The relationship of near-neighbor and random NAC values for different group sizes or spatial separation of *Alseis blackiana* individuals belonging to three diameter classes. A. 0–2 cm. B. 2–8 cm. C. > 8 cm. D. The NAC values for the three diameter classes versus distance among individuals being compared. Vertical bars represent ± one standard error.

had nearly converged with the random NAC value. The ratio between the near-neighbor NAC and the random NAC was 1.040 for group size 2, 1.025 at group size 20, and 1.002 at group size 60. The mean distance separating an individual at the center of a group from its near-neighbors was about 3 m for group size 2, 10 m for group size 20 and nearly 21 m for group size 60 (Fig. 2B).

For the largest diameter class (> 8 cm) groups of more than 40 individuals could not be examined because of the limited number of individuals. The near-neighbor and random NAC values for this diameter class were consistently higher than values for the same group size in the smaller diameter classes (Fig. 2C). Nearest neighbors had NAC values of 1.436 which decreased to approximately 1.386 at group size 20 and 1.393 at group size 40. The NAC value dropped sharply from group size 2 to group size 10 and then leveled off at values somewhat below the random values. The ratio between random and neighbor NAC values was 1.018 for group size 2 and decreased to 0.994 for group size 20 and to 0.998 for group size 40. The mean distance separating a central individual from its neighbors was 10 m for the nearest-neighbor comparisons and increased to 43 m when 30 individuals were compared (Fig. 2C).

NAC values were compared among the three diameter classes as a function of the mean distance separating individuals in a near-neighbor group from its central individual (Fig. 2D). The NAC values for the 0–2 cm and the 2–8 cm diameter classes decreased from approximately

1.417 and 1.400 to 1.364 and 1.338, respectively, as distance among neighbors increased (i.e. increasing group sizes). The curves for the two smaller diameter classes were nearly identical. Although mean distances were much larger for the largest diameter class and the NAC values were also higher, the shape of the curve was similar to that for the smaller diameter classes until approximately 20 m. Beyond 20 m there was no relationship between NAC and mean distance among individuals.

Brosimum alicastrum

The only data on genetic differentiation presently available for *Brosimum* comes from the original genetic surveys on the FDP by Hamrick & Loveless (1989). The level of genetic differentiation (F_{ST}) among the four sample sites on the FDP was 0.050 ($\chi^2_3 = 28.80$; $P < 0.001$). The mean F_{IS} value across all loci was 0.120 ($\chi^2_1 = 4.15$; $P < 0.05$). This F_{IS} value includes a Wahlund effect due to spatial and temporal genetic heterogeneity.

In order to compare recruitment due to localized versus long distance seed dispersal, the density of seedlings in 60 m × 60 m plots centered on six female individuals were compared to four 40 m × 40 m plots established 60 m away from known females. The density of seedlings around the maternal trees ranged from 0.0 to 1.83 seedlings per 100 m^2 with a mean of 0.99 (sd = 0.30) per 100 m^2. The range for the four 40 m × 40 m plots was 0.06 to 1.00 individuals per 100 m^2 with a mean of 0.47 (sd = 0.23) individuals per 100 m^2. The mean seedling densities of maternal and isolated plots are not significantly different due to high among plot heterogeneity in seedling numbers.

Platypodium elegans

Seed shadows of 16 mature trees averaged 1.36 (sd = 0.61) seedlings per 100 m^2. There was extensive heterogeneity among adult trees with no seedlings found associated with one adult, while 9.2 seedlings per 100 m^2 were found around the adult with the highest seedling density. No *Platypodium* seedlings were found in three 40 m × 40 m plots established in areas 60 m from adult trees.

Genetic structure in *Platypodium* was measured at two spatial scales. First, genetic relationships between established individuals from throughout the FDP were compared. We also compared the genetic relationships of seedlings within and among the 50 m × 100 m plots centered around 16 adults. The level of genetic diversity among the three collection sites located on the FDP (F_{ST}) was 0.051 ($\chi^2_2 = 18.87$, $P < 0.001$) (Table 4). The mean F_{IS} value across all loci was 0.092 (χ^2_1 = 1.60; $P < 0.25$) (Table 5). Comparisons of genetic heterogeneity among the six seedling shadows with more than 15 seedlings produced a F_{ST} value of 0.091 ($\chi^2_5 = 174.72$; $P < 0.001$). The mean F_{IS} value within the seed shadow plots was 0.012 ($\chi^2_1 = 0.14$; $P < 0.750$).

In the 0–2 cm diameter class located throughout the FDP the number of alleles in common among near-neighbors (1.530) was higher than when random individuals were compared (1.426) (Fig. 3A). NAC values decreased as group sizes increased to six individuals. Beyond group size 10 there was a gradual convergence of the neighbor values to the corresponding random values. The NAC ratio was 1.071 for group size 2, 1.031 at group size 20 and 1.001 at group size 60 (Table 6). The mean distance of neighbors from a central individual in the 0–2 cm diameter class was approximately 25 m for group size 2, increased to approximately 75 m for group size 10 and then increased to 275 m for group size 60 (Fig. 3A).

The NAC value for near-neighbors in the 2–15 cm diameter class was 1.505 for group size 2 and decreased to 1.415 for group size 20 (Fig. 3B). At group size 60 the NAC had dropped to 1.395, and had converged with the random comparisons. Random comparisons within this diameter class were equal to the random values for the 0–2 cm diameter class. The ratio between the near-neighbor NAC and the random NAC was 1.067 for group size 2 but only 1.017 at group size 20 (Table 6). The mean distance of neighbors from

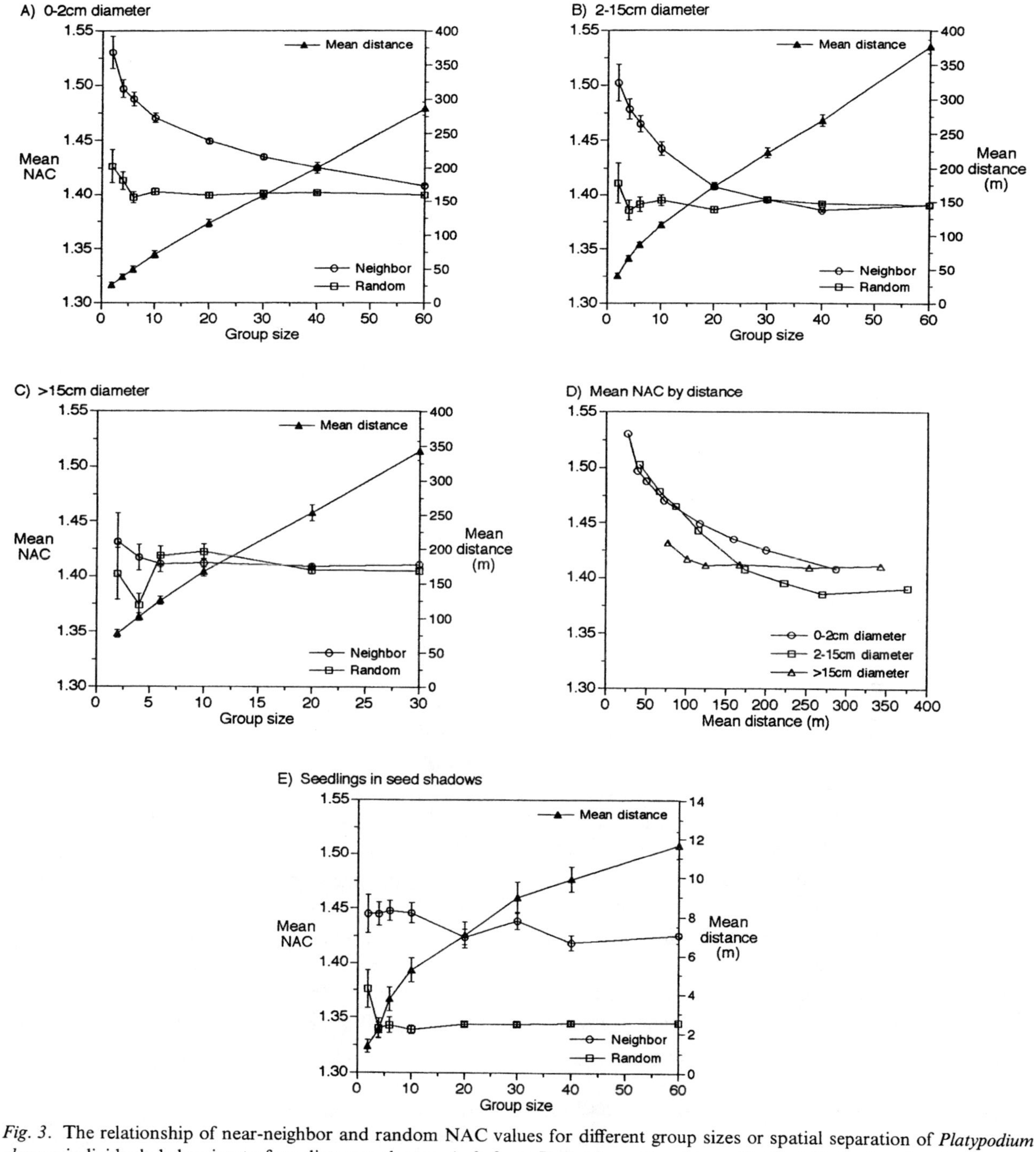

Fig. 3. The relationship of near-neighbor and random NAC values for different group sizes or spatial separation of *Platypodium elegans* individuals belonging to four diameter classes. A. 0–2 cm B. 2–15 cm C. > 15 cm D. The NAC values for the three diameter classes versus distance among individuals being compared. E. The NAC values for seedlings belonging to defined maternal seed shadows. Vertical bars represent ± one standard error.

a central individual was about 30 m for group size 2, 110 m for group sizes of 10 and 375 m for group sizes of 60 (Fig. 3B).

There were fewer group sizes available for the largest diameter class because of the limited number of individuals. Near-neighbor NAC values for this diameter class were consistently lower than values for the same group sizes in the smaller size classes (Fig. 3C). Nearest neighbors had NAC values of 1.430 which changed to approximately 1.424 for group size 10. The ratio between random and near-neighbor NAC values was 1.016 for group size two and changed to 1.001 at group size 20. The mean distance separating individuals from a central individual was 75 m for the nearest-neighbor comparisons and increased to 340 m at group size 30.

The NAC values of the 0–2 cm and the 2–15 cm diameter classes were similar to each other in having a close association with distance; as distance among neighbors increased (i.e. increasing group sizes) NAC values decreased from 1.525 to approximately 1.400 (Fig. 3D). In contrast, for the largest diameter class there was little relationship between NAC and distance.

NAC values were also calculated between individuals within the larger seed shadows (Fig. 3E). Near-neighbor comparisons were limited to seedlings belonging to the same seed shadow, while random comparisons were made between individuals selected from each of the seed shadows. The NAC values for the nearest-neighbor comparison (group size 2) was 1.446. At group size 20 and group size 60 the NAC values were 1.426 and 1.425 respectively. The NAC values across all group sizes for the random comparisons were much lower (1.345). There was little convergence towards the random NAC value by the neighbor comparisons. The ratio between the near-neighbor NAC and the random NAC was 1.049, 1.060, and 1.062 for seedlings with group sizes of 2, 20 and 60 respectively. The mean distance between neighbors was near 1 m for group size 2, 7 m for group size 20 and 11.8 m for group size 60. There was little relationship between distance and NAC within the seedling shadows (Fig. 3E).

Swartzia simplex var. *ochnacea*

The four sites analyzed on the FDP (Hamrick & Loveless 1989) consisted of individuals greater than 1cm DBH and were separated by approximately 200 m. The F_{ST} value among these four sites was 0.037 (Table 4) indicating that significant genetic heterogeneity ($\chi^2_3 = 73.04$; $P < 0.001$) occurs among different subdivisions of the FDP. The mean F_{IS} value within each site was 0.161 ($\chi^2_1 = 25.58$; $P < 0.001$) indicating a significant deviation from Hardy-Weinberg expectations (Table 5). This F_{IS} value includes, in addition to inbreeding effects, any spatial and temporal genetic heterogeneity that occurs within these collection sites.

A total of 686 individuals were sampled from the WH2 study site. The 0–2 cm diameter class consisted of 262 individuals while the 2–4 cm class and individuals larger than 4 cm included 266 and 158 individuals, respectively. There did not appear to be any distinct clumping of individuals (Fig. 1C). On WH2 different subdivision sizes were used for the three diameter classes to minimize the variance in F_{ST} and F_{IS} due to small sample sizes. For the 0–2 cm class, F_{ST} among the eight 30 m × 30 m subdivisions with more than 15 individuals (Table 4) was 0.021 ($\chi^2_7 = 11.00$; $P < 0.100$). The mean F_{IS} value was 0.103 ($\chi^2_1 = 2.78$, $P < 0.100$) (Table 5). The F_{ST} value for the eight 30 m × 30 m subdivisions with more than 15 individuals in the 2–4cm diameter class (Table 4) was 0.019 ($\chi^2_7 = 10.11$; $P < 0.100$). The mean F_{IS} for this diameter class was 0.010 ($\chi^2_1 = 0.03$; $P < 0.750$). To obtain adequate sample sizes in the largest diameter class (> 4 cm), WH2 was divided into six 36 m × 36 m subdivisions containing at least 14 individuals. The F_{ST} among the subdivisions was 0.031 ($\chi^2_5 = 9.80$; $P < 0.100$) while the mean F_{IS} within these subdivisions was 0.043 ($\chi^2_1 = 0.29$; $P < 0.500$). Values of F_{ST} (0.004; $\chi^2_2 = 5.82$; $P < 0.100$) and F_{IS} (0.069; $\chi^2_1 = 3.47$; $P < 0.100$) were also calculated among the three diameter classes. Since this F_{IS} was based on individuals from all sections of the plot it may contain a significant Wahlund effect.

The NAC among near-neighbors in the 0–2 cm diameter class (1.356) was higher than the NAC based on random comparisons (1.322) (Fig. 4A). The NAC values decreased rapidly as group size increased to 20. Above group size 20, the NAC

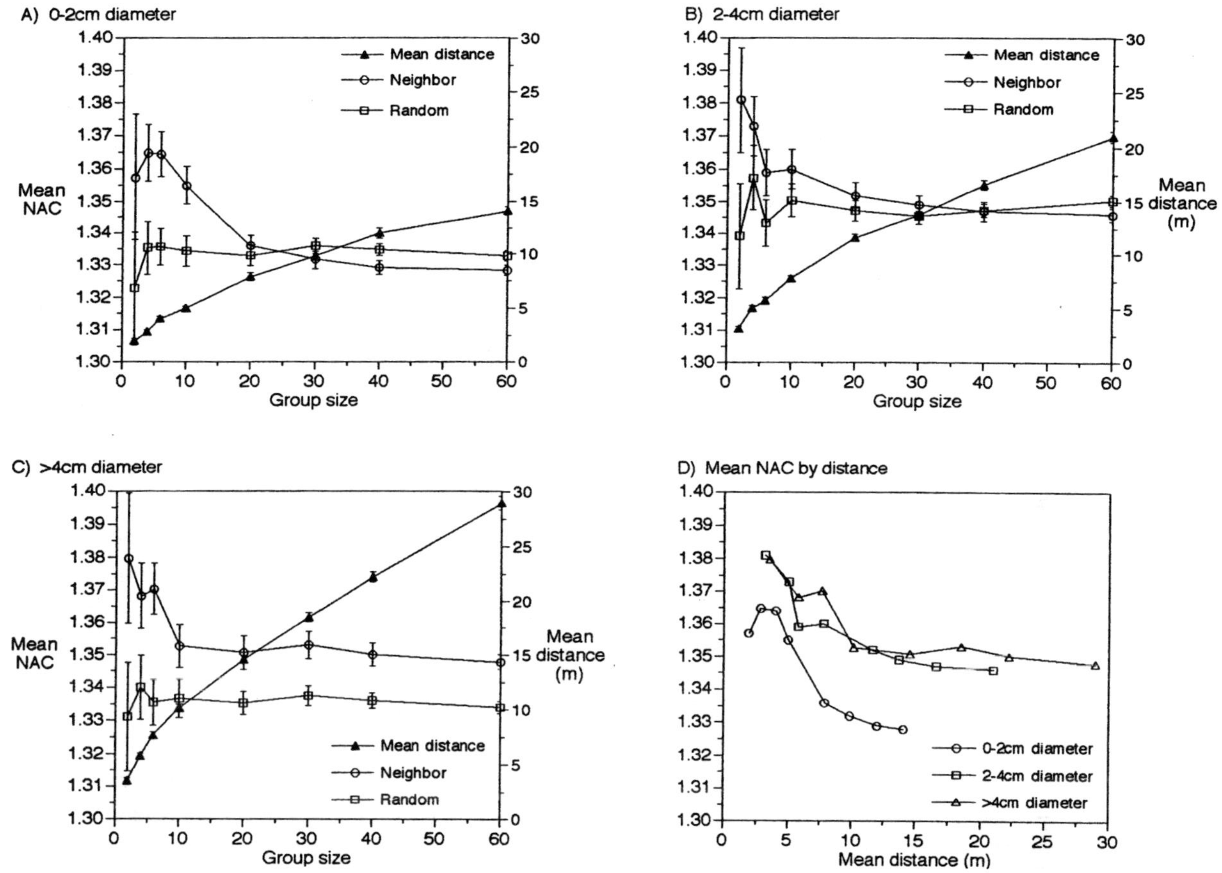

Fig. 4. The relationship of near-neighbor and random NAC values for different group sizes or spatial separation of *Swartzia simplex* individuals belonging to three diameter classes. A. 0–2 cm B. 2–4 cm C. >4 cm D. The NAC values for the three diameter classes versus distance among individuals being compared. Vertical bars represent ± one standard error.

among neighbors was equivalent to the random NAC. The ratio between the neighbor and the random NAC for group size 2 was 1.026 while the ratios for group sizes 20 and 60 were 1.003 and 0.998, respectively. The mean distance from a central individual to its near-neighbors was 2 m for group size 2 and increased steadily to 14 m for group size 60 (Fig. 4B).

The NAC for near-neighbor comparisons within the 2–4 cm diameter class was 1.381 for group size 2, decreased sharply to 1.359 for group size 10 then leveled off near the random NAC (1.350) after group size 20 (Fig. 4B). Ratios between near-neighbor and random NAC values ranged from 1.031 for group size 2, to 1.006 for group size 10 and 1.004 and 0.997 for group sizes 20 and 60, respectively. The mean distance among neighbors was about 3.5 m for group size 2, 12 m for group size 20 and 21 m for group size 60 (Fig. 4B). For the largest diameter class (>4cm) NAC values were highest (1.380) when nearest-neighbors were compared (Fig. 4C). At a group size of 10, NAC had decreased to 1.353. Beyond group size 10, NAC values level off, although they never approach the random values (1.335). The ratio between random and neighbor NAC values was 1.037 for group size 2 but decreased to 1.013 by group size 10. The mean distance separating near-neighbors was 4 m for group size 2 and steadily increased to 29 m by group size 60 (Fig. 4C).

The NAC values for the three diameter classes

were related to the mean distances among near-neighbors (Fig. 4D). The NAC for the smallest diameter class decreased from 2.5 m to approximately 10 m and then leveled off between 10 m and 15 m. The curves for the two larger diameter classes decreased between 2.5 m and 10 m and then leveled off. Fig. 4D also illustrates that NAC values of the 0–2 cm diameter class are somewhat lower than values for the two larger size classes.

Discussion

Significant spatial genetic heterogeneity was observed on the FDP for these tropical tree species. Significant or near significant levels of genetic heterogeneity were also found at much smaller spatial scales for the smaller diameter classes of *Alseis* and all diameter classes of *Swartzia*. The largest spatial genetic heterogeneity, however, was seen among the seedling shadows of *Platypodium* adults. This is not surprising since the seedling shadows should consist primarily of sibs. Although there was some overlap of seedling shadows (Fig. 1B) genetic heterogeneity among seedling plots was not greatly reduced since seedlings were primarily located near maternal individuals and away from plot margins.

The NAC analyses demonstrate that near-neighbors in the small and intermediate size classes of each species share more alleles than individuals located further apart. The most likely explanation for this observation is that spatially clustered individuals have at least one parent in common. Thus, even though there may be considerable mixing of seed shadows in the high density species (i.e. *Swartzia* and *Alseis*) near-neighbors tend to be more closely related than randomly paired individuals.

Species with similar densities but different seed dispersal mechanisms have somewhat different levels of fine-scale genetic structure. Although we couldn't compare the fine-scale genetic structure of *Platypodium* and *Brosimum*, the distribution of seedlings on the FDP indicates that the wind-dispersed *Platypodium* should have more genetic heterogeneity among seedling shadows than the more evenly distributed seedlings of the animal-dispersed *Brosimum*. The results of the NAC analysis for *Platypodium* confirms that near-neighbor pairs of seedlings within the seed shadow plots have a higher proportion of their alleles in common than expected by chance. We would predict that pairwise comparisons between the *Brosimum* seedlings should produce relatively lower NAC values due to the mixture of seedlings from different maternal trees.

At equal densities more structure should exist within populations of the wind-dispersed *Alseis* than for the bird-dispersed *Swartzia*. This is the case in the smaller diameter classes where *Alseis* has higher NAC ratios than *Swartzia* (Table 6). The NAC values for *Swartzia* drop more quickly with increasing group sizes, indicating either that patches of relatives contain fewer individuals or are more overlapping in *Swartzia* than in *Alseis*.

Comparisons of species with similar seed dispersal mechanisms but different densities are also consistent with expectations. At the spatial scale of the FDP the two species with lower densities, *P. elegans* and *B. alicastrum*, have higher F_{ST} values indicating that they have more genetic heterogeneity among collection sites separated by 100–200 m than the more continuously distributed *A. blackiana* and *S. simplex*. This heterogeneity may be due to the lower number of *Platypodium* and *Brosimum* adults that contribute genes to the different sections of the FDP.

At a smaller scale the existence of patches of related individuals is not surprising for the low density species (*Platypodium*) but is more unexpected for species with several adults per hectare (*Alseis* and *Swartzia*). Evidently small patches of related individuals exist in the smaller diameter classes of these high density species even though there must be considerable overlap of the seed shadows. In the two species with wind-dispersed seeds, the lower NAC ratios of *Alseis* relative to those of *Platypodium* indicate that patches of *Alseis* seedlings are derived from more than one maternal individual. The potential for *Alseis* seeds to move more than 100 m (Augspurger 1986) makes the mixture of seed shadows likely. The genetic structure seen in the seedling (0–2 cm) diameter class of *Alseis* carried over into the sapling diameter class (2–8 cm). The chief difference in the genetic structure of these two diameter classes was that the NAC values of the sapling class decreased at lower group sizes and that the mean distance between near-neighbors was greater. This is probably due to the loss of individuals from the family patches as the seedling cohorts thin. When *Platypodium* and *Alseis* reach the largest diameter class most of the genetic structure observed in the seedlings has been lost. This is almost certainly due to the disappearance of the patch structure in the large diameter classes of these species; as natural demographic processes occur, only one (or perhaps none) of the members of a seedling patch survive to the largest size class. As a result, the spatial distribution of individuals becomes more regular and genetic structure disappears.

A somewhat different picture is seen for *Swartzia*. In this species the largest size class (>4 cm) retains substantial fine-scale genetic structure. It is not clear why *Swartzia* behaves differently from *Platypodium* and *Alseis* but it may be due to the more even distribution of individuals within the three diameter classes. The lack of a 'J' shaped size distribution in WH2 may indicate that there is less mortality within seedling patches. As a result, the original family structure may be maintained in the larger diameter classes.

In *Alseis* and *Swartzia* the larger diameter classes have higher NAC values for both the near-neighbor and the random comparisons. There are at least two factors that could cause NAC to increase with size. First, if genetic diversity within the study site decreased in the older age classes both the neighbor and the random NAC values would increase. Second, if there was uniform selection for certain multilocus genotypes both NAC values would increase. For *Swartzia* and *Alseis* there is no indication that genetic diversity is lower or that there is uniform multilocus selection in the large diameter class. There is however, evidence that the larger diameter classes have a higher proportion of heterozygous individuals. Inbreeding coefficients (F_{IS}, Table 5) are largest in the smallest diameter class of both species. In *Alseis* the largest size class has the highest NAC and lowest F_{IS} values. For *Swartzia* the major difference in F_{IS} values comes between the smallest and the two larger diameter classes (Table 5). This is also where differences in NAC values occur. We conclude, therefore, that increases in heterozygosity associated with size has produced an increase in the NAC value.

The calculation of the number of alleles in common between pairs of individuals has proved to be a sensitive method to measure fine-scale genetic structure. The NAC value not only provides an estimate of genetic similarities between individuals but NAC can also be used to determine the number of individuals within a patch of related individuals and the distances between these individuals. When the number of individuals in any diameter class is large our protocol provides an accurate estimate of the difference in genetic similarity between near-neighbors and randomly chosen individuals.

There are, however, additional ways that the NAC procedure can be used to describe fine-scale genetic structure. In this paper we compared randomly chosen central individuals with their 1, 3, 5, 9 etc nearest-neighbors. By including the NAC comparisons from the smaller group sizes in the estimates of the NAC for the larger group sizes there is a carryover effect that overestimates mean patch sizes. Actual patch sizes could be better defined by comparing the central individual with its five nearest neighbors, then with its 6–10 nearest neighbors etc. Also, it would be useful to convert NAC ratios into measures of genetic relatedness. This was done by Surles *et al.* (1990) for open-pollinated families of *Gleditsia triacanthos* and *Robinia pseudoacacia*. Since the expected values for half- or full-sibs vary according to the genetic diversity in the population these values should be calculated separately for each population. Nevertheless, it appears that the NAC procedure has considerable potential as a measure of multilocus genetic structure in plant populations.

Concluding remarks

Considerable fine-scale genetic structure exists in these tropical tree populations. Furthermore, the magnitude and spatial distribution of genetic structure is related to the seed dispersal mechanisms and adult densities that characterize each species. Although it is dangerous to generalize from a sample of one species per density and seed dispersal category, our results indicate that species with wind-dispersed seeds and with lower densities develop more genetic structure in their seedlings than species with animal dispersed seeds or higher densities. The effects of seed dispersal on the establishment of genetically related near neighbors (i.e. half- and full-sibs) has implications for demographic and reproductive processes. For example, in species whose seed dispersal mechanisms promote the development of strong patch structure, competition for water, nutrients, and light will often be among related individuals. In addition, the spread of pathogens among susceptible seedling cohorts may be facilitated by the short distances separating related individuals (Augspurger & Kelly 1984). In species where fine-scale genetic structure established during seed dispersal persists into the adult generation (i.e. *Swartzia*), the likelihood of inbreeding will be increased (Hamrick & Loveless 1986). Biparental inbreeding should be lower in species where patch structure deteriorates with age (i.e. *Alseis* and *Platypodium*). Analyses of the breeding structure of tropical trees (e.g. Hamrick & Murawski 1990) coupled with analyses of fine-scale genetic structure should greatly enhance our understanding of how demographic and evolutionary processes act to produce the next generation of reproductive adults.

Acknowledgements

We wish to thank the Smithsonian Tropical Research Institute for the use of their facilities on BCI. Thanks are also due to Steve Hubbell and Robin Foster for all the help and encouragement they have given over the years. Sue Sherman-Broyles gave valuable technical assistance during the electrophoretic analyses. A. Schnabel and M.D. Loveless helped with the collection of *Alseis blackiana*. D. Santamaria, R. Peréz and C. Chung provided able field assistance. Funds were provided by a grant from the Mellon Foundation to J.L.H. and by NSF grants BSR 860083 and BSR 8918420.

References

Augspurger, C. K. 1983. Offspring recruitment around tropical trees: Changes in cohort distance with time. Oikos 40: 189–196.

Augspurger, C. K. & Kelly, C. K. 1984. Pathogen mortality of tropical tree seedlings: Experimental studies of the effects of dispersal distance, seedling density and light conditions. Oecologia 61: 211–217.

Augspurger, C. K. 1986. Morphology and dispersal potential of wind-dispersed diaspores of neotropical trees. Am. J. Bot. 73: 353–363.

Fleming, T. H. & Heithaus, E. R. 1981. Frugivorous bats, seed shadows, and the structure of tropical forests. Biotropica 13(Supl): 45–53.

Hamrick, J. L. & Godt, M. J. W. 1989. Allozyme diversity in plants. Pp. 43–63. In: Brown, A. H. D., Clegg, M. T., Kahler, A. L. & Weir, B. S. (eds.). Population Genetics, Breeding and Germplasm Resources in Crop Improvement. Sinauer Press. Sunderland, Mass.

Hamrick, J. L., Godt, M. J. W. & Sherman-Broyles, S. L. 1991. Factors influencing levels of genetic diversity in woody plant species. New Forests, in press.

Hamrick, J. L. & Loveless, M. D. 1986. The influence of seed dispersal mechanisms on the genetic structure of plant populations. Pp. 211–223. In: Estrada, A. & Fleming, T. H. (eds.), Frugivores and Seed Dispersal. Dr. Junk Publications, Dordrecht, The Netherlands.

Hamrick, J. L. & Loveless, M. D. 1989. The genetic structure of tropical tree populations: Associations with reproductive biology. pp. 129–146. In: Bock, J. H. & Linhart, Y. B. (eds.), The Evolutionary Ecology of Plants. Westview Press, Boulder, Colorado.

Hamrick, J. L. & Murawski, D. A. 1990. The breeding structure of tropical tree populations. Pl. Sp. Biol. 5: 157–165.

Harcombe, P. & Riggins, R. 1968. Observations on the pollination biology of *Swartzia simplex*. Organization for Tropical Studies Course Book, Reproductive Biology in Tropical Plant Ecology. July-August 1968.

Howe, H. F. 1989. Scatter- and clump-dispersal and seedling demography: Hypothesis and implications. Oecologia 79: 417–426.

Howe, H. F. 1990. Seed dispersal by birds and mammals: Implications for seedling demography. Pp. 191–218 In:

Bawa, K. S. & Hadley, M. (eds.) Reproductive Ecology of Tropical Forest Plants. Man and the Biosphere Series, Volume 7. UNESCO and Parthenon Publishing Group, Paris.

Howe, H. F. & Smallwood, J. 1982. Ecology of seed dispersal. Ann. Rev. Ecol. Syst. 13: 201–228.

Howe, H. F., Schupp, E. W. & Westley, L. C. 1985. Early consequences of seed dispersal for a neotropical tree (*Virola surinamensis*). Ecology 66: 781–791.

Hubbell, S. P. & Foster, R. B. 1983. Diversity of canopy trees in a neotropical forest and implications for conservation. Pp. 25–41 *In* Sutton, L. C., Whitmore, T. C. & Chadwick, A. C. (eds.) Tropical Rain Forest: Ecology and Management. Blackwell Scientific Pub. Oxford.

Hubbell, S. P. & Foster, R. B. 1990. Structure, dynamics and equilibrium status of old-growth forest on Barro Colorado Island. Pp. 542–564. In: Gentry, A. H. (ed) Four Tropical Rainforests. Yale University Press, New Haven, Connecticut.

Levin, D. A. 1981. Dispersal versus gene flow in plants. Ann. Mo. Bot. Gard. 88: 233–253.

Li, C. C. & Horowitz, D. G. 1953. Some methods of estimating the inbreeding coefficient. Am. J. Hum. Genet. 5: 107–117.

Loveless, M. D. 1991. Isozyme variation in tropical trees: Patterns of genetic organization and processes influencing genetic structure. New Forests, in press.

Loveless, M. D. & Hamrick, J. L. 1984. Ecological determinants of genetic structure in plant populations. Ann. Rev. Ecol. Syst. 15: 65–95.

McCauley, D. E. 1991. Genetic consequences of local population extinction and recolonization. Trends Ecol. Evol. 6: 5–8.

Slatkin, M. 1977. Gene flow and genetic drift in a species subject to frequent local extinction. Theor. Pop. Biol. 12: 253–262.

Surles, S. E., Arnold, J., Schnabel, A., Hamrick, J. L. & Bongarten, B. C. 1990. Genetic relatedness in open-pollinated families of two leguminous tree species, *Robinia pseudoacacia* L. and *Gleditsia triacanthos* L. Theor. Appl. Genet. 80: 49–56.

Wade, M. J. & McCauley, D. E. 1988. Extinction and recolonization: Their effects on the genetic differentiation of local populations. Evolution 42: 995–1005.

Workman, P. L. & Niswander, W. C. 1970. Population studies on southwestern Indian tribes II. Local genetic differentiation in the Popago. Am. J. Hum. Genet. 22: 24–49.

Wright, S. 1951. The genetic structure of populations. Ann. Eugen. 15: 323–354.

Wyatt, R. 1981. Components of reproductive output in five tropical legumes. Bull. Torr. Bot. Club 108: 67–75.

Vegetatio **107/108**: 299–318, 1993.
T. H. Fleming and A. Estrada (eds). Frugivory and Seed Dispersal: Ecological and Evolutionary Aspects.

Seed rain and advanced regeneration in a tropical rain forest

M. Martínez-Ramos [1] & A. Soto-Castro
Centro de Ecología, Universidad National Autónoma de México, Ap. Post. 70–275, México, D.F. 04510, México

Keywords: Tropical rain forest, Seed dispersal, Advanced regeneration, Seed and seedling ecology

Abstract

By comparing seed rain, seedling and sapling community structures we assessed the possible role played by vertebrate seed dispersal as a structuring factor in advanced regeneration of closed-canopied sites in the tropical rain forest of Los Tuxtlas, Mexico. Seed weight, initial morphology of seedlings and species abundance were also analyzed to determine if these traits influenced the probability of establishment in the shade.

About half of the seed species falling in five closed forest sites (25 × 25 m) during one year came from fruiting trees growing within the sites (local seeds) and half from fruiting trees found outside the sites (immigrant seeds). Seeds of liana and upper-canopy species were over-represented among immigrant seeds compared with seeds of understory tree species. This probably reflects the activity of frugivorous arboreal mammals, bats, and birds. Species with immigrant seeds had both a lower abundance and a narrower spatial distribution than locally produced seeds. Therefore, immigrant seeds showed higher diversity values than locally produced seeds.

Average seed size and the proportion of epigeous seedlings were similar in local and immigrant species. Under closed-canopied sites, factors affecting community organization seem to operate selectively, favoring the establishment of large-seeded, local abundant species in the advanced regeneration. However, the fact that some saplings of immigrant species were found in the plots suggest that a slow species infiltration may be occurring leading to a slow shift in the advanced regeneration species composition. We propose that the influence of seed dispersal on advanced regeneration structure depends on the disturbance history of the patches where seeds land.

Nomenclature: Scientific names and authorities follow Ibarra & Sinaca (1987)

Introduction

During the last two decades, much attention has been paid to understand the role played by canopy gaps in the regeneration processes of tropical rain forest tree communities (Denslow 1987; Oldeman & van Dijk 1991). These studies have indicated that seedling and sapling banks established before the opening of a gap are important to the gap-phase regeneration of the so-called non-pioneer or persistent species (Sarukhán *et al.* 1985; Brokaw 1985; Uhl *et al.* 1988; Raich & Christiansen 1989). However, we are only beginning to understand how this advanced regeneration originates and becomes structured. Since the persistent trees constitute the principal component of diversity, structure and dynamics in these systems (Denslow 1987), this problem is an important ecological issue.

Seeds are a major source of propagules for rain forest regeneration. Seed rain falling beneath a mature tree is composed of seeds produced by the tree itself (local seeds) and by other trees (immigrant seeds). Locally produced seeds represent potential advanced self-regeneration whereas those seeds arriving via dispersal agents represent a potential advanced immigrant-regeneration. Although we can imagine that both types of regeneration take place, we don't know what proportion of the seedling and sapling banks found in closed-canopied sites originate from these two seed sources. In its final consequences, self-regeneration maintains a floristic forest mosaic whereas immigrant-regeneration may homogenize forest composition if seeds are widely dispersed or may create floristic heterogeneity if seed dispersal is patchy. Then, forest structure must be the result of the combination of these two regeneration alternatives and in order to better understand tropical forest structure, we must know the contribution of each alternative.

In tropical rain forests the main seed vectors are vertebrate frugivores (Terborgh 1990). Therefore, frugivory is expected to be an important factor in the organization of advanced regeneration. Under closed-canopies sites, low light availability (Pearcy 1990; Chazdon 1988), physical damage (Clark & Clark 1989) and natural enemies (Augspurger 1984; Dirzo 1984; Clark & Clark 1984; Schupp 1988a, b) are important mortality factors operating on seed and seedling populations. The probability of any species becoming established in advanced regeneration depends on how well its members can withstand mortality risks imposed by an environment that changes as plants develop from seed to later life-cycle stages. A complex of traits, including the initial amount of maternal resources (Foster, 1986; Howe 1990; Hladik & Miquel 1990), antiherbivore defenses (Dirzo 1987), carbon balance (Pearcy 1990; Chazdon 1988) and seedling size (Ng 1978; Sarukhán *et al.* 1984), determines this probability (Primack 1990).

The action of mortality factors as part of controlling or regulating population mechanisms, may have very different consequences to the organization of advanced regeneration and to the structure of the whole forest community (Connell *et al.* 1984; Hubbell & Foster 1986, 1990a). If populations are regulated by density or frequency-dependent mechanisms, many species can persist in equilibrium. In contrast, if populations are controlled by density-independent mechanisms, diversity will be the result of a balance between local species extinction and colonization rates, ruled, to a great extent, by random events (Hubbell & Foster 1986). Under these non-equilibrium circumstances, we can expect an unpredictable advanced regeneration structure and composition. In any community, however, these two sets of forces are present and their relative importance may change through space and time (Hubbell & Foster 1990a; Hubbell & Foster 1990b).

In 1988 we began a long-term study whose aims were to identify patterns, processes and mechanisms involved in the organization of advanced regeneration in the Los Tuxtlas rain forest. This paper presents data obtained during the first year. By comparison of the coexisting seed rain, seedling and sapling community structures we explore: i) the contribution of immigrant seeds to the seed rain falling in closed canopy sites, ii) the contribution of immigrant seeds to the seedling and sapling communities in those sites and iii) the existence of community organization mechanisms during advanced regeneration. Particularly, we take up the following questions: To what degree do dispersed seeds contribute to the species richness of the seed rain? Have seeds of trees maturing at different canopy heights the same chance to be dispersed? What is the contribution of these tree groups to the diversity of the seed rain and advanced regeneration communities? How does the initial community structure, represented by the seed rain, change as individuals develop to further life-cycle stages (*i.e.* seedlings and saplings)? What factors affect such transitions and what is their possible role in the organization of the advanced regeneration community? Although our main concern is centered on the tree community because ecological knowledge available for lianas is very limited (Putz 1983), we decided to include this life form in this study. In concluding, we discuss the possible roles played by the relationship between seed dispersal and forest dynamics in forest community organization.

Material and methods

This study was carried out at the Los Tuxtlas tropical field station in south-eastern Mexico (longitude 95°04′ latitude 18°34′ N). This locality represents the northernmost neotropical rain forest (Dirzo & Miranda 1991). Average canopy height is about 25 m, and tree community diversity is relatively low compared with other tropical rain forests (Bongers *et al.* 1988). However, more than 250 tree species, and nearly 900 vascular plants, have been found in the 700 ha preserve (Ibarra & Sinaca 1987). Information on the climate, geology and general ecology of the locality can be found elsewhere (Gómez-Pompa & Del Amo 1985).

We established five 25 × 25 m permanent study plots in relatively flat, closed canopy forest sites in May 1988. Plot size represented forest areas similar to those affected by the largest treefall gaps at Los Tuxtlas (Popma *et al.* 1988). In each plot, all tree and liana species greater than 1 cm in diameter at breast height (dbh) were recorded and mapped. We define saplings as trees with a dbh 1–2 cm. Sixty four uniformly spaced points were located at each site. We identified and recorded the species of all tree and liana crowns found over each point.

To measure seed rain, we placed twelve 1 m^2 seed traps randomly on each plot. The trap material was collected monthly, from June 1988 to May 1989, and all apparently healthy seeds larger than 1 mm in length were identified and counted. We considered the seed rain community to be those species recorded in the traps during the year period. Average seed dry weight per species was also determined (G. Ibarra & M. Martínez-Ramos unpubl.).

Seedling community structure was studied at sixteen 1 m^2 permanent quadrats, randomly established within each study plot. In June 1988, all seedlings and saplings ≤ 1 m tall were identified, counted and mapped in each quadrat. Hereafter, these plants will be referred to as old-seedlings. Every three months, from June 1988 to May 1989, all new seedlings were recorded and marked in the quadrats. Hereafter, these seedlings will be referred as recruited-seedlings.

Data analyses

Although we recorded species whose seeds come from fleshy and dry fruits, in the present paper we limited the data analyses to those species whose fleshy fruits or arilate seeds suggest seed dispersal by animals. This was done in order to assess the role that frugivory has as an factor influencing advanced regeneration. Furthermore, wind dispersed species represented a minor fraction ($<15\%$) of the total number of species recorded in the seed rain and its omission did not change the results obtained in the present study (Martínez-Ramos & A. Soto, unpubl.). Seed, seedling and sapling species were classified as belonging to i) trees or lianas, and coming from ii) a local or immigrant origin. Local species were those produced by mature plants whose crowns were found within the plot or touching the plot borders, while immigrant species were those for which crowns of parent plants were completely outside the plots. Therefore, we assumed that immigrant species were brought in by animal dispersers.

Seed, seedlings and sapling species were assigned to the following tree groups: Low-Canopy (LC = tree species maturing at heights lower than 10 m), Mid-Canopy (MC = tree species maturing at heights lower than 20 m) or Upper-Canopy (UC = species maturing at height taller than 20 m). The L group included liana species.

To determine whether rates of seed dispersal differ among species found at different canopy heights, we constructed the following hypothesis: The probability of detecting at least one seed at a given point in the forest produced by species found at different canopy height positions depends only on the proportion in which these species are represented in the entire tree community. To evaluate the null hypothesis we used i) the community structure data (woody plants ≥ 1 cm dbh) available for a 1 ha plot at Los Tuxtlas forest (Bongers *et al.* 1988) and ii) an intensive check-list of seeds and fruits produced by trees and liana species at the same forest (G. Ibarra & M. Martínez-Ramos unpubl.). From these two sources, only species that produce fleshy fruits or arillate seeds were considered. Tree and liana species recorded in the 1-ha plot on the one hand, and in the check-list on the other hand, were placed into the LC, MC, UC and L groups. The resulting proportions from the two sources were averaged to yield the expected random proportions for each plant group. The null hypothesis was tested using Chi-square tests, comparing the local and immigrant seed, seedling and sapling communities. Because of the small sample sizes available in the majority of the study plots, the five plots were pooled for the statistical analyses. We considered local species as those having either seeds, seedlings or saplings and mature individuals in at least one of the five study plots.

We calculated species richness (S = number of species in the sample), diversity (H' Shannon-Wiener Index) and equitability (J = H'/Ln (S)), by combining all plots. The calculations were made for: i) each plant group, ii) the entire tree community, and iii) the tree plus liana community. An index of species similarity was computed in order to estimate the probability that species recorded at a given life-cycle stage (seed, recruited-seedling, old-seedling or sapling) were also present in the next stage. This index measures the proportion of species shared by two samples (Greig-Smith 1983). We calculated species similarity for all tree species and for local and immigrant species separately.

An indicator of the seedling shade-tolerance was obtained by using i) the mean dry seed weight per species (Foster 1986), and ii) the initial morphology of seedlings. Initial morphology was classified into two categories: seedlings with photosynthetic cotyledons (PC-seedlings) and seedlings with cotyledons, hypocotyl and/or endosperm functioning as reserve organs (RC-seedlings). The first category corresponds to the epigeal germination type and the second one to the hypogeal, semi-hypogeal and durian germination types proposed by Ng (1978) and Hladik & Miquel (1990). Elsewhere, it has been shown that RC-seedlings, mostly emerging from large seeds, have a significantly higher survivorship than PC-seedlings under closed canopy conditions (Martínez-Ramos 1991).

Results

Forest community structure in the study plots

Two of the five study plots were located in successional forest patches ('building' *sensu* Whitmore 1989), and were dominated by the pioneer tree *Cecropia obtusifolia* (Moraceae). These patches were disturbed by the opening of tree-fall gaps eight and twenty years ago, respectively. The younger patch showed the highest frequency of small trees (≤ 5 cm dbh) of the five study plots. The twenty year-old patch showed highest frequencies of medium-sized trees (10–20 cm dbh). The other three plots were localized in forest patches in mature phase (*sensu* Whitmore 1989). Two of these patches were dominated by *Nectandra ambigens* (Lauraceae), the most abundant upper-canopy tree at Los Tuxtlas. The third represented a diverse patch with trees of *Ficus yoponensis* (Moraceae), *Pseudolmedia oxyphyllaria* (Moraceae) and *Cordia megalantha* (Boraginaceae) as the species with highest total community basal area and crown cover (Martínez-Ramos 1991). The frequency of seedlings and small saplings (≤ 1 m in height) in the mature patches was significantly higher than in the two building forest patches ($\chi^2 = 17.2$, DF = 4, $P < 0.01$; Table 1).

Notwithstanding the above differences among plots, community diversity was quite similar and high in all of the plots (S range: 26–34, H'range: 2.73–2.89; J range = 0.84–0.87) (Table 1).

Table 1. Tree community structure in the five study plots at Los Tuxtlas tropical rain forest. Patch age indicates the time elapsed since the last treefall detected in the plot (accordingly to Martínez-Ramos *et al.* 1988a). Seedling and sapling density refers to plants ≤1 m in height. Tree density, S, H′ and J refer to plants with dbh ≥ 1 cm. All values are given for 25 × 25 m plots.

	Plots				
	1	2	3	4	5
Patch age (yrs)	>35	>35	20	>35	8
Seedling & sapling density	7,092	13,767	3,963	11,458	2,593
Tree density	69	60	62	51	81
Number of species (S)	31	33	30	25	31
Shannon index (H′)	2.51	2.89	2.73	2.63	2.55
Equitability (J)	0.84	0.86	0.84	0.87	0.85

Null hypothesis community

One hundred and eleven tree and liana species, producing fleshy fruits or arillate seeds, occurred on the 1-ha plot, and 174 tree and liana species occurred in the check-list. There were no significant differences in species frequency between these two sources of data when species were classified in LC, MC, UP and L plant groups $\chi^2 = 2.76$, DF = 3, $P > 0.50$). When both sources were averaged, percentages per plant group were: LC = 40.6, MC = 15.7, UC = 27.3, L = 16.

Seed rain, advanced regeneration and the null dispersal hypothesis

Seeds

A total of 160 species in 59 families was recorded in the plots over the course of the study. Fifty two per cent (88 species) were tree and liana species that produce fleshy fruits or arillate seeds. The rest were species with dry fruits or that belong to other life forms (palms and herbs).

Of the fleshy-fruited tree and liana species, 54.6% of the species were classified as arriving from outside of the plots (Table 2). This value varied among height groups. Lianas and UC trees showed the highest percentages and LC trees the lowest, both at an individual plot level (Table 2) and when all plots were pooled (Fig. 1). Immigrant liana seeds, and to a lesser extend those produced by UC and MC trees, showed a significantly higher species frequency than expected by chance (Fig. 1). When the liana group was omitted from the analysis, then LC species become under-represented whereas UC and MC species show even higher frequencies than expected by chance ($\chi^2 = 6.0$, DF = 2, $P < 0.05$). The species frequency per height group in local seeds did not differ from those expected by chance (Fig. 1).

Recruited-seedlings

A total of 56 species was recorded as recruited-seedlings in the five plots. Immigrant seedling species accounted for 48.2 % of the total (Table 2). Immigrant liana seedlings tended to be over-represented and immigrant LC seedlings under-represented (Table 2). Pooling all plots, species frequency of immigrant liana and UC seedlings were significantly higher than that expected by chance (Fig. 1). Species frequency per plant group in locally recruited-seedlings reflected the community structure of the forest (Fig. 1).

Old-seedlings

In the old-seedling community, 60 tree and liana species were found in the five study plots. Immigrant species accounted for a significantly lower percentage ($\chi^2 = 3.29$, $P < 0.10$) than locally produced seedlings to the total number of species (Table 2) and, in contrast to the seed rain and recruited-seedling communities, species frequencies of local and immigrant height groups were quite similar within each plot and when all plots were combined (Table 2). Species frequency of

Table 2. Seed rain, recruited-seedling and old-seedling community species richness under five closed sites in the Los Tuxtlas rain forest, Mexico. LC = low-canopy (mature trees < 10 m height), MC = mid-canopy (10.1–20 m height), UC = upper-canopy (> 20 m height), L = lianas. I = immigrant species, L = local species. The I/T ratio measures the percentage of immigrant species out of the total species richness.

	Number of species by plot											
	1		2		3		4		5		All plots	
	I	L	I	L	I	L	I	L	I	L	I	L
1) Seed rain												
LC	5	1	2	5	4	7	4	5	9	1	8	12
MC	9	2	4	4	3	2	2	3	5	3	9	6
UC	14	5	13	6	14	6	11	8	8	6	13	14
L	12	2	8	0	14	1	8	4	13	3	18	8
Total	40	10	27	15	35	16	25	20	35	13	48	40
I/T(%)	80.0		64.3		68.6		55.6		72.9		54.6	
2) Recruited-seedling												
LC	2	4	2	3	1	0	2	5	1	2	5	10
MC	3	1	2	2	2	2	2	1	1	0	4	4
UC	1	4	3	3	5	5	5	4	2	2	9	10
L	4	2	2	0	6	0	3	3	0	1	9	5
Total	10	11	9	8	14	7	12	13	4	5	27	29
I/T(%)	47.6		52.9		66.7		48.0		44.4		48.2	
3) Old-seedling												
LC	9	8	6	9	5	6	6	5	6	5	9	13
MC	5	2	4	1	5	2	3	3	3	2	6	5
UC	4	5	4	4	3	4	6	4	3	6	4	10
L	4	5	5	3	5	2	8	3	7	2	4	9
Total	22	20	19	17	18	14	23	15	19	15	23	37
I/T(%)	52.4		52.8		56.3		60.5		55.9		38.3	

immigrant and local old-seedlings reflected the community structure of the forest (Fig. 1).

Saplings

We pooled our data on tree saplings for statistical analysis. For the liana group we did not take any sapling data. A total of 21 tree species was recorded in the five study plots. Immigrant species were the minor component of species richness in the community (28.6% of the total; $\chi^2 = 3.86$, DF = 1, P = 0.05). In the UC tree group no immigrant species were found. Combining immigrant and local species, LC saplings showed the highest species frequency (42.9% of the total) whereas UC saplings showed the lowest (Fig. 1). These frequencies were not statistically different from those expected under a random distribution (Fig. 1).

Changes in community structure among life-cycle stages

Species richness

A reduction in the number of species occurred from the seed stage to later stages (see Fig. 3). As

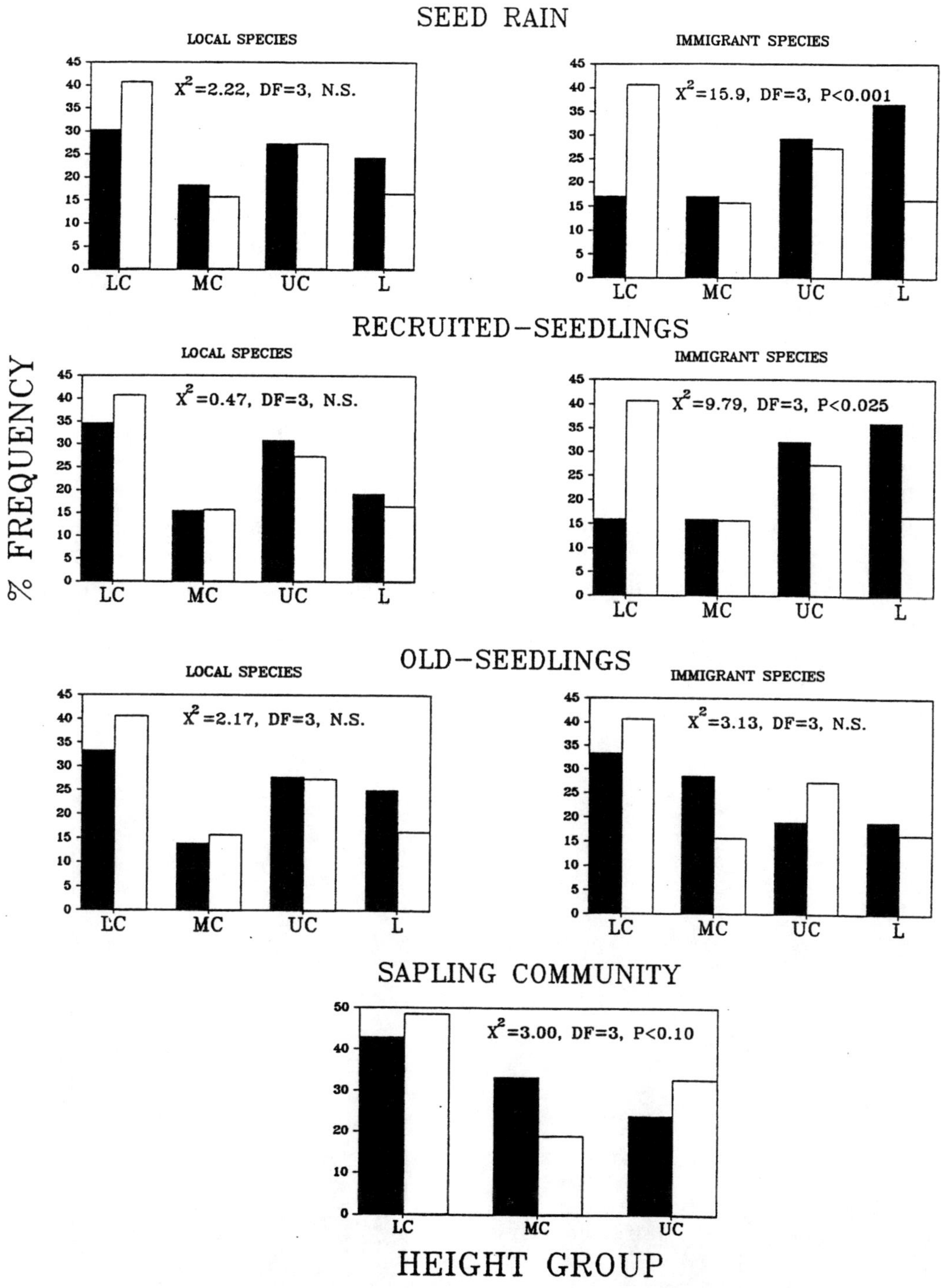

Fig. 1. Observed (filled bars) and expected (open bars) (under a dispersal null hypothesis) species frequencies per height group in local and immigrant species for: Seed rain, recruited-seedling, old-seedling and sapling communities in closed sites of Los Tuxtlas forest. Results of χ^2 analysis are indicated inside each graph. See text for further details.

a result, only 30 % of the total number of tree species recorded in the seed rain were also found in the sapling community (Fig. 2).

Species similarity between communities representing different life-cycle stages clearly showed lower values for immigrant species than for local ones (Fig. 2). This result suggests that transition probabilities among life-cycle stages are higher for locally produced seeds than for immigrant seeds.

Diversity: relative abundance among species

Associated with the decreasing species richness recorded from the seed rain to sapling communities, more abundant species suffered a strong thinning effect. Consequently community diversity and equitability increased along this sequence (Fig. 3). Immigrant species were less abundant compared to local species in all the life-cycle stages. A Chi-square analysis showed that local height groups in all life-cycle stages had a significantly higher frequency of common species (>4 individuals per species in the five plots) than did the immigrant groups ($P<0.01$). In terms of interplot frequency, immigrant species were also rare. Most immigrant species were found in only one of the five study plots. In contrast, most local species were widely distributed through the plots (Fig. 4).

The most abundant species had different community abundance positions in different life-cycle stages (Fig. 3). Small-seeded and light-demanding trees *Cecropia obtusifolia* (Moraceae), *Lunania mexicana* (Flacourtiaceae), *Ficus yoponensis* (Moraceae), and *Dendropanax arboreous* (Araliaceae), which were the most common species in the seed rain community, were rarely (*C. obtusifolia*, *D. arboreous*) or never recorded (*L. mexicana*, *F. yoponensis*) in later stages. This presumably reflects the inability of these species to establish in the shade. The reduction in abundance of these species increased diversity in the recruited-seedling community (Fig. 3). Other reasons appear to explain the changes in abundance that occurred in some persistent species. For example, the decrease in abundance of *Pseudolmedia oxyphyllaria* (Moraceae), a large seeded (average dry weight: 500 mg $seed^{-1}$) shade-tolerant tree and sixth in abundance in the seed rain community

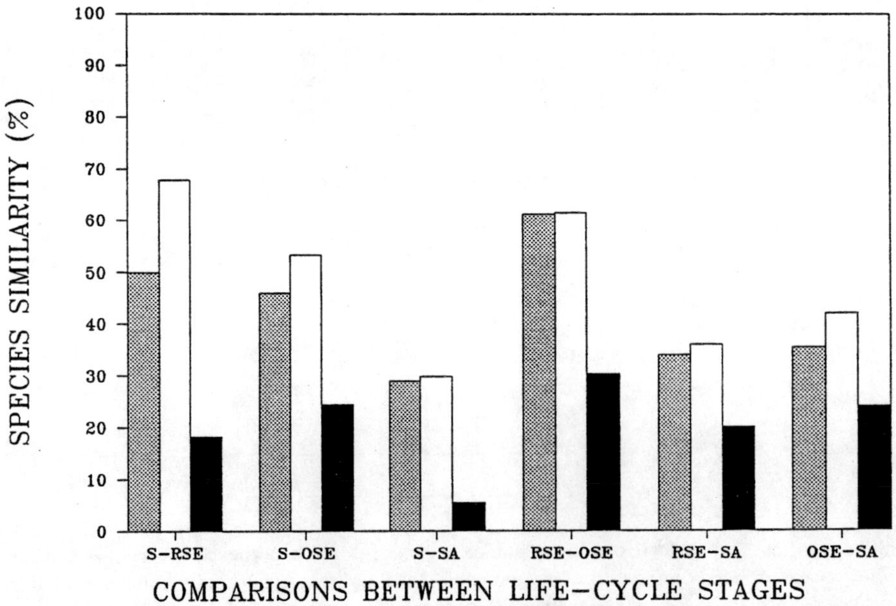

Fig. 2. Index of species similarity values obtained for pair-wise comparisons between seed rain (S), recruited-seedling (RSE), old-seedling (OSE) and sapling (SA) communities. Values for all species are gray bars, values for local species are white bars and values for immigrant species are black bars.

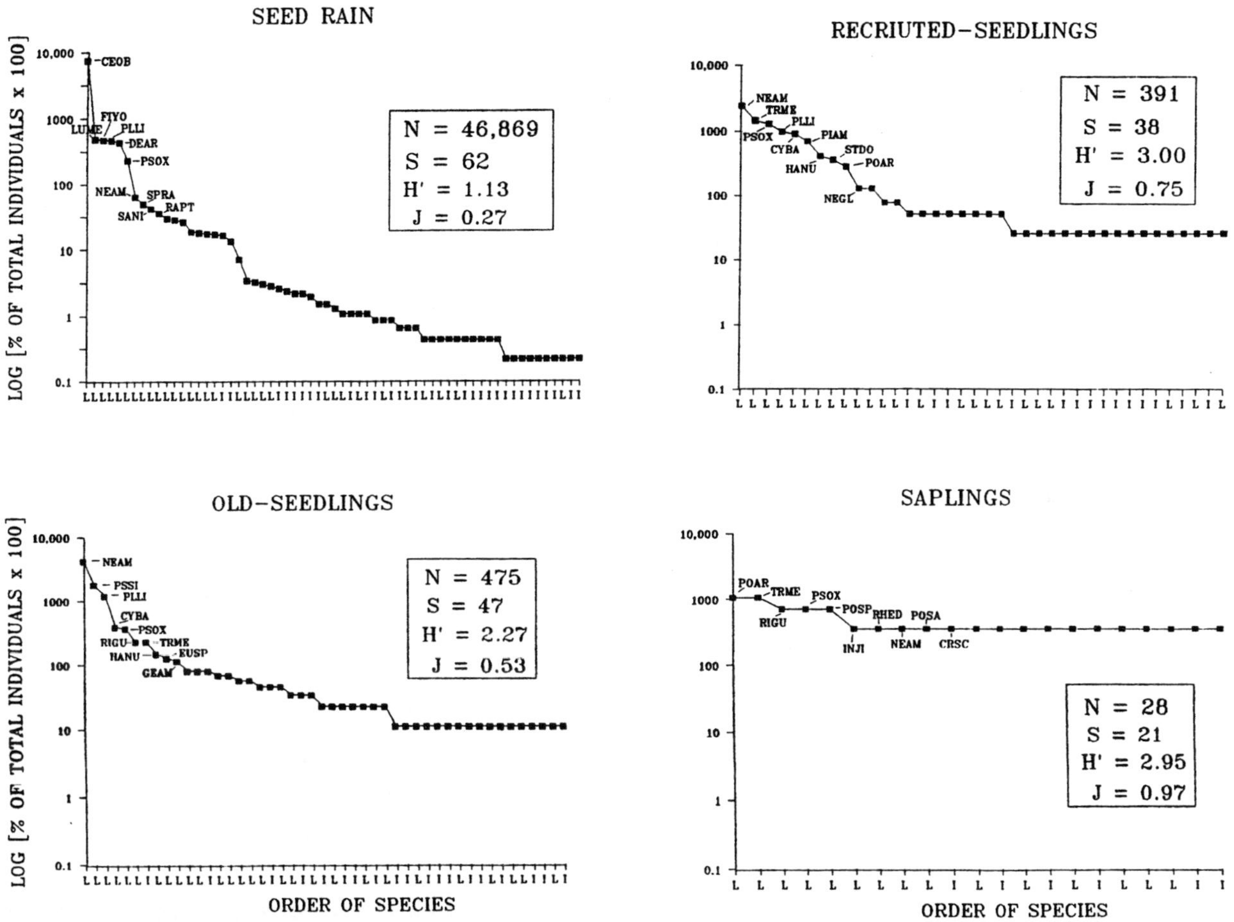

Fig. 3. Dominance-diversity curves for seed rain, recruited-seedling, old-seedling and sapling communities in closed forest sites at Los Tuxtlas. The number of individuals (N), species richness (S), diversity (H') and equitability (J) are given in each case. The ten most abundant species are indicated in each community. In the X-axes, L indicates local species and I the immigrant ones. Species codes correspond to the following tree species: CEOB (*Cecropia obtusifolia*), CRSC (*Croton schedianus*), CYBA (*Cymbopetalum baillonii*), DEAR (*Dendropanax arboreous*), EUSP (*Eugenia* sp), FIYO (*Ficus yoponensis*), HANU (*Hampea nutricea*), GEAM (*Genipa americana*), INJI (*Inga jinicuil*), LUME (*Lunannia mexicana*), NEAM (*Nectandra ambigens*), NEGL (*Nectandra globosa*), PIAM (*Piper amalago*), PLLI (*Pleuranthodendron lindenii*), POAR (*Poulsenia armata*), PSOX (*Pseudolmedia oxyphyllaria*), RAPT (*Randia pterocarpa*), RIGU (*Rinora guatemalensis*), RHED (*Rheedia edulis*), SANI (*Sapium nitidum*), SPRA (*Spondias radlkoferi*), STDO (*Stemadenia donnell-smithii*), TRME (*Trophis mexicana*), POSA (*Pouteria sapota*), POSP (*Pouteria* sp).

(Fig. 3), from the seed to seedling stages was mainly due to seed predation by rodents (Martínez-Ramos 1991). In comparison, *Nectandra ambigens* (Lauraceae), an upper-canopy tree which produces large crops of big seeds (average dry weight: 1800 mg seed-[1]) and seventh in abundance in the seed rain community (Fig. 3), suffered lower seed predation rates than *P. oxyphyllaria* (Martínez-Ramos 1991). Thus, differences in the rates of seed predation between these species may explain why in the seedling community the abundance position of *N. ambigens* was one order of magnitude higher than *P. oxyphyllaria* (Fig. 3). Supra-annual variability in fruiting behavior and high seedling survivorship of *Trophis mexicana* (Moraceae) (Martínez-Ramos 1991) may explain why this species was not found in the seed rain in the year studied, whereas it was one of the ten most abundant species in the seedling and sapling communities (Fig. 3).

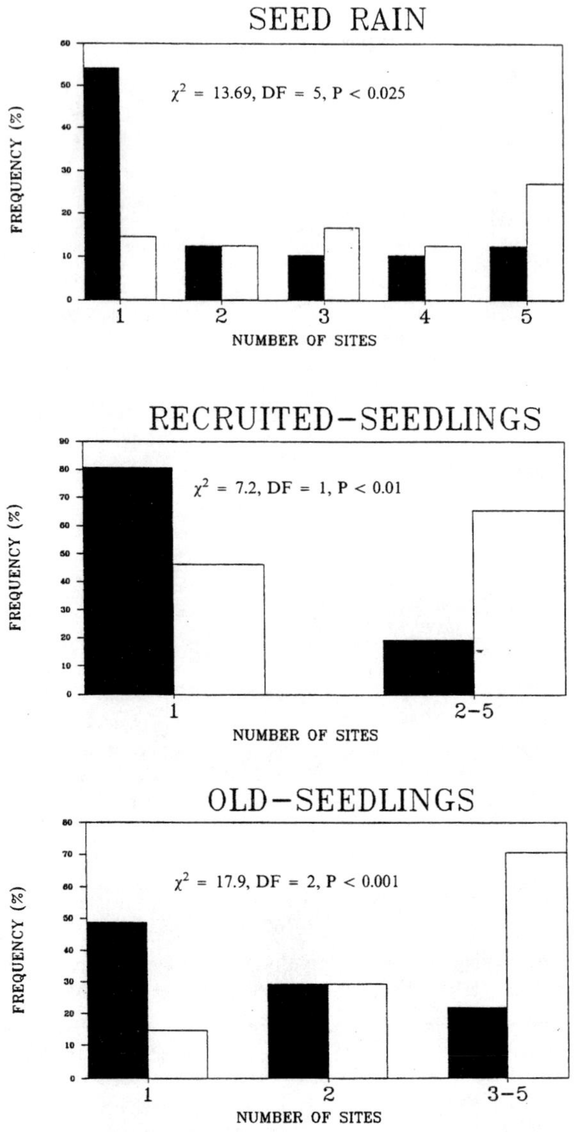

Fig. 4. Spatial frequencies of immigrant (filled bars) and local (open bars) species in the seed rain, recruited-seedling and old-seedling communities.

Pooling all height groups (including the lianas), immigrant seeds showed higher levels of diversity than locally produced seeds, and they were important components of the diversity observed in the seed rain community (Fig. 5a). However, in later life-cycle stages, the relative contribution of immigrant seed species to overall community diversity decreased (Fig. 5a). Diversity and equitability in the immigrant plants remained relatively constant among the different life-cycle stages but increased in the local plants (Fig. 5a, b). The small difference in diversity between local and immigrant tree species observed in the seedling and sapling communities (Fig. 5a) was due to an increase in equitability in the local species (Fig. 5b). Because most species in the sapling community were local species (Fig. 3), local saplings were more diverse than immigrant saplings (Fig. 5a).

When height groups were analyzed separately, few differences with respect to the above trends were observed. Lianas showed the highest diversity among all height groups (Fig. 5a). Among tree groups, UC seeds, seedlings and saplings showed the lowest diversity values while the highest ones were found in the LC groups (Fig. 5a). Within the MC group, immigrant communities had consistently higher diversity values than local ones (Fig. 5a).

Species transition between life-cycle stages and some of its possible determinants

From the above results it is apparent that immigrant propagules make an important contribution to the community structure observed in the seed rain and, to a lesser degree, to that of seedlings and saplings. Which seed species arriving at the forest soil eventually establish as seedlings and saplings? Here, we explore three life-history traits that can influence establishment under the shade conditions of the forest. Because of insufficient data for statistical analysis in the liana group, the following results include only tree species.

Seed size

It can be assumed that seed size is positively related to the amount of maternal investment per seed (Fenner 1985; Foster 1986). If larger maternal investments confer higher probabilities of seedling survival, we would expect that tree species producing large seeds have a higher chance of being represented in the seedling and later life-cycle stages. In fact, there was an increase in the average seed size per species from the seed to sapling communities ($F_{3,158} = 3.57$, $P < 0.01$);

however, the only significant difference detected was between the average seed size of species found in the seed rain and that of species found in the old-seedlings and saplings communities (Fig. 6a). A two-way analysis of variance did not detect a significant interaction effect on the average seed size either between life-cycle stages and the origin of propagules (*i.e.* immigrant *vs* local; $F_{3,154} = 0.72$, $P = 0.52$) nor between life-cycle stages and tree groups ($F_{6,150} = 0.70$, $P = 0.56$). Finally, there was no significant difference in mean seed size among either LC, MC and UC groups ($F_{2,156} = 0.44$, $P = 0.65$) nor between local and immigrant species ($F_{1,160} = 2.63$, $P = 0.11$).

Seedling morphology

Initial morphology of tree seedlings was the second attribute that we analyzed. We investigated the idea that, under shade conditions, seedlings having large maternal reserves (which we termed RC-species) would have higher survivorship probabilities than seedlings with epigeous, photosynthetic, cotyledons (PC-species). In our seed rain sample, species that procedure RC-seedlings had significantly larger seeds ($\bar{X} \pm SE = 2.81 \pm 0.11$ mg [log scale]) than those producing PC-seedlings ($\bar{X} \pm SE = 1.44 + 0.13$ mg) ($t = 7.33$, $DF = 60$, $P < 0.001$). Photosynthetic cotyledons should play an important role in seedling establishment in sunny conditions but not in shade conditions (Ng 1978; Hladik & Miquel 1990; Martínez-Ramos 1991). If so then we would expect the ratio RC-species/PC-species to increase from the seed to sapling communities. A contingence table analysis made with a log-linear model (MacCullagh & Nelder 1983) showed that the ratio RC-species/PC-species differs significantly among life-cycle stages ($\chi^2 = 13.36$, $DF = 3$, $P < 0.001$). The proportion of CR-species was not significantly different between the seed rain and recruited-seedling communities, but this proportion was significantly higher in the old-seedlings and sapling communities (Fig. 6b). The higher number of epigeous species in the seed rain (63 % of the total number of species) indicates that this type of seedling morphology predominates in the entire tree community at Los Tuxtlas. The fact that a very similar RC/PC ratio was detected in the seed rain and recruited seedling communities indicates that species producing one or the other type of seedling have similar probabilities of germination and recruitment as new seedlings. A contingency table analysis made with a log-linear model showed that the frequency of species producing RC- or PC-seedlings was not significantly different either with respect to tree canopy positions (LC, MC and UC groups; $\chi^2 = 2.13$, $DF = 2$, $P > 0.10$) or to origin of seeds (local or immigrant; $\chi^2 = 0.84$, $DF = 1$, $P > 0.10$).

Species abundance

Abundance may determine the probability that a given species reaches later life-cycle stages. If mortality probabilities are the same for all species, we can expect that rare (low abundance) species have a higher risk of extinction than abundant species. To explore this idea, seed rain, recruited-seedling and old-seedling communities were divided in four categories: 1) local species recorded in the life-cycle stage *i* and in the next *i* stage (*i.e.* remaining local species -RLS-), 2) local species whose individuals presumably died in stage *i* and therefore are not recorded in the stage *i* (*i.e.* extinct local species –ELS–), 3) immigrant species present in the *i* and *i* stages (*i.e.* remaining immigrant species –RIS–), and 4) immigrant species present only in the *i* stage (*i.e.* extinct immigrant species –EIS–). Next, each of these species groups were arbitrarily divided into two abundance categories: 1) 'rare' species, with ≤ 4 individuals and 2) 'abundant' species, with > 4 individuals, considering population sizes recorded in the five study plots. With the data available, only these two categories were amenable to statistical analysis. To determine whether species frequency in each of the above four groups was independent of species abundance, we constructed a null hypothesis by obtaining the proportion of rare and abundant species in the entire community for each of the life-cycle stages. These proportions were then multiplied by the species number recorded for RLS, ELS, Ris, and EIS groups to obtain the expected random frequencies of rare and abundant species within each

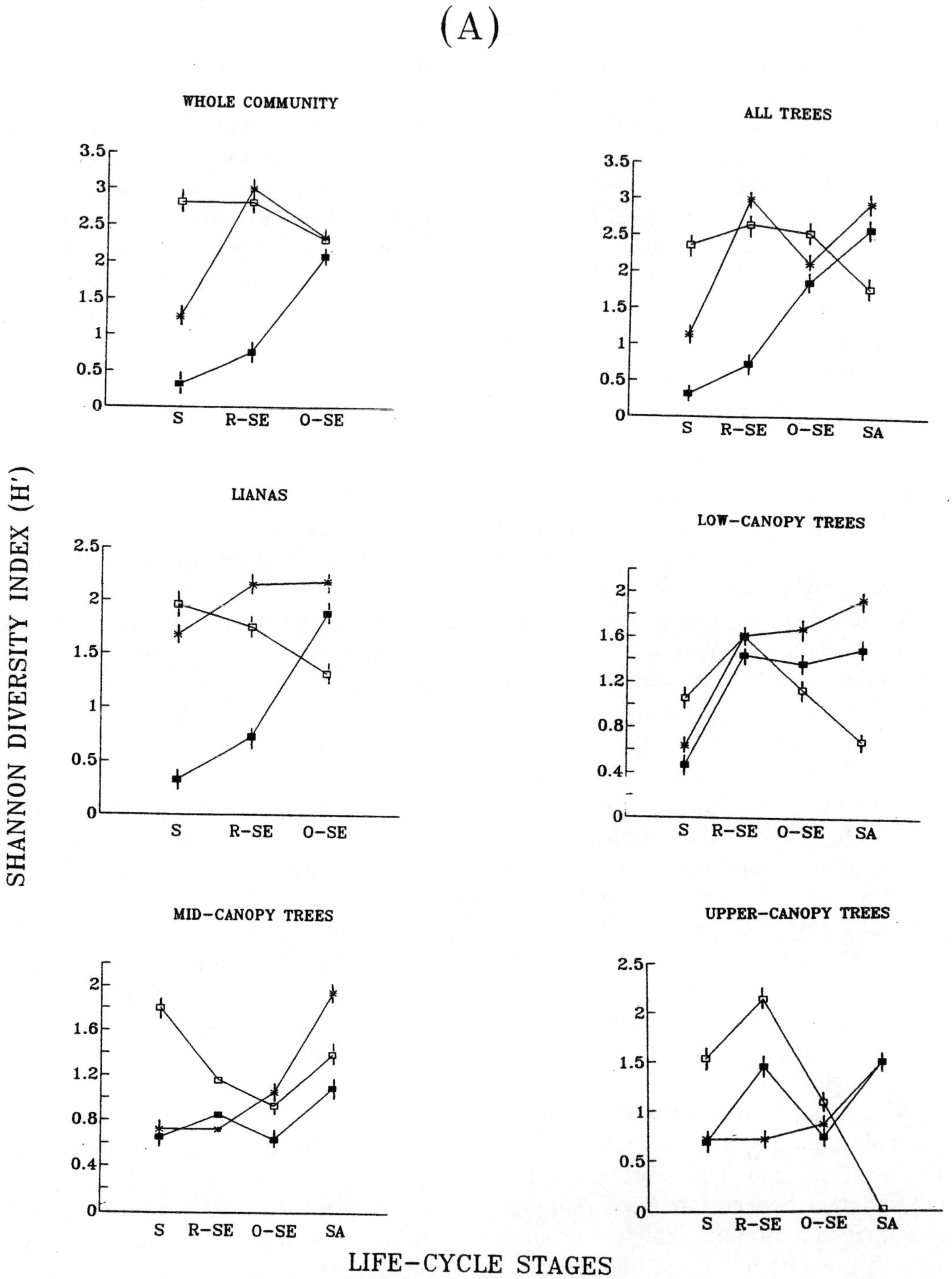

group. The observed and expected values were compared using Chi-square tests.

In the seed rain community (immigrant plus local species), remaining species were predominantly abundant; rare species were more likely to go extinct (Fig. 7). In the recruited-seedling community, extinct species were also predominantly rare. In the old-seedling community remaining

(B)

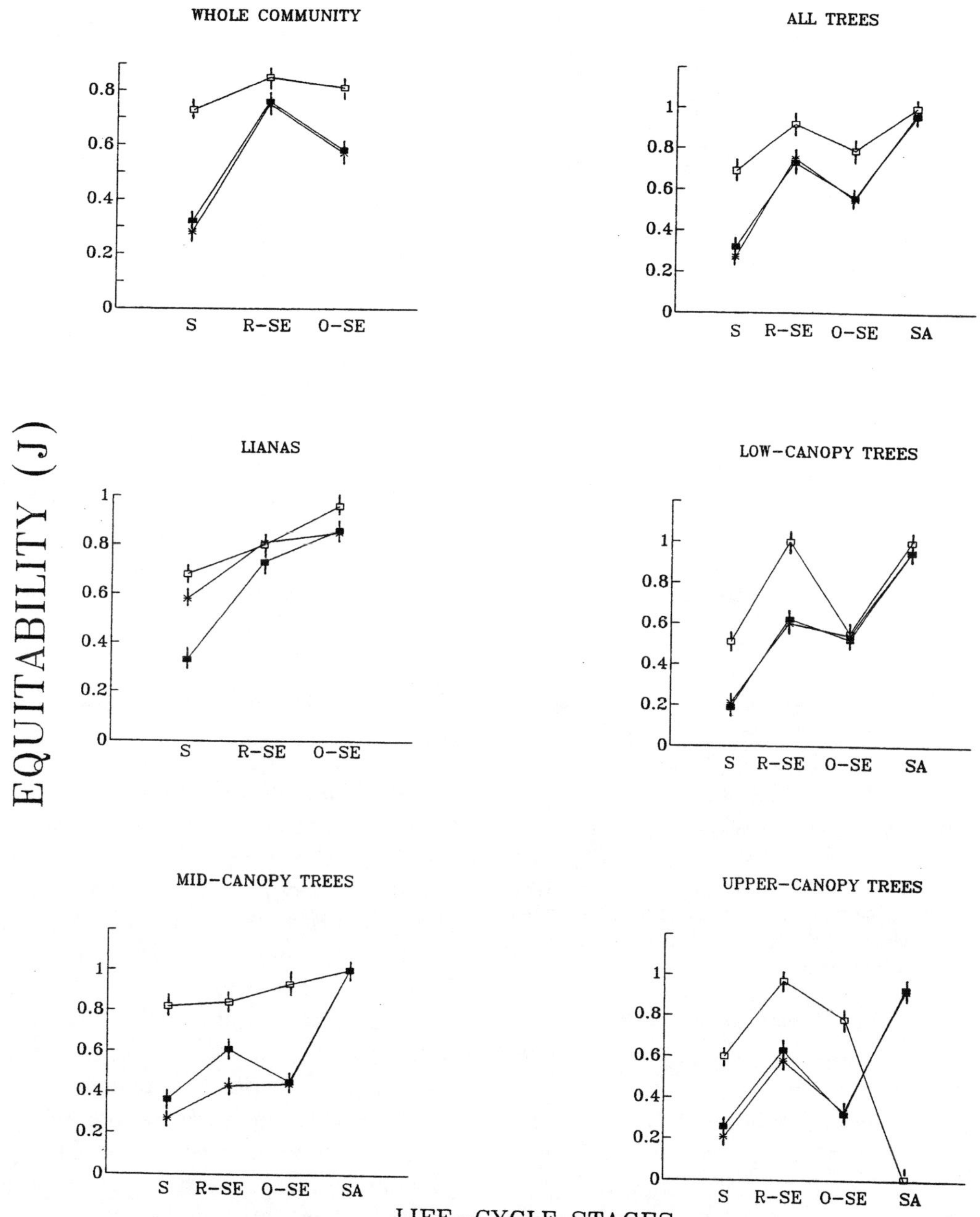

LIFE-CYCLE STAGES

Fig. 5. Diversity (A) and equitability (B) changes among life-cycle stages in the entire community (lianas plus trees), in the tree and liana communities, and in different canopy positions of trees. In each case the values for local, immigrant and all species are shown. S = seeds, R-SE = recruited seedlings, O-SE = old seedlings, SA = saplings. Vertical lines represent two standard errors. □ = immigrant species, ■ = local species, ★ = all species.

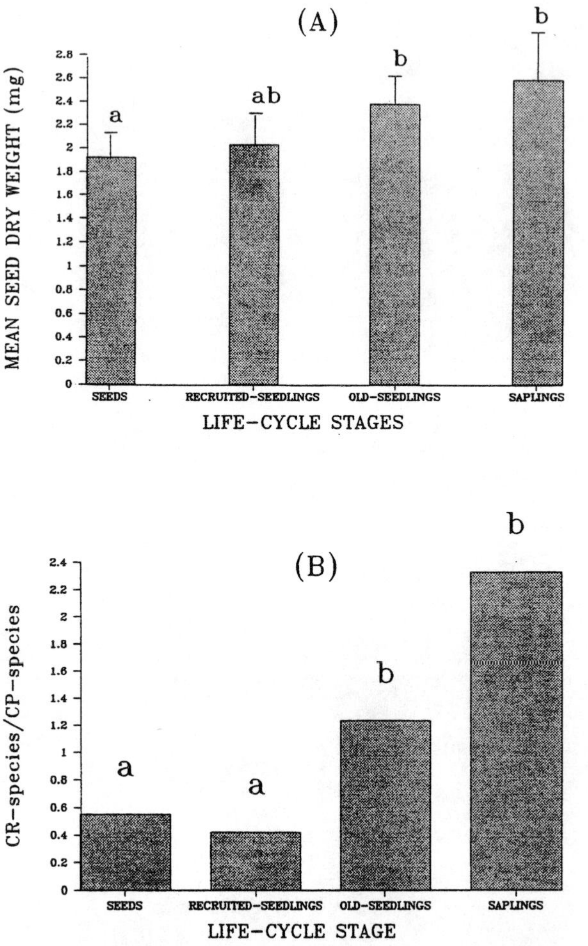

Fig. 6. Changes in (A) average seed weight (on a log-scale) and (B) the ratio of RC-species/PC-species among early life-cycle stages of the tree community found in closed forest sites at Los Tuxtlas; PC-species = species producing seedlings with photosynthetic cotelydons, RC-species = species producing seedlings with cotelydons, endosperm or hypocotyl functioning as reserve organs. Vertical lines indicates two standard errors. Bars that share the same letters are not statistically different ($P < 0.05$).

species were predominantly abundant. When local and immigrant species were analyzed separately, it was clear that the RLS group was significantly over-represented by abundant species in all life-cycle stages. In contrast, the EIS group included a significantly higher frequency of low abundance species in most life-cycle stages (Fig. 7).

Overall, the above results suggest that abundant species have higher probabilities of being represented in the next stage, and that probability was higher for local than for immigrant species. However, several rare species of immigrant origin were also recorded in the sapling community (Fig. 3). Therefore, other factors besides species abundance alone, seem to influence the species persistence over time.

Discussion

Three major points emerged from this study: i) Immigrant species are a significant component of diversity for both the seed rain and seedling communities in closed canopy sites, ii) contribution to the seed rain depends on position in the canopy: seeds of upper-canopy trees and liana species were the major components of the immigrant seed community, and iii) under closed-canopied sites, factors affecting community organization seem to operate selectively, favoring the establishment of large-seeded, local abundant species in the advanced regeneration.

Seed dispersal, forest community structure and seed rain

The estimated contribution of immigrant species to the seed rain based on only one year can be only a rough estimate, since this is an extremely variable phenomenon. Spatial and temporal changes in both the intensity of frugivory and the fruiting behavior of plants (Foster 1990b; Gautier-Hion 1990; Terborgh 1990) may effect the structure and composition of the seed rain falling in different places in the forest, and over time in a particular forest patch. The high interpatch variability detected in the proportion of seeds coming from an immigrant origin (Table 2) may be related to such sources of seed dispersal variation.

More interesting, however, is the over-representation of tree canopy and liana species in the seed rain. Assuming similar crop sizes among trees of different heights, this pattern could be the result of more intense seed dispersal rates in the upper and mid-canopy level of the forest, where most fruit-eating arboreal mammal and bird spe-

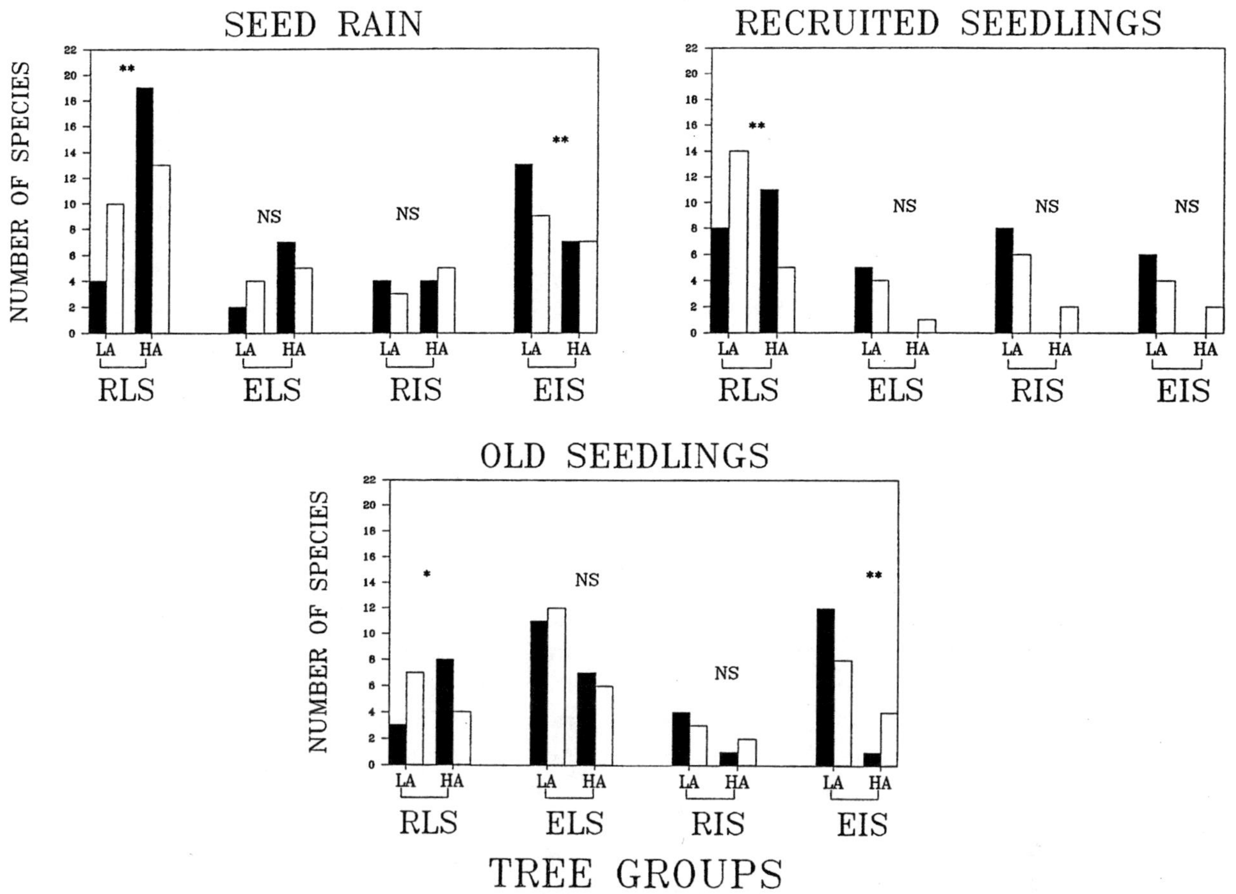

Fig. 7. Relationship between species abundance (LA = low abundant, HA = high abundant) and the frequency of remaining local species (RLS), extinct local species (ELS), remaining immigrant species (RIS), and extinct immigrant species (EIS) in three life cycle-stage communities. ** = $P<0.01$, * = $P<0.01$, *N.S.* = $P>0.05$ based on Chi-square contingency table analysis. Observed frequencies in filled bars, expected frequencies in open bars. In cases in which expected frequencies were less than 5, the appropriate correction to the Chi-square test was made (Sokal & Rolf 1981). See text for further details.

cies feed (Terborgh 1986). At Los Tuxtlas, large frugivorous mammals (such as monkeys, kinkajous and porcupines) and birds (such as toucans and parrots) rarely feed in the sparsely-distributed small understory trees (R. Dirzo pers. comm.). This suggests that a lack of dispersers is one reason why low-canopy trees were under-represented in the seed rain.

An alternative explanation may be related to differences in fecundity among tree species. We can expect average fecundities to increase from understory to upper-canopy trees due to larger crown sizes and better light conditions of the latter. The few fecundity schedules available for rainforest trees seem to support this assertion (Alvarez-Buylla & Martínez-Ramos 1992). Even with similar seed dispersal rates in understory and canopy tree groups, the larger fruit crops of upper-canopy trees could enable a larger number of seeds to be dispersed per unit area.

Lianas, which were the most frequent species represented as immigrant seeds, have smaller stem diameters than trees (Putz 1983). This might seem to contradict the above scenario. However, the rate of increase of crown-leaf area per unit of rate of increase in stem diameter is higher in lianas than in trees, and individual lianas may have extensive leaf crowns spreading over several trees (Putz 1983). It is therefore possible that liana fruit crops may reach comparable sizes to those of some canopy trees. This topic deserves further study.

Variability of seed rain over time and space

At Los Tuxtlas and in other rainforest localities (Foster 1990a; Gautier-Hion 1990), lianas, canopy trees and understory species differ in their timing of fruit production. At Los Tuxtlas and Barro Colorado Island, Panama, most canopy species produce fruits in the rainy season whereas several understory species (mostly dispersed by migratory birds) fruit in winter (G. Ibarra pers. comm.; Foster 1990a). An added factor could be the differences in seed fall timing between canopy and understory trees, which could lead to temporal changes in the seed rain composition.

The abundance and spatial distribution of fruiting species must also play a role in generating diversity in the seed rain. Because most of the species are rare and/or have a clumped distribution (M. Martínez-Ramos & E. Alvarez-Buylla, unpubl.; Hubbell & Foster 1983), we can expect considerable variation in the species composition of the seed rain falling at different forest locations. This expectation seems to be illustrated by the fact that most immigrant seed species were recorded in only one of the five study plots (Fig. 4).

Seed rain and the organization of advanced regeneration

With the exception of a few light-demanding trees (e.g. *Ficus yoponensis*, *Lunania mexicana*), most abundant species recorded as seeds were also present as seedlings. Even some seedlings of the gap-dependent tree *Cecropia obtusifolia* were encountered, presumably because of the great abundance of seed present (Fig. 3a). The light environment of the closed canopy sites studied did not limit germination and early development of most species.

In most species there was a strong numerical reduction in the transition from seed to seedling stage. Seeds experience a variety of mortality sources, including physiological disorders, predispersal predation, random events such as inhibitory germination by litter (Vázquez-Yanes *et al.* 1990), and post-dispersal predation or pathogen attack (Dirzo & Dominguez 1986; Howe 1990). If these mortality factors act with the same intensity on species independently of their abundance, rare species (at an extreme, those represented by only one individual) have a higher chance of becoming locally extinct. However, frequent-independent processes did not explain the many rare, mostly immigrant, species that were found in the old-seedling and sapling communities. *Poulsenia armata* (Moraceae) is particularly notable in this respect. Although an unimportant species in the seed rain and seedling communities, *P. armata* was the most abundant species in the sapling community. Another similar case was *Pouteria sapota* (Sapotaceae) (Fig. 3). In these species, rarity was associated with positive sapling establishment probabilities. Similar results have been found for some species in other neotropical rain forests (Hubbell & Foster 1990a).

Experimental evidence for the existence of regulatory forces operating on abundant seed and seedling populations has been documented for several species in the study plots used here (Martínez-Ramos 1991). For example, *Pseudolmedia oxyphyllaria* and *Nectandra ambigens*, two of the ten most abundant seed, seedling and sapling species in the study plots, demonstrated density-dependent changes in population numbers although through different processes. Under parent trees, seeds of *P. oxyphyllaria* are strongly preyed upon by rodents. However, the widely scattered seedlings indicate that seed dispersal may confer better survival probabilities to those seeds falling away from parent trees. This pattern is similar to that of *Virola surinamensis* in Panama (Howe 1990). In contrast, seeds of the upper-canopy tree *N. ambigens* suffer considerably lower seed predation than *P. oxyphyllaria*. Approximately half of the seeds falling under parent trees germinate there, forming high density populations of long-lived seedlings (see also Sarukhán 1978 and Córdova 1985); very few seedlings are found away from fruiting trees. Seed dispersal seems not to be important for seedling establishment (Córdova 1979), as in *Tetragastris panamensis* on Barro Colorado Island, Panama (Howe 1990). However, when *N. ambigens* seedling populations are crowded, mortality rates increase rapidly because of seedling competition and/or pathogen attack (Martínez-Ramos 1991). Certainly some inhibitory or regulatory mechanisms operate in seedling populations of *N. ambigens*, since this was the most abundant species in the seedling community (Fig. 3) but was poorly represented as saplings (Fig.3).

As a result of the thinning processes operating on abundant species, equitability and diversity increased in communities of late life-cycle stages. Within this trend, species that produce small seeds and PC-seedlings suffered the strongest numerical reduction, and most of them were not found at the sapling community. The increase in diversity from the seed rain to the sapling community was largely due to the elimination of these small seeded species. On the contrary, species that produce large seeds and RC-seedlings seem to have higher rates of survival and a higher probability of being represented in advanced regeneration. A higher seed and seedling mortality rate operating on the most abundant large-seeded species also helps to explain the higher diversity found in the old-seedling and sapling communities (Martínez-Ramos 1991).

In a community context, seed size and seedling morphology influence the survivorship probabilities of seedlings (Fenner 1985). Actually, these traits are correlated; species producing small seeds mostly give rise to epigeous seedlings that die at higher rates than seedlings with high maternal resources stored in specialized organs which are mainly produced by large seeded species (Ng 1978; Hladik & Miquel 1990). Light-demanding trees exhibit epigeous seedlings whereas most long-lived species produce large seeds that readily germinate in the shade (Ng 1978; Hladik & Miquel 1991; Vázquez-Yanes & Oroczo-Segovia 1984, 1990). High maternal resources enable seedlings to survive and grow even under reduced light conditions (Howe 1990). This is because large seeds give rise to robust seedlings (Ng 1978; Hladik & Miquel 1990; Howe & Ritcher 1982) which can withstand mortality risks imposed by physical and animal injuries (Fenner 1985; Foster 1986). These processes are consistent with the observation that large seeded, persistent species form the largest part of the seedling and sapling communities in closed canopy sites. Similar patterns of seedling recruitment have been found in an African rain forest (Hladik & Miquel 1990).

A mixture of physiological, population regulatory and random factors influence the organization of advanced regeneration. We suggest that species' biology (seed size, seedling morphology, shade-tolerance, plant-animal interactions), more than random events, organizes the community at the seedling and sapling levels in closed canopy sites.

Conclusions

Tropical rain forests are dynamic systems. From the information obtained in the present study we suggest links between the roles of gap dynamics and seed sources (local versus immigrant) in the regeneration processes of a tropical rain forest. The role of immigrant seeds may depend particularly on the type of patch where a seed lands. In forest patches with high canopy turn-over rates, such as those found along forest borders and in steep-sloped sites (Martínez-Ramos *et al.* 1988b), immigrant seeds will play an important role in generating within-patch diversity. In such sites, high turn-over rates do not favor the establishment of long-lived persistent trees. Therefore, advanced regeneration of these species will depend on immigrant seeds. Pioneer and late-secondary species may dominate the structure of the patches. However, because seedlings of these light-demanding trees do not establish in shade, seedlings of persistent species arriving through seed dispersal will constitute a diverse seedling community of low abundance. In fact, the patches studied found in the early successional stages have a significantly less abundant but more diverse seedling community than the patches found in the mature phase (Martínez-Ramos 1991).

At the other extreme, in forest patches with very low canopy turn-over rates in which long-lived species are established (Martínez-Ramos *et al.* 1988b), immigrant species will play a minor role in generating diversity. Long-lived tree species produce crops of large seeds (such as *Nectandra ambigens* at Los Tuxtlas) which in turn generate abundant seedling banks (Martínez-Ramos 1985). Because local species seem to have a better chance than immigrant species of recruiting into advanced regeneration, a self-regenerative pathway may predominate in these forest patches of low turn-over. As the time elapsed between disturbances increases, seedlings and saplings of local species eventually dominate advanced regeneration. When the canopy opens up, large seedlings and saplings may have competitive advantages over recently emerging seedlings in reaching a place in the canopy (Uhl *et al.* 1988). However, the observation that some immigrant species do get established as saplings also indicates that there may be a slow infiltration resulting in a slow shift in the species composition of advanced regeneration. In the long term this slow infiltration could lead to a floristic homogenization among forest patches.

Finally, in patches with an intermediate canopy turn-over rate we can expect a mixture of the two regeneration possibilities described above. Since rain forests are in reality shifting mosaics of vegetation patches, each one with a different disturbance history, the relative contribution of local and immigrant species to the seed rain community will change temporally and spatially. As this study suggests, understanding the role played by seed rain in community organization can be reached through an integrative approach, by studying the processes and mechanisms that govern the fate of seed populations as they advance from one life-cycle stage to the next.

Acknowledgements

This paper was possible through the generous assistance of many people. We particularly thank Santiago Sinaca who helped us with the identification of seeds and seedlings and to George Dyer, Teresa Hernández, Graciela García, Mario González, Steve Littman, Paulina Parlange, Eduardo Rincón-Gallardo, Maricruz Rosas, Jorge Rodríguez, and Saúl Segura whom worked with us in the field and/or in the laboratory. We thank the staff of Los Tuxtlas Tropical Field Station, Instituto de Biología, UNAM, for all the facilities provided. The manuscript was improved by helpful comments made by Elena Alvarez-Buylla, Fahkri Bazzaz Ted Fleming, Jim Hamrick, Collin C. Kelly, and Eugene Schupp. The project was supported by a grant from Consejo Nacional de Ciencia y Tecnologia of México. The final version of the manuscript was completed while MMR was a Bullard fellow in the Department of Organismic and Evolutionary Biology, Harvard University.

References

Alvarez-Buylla, E. & Martínez-Ramos, M; 1992. Demography and allometry of *Cecropia obtusifolia*, a neotropical pioneer tree – an evaluation of the climax – pioneer paradigm for tropical rain forests. J. Ecology 80: 275–290.

Augspurger, C. K. 1984. Seedling survival of tropical tree species: interactions of dispersal distance, light gaps, and pathogens. Ecology 65: 1705–1712.

Bongers, F., Popma, J., Meave, J. & Carabias, J. 1988. Structure and floristic composition of the lowland rain forest of Los Tuxtlas, Mexico. Vegetatio 74: 55–88.

Brokaw, N. V. L. 1985. Gap phase regeneration in a tropical forest. Ecology 66: 682–687.

Chazdon, L. R. 1988. Sunflecks and their importance to forest understory plants. Adv. Ecol. Res. 18: 1–63.

Clark, D. & Clark, D. 1984. Spacing dynamics of a tropical rain forest tree: evaluation of the Janzen-Connell model. Amer. Nat. 142: 769–788.

Clark, D. B. & Clark, D. A. 1989. The role of physical damage in the seedling portality of a neotropical rain forest. Oikos 55: 225–230.

Connell, J. H., Tracey, J. G. & Webb, L. J. 1984. Compensatory recruitment, growth, and mortality as factors maintaining rain forest tree diversity. Ecological Monographs 54: 141–164.

Córdova, B. 1979. Efectos de la densidad, la distancia al árbol y la depredación en el crecimiento y sobrevivencia de plántulas de *Nectandra ambigens* (Blake). Tesis de Profesional. Facultad de Ciencias, UNAM, México.

Córdova, B. 1985. Demografía de árboles tropicales. In: Gómez-Pompa, A. S. & Del Amo, S. (eds). Investigaciones sobre la regeneración de selvas altas en Veracruz, México Vol II. pp. 103–128. INIREB, Alhambra, México.

Denslow, J. 1987. Tropical rain forest gaps and tree species diversity. Ann. Rev. Syst. Ecol. 18: 432–451.

Dirzo, R. 1987. Estudios sobre interacciones planta-herbívoro en 'Los Tuxtlas', Veracruz. Rev. Biol. Trop. (Supl. 1) 35: 119–132.

Dirzo, R. & Dominquez, A. C. 1986. Seed shadows, seed predation and the advantages of dispersal. In: Estrada, A. & Fleming, T. H. (eds). Frugivores and seed dispersal. pp. 237–249. Dr. Junk Publishers, The Hague, the Netherlands.

Dirzo, R. & Miranda, A. 1991. Altered patterns of herbivory and diversity in the forest understory: a case study of the possible consequences of contemporary defaunation. In: Price, W. P., Lewinson, M., Wilson, G. & Woodruff, W. B. (eds). Plant-animal interactions: evolutionary ecology in tropical and temperate regions. pp. 273–447. John Wiley and Sons, Inc, USA.

Fenner, M. 1985. Seed ecology. Chapman and Hall. New York, USA.

Foster, S. A. 1986. On the adaptative value of large seeds for tropical moist forest trees: a review and synthesis. The Botanical Review 52: 269–299.

Foster, R. B. 1990a. Ciclo estaconal de caída de frutos en la isla de Barro Colorado. In: Leigh, E. G., Jr., Stanley, R. A. & Windsor, D. M. (eds). Ecología de un bosque tropical: ciclos estacionales y cambios a largo plazo. pp. 219–241. Smithsonian Tropical Research Institute, Panama.

Foster, R. B. 1990b. Hambruna en la Isla de Barro Colorado. In: Leigh, E. G., Jr., Stanley, R. A. & Windsor, D. M. (eds). Ecología de un bosque tropical: ciclos estacionales y cambios a largo plazo. pp. 271–283. Smithsonian Tropical Research Institute, Panama.

Gauthier-Hion, A. 1990. Interactions among fruit and vertebrate fruit-eaters in an African tropical rain forest. In: Bawa, S. & Hadley, M. (eds). Reproductive ecology of tropical forest plants. Man and the Biosphere Series. pp. 219–230. The Parthenon Pub. Group, Paris. France.

Gómez-Pomps, A. & Del Amo S. 1985. Investigaciones sobre la regeneración de selvas altas en Veracruz, México Vol. II. INIREB, Alhambra, México.

Greig-Smith, P. 1983. Quantitative plant ecology. Studies in ecology Vol. 9. Univ. Calif. Press, L. A., USA.

Hladik, A. & Miquel, S. 1990. Seedling types and plant establishment in an African rain forest. In: Bawa, K. S. & Handley, M. (eds). Reproductive ecology of tropical forest plants. Man and the Biosphere Series. pp. 191–218. The Parthenon Pub. Group, Paris, France.

Howe, H. F. 1990. Seed dispersal by birds and mammals: implications for seedling demography. In: Bawa, K. S. & Handley, M. (eds). Reproductive ecology of tropical forest plants. Man and the Biosphere Series. pp. 261–282. The Parthenon Pub. Group, Paris, France.

Howe, H. F. & Ritcher, W. M. 1982. Effects of seed size on seedling size in *Virola surinamensis*; a within and between tree analysis. Oecologia 53: 347–351.

Hubbell, S. P. & Foster, R. B. 1983. Diversity of canopy trees in a neotropical forest and implications for conservation. In: Sutton, S. L., Whitmore, T. C. & Chadwik, A. C. (eds). Tropical rain forest ecology and management. pp. 25–41. Blackwell Scientific Publications, Oxford, UK.

Hubbell, S. P. & Foster, R. B. 1986. Biology, chance and history and the structure of tropical rain forest tree communities. In: Diamond, J., Case, T. J. (eds). Community Ecology. pp. 314–329. Harper & Row Publications. New York, USA.

Hubbell, S. P. & Foster, R. B. 1990a. The fate of juvinile trees in a neotropical forest: implications for the natural maintenance of tropical tree diversity. In: Bawa, K. S. & Handley, M. (eds). Reproductive ecology of tropical forest plants. Man and the Biosphere Series. pp. 317–344. The Parthenon Pub. Group, Paris, France.

Hubbell, S. P. & Foster, R. B. 1990b. Structure, dynamics, and equilibrium status of old-growth forest on Barro Colorado Island. In: Gentry, A. H. (ed). Four neotropical rainforests. pp. 522–541. Yale University Press, New Haven, USA.

Ibarra, G. & Sinaca-Colín, S. 1987. Listados florísticos de México VII. Estacíon dc Biología Tropical Los Tuxtlas. Inst. Biol. UNAM.

MacCullagh, P. & Nelder, J. A. 1983. Generalized linear models. Chapman & Hall, London, UK.

Martínez-Ramos, M. 1985. Claros, ciclos vitales de los árboles tropicales y la regeneracíon natural de las selvas altas perennifolias. In: Gómez-Pompa, A. & Del Amo, S. 1985 (eds). Investigaciones sobre la regeneración de selvas altas en Veracruz, México Vol II. pp. 191–239. INIREB, Alhambra, México.

Martínez-Ramos, M., Alvarez-Buylla, E., Sarukhán, J. & Piñero, D. 1988a. Treefall age determination and gap dynamics in a tropical rain forest. J. Ecol. 76: 700–716.

Martínez-Ramos, M., Sarukhán, J. & Piñero, D. 1988b. The demography of trees in the context of forest gap dynamics: the case of *Astrocaryum mexicanum* at Los Tuxtlas tropical rain forest. In: Davy, D. J., Hutchings, M. J. & Watkinson, A. R. (eds). Plant population ecology. pp. 293–313.. Blackwell, Oxford, U. K.

Martínez-Ramos, M. 1991. Patrones, procesos y mecanismos en la comunidad de plántulas de una selva húmeda neotropical. Tesis doctoral, UNAM, México.

Ng, F. S. P. 1978. Strategies of establishment in Malayan forest trees. In: Tomlinson, P. B. & Zimmerman, H. (eds). Tropical Trees and Living Systems. pp. 129–162. Cambridge Univ. Press. Cambridge, USA.

Oldeman, A. A. R. & van Dijck, J. 1991. Diagnosis of the temperament of tropical forest trees. In: Gómez-Pompa, A., Whitmore, T. C. & Handley, M. (eds). Rain forest regeneration and management. Man and the biosphere series. pp. 21–65. The parthenon Pub. Group. Paris, France.

Pearcy, W. R. 1990. Sunflecks and photosynthesis in plant

canopies. Ann. Rev. Plant Physiol. Plant Mol. Biol. pp. 421–453.

Popma, J., Bongers, F., Martínez-Ramos, M. & Veneklaas, E. 1988. Pioneer species distribution in treefall gaps in neotropical rainforest: a gap definition and its consequences. J. Trop. Ecol. 4: 77–88.

Primack, R. B. 1990. Seed physiology, seed germination and seedling ecology. In: Bawa, K. S. & Handley, M. (eds). Reproductive ecology of tropical forest plants. Man and the Biosphere Series. pp. 233–236. The Parthenon Pub. Group, Paris, France.

Putz, F. E. 1983. Liana biomass and leaf area of a 'tierra firme' forest in the Rio Negro Basin, Venezuela. Biotropica 15: 185–189.

Raich, W. J. & Christensen, L. N. 1989. Malaysian dipterocarp forest: tree seedling and sapling species composition and small-scale disturbance patterns. Nat. Geog. Res. 5: 348–363.

Sarukhán, J. 1978. Studies on the demography of tropical trees. In: Tomlinson, P. B. & Zimmerman, H. (eds). Tropical trees as living systems. pp. 163–184. Cambridge Univ. Press., Cambridge, USA.

Sarukhán, J., Martínez-Ramos, M. & Piñero, D. 1984. The analysis of demographic variability at the individual level and ist population consequences. In: Dirzo, R. & Sarukhán, J. (eds). Perspectives on plant population ecology. pp. 141–165. Sinauer Ass. Inc., Sunderland Mass., USA.

Sarukhán, J., Piñero, D. & Martínez-Ramos, M. 1985. Plant demography: a community level interpretation. In: White, J. (ed). Studies in plant demography: a festschrift for John L. Harper. pp. 17–31. Acad. Press., London, UK.

Schupp, E. W. 1988a. Seed and early seedling predation in the forest understory and in treefall gaps. Oikos 51: 525–530.

Schupp, E. W. 1988b. Factors affecting post–dispersal seed survival in a tropical forest. Oecologia (Berlin) 76: 525–530.

Sokal, R. & Rohlf, R. 1981. Biometry. W. H. Freeman & Co. NY USA.

Terborgh, J. 1986. Community aspects of frugivores in tropical forests. In: Estrada, A. & Fleming, T. H. (eds). Frugivores and seed dispersal. pp. 371–384. Dr. Junk Publish., The Hague, Netherlands.

Terborgh, J. 1990. Seed and fruit dispersal-commentary. In: Bawa, K. S. & Handley, M. (eds). Reproductive ecology of tropical forest plants. Man and the Biosphere Series. pp. 181–190. The Parthenon Pub. Group, Paris, France.

Uhl, C., Clark, K., Dezzeo, N. & Maquirino, P. 1988. Vegetation dynamics in Amazonian treefall gaps. Ecology 69: 751–763.

Vázquez-Yanes, C. & Orozco-Segovia, A. 1984. Ecophysiology of seed germination in tropical humid forests of the world: a review. In: Medina, E., Mooney, H. A. & Vázquez-Yanes, C. (eds). Physiological ecology of plants of the wet tropics. pp. 37–49. Dr. Junk Publishers, The Hague, the Netherlands.

Vázquez-Yanes, C. & Orozco-Segovia, A. 1990. Seed dormancy in the tropical rain forest. In: Bawa, K. S. & Handley, M. (eds). Reproductive ecology of tropical forest plants. Man and the Biosphere Series. pp. 233–236. The Parthenon Pub. Group, Paris, France.

Vázquez-Yanes, C., Orozco-Segovia, A., Rincón, E., Sánchez-Coronado, M. E., Huante, P., Barradas, V. & Toledo, J. R. 1990. Light beneath the litter in a tropical forest: effect on seed germination. Ecology 71: 1952–1958.

Whitmore, T. C. 1989. Canopy gaps and two major groups of forest trees. Ecology 70: 536–538.

Vegetatio **107/108**: 319–338, 1993.
T. H. Fleming and A. Estrada (eds). Frugivory and Seed Dispersal: Ecological and Evolutionary Aspects.

Monitoring seed dispersal at isolated standing trees in tropical pastures: consequences for local species availability

S. Guevara[1,2] & J. Laborde[1]
[1]*Instituto de Ecología, Apartado Postal 63, 91000, Xalapa, Veracruz, México;* [2]*Laboratorio de Ecología, Facultad de Ciencias, Universidad Nacional Autónoma de México, 04510 México D. F., México*

Keywords: Frugivory, Landscape ecology, Seed deposition, Tropical Rain Forest

Abstract

The tropical rain forest landscape has been transformed to a mosaic composed of patches of crops, secondary vegetation and remnant forest fragments of different shapes and sizes. Isolation of patches and fragments is a critical issue in the maintenance of local species diversity. In this study we focus on the dispersal of propagules by birds to understand the movement of plants between landscape components. Seed deposition and the behavior of frugivorous birds were monitored at four isolated fig trees (*Ficus yoponensis* and *F. aurea*) in man-made pastures. Seed deposition was measured by trapping seeds under canopy trees for six months and by direct observation of bird visits to the four trees for one year. Seed deposition densities were 465, 614, 632 and 1097 seeds/m^2 accumulated over six months under each of the four trees. We recorded 8268 seeds of 107 species under the trees, among them, 6726 seeds (81%) were of 56 species dispersed by vertebrate frugivores. Seeds of tree species accounted for 26% of the total species. Seventy-three species of birds perched in the observed trees, and 3344 visits were made by 47 species of frugivores. Frugivorous birds occurred in two groups: resident species nesting in the pastures and resident species nesting elsewhere. Propagule exchange between landscape components is clearly influenced by the behavior of these two groups. Structure and dynamics of the landscape depend on plant species availability within the mosaic. This availability is high and suggests possibilities for the management of the local species diversity of tropical rain forests.

Nomenclature: follows Gómez-Pompa, A. & Sosa, V. 1991. Flora de Veracruz, Listado de Especies, Instituto de Ecología, A. C., Xalapa, México and Ibarra-Manríquez, G. & Sinaca, S. 1987. Listados Florísticos de Mexico VII, UNAM, México, for plants and A. O. U. 1983. Check list of North American birds, 6th edition. Allen Press, Inc. Kansas, USA, for birds.

Introduction

The original area of tropical rain forest (TRF) in Mexico is decreasing daily owing to deforestation. Originally, TRF in Mexico represented 1.6% of all neotropical rain forest, however Mexican humid forest has been drastically cleared in recent decades, resulting in its reduction to approximately one million hectares. This represents only 5% of its original area and consists of only a few tracts located in the states of Veracruz, Chiapas, Oaxaca, Campeche and Quintana Roo. The remaining TRF along the coastal lowlands of the Gulf of Mexico has been reduced to patches of irregular size and shape that are scattered among pastures and agricultural fields. The use of land for pasturing cattle has continued to gain economic importance and now accounts for 11% of the area of Mexico. This corresponds to 35% of all TRF area in Mexico (Toledo, *et al.* 1989). Remaining TRF is found on very steep hillsides, the peak of mountains and hills, along creeks and, rarely, in abandoned sites. The open fields are heterogeneous in vegetation structure and surrounded by living fences of trees of medium height. Also, open fields are crossed by riparian corridors of trees of moderate to tall heights and are interspersed with very large isolated standing trees (IST) whose number varies from pasture to pasture (Fig. 1).

Propagule dissemination seems to be a good estimator of the isolation of populations and species in these landscapes. Since a high proportion of the species of neotropical rain forests (Howe & Smallwood 1982) and local flora (Ibarra-Manríquez 1985) is dispersed by birds, the behavior and movements of frugivorous birds in the fragmented landscape are particularly important for the dispersal of seeds between and within different components of the landscape. Considering that many birds flying across open spaces look for perches (Wegner & Merriam 1979; McDonnel & Stiles 1983; Guevara *et al.* 1986), we decided to monitor seed dispersal around large trees isolated in pastures.

Floristic composition and structure of vegetation under the canopy of ISTs is clearly distinct compared to vegetation outside the canopy. Directly under most ISTs are several native rain forest species and late secondary succession species, which are uncommon in unshaded pastures (Guevara *et al.* 1986; Guevara *et al.* 1992). We hypothesized that the two most important factors which result in the greater abundance and richness of woody plants under ISTs are i) a more favorable environment for the germination and establishment of woody species and ii) a high number of the propagules of woody species beneath ISTs, resulting from seed deposition by frugivorous dispersers.

First hypothesis was tested by Guevara *et al.* (1992) through the analysis of the vegetation under *Ficus* spp. (n = 30) and *Nectandra* spp. (n = 20) trees. Their results showed clear floristic composition differences under the trees compared with open pastures. Most of the forest tree species dispersed by birds occurred frequently under IST canopies but were scarce in open pastures. These results complement this study.

To test the second hypothesis, we monitored seed deposition and the visitation frequency of frugivorous birds at four isolated trees in currently grazed pastures at Los Tuxtlas, Mexico. While in the canopy of ISTs, frugivorous birds regurgitate and defecate seeds ingested in adjacent vegetation spread over a wide area.

Our results show that several species of avian frugivorous are actively visiting ISTs throughout the year. A consequence of this is that a large number of seeds of zoochorous species are deposited under the canopy of ISTs in the middle of man-created pastures.

Study site

The study area is located in the volcanic mountain range of the Los Tuxtlas region in the southeastern area of the state of Veracruz, Mexico (between 18° 30'–18° 40' North latitude and 95° 03'– 95° 10' West longitude), near the 'Los Tuxtlas' Biological Research Station of the Universidad Nacional Autónoma de México (Fig. 1). The Los Tuxtlas range is covered with TRF with a closed canopy 30–35 m in height. Plant community structure is characterized by a relative dominance of palms, with *Astrocaryum mexicanum* in the understorey, *Pseudolmedia oxyphyllaria* (Moraceae) in the middle storey and *Nectandra ambigens* (Lauraceae) in the canopy (Bongers *et al.* 1988). Soils show poorly developed profiles and contain high amounts of organic material (Chizon 1984). They are classified as vitric andosols (FAO/UNESCO, 1975) and their texture is sandy loam. The area receives an average of 4900 mm of rainfall annually, with a relatively 'dry season' from March to May (monthly mean = 111.7 ± SD 11.7 mm), and a 'rainy sea-

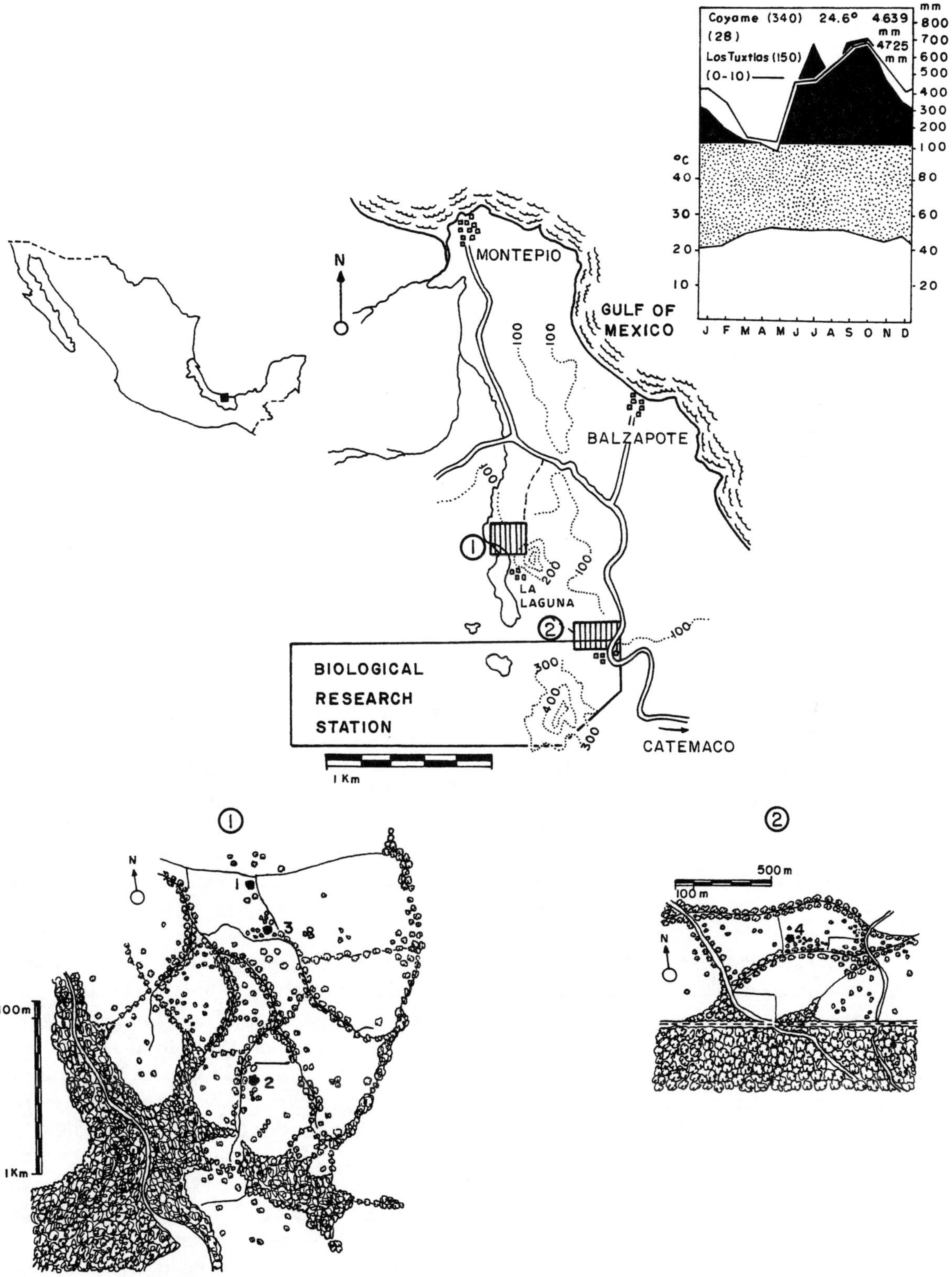

Fig. 1. Study site at Los Tuxtlas, Veracruz, Mexico and climatic diagram of Coyame (35 km from the study site).

son' from June to February (monthly mean = 486.2 ± SD 87.0 mm). Mean annual temperature is 27 °C (Lot-Helgueras 1976; Estrada *et al.* 1985).

Methods

Selection of trees

To identify the most important isolated trees species, we surveyed 13 pastures with a total area of 81.4 ha and determined the importance values, measured as the sum of relative abundance and relative frequency in pastures, of 265 individuals of 57 species. The results of this survey (Guevara *et al.* 1992), suggested that fig trees (*Ficus* spp.) are the most important. The infrutescences of *Ficus* are consumed by a large variety of frugivores, i.e. potential seed dispersers (Coates-Estrada & Estrada 1986; Bronstein & Hoffmann 1987; Van Dorp 1985). Four isolated trees in currently grazed pastures were selected for this study (Fig. 1). Two of these were 'free-standing' trees, *F. yoponensis*, Desv. (subgenus *Pharmacosycea*), hereafter refered to as tree 1 and 2. The other two were 'hemiepiphytic-strangler' trees, *Ficus aurea*, Nutt. (subgenus *Urostigma*), hereafter refered to as trees 3 and 4. Height of the trees were 26.9, 28.5, 24.1 and 24.6 m, and canopy cover was 669, 585, 415 and 335 m^2, respectively.

The syconia of *F. aurea* are relatively small (7–8 mm diameter), dark red in color and typically dispersed by birds (ornithochorous), while those of *F. yoponensis* are larger (13–18 mm diameter), yellow-green in colour and very aromatic. The latter are typically dispersed by mammals (mammalochorous). Fruiting phenology of the genus *Ficus* is asynchronous with year-round fruit production; a given tree may fruit two or more times annually (Milton *et al.* 1982; Windsor *et al.* 1989).

Observations of birds

Previous studies have shown that migratory birds can be important frugivores in secondary forest in the tropics (Martin 1985). North American migratory species are present in Los Tuxtlas region between September and April (Coates-Estrada & Estrada 1985). Many are frugivores which, in addition to resident birds, may visit IST in pastures during those months. Therefore, there were four observation periods, each lasting three consecutive days, at each tree. Observations were carried out when migratory birds were (1) present and (2) absent (May to August). This was done both when the selected trees were (3) fruiting and (4) when they were not fruiting.

All observations were made on days with favorable meteorological conditions for bird activity. Days with heavy rain and strong winds were avoided. Visits by all bird species to the study trees were recorded throughout the observation periods. Bird species, direction of arrival, time spent in the tree and direction of departure were recorded for each visit. Flight directions were classified as N, NE, E, SE, S, W, SW or NW with respect to the study tree. Morning observations began at sunrise and ended at 1030 h, when bird activity diminished markedly. Afternoon observations began at 1600 h and ended at sunset each day. Observations were made from a hidden position in neighbouring trees which provided a clear view of the canopy of the study tree. These were located 15 to 30 m from the study trees. Zoom binoculars (10–22 × magnification; objective lens diameter = 50 mm) were used for observations.

Exceptions to this methodology were: 1) Hurricane 'Hugo' and other meteorological disturbances made one observation period for tree 1 impossible (i.e. no migratory birds, tree not fruiting). Therefore we report the results of 15 three-day observation periods, performed between May 1988 and October 1989, totalling 247 hours of observation. 2) Tree 3 fruited when there were no migratory species in Los Tuxtlas. Therefore there are no data for this tree for the 'migratory species present, tree fruiting' condition.

χ^2 tests (Batschelet 1981) were used to determine if frequency of flight arrival direction was random. In order to assess the effect of nearby arboreal vegetation on flight arrival direction by birds, canopy cover was measured for all trees which were greater than 3 m in height, and located within a 75 m radius of the study trees. Arboreal vegetation within this radius was classified with respect to direction in the same manner as bird arrival flights. Spearman Rank Correlation was used to determine if the frequency of flight arrival directions and arboreal canopy cover in the vicinity of each study tree were related.

Seed deposition

To measure rates of seed deposition, 15 seed traps were randomly placed beneath the canopy of the four trees. Gaps in the tree canopy were avoided. Traps were conical with a diameter of 50 cm, a height of 51 cm, and were made of Nytall nylon mesh No. 6066 (pore size 280μ). Total seed trap area was 2.94 m^2 for each site. Each sampling site was enclosed by a barbed wire fence to prevent interference by cattle.

Seeds were collected from the traps monthly over a six month period (from 15 March to 14 September 1988). In the laboratory seeds were separated from litterfall, insect remains and trash and then sorted for identification and counting using a dissecting microscope (10×). Species identification was based on the Reference Seed Collection of the Laboratorio de Ecología, Facultad de Ciencias, UNAM and Estacíon de Biología Tropical Los Tuxtlas, UNAM.

Results

Birds

During 247 h of observation 73 species of birds in 17 families were seen perching on the four observation trees. Of these, 52 species were resident, 20 species were migratory and one was not determined (Table 1). The species list contains 15 primarily frugivorous birds and 32 facultative frugivores (total = 47 frugivorous species).

Table 1. Bird species observed perching in four Isolated Standing Trees (ISTs) of *Ficus* at Los Tuxtlas, Veracruz, México, and the number of visits they made to each tree.

Bird species	Family-Subf	DI	ST	HA	Number of visits to			
					Ficus yoponensis		*Ficus aurea*	
					Tr 1	Tr 2	Tr 3	Tr 4
Frugivores								
Amazona autumnalis (Linnaeus)	PSITT	Fs	r	MSO	2	6		55
Caryothraustes poliogaster (Du Bus de Gisignies)	EMB-Card	FI	r	MS		2		
Celeus castaneus (Wagler)	PICID	IF	r	MS				1
Coereba flaveola (Linnaeus)	EMB-Coer	NFI	lm	MS	1			
* *Columba flavirostris* (Wagler)	COLUM	Fs	r	SO			71	
Columbina inca (Lesson)	COLUM	Fs	r	O		2		
* *Cyanocorax morio* (Wagler)	CORVI	IFV	r	MSO	21	19	19	32
Cyanocorax yncas (Boddaert)	CORVI	IFV	r	S		2		
Dendroica petechia (Linnaeus)	EMB-Paru	IF	m	SO		4		
Dendroica spp.	EMB-Paru	IF	m	–	3	6		11
* *Dives dives* (Deppe)	EMB-Icte	IFs	r	SO	23	52	94	73
Euphonia affinis (Lesson)	EMB-Thra	FI	r	SO			2	
Euphonia hirundinacea Bonaparte	EMB-Thra	FI	r	MS	16	2	42	134
Habia sp.	EMB-Thra	IF	r	MS		2		
Icterus dominicensis (Linnaeus)	EMB-Icte	IF	r	S				2
Icterus galbula (Linnaeus)	EMB-Icte	IFN	m	MS	3	41	13	37
Icterus gularis (Wagler)	EMB-Icte	IF	r	SO	1	7	16	1

Table 1. Continued.

Bird species	Family-Subf	DI	ST	HA	Number of visits to			
					Ficus yoponensis		*Ficus aurea*	
					Tr 1	Tr 2	Tr 3	Tr 4
Icterus spurius (Linnaeus)	EMB-Icte	IFN	t	O	1		2	1
* *Megarhynchus pitangua* (Linnaeus)	TYR-Tyra	IVF	r	MSO	5	4	4	97
* *Melanerpes aurifrons* (Wagler)	PICID	IF	r	MS	149	96	233	144
Melanerpes pucherani (Malherbe)	PICID	IF	r	MSO			1	17
* *Myiodynastes luteiventris* Sclater	TYR-Tyra	IF	sr	MS	10	34	15	1
* *Myiozetetes similis* (Spix)	TYR-Tyra	IF	r	SO	19	41	15	16
Ortalis vetula (Wagler)	CRACI	FI	r	S				1
Pachyramphus aglaiae (Lafresnaye)	TYR-Tity	IF	r	SO		3	5	4
Piranga rubra (Linnaeus)	EMB-Thra	IF	m	MS	1		12	4
* *Pitangus sulphuratus* (Linnaeus)	TYR-Tyra	IFV	r	SO	23	3	30	34
* *Psarocolius montezuma* (Lesson)	EMB-Icte	IFV	r	MSO	3	11	2	243
Pteroglossus torquatus (Gmelin)	RAMPH	FIV	r	MSO	21	13		102
Ramphastos sulfuratus Lesson	RAMPH	FIV	r	MSO	2	20	8	44
Ramphocelus sanguinolentus (Lesson)	EMB-Thra	IF	r	MS		3	3	
Saltator atriceps (Lesson)	EMB-Card	FI	r	SO		20	5	
Saltator maximus (Müller)	EMB-Card	FI	r	SO	3		19	
* *Scaphidura oryzivora* (Gmelin)	EMB-Icte	IFs	r	O				127
Sturnella magna (Linnaeus)	EMB-Icte	IFs	r	O				1
* *Thraupis abbas* (Deppe)	EMB-Thra	IF	r	SO	50	22	97	211
Thraupis episcopus (Linnaeus)	EMB-Thra	FI	r	SO			1	1
* *Tityra inquisitor* (Lichtenstein)	TYR-Tity	FI	r	MS	39	23	11	5
* *Tityra semifasciata* (Spix)	TYR-Tity	FI	r	MSO	22	71	62	33
Tolmomyias sulphurescens (Spix)	TYR-Elea	IF	r	MS	1		2	
Trogon violaceus Gmelin	TROGO	IF	r	MS	1	4		23
Turdus assimilis Cabanis	MUS-Turd	IF	lm	MS			1	7
Turdus grayi Bonaparte	MUS-Turd	IF	r	MS	5	4	16	99
Tyrannus melancholichus Vieillot	TYR-Tyra	FI	r	O			2	
Tyrannus tyrannus (Linnaeus)	TYR-Tyra	IF	t	O		32		
Vireo flavifrons Vieillot	VIREO	IF	m	SO	3			2
Vireo (aff) *solitarius*	VIREO	IF	m	MS		1		
Non frugivores								
Basileuterus culicivorus (Deppe)	EMB-Paru	I	r	MS		9		
Basileuterus rufifrons (Swainson)	EMB-Paru	I	r	SO	6	24		
Buteo nitidus (Latham)	ACCIP	VI	r	SO		1	2	4
Buteogallus anthracinus (Deppe)	ACCIP	VI	r	MSO			1	
* *Campylorhynchus zonatus* (Lesson)	TROGL	I	r	SO		3	12	74
Contopus cinereus (Spix)	TYR-Fluw	I	r	MS	7			
* *Crotophaga sulcirostris* Swainson	CUCUL	I	r	SO	27	5	6	1
Dryocopus lineatus (Linnaeus)	PICID	I	r	SO	2	1	2	
Empidonax flaviventris (Baird & Baird)	TYR-Fluv	I	m	MS	1		1	
Empidonax minimus (Baird & Baird)	TYR-Fluv	I	m	S		2		
Empidonax sp.	TYR-Fluv	I	–	–			1	
Geothlypis poliocephala Baird	EMB-Paru	I	r	O				1
* *Glaucidium brasilianum* (Gmelin)	STRIG	VI	r	MSO			2	
Herpetotheres cachinnans (Linnaeus)	FALCO	V	r	SO		1		
Lepidocolaptes souleyetii (Des Murs)	DENDR	I	r	MS		2		2

Table 1. Continued.

Bird species	Family-Subf	DI	ST	HA	Number of visits to			
					Ficus yoponensis		*Ficus aurea*	
					Tr 1	Tr 2	Tr 3	Tr 4
Mniotilta varia (Linnaeus)	EMB-Paru	I	m	MS		2		3
Myiarchus sp.	TYR-Tyra	I	r	MS	12	12		
Oporornis formosus (Wilson)	EMB-Paru	I	m	MS				5
Parula americana (Linnaeus)	EMB-Paru	I	m	O	2		1	
Piaya cayana (Linnaeus)	CUCUL	I	r	MS		1		
Piculus rubiginosus (Swainson)	PICID	I	r	MS				1
Polioptila caerulea (Linnaeus)	MUS-Sylv	I	m	MS	5	4	16	2
Setophaga ruticilla (Linnaeus)	EMB-Paru	I	m	MS		1		
Vireolanius pulchellus Sclater & Savin	VIREO	I	lm	M			1	
Wilsonia citrina (Boddaert)	EMB-Paru	I	m	MS			2	
Wilsonia pusilla (Wilson)	EMB-Paru	I	m	MSO	33	15	6	17
Non-identified visits					2	16	5	21
Subtotals	Frugivores	Visits			428	550	803	1563
		Species			26	30	29	32
	Non frugivores	Visits			95	83	53	110
		Species			9	15	13	10
	Total	Visits			523	633	856	1673
		Species			35	45	42	42

* Bird species nesting in ISTs (any tree species) in pastures. *Columba flavirostris* was the only species which constructs its nest directly in one of the target trees (Tree 3).
Family (fiver letter abbreviations or FAM-Subfamily abbreviation).
DI = Diet (F = Fruit, potential seed dispersers; I = Invertebrates; V = Vertebrates; N = Nectar; s = seeds, potential seed predators).
ST = Status (r = resident; m = Nort American migratory; lm = local migratory; sr = summer resident; t = transitory).
HA = Habitat (M = Mature Rain forest; S = Secondary forest; O = Open areas).
Diet, status and habitat data are from Coates-Estrada & Estrada (1985). Diet data are in order of feeding preference.
Nomenclature follows 'Check-list for North American Birds' (A.O.U. 1983, 6th edition).

Twenty two species were insectivores and four species consumed vertebrates. Bird species in the latter two categories do not consume fruit and therefore do not disperse seeds. The frugivorous birds observed represent 32% of the total frugivorous avifauna reported for the Los Tuxtlas Biological Station, not including aquatic species (Coates-Estrada & Estrada 1985).

There was a total of 3,685 visits to the four trees by all bird species. Of these, 91% were made by the 47 frugivorous species (Table 1). Within this guild the facultative frugivores were most numerous and visited the trees most often (Fig. 2a). The frugivorous birds spent a total of 409.5 h perching in the trees. The time spent during each visit varied widely (range: <1–50 min). Fifty percent of the visits lasted between 2 and 10 min. The average time spent during each visit at *F. aurea* trees was 7.7 ± 9.9 S. D. min while the average time spent et *F. yoponensis* was 6.1 ± 7.4 min.

Thirteen species of resident frugivorous birds had nests in ISTs in the pastures and made 2,409 visits (72% of the frugivore visits). Twenty-two of the resident species did not have nests in the pasture and only made 749 visits (22.4% of the frugivore visits). The 11 migratory species made 186 visits (5.6%, Fig. 2b).

Visitation frequencies to ISTs by frugivorous

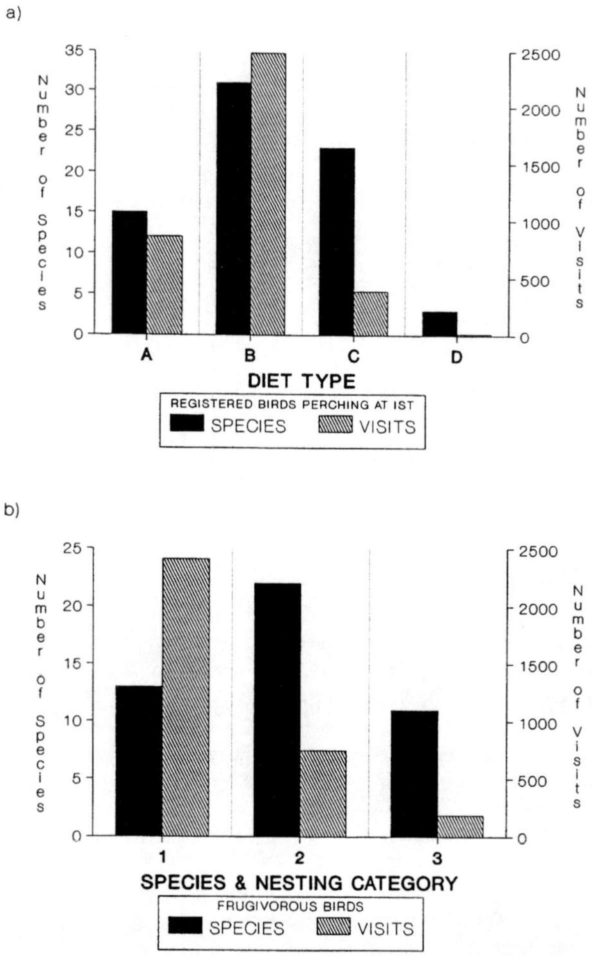

Fig. 2. Number of species and number of visits for birds observed perching in 4 ISTs of *Ficus* spp. in pastures at Los Tuxtlas, Ver. Mexico. Results are grouped in bird species categorized by diet type (a) and for frugivorous birds by nesting category (b). (a) Bird species which; A) primarily eat fruit, B) had a mixed diet of animal origin and fruit, C) eat invertebrates (mainly insects) and D) eat vertebrates. Species in categories 'A' and 'B' are considered as 'Frugivorous birds' in this study, while those in 'C' and 'D' do not eat fruit. Species diets are based on Coates-Estrada & Estrada (1985), see Table 1. (b) Frugivorous birds of resident species which; 1) were observed nesting in ISTs (trees of any species), 2) were not observed nesting in ISTs and of 3) North American migratory species which do not nest in the Los Tuxtlas region.

birds were highly variable. In 14 of the 15 observation periods the number of visits per hour to the study trees by all species of frugivorous birds ranged from 5 to 18. During one observation period at tree 4, a visitation rate of 80 visits/hour was recorded. This occurred in March 1989 when this tree had copious fruit and fruit availability in the neighboring vegetation was extremely limited. Our results show that ISTs are continuously visited by frugivorous birds throughout the year when metereological conditions are favorable for bird activity. However, despite clear differences in fruit morphology between the two species of *Ficus* and the offering of a feeding reward when the tree is fruiting, there was no significant relationship between the frequency of visits to ISTs by frugivorous birds and the presence of ripe fruit in the tree ($F_{(1, 11)} = 2.251$; $P = 0.162$) or the species of *Ficus* involved ($F_{(1, 11)} = 1.523$; $P = 0.243$).

Visitation frequency for resident birds which were nesting in the pasture was generally highest when the observed tree was fruiting, as in case of both *F. aurea* trees (Fig. 3). Although visitation frequency by birds in this category generally decreased during the times of the year when migratory birds were present, this decrease was not coincident with an increase in visits by migratory bird species. In 14 of the 15 observation periods, both when the trees were fruiting and not fruiting, visitation frequency by species of resident birds not nesting in the pasture was much less than that of species of resident birds with nests in the pastures. Visitation frequencies of migratory species were very low in seven of the nine observation periods when they were in the region (Fig. 3).

Tree 4 was an exception to these general trends for the observation period when this tree was fruiting and migratory species were in the region for reasons noted above. During one observation period at tree 2, a mixed flock containing many birds of the migratory species *Icterus galbula* and *Tyrannus tyrannus* was foraging for insects and fruit in the vicinity of the observed tree. This explains the relatively high frequency of visits to tree 2 by migratory birds during that observation period.

The total number of frugivorous species recorded for all observations at each tree ranged between 26 to 32 species (Table 1) . The average number of species per observation period (of three consecutive days) was: 13 ± 2.6 S. D., 17.5 ± 2.1, 15.5 ± 1.0 and 18.7 ± 6.2, respectively for trees

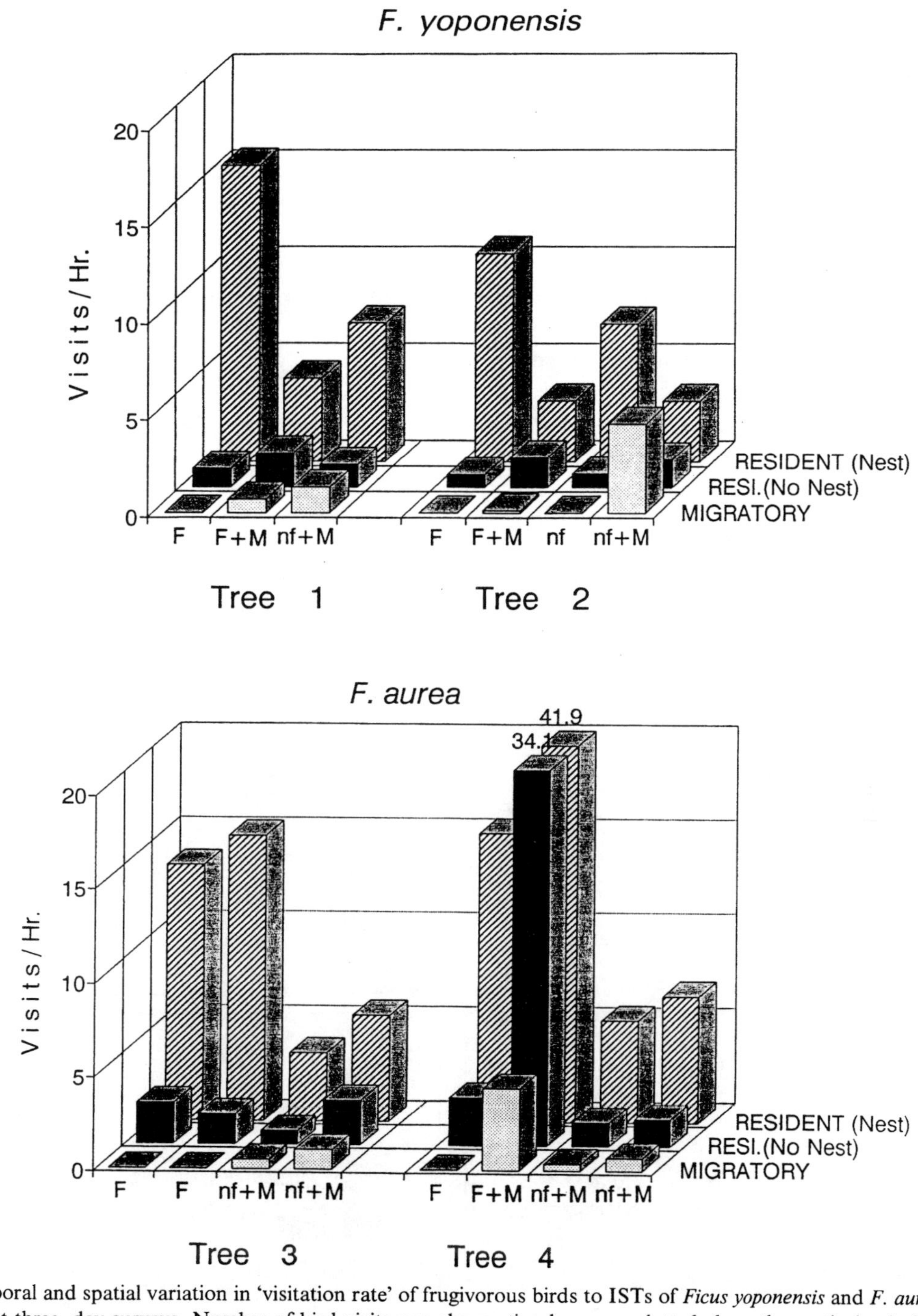

Fig. 3. Temporal and spatial variation in 'visitation rate' of frugivorous birds to ISTs of *Ficus yoponensis* and *F. aurea* as detected in 15 different three–day surveys. Number of bird-visits per observation hour are plotted along the vertical axis for each survey (showed along X axis). Surveys were carried out at different times of the year, when the surveyed tree was bearing ripe fruits (F) or when there were no fruit (nf) on the tree as well as when North American migratory species were present (M) in the study area or when absent. Species are grouped according to 3 nesting categories; resident species for which at least one functional nest in IST was observed, resident species which do not nest in IST and N. Amer. migratory species which do not nest in the Los Tuxtlas region.

1–4. Differences between trees were not significant ($F_{(3, 11)} = 1.643$, $P = 0.236$).

Bird species similarity was highest between *F. yoponensis* tree 1 and *F. aurea* trees 3 and 4, (Sorensen Similarity Index = 0.73 and 0.83 respectively), followed by the two individuals of *F. aurea* (S. S. Index = 0.72); *F. yoponensis* tree 2 showed the lowest similarity values (S. S.Index = 0.68, 0.64, and 0.65) with trees 1, 3 and 4 respectively.

Frequency of flight arrival direction was non-random ($\chi^2 = 272.9$ $P \ll 0.001$; 355.5 $P \ll 0.001$; 293.9 $P \ll 0.001$ and 244.3 $P \ll 0.001$) at each of the four study trees (Fig. 4). This variable was correlated with the arboreal canopy cover within a 75 m radius of the *F. yoponensis* study trees (Spearman Rank Correlation; rs = 0.76 $P < 0.05$ and rs = 0.89 $P < 0.01$, tree 1 and 2, respectively). Despite the non-random frequency of arrival directions, this correlation was not significant for either *F. aurea* tree, (SRC; rs = −0.1190 and rs = 0.6190 for trees 3 and 4, respectively).

Median body size of frugivorous birds observed at ISTs did not differ from the median size of the entire frugivorous avifauna at the Los Tuxtlas Biological Station (Fig. 6).

Seed deposition

Total seed capture between March and September 1988 was 8,268 seeds. Mean cummulative

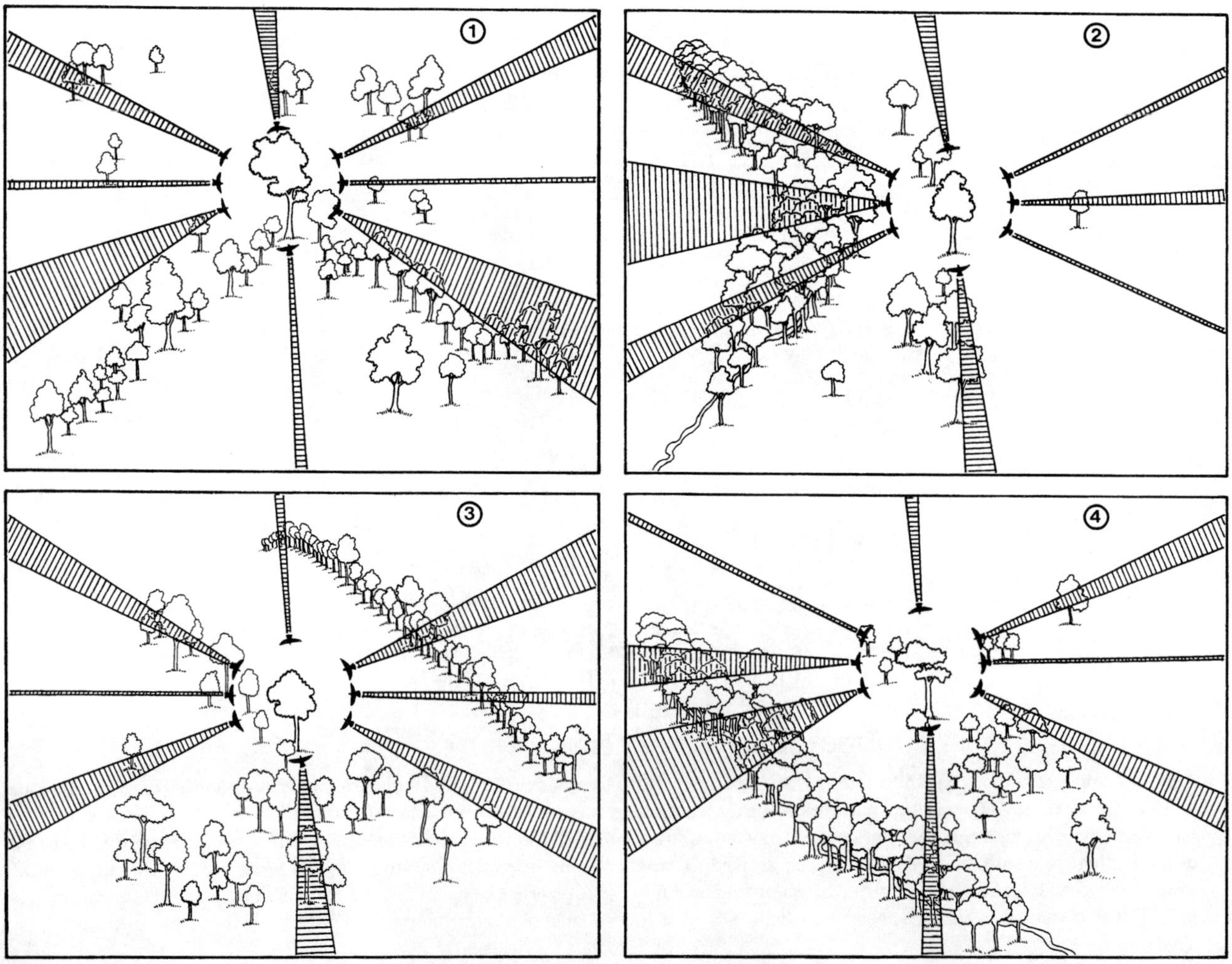

Fig. 4. Bird arrival directions for *Ficus yoponensis* and *F. aurea* trees. Frequency is shown by the width of the flight strip.

seed density over the entire sampling period was 710 ± 239 seeds/m^2 beneath the canopies of the four ISTs. One hundred and seven species (including 13 morphospecies) were captured in the 60 seed traps from the four trees. Number of species recorded under each tree was 59, 65, 68 and 57 for trees 1–4 (Table 2).

Table 2. Number of seeds of plant species captured in seed traps beneath four ISTs of *Ficus* spp. at Los Tuxtlas, Veracruz, México, between March 15 and September 14 in 1988.

Plant species	Family	GF	DM	Number of seeds beneath			
				Ficus yoponensis		*Ficus aurea*	
				Tr 1	Tr 2	Tr 3	Tr 4
Acalypha diversifolia Jacq.	EUPHO	S	–	5	4		1
Aechmea bracteate (Sw.) Griseb	BROME	E	Z	14	32	8	24
Aechmea nudicaulis (Sw.) Griseb	BROME	E	Z			1	2
Anthurium scandens (Aublet) Engl.	ARACE	EH	Z				1
Blechum brownei (L.) Ant. Juss.	ACANT	H	–	1		11	1
Borreria sp. (1)	RUBIA	H	–		30		
Borreria sp. (2)	RUBIA	H	–		1		
Bursera simaruba (L.) DSarg.	BURSE	T	Z	20	9	28	7
Casearia sylvestris Sw.	FLACO	T	Z	1			
Cecropia obtusifolia Bertol.	MORAC	T	Z	499	700	497	1030
Cedrela odorata L.	MELIA	T	A		1	1	
Chaptalia nutans (L.) Polak	COMPO	H	A			1	
Cissus microcarpa Vahl	VITAC	V	Z		1	8	1
Clibadium arboreum J. D. Smith	COMPO	S	–		1		
Clusia sp.	GUTTI	ET	Z				3
Conostegia xalapensis (Kunth in H.B.K.) Don ex DC	MELAS	S	Z	232	21	232	17
Cordia spinescens L.	BORAG	H	Z	164	132	20	1
Cordia stenoclada I. M. Johnston	BORAG	T	Z			1	
Coussapoa purpusii Standley	MORAC	ET	Z	4	7	10	14
Cupania glabra Sw.	SAPIN	T	Z	4	1	1	
Cymbopetalum baillonii R. E. Fries	ANNON	T	Z		1	1	
Cyperus tolucensis Kunth in H.B.K.	CYPER	H	–	11			1
Dendropanax arboreus (L.) Decne. & Planchon	ARALI	T	Z			2	
Drymonia strigosa (Oersted) Whieler	GESNE	V	–		1		9
Epiphyllum crenatum (Lindley) G. Don	CACTA	E	Z		1		
Eugenia capuli (Cham & Schldl.) Berg.	MYRTA	T	Z	2	2	33	1
Eupatorium araliifolium Less	COMPO	H	A	1	8	4	2
Eupatorium galeottii Robins	COMPO	T	A	1	22		
Eupatorium macrophyllum L.	COMPO	H	A				1
Eupatorium sp. (1)	COMPO	S	A	3	26	1	2
Eupatorium sp. (2)	COMPO	H	A	13	29	16	10
Eupatorium sp. (3)	COMPO	H	A	1	3	1	2
Eupatorium sp. (4)	COMPO	H	A		1	3	
Euphorbia heterophylla L.	EUPHO	H	Z	25	24	1	
Ficus (*Pharmacosycea*) spp.	MORAC	T	Z	*	*	6	22
Ficus (*Urostigma*) spp.	MORAC	T	Z	358	204	*	*
Forsteronia viridescens Blake	APOCI	V	A		1		1
Gouania lupuloides (L.) Urban	RHAMN	V	A		1	2	
Guarea glabra Vahl	MELIA	T	Z				1

Table 2. Continued.

Plant species	Family	GF	DM	Number of seeds beneath			
				Ficus yoponensis		*Ficus aurea*	
				Tr 1	Tr 2	Tr 3	Tr 4
Hampea nutricia Fryx.	MALVA	T	Z	1			
Heliconia uxpanapensis Gutiérrez	MUSAC	H	Z	1			
Heliocarpus sp.	TILIA	T	A		7	2	3
Hyptis sp.	LABIA	H	–	14	3	1	2
Juanulloa mexicana (Schldl.) Miers	SOLAN	ES	Z				4
Lantana camara L.	VERBE	H	Z	2			1
Liabum discolor (Hook. & Arn.) Benth. & Hook	COMPO	H	A			1	
Mascagnia vaccinifolia Niedz	MALPI	V	A		4	7	1
Mikania leiostachya Benth.	COMPO	H	A		33	4	3
Myriocarpa longipes Liebm	URTIC	S	Z	2	9	23	
Neurolaena macrocephala Sch. Bip. ex Hemsl.	COMPO	S	A	4	4	6	1
Oeropanax sp.	ARALI	ET	Z	23	49	12	34
Orthion oblanceolatum Lundell	VIOLA	T	–	493[a]			2
Parathesis lenticellata Lundell	MYRSI	T	Z			19	
Passiflora helleri Peyr.	PASSI	V	Z		1	1	44
Phyllanthus compressus Kunth in H.B.K.	EUPHO	H	–	5	9	1	
Phytolacca rivinoides Kunth & Bouché	PHYTO	H	Z				1
Piper aequale Vahl	PIPER	S	Z			1	
Piper amalago L.	PIPER	S	Z	2			2
Piper auritum Kunth in H.B.K.	PIPER	S	Z	312	68	10	15
Piper hispidum Sw.	PIPER	T	Z	485	93	92	262
Piper sanctum Schldl. ex Miq.	PIPER	S	Z	3	4	1	
Piptocarpha chontalensis (L.) Urban	COMPO	S	A	3	4		1
Platymicium pinnatum (Jacq.) Dugand	LEGUM	T	A		1		
Psychotria chiapensis Standley	RUBIA	T	Z		1	1	
Rhipsalis bartletti Clover	CACTA	E	Z	1	8		5
Rhynchospora radicans (Schldl. & Cham.) Pfeiffer ssp. *radicans*	CYPER	H	–	23	2	2	
Robinsonella mirandae Gómez-Pompa	MALVA	T	A	9	2	35	79
Rollinia jimenezii Saff.	ANNON	T	Z				4
Sapium nitidum (Monach.) Lundell	EUPHO	T	Z	1	1	2	2
Saurauia yasicae Loes	ACTIN	T	Z	2	5		
Scleria sp.	CYPER	H	Z	12	7	1	1
Siparuna andina (Tul.) A. DC.	MONIM	S	Z		2		1
Solanum rugosum Dunal	SOLAN	H	Z			2	
Solanum schlechtendalianum Walp.	SOLAN	S	Z	10	8	5	
Solanum sp. (1)	SOLAN	H	Z	61		6	14
Solanum sp. (2)	SOLAN	H	Z	162	23	2	3
Souroubea loczyi (V. Richter) Room	MARCG	ES	Z			1	1
Stemmadenia sp.	APOCY	T	Z	3		4	1
Syngonium podophyllum Schott	ARACE	EH	Z	1		1	
Tetrorchidium rotundatum Standley	EUPHO	T	Z	6	19	14	8
Trema micrantha (L.) Blume	ULMAC	T	Z	12	12	5	7
Trichilia breviflora Blake & Standley	MELIA	T	Z	1			
Verbesina turbacensis Kunth in H.B.K.	COMPO	H	A		8	1	
Vernonia patens Kunth in H.B.K.	COMPO	H	A	5	14	6	10
ASCLEPIADACEAE	ASCLE	–	A			1	

Table 2. Continued.

Plant species		Family	GF	DM	Number of seeds beneath			
					Ficus yoponensis		*Ficus aurea*	
					Tr 1	Tr 2	Tr 3	Tr 4
COMPOSITAE (1)		COMPO	–	A	4	11	2	15
COMPOSITAE (2)		COMPO	–	A		13		2
COMPOSITAE (3)		COMPO	–	A			1	1
COMPOSITAE (4)		COMPO	–	A	2	23	6	154
COMPOSITAE (5)		COMPO	–	A	1	6	157	4
EUPHORBIACEAE		EUPHO	–	Z	1			
SOLANACEAE (1)		SOLAN	–	Z	3	55	3	
SOLANACEAE (2)		SOLAN	–	Z			2	
SOLANACEAE (3)		SOLAN	–	Z	183		22	
Not identified (n = 13 spp.)		–	–	–	20	35	16	18
Subtotals	Zoochorous	Seeds			2613	1500	1079	1534
		Species			35	30	38	32
	Non zoochorous	Seeds			619	308	289	326
		Species			24	35	30	25
	Total	Seeds			3232	1808	1368	1860
		Species			59	65	68	57

Family (five leter abbreviation).
GF Growth form (T = Tree; S = Shrub; V = Vine; E = Epiphyte & Hemiepiphyte; H = Herb).
DM = Dispersal mode (Z = Zoochory; A = Anemochory).
[a] A fruiting individual of *Orthion oblanceolatum* was located less than 3 m from some of the seed traps beneath Tree 1.
* Seeds of *Ficus* were identified only to subgenus and those seeds of the same subgenus as the sampled IST were not counted.
Nomenclature follows Gómez-Pompa & Sosa (1991) and Ibarra-Manríquez & Sinaca (1987).

Seeds dispersed by birds or bats were abun dant under the four ISTs. A total of 6,726 seeds (81% of total capture) belonging to 56 zoochorous species (52%) was recorded during the sampling period. For the zoochorous species a mean of 1,681 $\pm$ 526.3 S. D. seeds per tree and a mean of 34 $\pm$ 3.1 species per tree were recorded. The most abundant species were *Cecropia obtusifolia*, *Conostegia xalapensis*, *Cordia spinescens*, *Oreopanax* sp., *Piper auritum*, *P. hispidum* and *Ficus* spp. (Table 2). Except for *Piper* species, the fruits of these plants are commonly eaten by several of the frugivorous birds that visited the trees.

Species abundance curves (Fig. 5) show that a greater number of zoochorous than anemochorous species were deposited under each of the four trees. The number of seeds was higher for zoochorous than anemochorous species; most zoochorous species were represented by more than three seeds.

Tree species (primary, secondary and gap species) were the most common life forms among the seeds and accounted for 26% of the species and 62% of the total seeds (Table 3). Most of the seeds of tree species were of zoochorous species commonly dispersed by frugivorous birds. The scarcity of herb seeds can be explained by their dispersal mode and the shape and height of the traps.

The median size of zoochorous seeds captured under ISTs was significantly smaller than that of all rain forest seeds that commonly ripen their

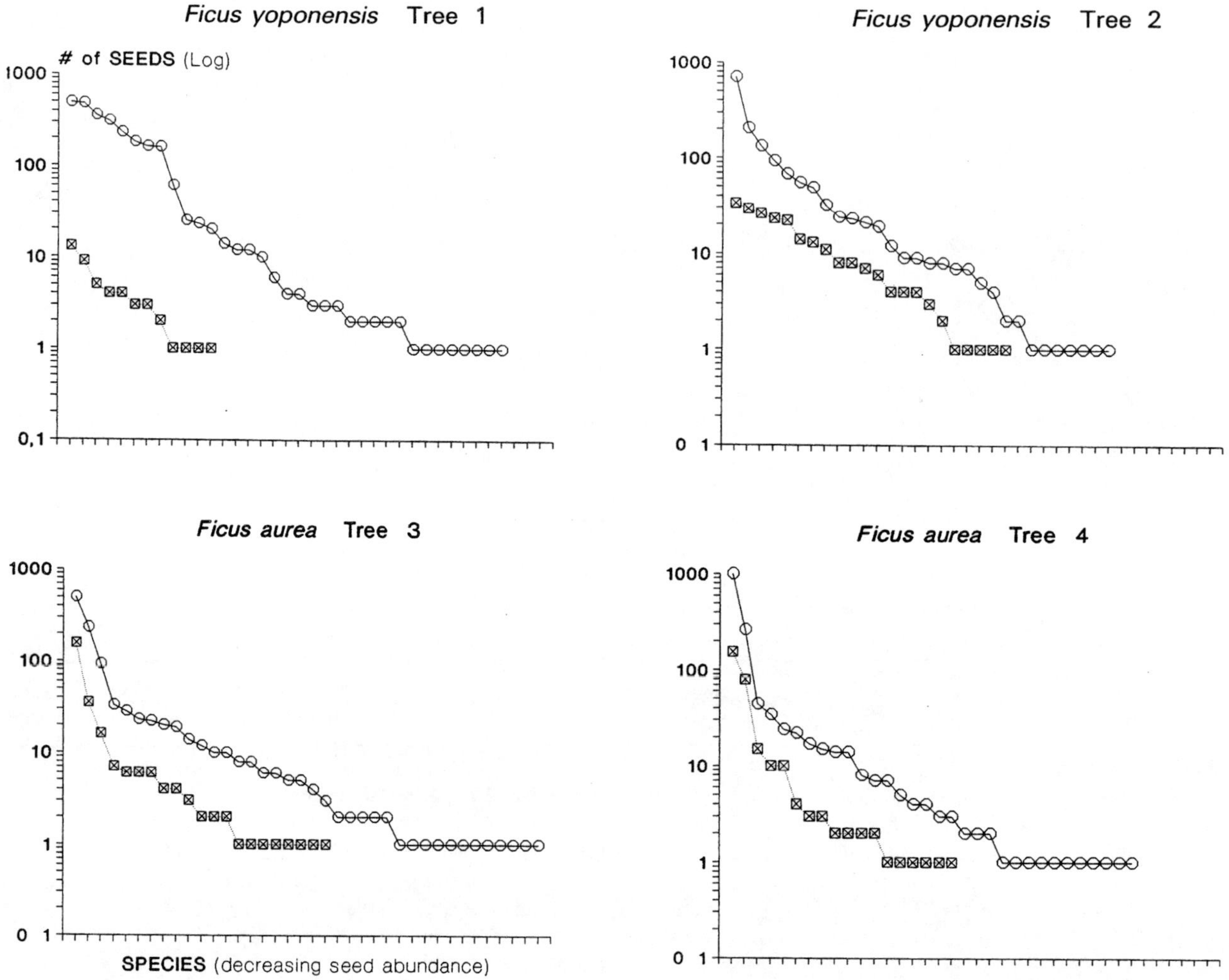

Fig. 5. Number of seeds (Logarithmic) for species captured in seed-traps beneath four ISTs of *Ficus* spp. in pastures of Los Tuxtlas between March 15 and September 14 1988. Captured species are subdivided as wind dispersed (Anemochorous, squares in the graph) and animal dispersed (Zoochorous, circles) species and they are ordered along the horizontal axis in decreasing seed abundance.

fruit during the months included in the sampling period (Fig. 6).

Fate of seeds

A previous study in currently grazed pastures in the same area showed that seeds of woody species deposited under the canopy of ISTs are successfully established (Guevara *et al.* 1992). For each of 50 ISTs, two 4-m^2 quadrats were sampled under the canopy and 5 m outside the canopy in the open pasture. The mean number of plant species per quadrat under canopies was significantly higher than open pasture sites. Vegetation under the canopy was structurally and floristically different from open pasture. Compositional differences were related to a higher proportion of zoochorous and rainforest species at canopy sites. Structural differences were related to a higher number and density of tree species recruits. The mean density of tree species (present mostly as seedlings and saplings) for the canopy sites was seven times greater than the mean density recorded for the open pasture sites (Table 4).

Table 3. Distribution of plant growth forms among the species captured in seed traps beneath four ISTs of *Ficus* at Los Tuxtlas, Veracruz, México.

Growth form	*Ficus yoponensis*		*Ficus aurea*		Total
	Tr 1 # of species (%) **# of seeds (%)**	Tr 2	Tr 3	Tr 4	
Tree	17 (29) **1898 (59)**	17 (26) **1081 (60)**	18 (27) **744 (55)**	14 (25) **1429 (77)**	28 (26) **5152 (62)**
Shrub	10 (17) **576 (18)**	11 (17) **151 (8)**	8 (12) **279 (20)**	8 (14) **40 (2)**	13 (12) **1046 (13)**
Vine	0 **0**	6 (9) **9 (1)**	4 (5) **18 (1)**	5 (9) **56 (3)**	6 (6) **83 (1)**
Epihyte & hemiepiph.	5 (8) **43 (1)**	5 (8) **97 (5)**	6 (9) **33 (3)**	9 (16) **88 (5)**	11 (10) **261 (3)**
Herb	16 (27) **501 (15)**	16 (25) **327 (18)**	19 (28) **84 (6)**	15 (26) **53 (3)**	26 (24) **965 (12)**
Unidentified growth form	11 (19) **214 (7)**	10 (15) **143 (8)**	13 (19) **210 (15)**	6 (10) **194 (10)**	23 (22) **761 (9)**
Total per site					
Species	59 (100)	65 (100)	68 (100)	57 (100)	107 (100)
Seeds	**3232 (100)**	**1808 (100)**	**1368 (100)**	**1860 (100)**	**8268 (100)**

Table 4. Floristic composition of pastures in Los Tuxtlas, Veracruz, México. Analysis from sampling 50 ISTs (30 of *Ficus* spp. and 20 of *Nectandra* spp.). For each sampled tree, two 2 m × 2 m quadrats were located; one under the tree-canopy and the other 5 m outside the canopy in the open pasture. From Guevara *et al.* 1992.

Pasture vegetation	Under IST canopy	Open pasture
Species richness	193	107
Herb species (%)	84 (43.5)	65 (60.7)
Woody species (%)	109 (56.5)	42 (39.3)
Zoochorous species (%)	97 (50.3)	40 (37.4)
Mean density of tree saplings/m^2 (SD)	1.4 (0.7)	0.2 (0.3)

Discussion

ISTs in tropical pastures are excellent sites for monitoring seed dispersal. In one year 4 fig trees of 2 species were visited by 23% of the total local avifauna, of which 63% were frugivorous species. Under the same four trees a high number of seeds belonging to 56 zoochorous species were deposited during six months. This abundant seed input under the trees can be related to the observation that most seed defecation and regurgitation by avian frugivores occurs when birds are perched or immediately after they take off, rather than during flight (McDonnel & Stiles 1983; Stiles & White 1986; Charles-Dominique 1986). Seed deposition is also related to time spent perching in the trees. Our data agree with results of Bronstein & Hoffman (1987) for fig species (1.9–19.7 min with a mode of 3.3–8.5 min) and with data of Wheelwright (1991) for bird dispersers of lauraceous trees (5.5–11.7 min) as well as Pratt & Stiles' (1983) data for various tree species in Asia (4.5–12.5 min).

Frugivory and seed deposition

Isolated *F. yoponensis* and *F. aurea* trees were visited by 36 and 39 species of avian frugivores,

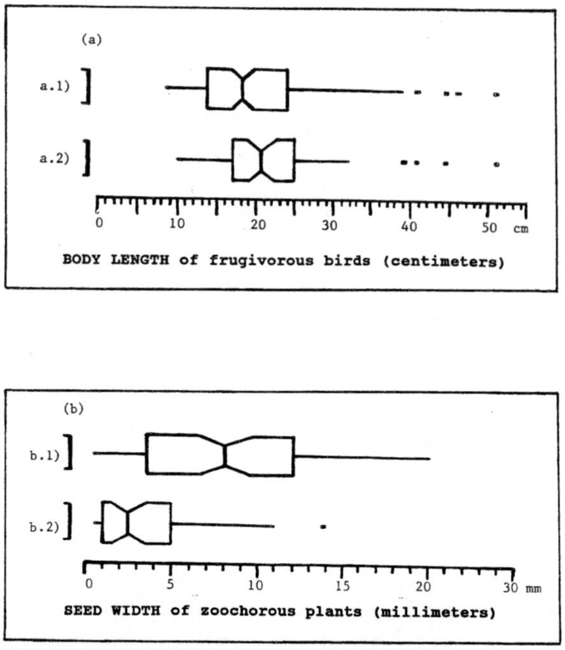

Fig. 6. Size distribution of frugivorous birds and zoochorous seeds. Notched box & whisker plot, which display a statistical summary, dividing the data in four areas of equal frequency. The boxes contain 50% of the data and the vertical line which divides it is the 'median' the notched portion of the box is the 'median' confidence interval. Each plot displays the size distribution of: (a) Body lenghts (cm) of frugivorous birds comparing; a.1) species reported for 'Los Tuxtlas', (n = 145, aquatic species are not included) data from Coates-Estrada & Estrada (1985), with a.2) species observed perching on any of the 4 IST of *Ficus* (n = 47). (b) Seed widths (mm) of zoochorous species comparing; b.1) rain forest species whose fruit commonly ripen between March and September (n = 73) data are from Ibarra-Manríquez 1985, with b.2) species captured in seed-traps under four isolated trees of *Ficus* spp. between March 15 and September 14 in 1988 (n = 38).

respectively. This represents as many or even more visitors than those reported in previous studies of fig trees in TRFs of Costa Rica and Los Tuxtlas (Bronstein & Hoffman 1987; Coates-Estrada & Estrada 1986; Van Dorp 1985). This is particularly interesting because we were studying isolated trees in pastures which have lower chances of being visited by birds than fig trees in the rain forest (Bronstein & Hoffman 1987). Figs produce huge crops of fruits (two times each year for *F. yoponensis* and continuously for *F. aurea*, Milton *et al.* 1982; Windsor *et al.* 1989) that are attractive to many species of frugivorous vertebrates. As keystone resource species (Terborgh 1986), fig trees are the most common and appealing perching trees in the pastures of Los Tuxtlas region and elsewhere in Central America (Janzen 1979; McKey 1989).

Bird species that move seeds across pastures can be divided into two groups: those nesting in the pastures and those not nesting in the pastures. The first group includes birds nesting in pastures that are also common in TRF, e.g. the Golden-fronted Woodpecker (*Melanerpes aurifrons*), the Brown Jay (*Cyanocorax morio*) and the Masked Tityra (*Tityra semifasciata*). They are important in the movement of seeds within the pastures, but they are particularly important in the movement of seeds from forest fragments to pastures. This group also includes species which nest in pastures that prefer open or disturbed areas and that rarely move into TRF fragments. Among these birds, the most important visitors to ISTs were the Great Kiskadee (*Pitangus sulphuratus*), the Boat-billed Flycatcher (*Megarhynchus pitangua*) and the Melodious Blackbird (*Dives dives*). These birds consume fruits from the edges of the rain forests, secondary forests, riparian zones, living fences and from isolated trees.

The second group of seed dispersers contains species of frugivorous birds such as the toucans (*Ramphastos sulfuratus* and *Pteroglossus torquatus*) as well as the Clay-colored Robin (*Turdus grayi*), that do not nest in pastures. These avian frugivores have a very broad fruit diet that contains a high proportion of primary rain forest species (Van Dorp 1985). The main feeding and reproductive (nesting site) areas of these species are mature rain forest tracts (fragments), and they perch on ISTs in order to cross the pastures which separate forest fragments. Therefore, these birds are important in the movement of seeds between rain forest fragments and when they eat fruit at ISTs in pastures, they may reintroduce their seeds into the forest.

Migratory species of frugivorous birds are not important visitors to ISTs in pastures. Even during the season when they are present in the Los Tuxtlas region (September to April) most visits to our trees were made by resident species. Perhaps the most important disperser of seeds among the migratory species is the Summer Tanager (*Piranga rubra*) which consumes the fruits of several plants at Los Tuxtlas and which was observed at three of the four study trees (Table 1).

The high visitation frequencies of frugivorous bird species and their high numbers suggest a high degree of propagule movement across the pastures. Although we sampled only for a six month period, we trapped 6726 seeds of 56 zoochorous species under the four sampled ISTs, which demonstrates the high level of activity of frugivores in the pastures and their low degree of dietary specialization.

Considering all recorded plant species (regardless of dispersal mode), total richness in seed deposition under isolated trees (107 species) is comparable to that which occurs beneath the rain forest canopy and in secondary forests in Costa Rica – 75 and 48 species, respectively (Young *et al.* 1987). Number of seeds deposited under isolated trees (701 seeds/m^2, range: 460–1096 seeds/m^2) is also similar to the values obtained in the Costa Rican study (490–2090 seeds/m^2).

Flight behaviour of birds

Bird flight direction followed distinct pathways, as seen in the frequency of arrival at each tree (Fig. 4). As we have demonstrated, arrival directions at ISTs are not random but are influenced by the arboreal vegetation in the vicinity of the trees. In both *F. aurea* trees flight arrival directions were not correlated with the arboreal vegetation within a 75 m radius of the study tree, but arboreal elements outside this radius were used by the birds to reach the study trees. In agreement with Wegner & Merriam (1979) for temperate areas, we found that the movement of frugivorous birds in open areas (tropical pastures) is governed by the presence and spatial arrangement of forest elements such as riparian corridors, living fences or other ISTs. Nevertheless, there was no difference in the number of zoochorous species and seeds deposited under *F. yoponensis* tree 1 (the most isolated tree), and the least isolated *F. aurea* tree 4. This reflects the negative correlation between the distance of the seed source from the tree and seed dispersion and deposition under suitable perches by frugivorous birds (McClanahan 1986; McClanahan & Wolfe 1987). The flight paths of birds depend not only on the spatial arrangement of trees in the pasture but also on the availability of fruit in trees of the same or neighboring species as well as on the location of nests (Bronstein & Hoffmann 1987).

Dispersal of rain forest species in pastures

At least 13 species of mature forest trees, including *Cymbopetalum baillonii*, *Dendropanax arboreus*, *Guarea glabra* and *Orthion oblanceolatum*, were dispersed by birds under the four ISTs. Resident bird species nesting in pastures, as well as species not nesting in pastures are actively dispersing plant species distances greater than 300 m from the seed source.

Frugivorous birds perching on fruiting trees in the middle of the pastures deposit seeds of different species under the tree canopy, and also ingest their fruits and seeds, carrying them to other trees and perhaps even to the next forest fragment. In the fragmented landscape, remaining TRF fragments are important seed sources of rain forest species. Outside the forest fragments, standing trees that are remnants of the TRF canopy are also seed sources, but in addition they are recruitment foci for rainforest species in man-created pastures.

Germination and establishment of rain forest species

Trees in the middle of the pasture offer safe sites for germination and establishment of primary and secondary species (trees, shrubs and herbs) as is demonstrated by Guevara *et al.* (1992). They reported that 193 plant species (109 woody and 84 herbaceous) were established under fifty ISTs. Among the reported plant species, 53 were primary forest tree species and 51 primary forest climbers, epiphytes or herbs. It is also possible to find new recruits of tree and shrub species growing in open conditions, as is the case for *Stemmadenia donnel-smithii*, *Piper hispidum* and *Cupania glabra*, but they are clearly clumped around ISTs, and their respective densities are significantly higher under the canopy of ISTs than outside the canopies (Guevara *et al. op. cit.*).

Our seed deposition data show higher numbers of colonizing (pioneer) tree species with small seeds and dormancy (e.g. *Cecropia obtusifolia*, *Heliocarpus appendiculatus*, *Trema micrantha*), than mature forest species with larger seeds. However, in the vegetation already established under ISTs, Guevara *et al.* (*op. cit.*) found that mature forest species with relatively large seeds lacking dormancy (e.g. *Cymbopetalum baillonii*, *Guarea glabra*, *Poulsenia armata*) were more numerous and frequent than gap colonizer species. These results suggest that despite a relatively low number of seeds dispersed across pastures, rain forest species with relatively large seeds germinate and establish successfully under ISTs.

Propagule migration between landscape patches

The components of the landscape exchange seeds very readily. In the present-day landscape at Los Tuxtlas, open pastures are invaded by species of forest trees and borders of forest fragments receive the seeds of grasses, forbs and herbs from pastures and secondary vegetation (Guevara 1986). Under ISTs in pastures, zoochorous species outnumbered the anemochorous species in terms of species number and individual abundances (Fig. 5). Nevertheless, the isolation of sites and the distance to seed sources should create a floristic gradient around ISTs. Species restricted to forest fragments should decrease and the species found in pastures as part of riparian corridors, living fences or as ISTs, should increase, as is the case for *Cymbopetalum baillonii*, *Guarea* spp., etc.

Demonstration of the consequences of seed dispersal across the pastures for forest species is difficult. Birds are selecting the species, handling the seeds and depositing them in patchy microhabitats (IST). This group of frugivorous birds selects, consumes and deposits the seeds of a subset of mature forest, secondary and ruderal species of trees, shrubs and herbs. The survival of zoochorous tree species depends on pasture management practices. Browsing eliminates the most palatable species such as *Brosimum alicastrum* and *Nectandra* spp. Periodic cutting of the non-palatable species under ISTs increases mortality rate of species such as *Zanthoxyllum kellermanii*. Species which survive and reproduce usually are secondary species which have become established under the perimeter of the canopy of ISTs or in open conditions such as *Bursera simaruba*, *Piper hispidum*, *Acacia cornigera* and those non-palatable species which are able to resprout from stumps such as *Tabaernomontana* sp., *Stemmadenia donnel-smithii* and *Sapium* sp.

The structure and dynamics of the TRF-pasture landscape depends on the species availability within the mosaic. Our results show differential plant and bird species availability in this landscape, demonstrating that the forest fragments are not as isolated as commonly though. In the fragmented agricultural landscape of the humid tropics, there are some processes and mechanisms which promote the consumption, dispersal and deposition of fruits and seeds as well as the existence of safe sites for germination, establishment, survival and breeding of species now considered as restricted only to forest fragments.

Conditions favorable for the germination and establishment of TRF tree species, previously believed to occur only in TRF or TRF fragments, also occur beneath ISTs located in man-created pastures. This strongly suggests that the 'bottleneck' (Gómez-Pompa *et al.* 1972; Kellman 1980; Uhl *et al.* 1988) for TRF regeneration in agricultural landscapes such as the one at Los Tuxtlas, is neither a problem of seed dispersal across extensive open areas nor is it one of establishment. The current practice of periodically cutting vegetation below ISTs in active pastures and cattle browsing are more likely responsible for the lack of TRF tree species establishment away from forest fragments. Therefore, the conservation of TRF species could be promoted even in very fragmented and heterogeneous agricultural landscapes under proper management practices.

The concept of the primary forest of the humid tropics as fragile (Gómez-Pompa *et al.* 1972) is based in part on the limitation imposed by poorly dispersed, short-lived seeds. This concept should be revised in light of the evidence of tree species seed dispersal and deposition, as well as their germination and establishment (Guevara *et al.* 1992), under isolated trees in pastures. This could be the key to reducing regeneration time in abandoned fields. This resiliency suggests the possibility for the development of management practices that take advantage of the useful primary species which could recolonize pastures profusely. However, in order to quantify the demographic consequences of seed dispersal under ISTs, long term studies are necessary to determine whether TRF tree species which establish beneath ISTs can mature and reproduce *in situ*.

Acknowledgements

This research was funded by Direccíon General de Invetigacíon Cientifica y Superacion Académica, Secretaría de Educación Pública and UNESCO Tropical Rain Forest Project. We thank Luis Juan Argüero, Ceferino Escobar and Santos Tepo for land and facilities; Raúl Ortiz for analyzing the seed trap data; Patricia Moreno-Casasola, Bianca Delfosse, and Raúl Ortiz for assistance in the field; Guillermo Ibarra-Manríquez and Santiago Sinaca for seed and plant identification; Patricia Moreno-Casasola, Bianca Delfosse and Theodore H. Fleming for their critical reviews of the manuscript.

References

Bathschelet, E. 1981. Circular statistics in Biology. Academic Press, London NWT.

Bongers, F., Popma, J., Meave del Castillo, J. & Carabias, J. 1988. Structure and floristic composition of the lowland rain forest of Los Tuxtlas, Mexico. Vegetatio 74: 55–80.

Bronstein, J. L. & Hoffmann, K. 1987. Spatial and temporal variation in frugivory at a neotropical fig, *Ficus pertusa*. Oikos 49: 261-268.

Charles-Dominique, P. 1986. Inter-relations between frugivorous vertebrates and pioneer plants: *Cecropia*, birds and bats in French Guyana. In: Estrada, A. & Fleming, T. H. (eds.), Frugivores and seed dispersal. Dr. W. Junk Publishers, Dordrecht, Netherlands.

Chizon, S. E. 1984. Relacíon suelo-vegetacíon en la Estación de Biología Los Tuxtlas, Ver. (Un análisis de la distribución de los diferentes tipos de suelo en relación con la cubierta vegetal que soporta). Tesis, ENEP-Zaragoza, UNAM, Mexico.

Coates-Estrada, R. & Estrada, A. 1985. Lista de las aves de la Estacíon de Biología Los Tuxtlas, Inst. Biol. UNAM., Mexico, D. F.

Coates-Estrada, R. & Estrada, A. 1986. Fruiting and frugivores at a strangler fig in the tropical rain forest of Los Tuxtlas, Mexico. Journal of Tropical Ecology 4: 349–357.

Estrada, A., Coates-Estrada, R. & Martínez-Ramos, M. 1985. La Estacíon de Biología Tropical Los Tuxtlas: un recurso para el estudio y conservación de las selvas del trópico húmedo. In: A. Gomez-Pompa and S. del Amo (eds.) Investigaciones sobre la regeneración de selvas altas en Veracruz, Mexico. Vol. II. pp. 395–400. INIREB-Alhambra, Mexico, D. F.

Estrada, A., Coates-Estrada, R. & Vázquez-Yanes, C. 1984. Observations on fruiting and dispersers of *Cecropia obtusifola* at Los Tuxtlas, Mexico. Biotropica, 16: 315–318.

FAO/UNESCO. 1975. Soil maps of the world Vol. III Mexico and Central America. FAO/UNESCO, Paris.

Gómez-Pompa, A., Vázquez-Yanes, C. & Guevara, S. 1972. The tropical rain forest: a non-renewable resource. Science 177: 762–765.

Guevara, S. 1986. Plant species availability and regeneration in Mexican tropical rain forest. Acta Universitatis Upsaliensis. Comprehensive Summaries of Uppsala Dissertations from the Faculty of Science. 48.

Guevara, S., Purata, S. & Van der Maarel, E. 1986. The role of remnant trees in tropical secondary succession. Vegetatio 66: 74–84.

Guevara, S., Meave del Castillo, J., Moreno-Casasola, P. & Laborde, J. (1992). Floristic composition and vegetation structure under isolated trees in neotropical pastures. Journal of Vegetation Science 3: 655–664.

Howe, H. F. & Smallwood, J. 1982. Ecology of seed dispersal. Ann. Rev. Ecol. Syst. 13: 201–228.

Ibarra-Manríquez, G. 1985. Estudios preliminares sobre la flora leñosa de la Estacíon de Biología Tropical Los Tuxtlas, Veracruz, Mexico. Tesis Profesional, Facultad de Ciencias UNAM, Mexico.

Janzen, D. H. 1979. How to be a fig? Ann. Rev. Ecol. Syst. 10: 13–51.

Kellman, M. 1980. Geographic patterning in tropical weed communities and early secondary successions. Biotropica 12 (suppl. Tropical Succession): 34–39.

Lot-Helgueras, A. 1976. La Estacíon de Biología Tropical Los Tuxtlas: pasado, presente y futuro. In: A. Gómez-Pompa *et al.* (eds.) Investigaciones sobre la regeneración de selvas altas en Veracruz, Mexico. CECSA, CNEB, INIREB, Mexico, D. F.

McDonnel, M. J. & Stiles, E. W. 1983. The structural complexity of old field vegetation and the recruitment of bird dispersed plant species. Oecologia 56: 109–116.

Martin, T. E. 1985. Selection of second-growth woodlands by frugivorous migrating birds in Panama: an effect of fruit size and plant density? Journal of Tropical Ecology 1: 157–170.

McClanahan, T. R. 1986. The effect of seed source on primary succession in a forested ecosystem. Vegetatio 65: 175–178.

McClanahan, T. R. & Wolfe, R. W. 1987. Dispersal of ornitochorous seeds from forest edges in central Florida. Vegetatio 71: 107–112.

McKey, D. 1989. Population biology of figs: Applications for conservation. Experientia 45: 661–673.

Milton, K., Windsor, D. M., Morrison, D. W. & Estribi, M. 1982. Fruiting phenologies of two neotropical *Ficus* species. Ecology 63: 752–762.

Pratt, T. K. & Stiles, E. W. 1983. How long fruit eating birds stay in the plants where they feed: implications for seed dispersal. American Naturalist 122: 797–805.

Stiles, E. W. & White, D. W. 1986. Seed deposition patterns: influence of season, nutrients and vegetation structure. In: Estrada, A. & Fleming, T. H. (eds.). Frugivores and seed dispersal. Dr W. Junk Publishers, Dordrecht, Netherlands.

Terborgh, J. 1986. Keystone plant-resources in the tropical forest. In: M. E. Soule (ed.). Conservation Biology. The Science of Scarcity and Diversity. Sinauer, Sunderland, Mass.

Toledo, V. M., Carabias, J., Toledo, C. & González-Pacheco, C. 1989. La producción rural en Mexico: alternativas ecológicas. Fundación Siglo Veintiuno. Mexico.

Uhl, C., Buschbacher, R. & Serrâo, E. A. S. 1988. Abandoned pastures in eastern Amazonia. I. Patterns of plant succession. J. Ecol. 76: 663–681.

Van Dorp, D. 1985. Frugivoria y dispersión de semillas por aves. En: Gómez-Pompa, A. & del Amo, S. (eds.). Investigaciones sobre la regeneración de selvas altas en Veracruz, Mexico. Vol II. INIREB, Alhambra Mexicana, Mexico.

Wegner, J. F. & Merriam, G. 1979. Movements by birds and small mammals between a wood and adjoining farmland habitats. J. Appl. Ecol. 16: 349–358.

Wheelwright, N. T. 1991. How long do fruit-eating birds stay in the plants where they feed? Biotropica 23: 29–40.

Windsor, D. M., Morrison, D. W., Estribi, M. A. & de León, B. 1989. Phenology of fruit and leaf production by 'strangler' figs on Barro Colorado Island, Panamá. Experientia 45: 647–653.

Young, K. R., Ewel, J. J. & Brown, B. J. 1987. Seed dynamics during forest succession in Costa Rica. Vegetatio 71: 151–173.

Vegetatio **107/108**: 339–349, 1993.
T. H. Fleming and A. Estrada (eds). Frugivory and Seed Dispersal: Ecological and Evolutionary Aspects.

The role of seed dispersal in the natural regeneration of rain forest after strip-cutting in the Peruvian Amazon

David L. Gorchov[1], Fernando Cornejo[2], Cesar Ascorra[3] & Margarita Jaramillo[4]
[1]*Department of Botany, Miami University, Oxford, OH 45056, USA;* [2]*Casilla 11, Chaclacayo, Lima 8, Peru;* [3]*Museo de Historia Natural 'Javier Prado,' Casilla 140434, Lima 14, Peru;*[4]*Av. Mariscal Caceres 2022, Iquitos, Peru*

Keywords: Bats, Birds, Forest regeneration, Peru, Seed dispersal, Sprouting, Tropical rain forest

Abstract

Seed dispersal and forest regeneration were studied on a 30 × 150 m strip cleared by 'strip-cutting,' a system of forest management designed for sustained yield (Hartshorn 1989), in high terrace rain forest in the Department of Loreto, Peru. After one year the strip was dominated by seedlings of a few bat- and bird- dispersed pioneer tree species (*Cecropia* spp., Melastomataceae, and *Alchornea triplinervia*); stump sprouts from cut trees and saplings that survived the clearing were less numerous. The density of saplings (> 2 m in height) surviving the clearing was 903 per hectare; 94% of these survived the subsequent 18 months. About 30% of 417 stumps (> 7.5 cm diameter at breast height) resprouted within 3 months, with an additional 10% sprouting in the subsequent 10 months. Sprouting frequency was greater for small stumps than large and varied greatly among plant families. Seed deposition over this year was much lower in the interior of the strip, both in species richness and numbers of seeds, than within the forest; strip edges were intermediate in richness and number. The decline in seed input from forest to edge to strip, both in species and in numbers of seeds, was most pronounced for bird-dispersed taxa (primarily Melastomataceae); bat- and wind- dispersed taxa were more evenly distributed. The similarity in bat species composition between the strip and nearby primary forest was higher than the similarity in bird species composition between these habitats, reflecting a failure of many forest bird species to venture into the strip. The predominance of *Cecropia* spp. and other pioneers of minimal commercial value in the regeneration question the sustainability of strip-cutting. Subsequent succession and future tree species composition on the cleared strip will depend not only on the survivorship and growth of sprouts, survivors, and seedlings, but also on responses of different seed-dispersing animal taxa to changes in the species composition and structure of the vegetation in the strip.

Introduction

Natural forest management for sustained yield is considered an alternative to deforestation in tropical rain forests (Mergen & Vincent 1987; Buschbacher 1990). One system of forest management for sustained timber yield that holds great promise is the strip-cut system (Hartshorn 1989; Anonymous 1990; Buschbacher 1990). In this system long, narrow (30–40 m) strips are clearcut and all timber is extracted. This width was selected to simulate the diameter typical of natural treefall gaps, which are the principal sites of natural regeneration of many valuable light tropical hardwoods (Hartshorn 1980, 1989). Animal, rather than mechanical, traction is used in order to minimize compaction of the soil. No seedlings are planted in the strips; rather natural regeneration from seeds and stump sprouts is permitted. Silvicultural treatments may be applied to improve the representation of desirable tree species. In a forest managed according to this scheme, a small percentage of the surface area will be cut each year, for the first 30–40 years. After this time, timber would be extracted from the regenerated forest in the oldest strips. The first implementation of this strip-cut system has been in the Palcazu Valley in Peru, as part of a joint United States Agency for International Development (AID) and Peru Instituto Nacional de Desarrollo (INADE) development project (Hartshorn 1989; Anonymous 1990).

Before the strip system is implemented on a large scale, it is necessary to evaluate its potential for long-term sustainability. A key factor in the determination of sustainability is the species composition of the forest that regenerates after logging. That species composition may be shaped by four modes of regeneration: advance regeneration (surviving seedlings and saplings), sprouting, germination of dormant seeds in the seed bank, and dispersal of seeds from the forest. We are investigating the contribution of each mode, with particular focus on seed dispersal, to forest regeneration after strip cutting in a lowland rainforest site in the Peruvian Amazon. The seed dispersal research is further focused on the role of birds and bats, as these are among the principal dispersal agents of Neotropical rain forest trees (Howe & Smallwood 1982). By investigating the spatial and temporal patterns of seed dispersal we hope not only to help evaluate the sustainability of strip- cutting, but also to suggest changes to parameters such as strip width and season-of-cutting that may increase the abundance of desired tree species. In this paper we report on major trends observed during the first year of our study.

Methods

Study site and climate

This research was done at the Centro de Investigaciones Jenaro Herrera (CIJH) (4° 55′ S, 73° 40′ W), a field station operated by the Instituto de Investigacion de la Amazonia Peruana (IIAP) and located on the Ucayali River approximately 140 km south of Iquitos. Long-term records show a mean annual precipitation of 2800 mm, with a relatively dry period from June to September (Spichiger et al. 1989).

Two strips, each 30 by 150 m, were cut in primary lowland rain forest on high terrace. A few large trees of selected species had been removed from this forest approximately 15–20 years earlier, but its canopy was intact. Soils were sandy clay loam with pH 3.6–4.0 (unpubl. data). All analyses reported here refer to the first of these strips, cleared at the end of the rainy season (April – May 1989). This strip was on level topography, with its long axis running north–south. The families best represented in the canopy were Lecythidaceae, Leguminosae, and Sapotaceae; Burseraceae were particularly numerous in the midstory.

A reference collection of seeds was developed from plant specimens with ripe fruit collected from the forest and from disturbed areas. Plant identifications were made by Fernando Cornejo and Rodolfo Vasquez. Voucher specimens will be deposited at the Herbarium Herrense at CIJH at the completion of this project.

Clearing

Lianas and understory palms were cut in December 1988; trees greater than 5 cm dbh were cut 24 April–9 May 1989, beginning with the smallest diameter classes and progressing to larger classes. Wherever possible we avoided cutting trees smaller than 5 cm dbh. Directional felling was used to ensure that all crowns landed within the 30 m strip. Six large trees located near the edge of the strip but leaning out were not felled. Botanical specimens were collected from each felled tree. Stumps were tagged with the number used for the botanical specimen.

Eight of the large trees were sawn into boards within the strip using a 'tableadora' consisting of a chainsaw mounted on a carriage. Some of the other wood was processed as roundwood, 'sinchinas,' and charcoal (Cornejo & Gorchov, in prep.). All forest products were removed by human labor; neither machines nor draft animals entered the strip. Unused logs were removed from the strip in order to simulate site conditions that would predominate after the more complete utilization planned for this management system (Anonymous 1990). These logs were deposited in the adjacent forest.

Regeneration

Small trees (< 5 cm dbh but > 2 m in height) that survived the clearing were tagged and measured in June 1989, shortly after the strip was cleared, and again in November 1990. The northernmost 15 m and southernmost 15 m were excluded from this inventory because these areas were subject to a different environment (e.g. more shade) than the rest of the strip.

Twenty transects, each 0.5 m wide and extending across the strip (30 m) were inventoried in July 1989 and July 1990. In the first inventory all stems that had survived the clearing, no matter how small, were marked with a spot of red paint, so that this advance regeneration could later be distinguished from new seedlings. In the second inventory, we counted all stems taller than 50 cm; only stems classified as trees are analyzed here. Tree stems were classified as stump sprouts (even small diameter stems were examined for evidence of cutting and resprouting), advance regeneration (present in 1989 inventory but not sprouts), or seedlings. Only the tallest stem was counted on stumps with multiple sprouts.

In August 1989 and again in June 1990 we censused stumps of trees > 7.5 cm dbh for the number of sprouts and height of the highest sprout.

Seed traps

Ninety six seed traps were placed within the strip on 15 June 1989, shortly after clearing was completed. Each trap consisted of a nylon cloth supported by a plastic frame measuring 50 cm × 50 cm standing 40 cm above the ground. Traps were placed in six columns of 16 traps each; columns were located 2.5, 7.5, and 12.5 m from the west edge of the strip, and 2.5, 7.5, and 12.5 m from the east edge (Fig. 1). Two additional columns of 16 traps each were placed in the forest 25 m east and 25 m west of the strip (Fig. 1). Within each column, adjacent traps were 7.5 m apart.

At semi-monthly intervals for 12 months the cloths were collected and replaced by clean cloths. The contents of each cloth were carefully checked for seeds and fruits, which were identified with the aid of the reference collection. When intact fruits were recovered the number of seeds contained within was counted or was estimated using the average from a sample of fruits of that species. In order to expedite the detection and counting of seeds less than 1 mm in diameter, we passed the contents of the forest and edge traps through a fiberglass screen with 1 mm square openings and separately inspected the fractions that did and

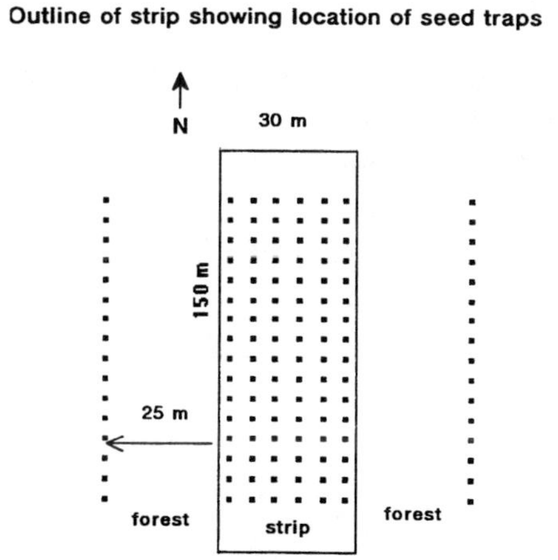

Fig. 1. Location of seed traps in and near the 30 × 150 m logged strip. Eight columns of 16 traps each were positioned in June 1989: two columns in the forest (25 m west and 25 m east of the strip) and six columns within the strip (2.5 m, 7.5 m, and 12.5 m from the west edge and the same distances from the east edge).

did not pass. Beginning with the fifth semi-monthly collection we mixed the material that passed the screen and then inspected a randomly chosen half. Preliminary trials revealed that nearly all species present in the entire sample were detected in the subsample, because when very small seeds were found they were usually in large numbers. Seed counts from the inspected half were then doubled to approximate their abundance in the entire trap.

We identified any fruiting plants above each seed trap. This enabled us to distinguish between fruits and seeds that fell from the parent plant from those that were truly dispersed, and analyze the latter data separately.

Species were determined to be dispersed by birds, bats, or both based on the presence of their seeds in fecal samples obtained from captured animals (below). Species were considered wind-dispersed if their diaspores contained wings or plumes or if seeds dispersed passively from capsules that remained on the plant.

Bird and bat capture

In order to determine the dispersal mode of the seeds reaching the strip, and to document seasonal and successional changes in the use of the strip by potential dispersal agents, we netted birds and bats in the strip monthly from June 1989 through May 1990. To determine the broader habitat use of animals that visited or colonized the strip, we netted both taxa in primary forest (monthly) and a strip of approximately 15 year-old secondary forest 1 km from the strip (for five of the 12 months). Bats were also netted in agroforestry plantations and open areas near the CIJH buildings (six months). Nylon mist-nets were set for bats on one to three nights and for birds on one to four days per site per month. In the strip nets were set primarily along the perimeter. In the forest sites nets were primarily set along (for birds) or across (for bats) trails.

Captured animals were kept in cloth bags for at least one hour, identified, and released. Each bag was inspected for fecal material; if present this was carefully inspected for insect parts and intact seeds. Seeds found in feces were identified with the aid of the reference collection.

Bat and bird species composition in the different habitats was compared using the Index of Similarity $I = 2C / A + B$ where C equals the number of species shared between the two habitats and A and B are the numbers of species in each of the two habitats (Krebs 1985). This index ranges from 0 (no overlap) to 1.0 (complete overlap).

Results

Regeneration

Fourteen months after clearing, 743 arborescent stems taller than 0.5 m were found on the 20 0.5 m wide transects across the strip, equivalent to 2.48/m^2 or 24,800/ha. Most of these (72%) were classified as seedlings, 19% as stump sprouts, and 9% as advance regeneration; seedlings predominated in all size classes (Table 1).

Table 1. Mean ± standard error number of advanced regeneration (survivors), new seedlings, and stump sprouts of tree taxa, grouped by stem height category, in 20 15 m^2 (30 m × 0.5 m) transects in the strip, July 1990. Only the tallest sprout was counted on each stump.

Regeneration	Height category (m)				Total
Class	0.5–1.0	1.0–1.5	1.5–2.0	>2.0	
Adv. regeneration	1.2 ± 0.3	0.6 ± 0.2	0.4 ± 0.1	1.3 ± 0.3	3.4 ± 0.5
Seedlings	9.6 ± 0.8	6.0 ± 0.7	4.5 ± 0.6	6.8 ± 0.9	26.8 ± 1.6
Stump sprouts	3.0 ± 0.3	2.0 ± 0.3	0.9 ± 0.2	1.0 ± 0.2	6.9 ± 0.5
Total	13.8 ± 0.9	8.6 ± 0.8	5.8 ± 0.7	9.0 ± 1.1	37.2 ± 1.8

The majority of the 537 tree seedlings belonged to a few taxa of pioneers: four *Cecropia* species (45%), several species of Melastomataceae (11%), and *Alchornea triplinervia* (Euphorbiaceae) (8%). In contrast, the advance regeneration and stump sprouts included a diversity of mature forest trees. The family best represented in the advance regeneration was Burseraceae. Among sprouting stumps, the most numerous families were Lecythidaceae (especially *Eschweilera bracteosa*), Leguminosae (especially *Inga* spp.), Melastomataceae, Burseraceae (especially *Protium* spp.), and Meliaceae (especially *Guarea* spp.).

Within the central 120 meters of the strip, 325 small trees (>2 m in height) survived the felling of the larger trees, representing a density of 903 trees per hectare. Of these, 307 (94%) were alive when recensused in Nov. 1990 (18 months after clearing), 7 (2%) were dead, and 11 (3%) were not found and presumed dead.

Of 417 stumps of trees >7.5 cm dbh, 129 (31%) had one or more living sprouts three months after cutting (August 1989). Ten months later, an additional 56 stumps had sprouted, whereas only 14 that had sprouts in the earlier census lacked them in the latter, raising the total to 171 (41%). Likelihood of sprouting was related to tree diameter: smaller trees (7.5–15 cm dbh) were more than twice as likely to sprout as larger trees (> 30 cm dbh) (Fig. 2). Among the best represented plant families, frequency of sprouting was highest among trees in the Vochysiaceae, Lecythidaceae, Leguminosae, and Lauraceae (Table 2).

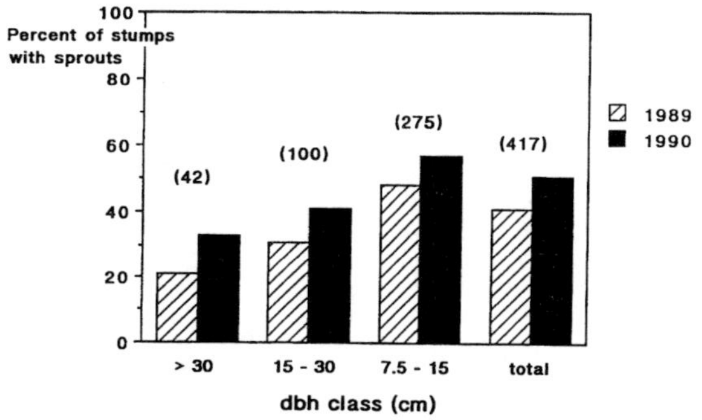

Fig. 2. Percentage of stumps in the logged strip with one or more sprouts three months (August 1989) and 13 months (June 1990) after clearing, for three diameter classes and for all stumps >7.5 cm dbh. Dbh refers to diameter of the tree when it was cut, not of the stump or the sprout. Number of stumps in each diameter class is given in parentheses.

Table 2. Number of stumps (>7.5 cm dbh) in the strip, number and percentage of these stumps with sprouts in June 1990, and net gain in this number since August 1989 (3 months after clearing), for the 12 plant families with the most stumps.

Family	Number of stumps		Net gain '89 to '90
	Total	Sprouting	
Lecythidaceae	59	43 (73%)	11
Burseraceae	38	10 (26%)	6
Leguminosae	37	19 (51%)	1
Sapotaceae	32	13 (41%)	1
Euphorbiaceae	31	7 (23%)	4
Myristicaceae	28	5 (18%)	2
Moraceae	21	4 (19%)	0
Lauraceae	20	10 (50%)	2
Melastomataceae	20	0 (0%)	0
Rubiaceae	11	3 (27%)	−3
Vochysiaceae	11	9 (82%)	0
Annonaceae	10	2 (20%)	0
Other	99	46 (46%)	18
Total	417	171 (41%)	42

Seed rain

Seeds of 115 species have been identified to date from the 128 seed traps over the first year after clearing. The families represented by the most species are Melastomataceae (19 spp.) and Moraceae (13 spp.).

Based on diet records, 20 of these species were dispersed by bats. The most frequently trapped bat-dispersed species were *Cecropia ficifolia*, *C. membranacea*, *C. distachya* (Moraceae), *Piper aduncum* (Piperaceae), and *Vismia* spp. (Hypericaceae).

Of 23 species recovered in the traps that were dispersed by birds, about half are in the Melastomataceae, including three of those most frequently trapped: *Miconia minutiflora*, *M. poeppigii*, and *Ossaea cucullata*. Other common bird-

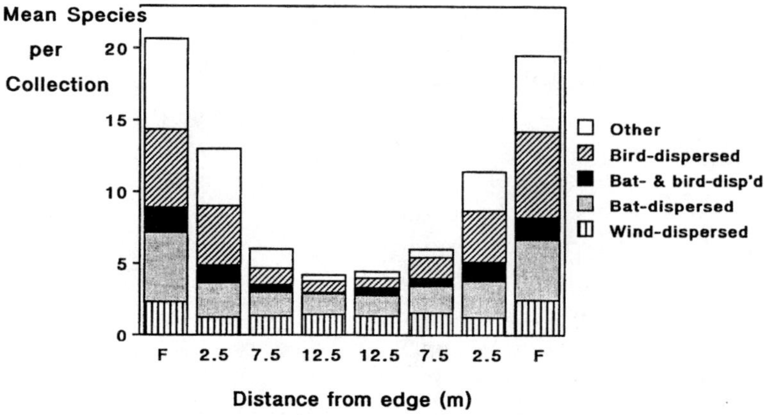

Fig. 3. Mean number of plant species represented by seeds in seed traps per semi-monthly collection, over the first year after clearing (June 1989–June 1990). Each bar represents a column of 16 traps (total area 4 m^2) located in the forest (F) 25 m from the strip or within the strip, 2.5 m, 7.5 m, or 12.5 m from the forest. Left to right on the graph corresponds to west to east (Fig. 1). Plant species are categorized by dispersal mode.

dispersed taxa were *Souroubea guianensis* (Marcgraviaceae), *Coussapoa parvifolia* (Moraceae), and *Alchornea triplinervia* (Euphorbiaceae). Three frequently trapped genera included bird- and bat-dispersed taxa with seeds that could not be distinguished: *Marcgravia* (Marcgraviaceae), *Philodendron* (Araceae), and *Ficus* (Moraceae).

Based on fruit and seed morphology we determined that 20 other species were wind dispersed. The two most frequent wind- dispersed species were *Adelobotrys marginata* (Melastomataceae) and *Ladenbergia magnifolia* (Rubiaceae).

The dispersal mode of the remaining 49 species has yet to be determined; they are grouped as 'other' in these analyses. Most (25) of these species are probably bird-dispersed, based on morphology and taxonomic affiliation. Other species have morphologies suggesting dispersed by arboreal or terrestrial mammals, or by both mammals and birds; for some species the dispersal mode is unknown.

The average number of species per semi-monthly collection was highest for the columns of traps in the forest, intermediate for the traps near the edge, and lowest in traps in the interior of the strip (Fig. 3). This pattern was most pronounced for bird-dispersed species. Wind-dispersed species were encountered at similar frequencies in all positions, while the spatial pattern for bat-dispersed species was intermediate (Fig. 3).

Twenty of the species with seeds in the traps had a fruiting individual located above one or more of these traps. Seeds recovered from traps below these fruiting conspecific adults were scored as 'fallen' rather than 'dispersed' in the following analysis.

The density of dispersed seeds was much greater in the forest than in the interior of the strip, 7.5 m or 12.5 m from the edge; traps 2.5 m from the edge showed intermediate density (Fig. 4). Fallen seeds showed a less regular pattern (Fig. 4), due primarily to numerous fruits fall-

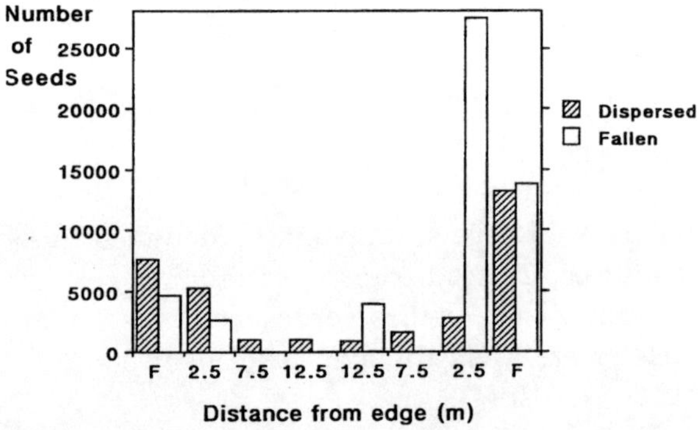

Fig. 4. Total number of dispersed and fallen seeds recovered from seed traps over the first year after clearing. Columns of seed traps are labelled as in Fig. 2. See text for definitions.

Table 3. Density of seeds of six taxa deposited in and near the strip over the first year after clearing. At each position seeds were collected from 32 traps, each 0.25 m^2. Forest traps were 25 m from the edge of the strip. Densities are per m^2.

Dispersal mode & species	Total seeds	Position			
		Strip, distance from edge			Forest
		12.5 m	7.5 m	2.5 m	
Bird					
Ossaea cucullata	1528	0.62	5.88	12.12	172.38
Miconia poeppigii	540	0.62	0.62	21.00	45.25
Bat					
Cecropia ficifolia	843	23.88	25.38	31.50	24.62
Piper aduncum	724	14.62	5.62	29.25	41.00
Wind					
Adelobotrys marginata	2666	70.88	102.25	32.62	127.50

ing from an epiphytic *Philodendron* directly onto one of traps near the east edge and to fruits falling from an fruiting individual of *Phytoloacca rivinoides* to a trap 12.5 m from the east edge.

The decline in seed input from forest to edge to strip interior was much greater for bird-dispersed seeds than for seeds dispersed by bats or wind. This pattern is illustrated by the density of the seed rain at each distance for five plant species well-represented in the traps and free of the confouding presence of a fruiting individual above one of the traps. For one bird-dispersed melastome, *Miconia poeppigii*, the seed rain near the edge of the strip was half that in the forest, while for another, *Ossaea cucullata*, it was more than an order of magnitude lower (Table 3). For both melastomes, the seed rain near the center of the strip was about two orders of magnitude lower than that in the forest. In contrast, one bat-dispersed species, *Piper aduncum*, showed less than an order of magnitude decline in seed density from forest to edge to center, and another, *Cecropia ficifolia*, showed little difference in seed rain among locations (Table 3.) The most common wind-dispersed species, *Adelobotrys marginata*, dispersed seeds to similar densities in the forest and strip interior, but to lower density at the edge (Table 3).

Bat and bird species composition

A total of 292 bats belonging to 33 species were netted in the strip; over all four sites 1214 bats (49 species) were netted. The species composition of bats netted in the strip was more similar to that of the nearby primary forest (24 species shared, $I = 0.79$) than it was to the secondary forest or open areas (Table 4). For all pairs of sites, the similarity in bat species netted was high (≥ 0.59).

The five most frequently netted bats, both in the strip and overall, were frugivores: *Carollia per-*

Table 4. Number of bat species netted in each habitat and index of similarity between habitats, June 1989–May 1990.

Site	# Captures	# Species	Similarity with:		
			Secondary forest	Strip	Open areas
Primary forest	422	28	0.67	0.79	0.64
Secondary forest	213	26		0.71	0.59
Strip	292	33			0.66
Open areas	287	28			
Total	1214	49			

Table 5. Number of bird species netted in each habitat and index of similarity between habitats, June 1989–May 1990.

Site	# Captures	# Species	Similarity with:	
			Secondary Forest	Strip
Primary forest	206	55	0.58	0.54
Secondary forest	132	38		0.53
Strip	150	45		
Total	488	79		

spicillata, C. brevicauda, Artibeus lituratus, A. jamaicensis, and *Phyllostomus hastatus*. The seeds most frequently obtained from fecal samples of the two *Carollia* species were *Vismia* spp. and *Piper* spp. The other three bat species, which are larger, most frequently had seeds of *Cecropia* spp. in their feces, including the four *Cecropia* species which regenerated in the strip.

We netted 150 birds (45 species) in the strip; 488 birds (79 species) over all three sites. Only 27 of the 45 species were also captured in the nearby primary forest (I = 0.54). A comparable level of similarity (I = 0.53) was found between the strip and the secondary forest (Table 5).

Frugivores (Columbidae, Pipridae, Thraupidae, and Turdidae) accounted for a similar proportion of captures (27%–29%) at all three sites. The fruit-eating birds most frequently captured in the strip were the White-crowned Manakin *Pipra coronata*, the Blue-crowned Manakin *Pipra pipra*, and the Fulvous-crested Tanager *Tachyphonus surinamus*; these species were also frequently netted in primary and secondary forest. A wide diversity of seeds was recovered from the feces of each of these bird species; the plant family best represented was the Melastomataceae.

Discussion

Regeneration mode

Some of the regenerating trees classified as seedlings may have actually been present before the clearing but their paint mark was lost. Studies that do not mark advance regeneration (e.g. Hartshorn 1989) probably group such plants as 'seedlings.' Other 'seedlings,' both in our study and others, may be sprouts from very small diameter stems that were broken near ground level during clearing, and which subsequently grew so as to obscure the break. The caveat that not all 'seedlings' actually germinated after the clearing applies not only to the following discussion but to other studies using similar methods.

The density of tree seedlings ≥ 50 cm in height we observed 14 months after clearing (17,900/ha) was about 65% higher than the density reported for two strip cuts in the Palcazu Valley censused after 2 and 2.5 years (Hartshorn 1989). Further sampling at our site will reveal whether stems thin to a comparable density in the second year. At both sites, the high stem density indicates a great potential for silvicultural treatments to favor desired tree species, although it does not obviate concern about their low abundance.

One of the tree species represented by the most seedlings in the strip 14 months after clearing, *Cecropia membranaceae*, had seeds present in the seed bank (unpubl. data). However, this species also was dispersed by bats into the strip during the first months after clearing, so the relative importance of the seed bank vs. newly dispersed seeds to this species' regeneration is not known. In a tropical forest in Mexico the seed bank of *C. obtusifolia* is rapidly depleted by predators but continually replenished by the seed rain (Alvarez-Buylla and Martinez-Ramos 1990). The other tree species with seedlings in the strip were not represented in the seed bank, hence their presence is attributable to the seed rain.

Whereas pioneer species of little commercial value predominated among seedlings, the advance regeneration and sprouting stumps included a diversity of commercially valuable species characteristic of later successional stages. Sprout density at 14 months (4600 sprouting stumps/ha) was about twice that in the Palcazu strips at 2 and 2.5 years (Hartshorn 1989). There is no indication that half of these sprouting stumps will die over the next year, as only 3% died between August

1989 and June 1990, far fewer than the number of inactive stumps that sprouted during this period. Nearly all stumps had multiple sprouts; if more than one survive this may lead to a bushy growth form and low timber yield. We have initiated experiments on thinning stems of common species to a single sprout to determine the effect on survivorship and growth.

The high density of saplings that survived the clearing operation and the first year of succession indicates that advance regeneration may play an important role in forest regeneration. For those commercially valuable tree species with poorly dispersed seeds and low sprouting frequency, these saplings may represent the only way to regenerate in the logged strip. Cutting these saplings during the systematic clearing of the understory before the felling operation (Anonymous 1990) eliminates this advance regeneration. Continued research on the survivorship and growth of advance regeneration, stump sprouts, and seedlings will elucidate the contribution of each mode to the regeneration of the forest.

Dispersal agents and spatial pattern

The tendency for bird-dispersed seeds to disperse within the forest or to the strip edge, while bat- and wind-dispersed seeds tend to disperse relatively evenly to forest, edge, and strip interior, is apparent both in the number of species of each dispersal mode represented in traps (Fig. 3) and in the number of seeds of frequently encountered species (Table 3). The paucity of wind-dispersed *Adelobotrys* seeds in traps near the forest edge, compared to those in the forest and strip interior, suggests that air patterns in clearings may disfavor seed deposition near the edge.

The tendency for bats, rather than birds, to disperse seeds to clearings has been reported for other Neotropical sites. In French Guiana, Charles Dominique (1986, see also De Foresta et al. 1984) found that in the forest and under large trees in clearings more seeds were dispersed during the day (by birds) than at night (by bats), whereas in clearings (5–10 m from the forest) and the interior of a gap, deposition of seeds occurred only at night. They attributed this pattern to the tendency for birds to defecate while perched and for bats to defecate in flight. Similarly, Thomas et al. (1988) found that for a savanna tree of *Muntingia calabura* (bird- and bat-dispersed) in Costa Rica, the proportion of seed-bearing feces found beneath the tree, as opposed to 5 and 15 m away, was greater during the day than at night. Those authors suggested that birds remained in the tree while processing fruits whereas bats carried fruits to feeding roosts and occasionally defecated in flight. Differences in the likelihood of transport from forest to open areas between bat- and wind-dispersed taxa on one hand and bird- and mammal- dispersed taxa on the other was proposed as the explanation for the predominance of bat- and wind-dispersed species in the early stages of primary succession on river meanders in southeastern Peru (Foster et al. 1986).

The high deposition of bird-dispersed seeds in traps 2.5 m from the edge may be attributable to defecation by birds while perched; many of the edge traps were located beneath branches of forest trees. Perch availability may be a major determinant of where seeds are deposited by birds within the strip, as it is in old fields in eastern North America (McDonnell & Stiles 1983). To test whether traps with more cover have greater input of bird- dispersed seeds we recorded the branch density at various heights and the light level above each trap; these data remain to be analyzed.

Although the movement patterns of birds and bats may change with the vegetation structure later in succession, the first year's results lead us to predict that increasing the width of the logged strip would result in an increase in the ratio of bat- and wind-dispersed species to bird-dispersed species in the regeneration. Narrower strips should favor bird-dispersed species, as well as decrease the predominance of pioneers; however, the large crown diameters of emergent trees means that they could not be felled completely within narrower strips.

Dispersers and succession

The greater similarity in bat vs. bird species composition among sites reflects the tendency for many bats to travel long distances and traverse diverse habitats during foraging (Heithaus & Fleming 1978). Capture of a bat in a particular habitat does not necessarily indicate it was foraging there, particularly when the capture is on a forest trail. For bats, the fact that the strip's community was more similar to that of nearby primary forest than to more distant sites with early successional vegetation, hints that many forest bats venture into the clearing. On the other hand, the abundance of seeds of pioneer *Cecropia* spp. in the seed traps indicates that at least some individuals of the large frugivorous bats that disperse these seeds visited the strip after foraging in early successional areas.

The bird community using the strip in the first year after clearing was as similar to that of nearby primary forest as to that of secondary forest, but we predict that as the strip's plant community develops into secondary forest, the bird community will converge toward that typical of secondary forest. We also predict that the number and diversity of frugivorous birds utilizing the strip will increase as the forest regenerates, and this will lead to an increase in the deposition of bird-dispersed seeds. Over the first year, however, capture rates have not increased.

Concluding remarks

We should note that our results apply only to strips set in a matrix of primary forest. In a forest managed by the strip clear-cut system, up to half the area will eventually be cleared; strips cleared late in the first rotation and in subsequent rotations will be bordered on at least one side by young forest. Species that thrive in the strips and reproduce within the cutting cycle (30–40 years), such as *Cecropia* spp., *Vismia* spp, and *Alchornea triplinervia*, should increase in abundance in such a managed forest. However, such pioneer trees are generally of little or no timber value. Tree species that do not regenerate in the strips or have long generation times are susceptible to extinction. Communities of seed-dispersing animals will shift in favor of species able to use the new habitats and fruit resources.

Although the movement patterns of dispersal agents and their deposition of seeds of species deemed valuable may be shaped by modifying parameters such as strip width or cutting date, the basic premise of natural regeneration relies on the presence of seed dispersal agents. Some disperser taxa may become locally extinct in a managed forest due to hunting or destruction of roost sites (bats), but their dispersal services may be replaced by other frugivorous taxa (Hallwachs 1986). Of potentially greater impact is the extinction of many disperser species due to the loss of a 'keystone' or 'pivotal' plant species utilized during seasons of resource scarcity (Howe 1984, Terborgh 1986). If such plant species are not capable of regenerating and reproducing in logged strips, it is crucial that they be included in parcels protected from logging or that they be spared from felling.

Acknowledgements

Cecilia Fox assisted with ornithological field research, Rodolfo Vasquez assisted in plant identification, and Jorge Palmeirim, Susan Mazer, and participants of the Symposium provided helpful comments on an earlier version of this manuscript. The Instituto de Investigacion de la Amazonia Peruana generously provided the use of the Centro de Investigaciones Jenaro Herrera field station. This research was supported under grant 7.228, Program in Science and Technology Cooperation, Office of the Science Advisor, U.S. Agency for International Development to DLG, FC, and John Terborgh.

References

Alvarez-Buylla, E. R. & Martinez-Ramos, M. 1990. Seed bank versus seed rain in the regeneration of a tropical pioneer tree. Oecologia 84: 314–325.

Anonymous. 1990. Manejo de bosques naturales de la selva alta del Peru. Un estudio de caso del Valle del Palcazu.

Documento tecnico. INADE-APODESA, AID, Ronco Consulting Corp., & Centro Cientifico Tropical, Lima.

Buschbacher, R. J. 1990. Natural forest management in the humid tropics: ecological, social, and economic considerations. Ambio 19: 253–258.

Charles Dominique, P. 1986. Inter-relations between frugivorous vertebrates and pioneer plants: *Cecropia*, birds and bats in French Guyana. In: Estrada, A. & Fleming, T. H. (eds), Frugivores and seed dispersal. Dr W. Junk, Dordrecht.

De Foresta, H., Charles Dominique, P., Erard, Ch., & Prevost, M. F. 1984. Zoochorie et premiers stades de la regeneration naturelle apres coupe en foret guyanaise. Rev. Ecol. (Terre Vie) 39: 369–400.

Foster, R. B., Arce B., J., & Wachter, T. S. 1986. Dispersal and the sequential plant communities in Amazonian Peru floodplain. In: Estrada, A. & Fleming, T. H. (eds), Frugivores and seed dispersal. Dr W. Junk, Dordrecht.

Hallwachs, W. 1986. Agoutis (*Dasyprocta punctata*), the inheritors of guapinol (*Hymenaea courbaril*: Leguminosae). In: Estrada, A. and Fleming, T. H. (eds), Frugivores and seed dispersal. Dr W. Junk, Dordrecht.

Hartshorn, G. S. 1980. Neotropical forest dynamics. Biotropica 12 (Suppl.): 23–30.

Hartshorn, G. S. 1989. Application of gap theory to tropical forest management: natural regeneration on strip clear-cuts in the Peruvian Amazon. Ecology 70: 567–569.

Heithaus, E. R. & T. H. Fleming. 1978. Foraging movements of a frugivorous bat, *Carollia perspicillata* (Phyllostomatidae). Ecol. Monogr. 48: 127–143.

Howe, H. F. 1984. Implications of seed dispersal by animals for tropical reserve management. Biol. Cons. 30: 261–281.

Howe, H. F. & Smallwood, J. 1982. Ecology of seed dispersal. Ann. Rev. Ecol. Syst. 13: 201–228.

Krebs, C. J. 1985. Ecology: the experimental analysis of distribution and abundance. 3rd ed. Harper and Row, New York.

McDonnell, M. J. & Stiles, E. W. 1983. The structural complexity of old field vegetation and the recruitment of bird-dispersed plant species. Oecologia 56: 109–116.

Mergen, F. & J. R. Vincent (eds.) 1987. Natural Management of Tropical Moist Forests. Yale Univ., New Haven, CT.

Spichiger, L., Meroz, J., Loizeau, P.-A., & Stutz de Ortega, L. 1989. Contribucion a la Flora de la Amazonia Peruana. Los Arboles del Arboretum Jenaro Herrera. Vol. 1. Conservatoire et Jardin Botoniques de Geneve, Switz.

Terborgh, J. 1986. Community aspects of frugivory in tropical forests. In: Estrada, A. & Fleming, T. H. (eds) Frugivores and seed dispersal. Dr W. Junk Publishers, Dordrecht.

Thomas, D. W., Cloutier, D., Provencher, M., & Houle, C. 1988. The shape of bird- and bat-generated seed shadows around a tropical fruiting tree. Biotropica 20: 347–348.

Vegetatio **107/108**: 351–362, 1993.
T. H. Fleming and A. Estrada (eds). Frugivory and Seed Dispersal: Ecological and Evolutionary Aspects.

Spatial scale and dispersion pattern of ant- and bird-dispersed herbs in two tropical lowland rain forests

C. C. Horvitz & J. Le Corff
Department of Biology, University of Miami, Coral Gables, FL 33124, USA

Keywords: Clark & Evans Index of aggregation, Clumping, Costa Rica, Nearest neighbor, Understory herbs

Abstract

The population level correlates of contrasting dispersal syndromes in closely related species are largely unknown. A family of tropical understory herbs, the Marantaceae, provides the opportunity for comparative studies as it contains many species with contrasting dispersal syndromes. As part of a study of the comparative population biology of ant- and bird-dispersed species, we test the hypothesis that spatial scale and dispersion pattern are related to dispersal type, proposing that bird-dispersed species will have a larger spatial scale than ant-dispersed species, and that, among bird-dispersed species, the scale will vary among three distinct dispersal types. We also propose that ant-dispersed species will show a more clumped dispersion pattern than the bird-dispersed species. Two types of spatial scale are examined: the amount of space occupied by individuals (maximum and actual) and the spacing among individuals within populations. The maximum size of plants showed a trend in the predicted direction. However, this trend was only of marginal statistical significance. The actual distribution of plant sizes, as measured by total leaf area and classified logarithmically, differed significantly among the dispersal types in the predicted direction. Spacing among individuals as measured by nearest neighbor distances also varied significantly by dispersal type, but not entirely in the predicted direction. Dispersion pattern analysis indicated that most study species in most populations had significantly clumped spatial patterning. The aggregation index varied 20-fold among study plots, but did not vary significantly by dispersal type. We conclude that while spatial scale was related generally to type of dispersal, dispersion pattern was not.

Nomenclature: follows Hammel (1984).
This paper was presented orally at the II Simposio – Taller de Frugivoría y Dispersión de Semillas, Los Tuxtlas, June 1991. This is Contribution No. 378 to the program in Tropical Biology at the University of Miami.

Introduction

How seed dispersal distances relate to spatial scale and dispersion patterns of plant populations is not well-known. A common intuition is that seed dispersal distances have evolved with respect to some expected spatial distribution of recruitment sites (Janzen 1970, 1971; Platt & Hermann 1986). It follows that seed dispersal distances should be positively correlated with inter-individual spacing in populations. Thus, plants with long distance dispersal should be characterized by more widely spaced individuals than plants with short distance dispersal. Size is likely to be correlated as well, giving the following suites of spatial scale and seed dispersal characters: long-distance dispersal, widely spaced individuals, and large size vs. short-distance dispersal, more closely spaced individuals and small size.

The importance of spatial scale in understanding ecological systems has recently been emphasized in theoretical and empirical word (see Wiens *et al.* 1986; O'Neill 1989), nevertheless the concept of spatial scale remains abstract. Spatial scale refers to relative spatial dimensions, and may include many different kinds of parameters, such as those that describe the size of individuals, or the average spacing among individuals within a patch or the average spacing among patches (Wiens *et al.* 1986). The units of scale will be units of length (or area or volume). The appropriate spatial scale for a study may be the radius (or area) of a circle (or volume of a sphere) you would need to draw around an individual that includes 'enough' of its environment to understand the *relevant* pattern of variability that an organism must face. Powell (1989) explained spatial scale: 'Intuitively...[it] is the distance one must travel before some quantity of interest changes.' These definitions emphasize that spatial scale is defined relative a particular question and that variation through space is an underlying component of the concept. It doesn't address whether there is such a quantity as the 'natural' spatial scale of a species or a phenomenon.

A related, but distinct, spatial concept is dispersion pattern. Dispersion pattern refers to whether individuals are randomly spaced, clumped or evenly spaced (Krebs 1989). Scale and dispersion have become entangled because the detection of a particular dispersion pattern depends upon the size (scale) of the study plot. It must be pointed out that any organism that occurs in populations and is not ubiquitous on the face of the earth can be considered to be clumped at some, sufficiently large, spatial scale. At such an extreme level, dispersion pattern ceases to be interesting. It seems more interesting to ask first, what is a biologically interesting spatial scale, and then to ask whether, at that scale, there is evidence for nonrandom spacing. One biologically interesting spatial scale is the average distance between individuals within a population: the average distance to nearest neighbors.

The question of spatial dispersion pattern is a separate issue from the question of spatial scale. In a recent model, Howe (1989) emphasized that seed rain may occur either by single seeds or by clumps of multiple seeds, depending upon how many seeds are carried/ingested or dropped/evacuated at once by animals. Single-seeded fruits are likely to have large seeds and relatively short distance dispersal as compared to multiple-seeded fruits. If the number of seeds per fruit and size of seeds are related to dispersal distance, then long-distance dispersal should be associated with significantly clumped seedlings, while short distance dispersal would be associated with randomly or evenly spaced seedlings. In Howe's model, however, such dispersion patterns are not expected to continue beyond the seedling stage; as no more than one of the multiple seeds in a clump will successfully be recruited.

If the number of seeds per fruit and size of seeds are not associated with dispersal distance, then dispersal distances may not even be correlated with spatial dispersion patterns of seedlings. No general predictions can be made; the natural history and ecology of each system must be considered. In comparing closely related species (in which seed size and seed number per fruit are quite similar) that vary in dispersal vector, the behavior of the animals may suggest the likely dispersion patterns of the seed rain and the subsequent dispersion pattern of recruited plants. In particular, ant-dispersed species should be clumped and bird-dispersed species more randomly spaced, simply as a result of the foraging behavior of the animals. Ants move relatively short distances and concentrate prey items in their nests; birds move much farther are quite likely to regurgitate or defecate seeds in non-specific locations following feeding bouts.

The population level correlates of differing dispersal syndromes in closely related plant species have only rarely been studied (Charles-Dominique 1986; Davidson & Morton 1984; O'Dowd and Gill 1986; Platt & Hermann 1986). A family of tropical understory herbs, the Marantaceae, provides the opportunity for comparative studies as it contains many species with contrasting dispersal syndromes (Horvitz & Beattie 1980; Horvitz & Schemske 1986abc; Horvitz 1991; O'Daniel 1987). Some species are adapted for short-distance dispersed (by ants) and others for relatively long-dispersal (by birds). Among the bird-dispersal species, three dispersal types have been described and dispersal distances are likely to vary among them (Horvitz 1991). Bird-dispersed species have mature seeds that are displayed conspicuously on the plant using two contrasting colors, in a way likely to attract avian frugivores (Janson 1983; Wheelwright & Janson 1985; Wilson & Melampy 1983). Ant-dispersed species have seeds that are scattered on the forest floor with a white, oily aril and relatively dark seed body, in a way likely to attract litter-foraging ants (Beattie 1985; Handel 1981; Hanzawa *et al.* 1988; Thompson 1981). Among the bird-dispersed species, the three morphological types are distinguished by the degree of branching of the infructescences, the seed size and the production of clonal propagules (Horvitz 1991).

A theoretical model of metapopulation dynamics for *Calathea* (Horvitz & Schemske 1986a) suggested that these understory plants may partition forest niches by utilizing different light environments (*sensu* Denslow 1980, 1987; Smith 1987) and field work indicated that these niches were associated with dispersal type (Horvitz 1991). Seedlings of the ant-dispersed species were found in relatively darker habitats. Seedlings of the highly-branched bird-dispersed species were mostly restricted to the lightest habitats and the seedling-sized plants of the other two bird-dispersed types were found in a variety of habitats, mostly intermediate (Horvitz 1991). That the species with large seeds or clonal propagules were more shade-tolerant is consistent with the general literature (Foster & Janson 1985; Caswell 1985). Since light gaps are relatively unpredictable in time and space in forests, long-distance dispersal should be favored in those species requiring gaps for recruitment; and shorter distance dispersal should be favored in those species able to recruit in shady habitats. Dispersal distance should be related to shade-tolerance because recruitment sites for shade-tolerant species will be more common (and on average closer) than recruitment sites for shade-intolerant species. We propose that spatial scale will be associated with dispersal type because of the expected association of dispersal type with dispersal distance.

In this paper, we test the hypotheses that bird-dispersed species will have a larger spatial scale than ant-dispersed species, and that, among the bird-dispersed species, the scale will vary among the three dispersal types: the largest scale will be for *bird* type, followed by *bird w/large seeds* and then *bird w/clonal propagules* (see Table 1 for a description of dispersal types from Horvitz 1991). Two types of spatial scale are examined: the amount of space occupied by individuals and the

Table 1. Summary of dispersal syndromes of some Costa Rican Marantaceae (reprinted by permission from Horvitz 1991).

Type	Infructescence		Display	Seed wt, mg	Reward
	ht, m	Branched,			% dry wt, (aril/aril + seed)
Bird	2–3	yes	Seeds sticking out of flattened bracts	500–1000	7–17
Bird w/large seeds	1–2.5	no	Seeds showing in waxy fruit capsule	2500–3000	8
Bird w/clonal propagules	1–1.5	no	Seeds embedded in infructescence	500–1000	5–6
Ant	0.1–0.3	no	White arils against dark seeds, forest floor	<500	5

spatial relationships among individuals in a population. With respect to dispersion pattern, we propose that ant-dispersed species will be more clumped than the bird-dispersed species.

Materials and methods

Study species and sites

For each dispersal type, we chose two or three species, a total of nine species (Table 2). Of these, eight species were in the genus *Calathea* and one was in the genus *Pleiostachya*. These plants share similar ecological habits (Hammel 1984; Kennedy 1977; 1978; 1983). Study sites were two lowland rain forests in Costa Rica, one on the Atlantic slope, La Selva Biological Station (Heredia Province), and one on the Pacific slope, Corcovado National Park (Osa Peninsula) (Hartshorn 1983). Four of the species were studied in both forests; but four were studied only at La Selva and onc was studied only at Corcovado, due to their extremely low abundance or absence at the alternate site. The study was based on plants chosen for a long term demographic study. The goal was to mark and map at least 100 individuals per forest per species. The number of study plots to achieve this goal varied from one to nine, as those plants that occurred only patchily or in low densities required more study plots. A total of 1763 plants were studied. Analysis of individual plant sizes utilized all these plants. Of these, 1609 individuals occurred in sufficient numbers within particular study plots to enable a spatial analysis of a given species within these plots.

Table 2. Study species: acronyms and number of individuals used for the size analyses at each forest. (S = La Selva; C = Corcovado).

Species	Acronym	Forest	Number of plants (size analyses)
Dispersal type: bird			
Calathea lasiostachya	LAS	S	117
C. lutea	LUT	C	116
	LUT	S	86
Pleiostachya pruinosa	PLE	C	66
	PLE	S	117
Dispersal type: bird w/large seeds			
C. gymnocarpa	GYM	S	152
C. inocephala	INO	C	180
	INO	S	115
Dispersal type: bird w/clonal propagules			
C. marantifolia	MAR	C	166
	MAR	S	172
C. donnell-smithii	DON	C	265
Dispersal type: ant			
C. cleistantha	CLE	S	114
C. hammelii	HAM	S	97

For each species at each study plot, all individuals were tagged and mapped during the 1990 rainy season (Jul-Oct), using either an XY-grid (laid out in the forest) or using a compass and meter tape to obtain clusters of relative polar coordinates that were later converted to an XY-coordinate system for each plot. The latter was the only practical solution for those species occuring at low densities. The boundaries of each study plot were established and marked with plastic tent stakes as permanent markers, to assure censusing of all individuals within a defined area.

Occupation of space by individuals

To determine the space occupied by individual plants, we measured the length of each leaf blade of each study plant during the rainy season census of 1990. From these measurements two indices of plant size were obtained: (1) the maximum attainable size as reflected by the maximum size of leaves and (2) the distribution of realized sizes as reflected by the total leaf area of plants.

The maximum attainable size of a plant is a spatial measurement that could reflect a limit to the spacing among individuals in a population. These monocot herbs produce sequentially larger and larger leaves during favorable growing periods; each species at each forest site has its own characteristic geometry and maximum size. We investigated whether maximum leaf size varied by dispersal type with a nonparametric Kruskal-Wallis test (SAS 1988).

The area of each leaf was calculated by measuring the length of each leaf of the individuals in the study and using a regression equation to predict leaf area from leaf length. To obtain the regression equation for each study species in each forest, we traced many leaves (n = 38 to n = 116), chosen to span the natural range of leaf sizes (with replication), during Jan-Jul 1990. These tracings were measured and cut out so that the area of the tracings could be determined with a leaf area meter. Regression analyses of the square root of leaf area on leaf length were performed on each species for each site; all were highly significant and indicated that leaf length is an excellent predictor of leaf area, explaining 96 to 99% of the variance (Table 3).

The leaf areas of whole plants were calculated by summing the areas of all the leaves of a plant. To study the distribution of plant sizes, we classified the plants into five categories, based on logarithmic midpoints for the classification intervals (10, 100, 1000, 10000, 100000 sq cm). We tested whether size distributions were independent of dispersal type with a G-test (Sokal & Rohlf 1981).

Spatial scale of plants in populations

To determine the spacing among individuals within populations, we used the position of each plant on the XY-grids of each study plot to calculate distances between plants. Only plots including at least 5 individuals of a particular species were utilized for the spatial analysis of that species. For each plant at each study plot, the distance to its nearest conspecific neighbor was determined. For each plot, the mean of these distances was calculated as an index of the spatial scale of the population. We tested whether the mean distances to nearest neighbors varied by dispersal type with the nonparametric Kruskal-Wallis test (SAS 1988).

Analysis of dispersion patterns of plants in populations

We then used the mean nearest neighbor distances to investigate spatial aggregation in each of the study plots with respect to ist own spatial scale, calculating the Clark & Evans Index of Aggrega-

Table 3. Regression of leaf area on leaf length. The model that was tested was the regression of the square root of the area on the length (PROC GLM, SAS 1988). $P<0.0001$ for all analyses.

Species	Forest	N (no. of leaves)	Slope	Intercept	r^2
Dispersal type: bird					
LAS	S	108	0.506	− 2.914	0.99
LUT	C	38	0.619	+ 1.464	0.99
LUT	S	92	0.641	+ 2.06	0.99
PLE	C	69	0.542	− 1.487	0.99
PLE	S	77	0.542	− 0.614	0.98
Dispersal type: bird w/large seeds					
GYM	S	99	0.656	+ 0.006	0.99
INO	C	60	0.608	− 1.027	0.99
INO	S	67	0.637	− 2.280	0.99
Dispersal type: bird w/clonal propagules					
MAR	C	63	0.635	− 1.424	0.99
MAR	S	80	0.568	− 0.928	0.97
DON	C	41	0.597	− 1.401	0.98
Dispersal type: ant					
CLE	S	116	0.456	+ 1.415	0.97
HAM	S	77	0.487	− 0.104	0.96

tion, modified by Donelly, (Krebs 1989) as follows:

$$R = r_A / r_E ,$$

where r_A = the observed mean distance to the nearest neighbor and r_E = the expected mean distance to the nearest neighbor. Under the null hypothesis of random spacing, the expected mean distance to the nearest neighbor is given by

$(1/2\sqrt{\text{density}})$ + a correction for the lack of a boundary strip.

The correction factor is necessary when the study plot is not bounded by an extra strip of land that could include neighbors of plants near the boundary of the study plot. It is given by:

$$(0.051 + 0.041/\sqrt{N}) \times (L/N) ,$$

where N = the number of individuals in the plot and L = the length of the boundary. The interpretation of the values of this index of aggregation are:

$R = 1$ means random spacing
$R < 1$ means clumping
$R \to 0$ means very clumped
$R > 1$ means even spacing
$R \to 2.15$ means very even spacing

To test the statistical significance of each R, we calculated:

$$Z = (r_A - r_E)/\text{standard error of } r_E ,$$

where the standard error of the expected is given by:

$$\frac{\sqrt{(0.07A + 0.037L\sqrt{(A/N)})}}{N}$$

in which A = area of the study plot and L = the length of the boundary and N = the number of individuals (Krebs 1989). We tested whether aggregation varied by dispersal type with the non-parametric Kruskal-Wallis test (SAS 1988).

Results

Occupation of space by individuals

The maximum attainable size of plants as measured by maximum leaf length showed a trend in the predicted direction. The mean for the *bird* type of dispersal was 2.4 times that of the *ant* type, while the other two dispersal types had intermediate mean maximum leaf sizes, 1.8 times and 1.3 times that of the *ant* type, for the *bird w/large seeds* and *bird w/clonal propagules*, respectively. However, this trend was only of marginal statistical significance (see legend Fig. 1). On a per species per forest basis, maximum leaf sizes varied nearly five-fold, with *C. lutea* (*bird* type of dispersal) having the largest leaf blades and *C. donnell-smithii* (*bird w/clonal propagules*) having the smallest (Fig. 1). If *C. lutea* had not been part of this study, there would have been little resolution among the three bird types of dispersal, although the difference between the bird types is pooled as a group and the *ant* type would still be evident. *C. donnell-smithii*, is exceptionally small, being much closer to the *ant* type in this character (Fig. 1).

The actual distribution of plant sizes, as measured by total leaf area and classified logarithmically, differed significantly among the dispersal

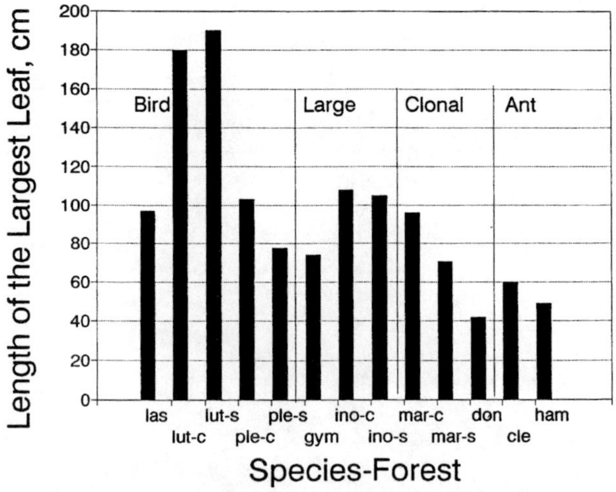

Fig. 1. The largest observed leaf size (the largest leaf of each plant was found and then the largest leaf of all the plants was found) is shown for each study species at each forest. Species are grouped by dispersal type. Maximum leaf size varied by dispersal type, but the difference was only marginally significant (Kruskal-Wallis Nonparametric Test: Chi-Square = 7.23, df = 3, $P < 0.065$). Means for each dispersal type (bird, bird w/large seeds, bird w/clonal propagules, and ant) were respectively: 129.6, 95.7, 69.7, 54.5 cm.

types (Table 4). The *bird* type had more plants than expected in the three largest size categories and proportionately more plants in the largest size class than any other dispersal type. The *bird w/large seeds* type had more plants than expected in the two largest size categories and proportionately more plants in the second-to-largest size class than any other dispersal type. The *ant* type was the only type to have more plants than expected in the smallest size category in which it had proportionately more plants than any other dispersal type. These results (Table 4) indicate general agreement with the hypothesis that the bird-dispersed species are larger. This trend is graphically depicted in Fig. 2; darker fill patterns are used to represent larger plants and the figure is darker to the left (bird dispersal) and lighter to the right (ant dispersal).

In spite of these general patterns, the individual species varied considerably (Fig. 2). *C. lutea* and *C. inocephala* were essentially the only species with plants in the largest size category, while all the species had at least some individuals in the second largest category. One of the *ant* dispersed species, *C. hammelii*, had a fairly large proportion of plants in the second largest category, because, even though the leaves are small (Fig. 1), plants of this species make many leaves. It also had a size distribution remarkably similar to that of *C.*

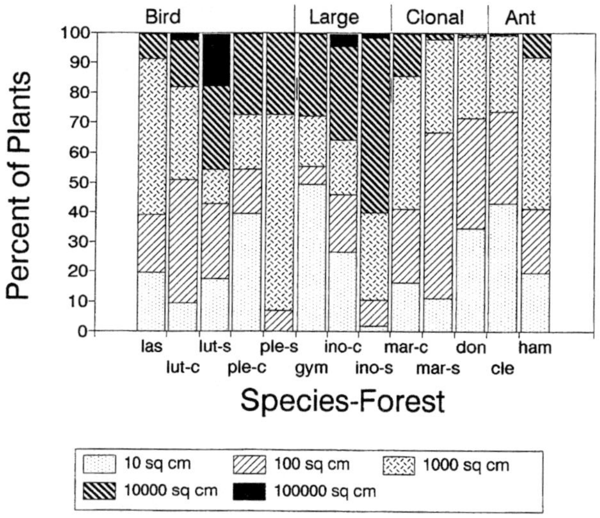

Fig. 2. The relative frequency of plants in each size class is shown for each study species at each forest. Size classes are determined by the logarithm of plant leaf area, midpoints of class intervals are given in the legend. The relative proportions are shown by the lengths of the distinct pattern-bars. Smaller plants are indicated by lighter fill patterns in the bars. Species are grouped by dispersal type. See Table 4 for statistical analysis of variation in size distribution by dispersal type.

Table 4. Percent of plants in each size class (logarithmic leaf area classification) for each dispersal type. The dispersal types differed significantly in size distributions. $G = 361.2$; df = 12; $P < 0.001$. Each entry in the table is marked to indicate whether the frequency in that cell is greater (+), smaller (−) or the nearly the same (≈) as expected under the null hypothesis that sizes are independent of dispersal type.

Dispersal type	Number of plants	Percent of plants in each leaf area (sq cm) class (midpoints)				
		10	100	1,000	10,000	100,000
Bird	502	14.9 −	22.1 −	39.0 +	20.3 +	3.6 +
Bird w/large seeds	447	28.0 +	12.1 −	20.8 −	36.9 +	2.2 +
Bird w/clonal prop.	603	22.9 ≈	39.0 +	33.0 +	5.1 −	0 −
Ant	211	32.2 +	26.5 ≈	37.0 +	4.3 −	0 −
Total	1763	23.0	25.9	32.1 (expected)	17.4	1.5

lasiostachya, a species with a *bird* type of dispersal that makes larger, but fewer leaves, but had relatively few large plants compared to other species of its dispersal type. The two species of the *bird w/clonal propagules*, *C. marantifolia* and *C. donnell-smithii*, were more similar in size distributions to the ant-dispersed species than to most of the species of the other types of bird dispersal (Fig. 2).

Spatial scale of plants in populations

There was tremendous variation in the distance of plants from one another. Densities, averaged over all study plots of each species within each forest, varied 72-fold, from only 0.016 to as many as 1.16 plants per sq m (Table 5). Mean distances to nearest neighbors compared among all study plots varied 27-fold, from 30 to 800 cm. Averaging over all study plots of each species within each forest, the mean of mean distances to nearest neighbors varied nearly 14-fold, from 42 to 584 cm (Table 5, Fig. 3). Mean nearest neighbor distances varied significantly by dispersal type (see legend Fig. 3), but not entirely in the predicted direction. When averaged by dispersal type, the ant-dispersed group had the nearest 'nearest

Table 5. Summary of nearest neighbor analyses performed on each study plot. Only those plots where >5 individuals of a species ocurred were used for that species. R = index of aggregation (Clark and Evans, modified by Donelly). All showed significant ($P<0.05$) clumping except as noted below. S = La Selva; C = Corcovado.

Species		Forest (no. of plots)	Average per study plot				
			Plot size (sq m)	No. of plants	Density (plants per sq m)	$\bar{x}$ Dist. to nearest neighbor, (cm)	R
Dispersal type: bird							
LAS	S	(5)	576	21.6	0.038	140.2	0.46[1]
LUT	C	(3)	278	35.0	0.126	75.3	0.38
LUT	S	(1)	1950	85.0	0.043	181.0	0.72
PLE	C	(6)	325	8.3	0.026	275.2	0.69[1,2]
PLE	S	(1)	400	108.0	0.270	116.0	0.36
Mean for this type						179.7	0.54
Dispersal type: bird w/large seeds							
GYM	S	(1)	1250	152	0.122	95.0	0.64
INO	C	(9)	319	19.4	0.061	126.7	0.49[1]
INO	S	(7)	706	11.0	0.016	584.3	1.23[3]
Mean for this type						313.2	0.81
Dispersal type: bird w/clonal propagules							
MAR	C	(6)	356	26.5	0.074	140.0	0.56[1]
MAR	S	(1)	165	113.0	0.068	131.0	0.65
DON	C	(4)	165	66.5	0.403	43.25	0.49
Mean for this type						104.0	0.54
Dispersal type: ant							
CLE	S	(3)	34	38.0	1.118	46.0	0.81[4]
HAM	S	(2)	42	48.5	1.155	42.0	0.71
Mean for this type						44.4	0.77

[1] One of these study plots had random spacing.
[2] One of these plots had significantly even spacing.
[3] Five of these plots had random spacing; two had even spacing.
[4] Two of these plots had random spacing.

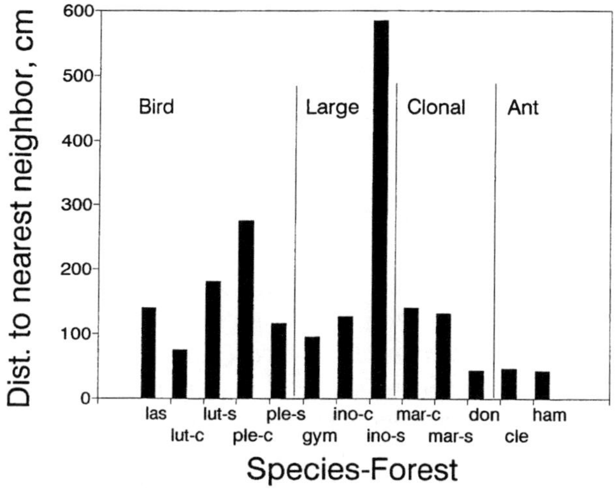

Fig. 3. The mean of the mean distances to nearest neighbors, averaged over all study plots, is shown for each study species at each forest. Species are grouped by dispersal type. The mean distance to nearest neighbors varied significantly by dispersal type (Kruskal-Wallis Nonparametric Test: Chi-Square = 12.9, df = 3, P < 0.005). See Table 5 for means by dispersal type.

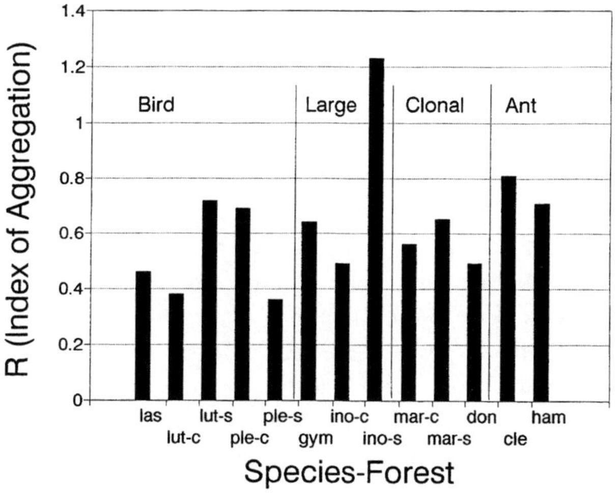

Fig. 4. The mean of R, the index of aggregation, averaged over all study plots, is shown for each study species at each forest. Species are grouped by dispersal type. The index of aggregation did not vary significantly by dispersal type (Kruskal-Wallis Nonparametric Test: Chi-Square = 6.2, df = 3, P > 0.10). See Table 5 for means by dispersal type.

neighbors', as predicted, but the type with the most distant 'nearest neighbors' was the *bird w/large seeds*, not the predicted *bird* type (Table 5). It was the behavior of a single species at a single forest that appears responsible for this large deviation from the predicted pattern. *C. inocephala*, a *bird w/large seeds* type of species, had exceptionally sparse distribution at the La Selva forest (Fig. 3), contributing the most to this pattern. Another interesting exception by an individual species was that *C. donnell-smithii*, a *bird w/clonal propagules* type of species, had nearly the same mean distance to nearest neighbors as the ant-dispersed species, very unlike any other bird-dispersed species.

Analysis of dispersion patterns of plants in population

The analysis of dispersion pattern made for each study plot indicated that 69.4% of all plots (n = 49 study plots) had a significantly clumped spatial pattern. Only 6.1% had a significantly even spatial pattern, while the remaining 24.5% of plots did not differ significantly from random spacing (Table 5). The aggregation index varied 20-fold from 0.07 to 1.43, but did not vary significantly by dispersal type (see legend Fig. 4). The population with the strongest clumping (R = 0.07) was one of the study plots for *Pleiostachya* at Corcovado, while another study plot of the same species at the same forest had one of the most even dispersions observed in this study (R = 1.42). Averaging over all study plots for each species within each forest, we found that all species at both forests tended to have clumped distributions (R < 1, Fig. 4) with the notable exception of *C. inocephala* at the La Selva forest. There was only a three-fold variation in the value of the average aggregation index (Fig. 4, Table 5). We point out especially that there was no trend for the ant-dispersed species to be more clumped than the bird-dispersed species.

Discussion

This paper grew directly out of the logistical difficulties that we encountered when we began to map and measure bird-dispersed *Calathea's*. The field experience impressed upon us how different the spatial scale of these populations were from those that we had previously studied, which had all been ant-dispersed species (Horvitz & Schemske 1986a and unpublished data, LeCorff, unpublished data). The analyses presented here serve to quantify and to clarify how spatial patterning in populations may be related to seed dispersal types. We found that bird-dispersed species exhibited larger spatial scale than ant-dispersed species, both with respect to the occupation of space by individuals and with respect to the spacing among plants in populations. Dispersion pattern, however, was independent of dispersal.

The resolution among the different types of bird-dispersed plants varied according to the spatial scale index studied. Maximum leaf size varied between the ant- and bird-dispersed species, but the bird-dispersed species showed less resolution by type among themselves, with one species, *C. lutea*, having leaves twice as large as the others. Relative size distributions varied more by dispersal types. Size may be important both in terms of display height and seed production. The larger species produce many more seeds than the smaller species. The mean distances to nearest neighbors also varied by dispersal types.

Despite these general patterns, there was considerable variation in spatial scale among species and for some species between forests. For example, *C. inocephala*, while similar in maximum sizes between the two forests, differed in relative leaf areas, mean nearest neighbor distances and dispersion patterns; seedlings being rare and plants being very widely spaced and randomly to evenly dispersed at La Selva, in contrast with Corcovado, *C. lutea* was similar in maximum sizes and size distributions between the two forests, but differed in mean nearest neighbor distances and dispersion patterns, plants being more closely spaced and more clumped at La Selva compared to Corcovado.

One intriguing pattern with respect to an individual species that emerged was that *C. donnell-smithii* showed maximum leaf size, size class distribution and nearest neighbor distances that were more similar to ant-dispersed species than bird-dispersed species. Of all the plants previously classified as bird-dispersed, it has the least showy infructescence. It is possible that mature seeds may be shed on the forest floor and that ants may be more important in the dispersal of this plant than has previously been assumed (Horvitz 1991).

An unexpected result of this study was that, despite the far more local movements of ants as foragers compared to birds, the ant-dispersed plants showed no more clumping than the bird-dispersed species. In fact, they showed less clumping. This result indicates that although the spatial scale is reduced for ant-dispersed species, the dispersion of plants relative to their own scale is not affected. These results emphasize that spatial scale is a separate issue from dispersion pattern. One should choose the spatial scale of interest for a particular organism with respect to a particular issue. Then one can ask whether, *at that scale*, individuals are randomly spaced. In this case, the scale of interest was the average distances between individuals within populations.

What is the biological significance of these results? The clumping of the bird-dispersed plants could result from several possible causes. One is the dropping of undispersed seeds beneath the mother plant. Another is the dropping of many seeds in a single seed load (regurgitated or defecated) or by returning several times to one perch with several seed loads. Another is the rarity of recruitment microhabitats; the plants can only make it in certains places and there is room for several plants at each recruitment patch. Further field studies may clarify these issues. The significance of our results with respect to density dependent mortality of plants (Augspurger 1984; Howe 1989; Janzen 1970, 1971) will require further experimental work. A longitudinal study of how spatial patterning changes from seedling stages to reproductive stages would clarify the biotic factors that may be responsible for producing the observed spatial distributions.

These analyses seem to confirm the general intuition that seed dispersal distances may have evolved with respect to some expected spatial distribution of recruitment sites. If recruitment is successful in the maternal habitat, then short-distance dispersal to nearby patches will be favored, while long-distance dispersal to far away patches will be favored only if recruitment success requires a habitat different from the mother plant's. This line of thought traditionally makes reference to the expected spatial distribution of particular characteristics of the abiotic environment, (e.g. shadiness vs. openness) or of the biotic environment (e.g. abundant herbivores vs. scarce herbivores). Another view (O'Neill 1989) is that short-distance dispersal is used by organisms that are good competitors, while long-distance dispersal is used by organisms that are not good competitors, but who use 'scale' as a way to persist in the environment. This latter view is different in that it does not emphasize any particular abiotic or biotic characteristics of near and far patches other than their distances; this view depends only on the general notion that distance confers escape from competition in a heterogeneous environment, (where the heterogeneity is only loosely defined). Currently, we are testing the hypothesis that *Calathea's* use long-distance dispersal to increase their probability of encountering light gap environments as recruitment sites. We are not investigating the idea that *Calathea's* may be using long-distance dispersal to escape predators, herbivores or competition, but future research may address these issues.

Acknowledgements

We thank Marcos Molina, Pamela Phillips, Devon Graham, and Rick and Jean Seavey for assistance in the field. We thank the staffs of the La Selva Biological Station and Corcovado National Park for logistical support. We thank Marcos Molina, Hortensia Quevedo and Tara Quevedo for data entry assistance. We thank Tara Quevedo especially for the conversion of clusters of polar coordinates into xy-maps. We thank Doug Levey for critical comments on an earlier draft ideas in this paper. The work was supported by NSF grant BSR-8906637 to C. Horvitz.

References

Augspurger, C. K. 1984. Seedling survival of tropical tree species: Interactions of dispersal distance, light-gaps, and pathogens. Ecology 65: 1705–1712.

Beattie, A. J. 1985. The evolutionary ecology of ant-plant mutualisms. Cambridge University Press. Cambridge, U.K.

Caswell, H. 1985. The evolutionary demography of clonal reproduction. Pages 187–224. In: Jackson, Buss and Cook, editors. Population Biology and Evolution of clonal organisms. Yale University Press. New Haven, CT.

Charles-Dominique, P. 1986. Inter-relations between frugivorous vertebrates and pioneer plants: *Cecropia*, birds and bats in French Guyana. Pages 119–136. In: Estrada, A. E. & Fleming, T. H., editors. Dr. W. Junk Publishers, Inc. Dordrecht. The Netherlands.

Davidson, D. W. & Morton, S. R. 1984. Dispersal adaptations of some *Acacia* species in the Australian arid zone. Ecology 65: 1038–1051.

Denslow, J. S. 1980. Gap partitioning among tropical rainforest trees. Biotropica 12 (Supplement): 47–55.

Denslow, J. S. 1987. Tropical rainforest gaps and tree species diversity. Ann. Rev. Ecol. Syst. 18: 431–451.

Foster, S. A. & Janson, C. H. 1985. The relationship between seed size and establishment conditions in tropical woody plants. Ecology 66: 773–780.

Hammel, B. 1984. Systematic treatment of the Cyclanthaceae, Marantaceae, Cecropiaceae, Clusiaceae, Lauraceae and Moraceae for the Flora of a Wet Lowland tropical forest, Finca La Selva, Costa Rica. Doctoral Dissertation. Duke University. North Carolina. USA.

Handel, S. N., Fisch, S. B. & Schatz, G. E. 1981. Ants disperse a majority of herbs in a mesic forest community in New York State. Bull. Torrey Bot. Club 32: 151–163.

Hanzawa, F. M., Beattie, A. J. & Culver, D. C. 1988. Directed disperal: Demographic analysis of an ant-seed mutualism. Am. Nat. 131: 1013.

Hartshorn, G. S. 1983. Plants. Pages 118–157. In: Janzen, D. H., editor. Costa Rican Natural History. University of Chicago Press. Chicago, IL. USA.

Horvitz, C. C. & Beattie, A. J. 1980. Ant dispersal of *Calathea* (Marantaceae) seeds by carnivorous ponerines (Fomicidae) in a tropical rain forest. Am. J. Bot. 67: 321–326.

Horvitz, C. C. & Schemske, D. W. 1986a. Seed dispersal and environmental heterogenity in a neotropical herb: a model of population and patch dynamics. Pages 169–186. In: Estrada, A. E. & Fleming, T., editors. Frugivores and Seed Dispersal. Dr. W. Junk. Dordrecht. The Netherlands.

Horvitz, C. C. & Schemske, D. W. 1986b. Seed dispersal of a neotropical myrmecochore: variation in removal rates and dispersal distance. Biotropica 8: 319–323.

Horvitz, C. C. & Schemske, D. W. 1986c. Ant-nest soil and seedling growth in a neotropical ant-dispersed herb. Oecologia 70: 318–320.

Horvitz, C. C. & Schemske, D. W. 1988. An experimental

analysis of the cost of reproduction in a neotropical herb. Ecology 69: 1741–1745.

Horvitz, C. C. 1991. Light environments, stage structures and dispersal syndromes of Costa Rican Marantaceae. Pages 463–485. Chapter 31. In: Huxley, C. & Culter, editors. Ant-plant Interactions. Oxford University Press.

Howe, H. F. 1989. Scatter- and clump-dispersal and seedling demography: hypothesis and implications. Oecologia 79: 417–426.

Janson, C. H. 1983. Adaptation of fruit morphology to dispersal agents in neotropical rainforest. Science 219: 187–189.

Janzen, D. H. 1970. Herbivores and the number of tree species in tropical forests. Am. Nat. 104: 501–528.

Janzen, D. H. 1971. Seed predation by animals. Ann. Rev. Ecol. and Syst. 2: 465–492.

Kennedy, H. 1977. An unusual flowering strategy and new species in *Calathea*. Bot. Notiser. 130: 333–339.

Kennedy, H. 1978. Systematics and pollination of the "closed-flowered" species of *Calathea* (Marantaceae). Univ. Cal. Pub. Bot. 71: 1–90.

Kennedy, H. 1983. *Calathea insignis* (Hoja Negra, hoja de sal, Bijagua, Rattlesnake Plant). Plants Species Accounts. Pages 204–206. In: Janzen, D. H., editor. Costa Rican Natural History. University of Chicago Press. Chicago. IL. USA.

Krebs, C. J. 1989. Ecological Methodology. Harper & Row, New York, NY. USA;

Levey, D. J. 1988a. Tropical wet forest treefall gaps and distributions of understory birds and plants. Ecology 69: 1076–1089.

O'Daniel, D. 1987. Seed dispersal and seed predation in two species of *Calathea* in a Costa Rican rainforest. Master's Dissertation. University of Texas at Austin.

O'Dowd, J. & Gill, A. M. 1986. Seed dispersal syndromes in Australian *Acacia*. Pages 87–121. In: Murray, D. R., editor. Seed dispersal. Academic Press. Australia.

O'Neill, R. V. 1989. Perspectives in hierarchy and scale. Pages 140–156. In: Roughgarden, J., May, R. M. & Levin, S. A., editors. Perspectives in ecological theory. Princeton University Press.

Powell, T. M. 1989. Physical and biological scales of variability in lakes, estuaries and the coastal ocean. Pages 157–173. In: Roughgarden, J., May, R. M. & Levin, S. A., editors. Perspectives in ecological theory. Princeton University Press.

Platt, W. J. & Hermann, S. M. 1986. Relationship between dispersal syndrome and characteristics of populations of trees in a mixed-species forest. Pages 309–322. In: Estrada, A. E. & Fleming, T. H. Frugivores and seed dispersal. Dr. W. Junk Publishers. Dordrecht. The Netherlands.

SAS. 1988. Statistical Analysis Software. Version 6.03. SAS Institute, Inc. Cary, North Carolina. USA.

Schemske, D. W. & Brokaw, N. V. L. 1981. Treefalls and the distribution of understory birds in a tropical forest. Ecology 62: 938–945.

Smallwood, J. 1982. The effect of shade and competition on emigration rate in the ant *Aphaenogaster rudis*. Ecology 63: 124–134.

Smith, A. P. 1987. Respuestas de hierbas del sotobosque tropical a claros ocasionados por la caida de arboles. Rev. Biol. Trop. (Suppl. 1): 111–118.

Sokal, R. R. & Rohlf, F. J. 1981. Biometry. Freeman, W. H. & Company. New York.

Thompson, J. N. 1981. Elaiosomes and fleshy fruits: phenology and selection pressures for ant-dispersed seeds. Am. Nat. 117: 104–108.

Wheelwright, N. T. & Janson, C. H. 1985 Colors of fruit displays of bird-dispersed plants in two tropical forests. Am. Nat. 126: 777–799.

Wiens, J. A., Addicott, J. F., Case, T. J. & Diamond, J. 1986. Overview: the importance of spatial and temporal scale in ecological interactions. Pages 145–153. In: Diamond, J. & Case, T. J., editors. Community ecology. Harper and Row, New York.

Willson, M. F. & Melampy, M. N. 1983. The effects of bicolored fruit displays on fruit removal by avian frugivores. Oikos 41: 27–31.

Vegetatio **107/108**: 363–374, 1993.
T. H. Fleming and A. Estrada (eds). Frugivory and Seed Dispersal: Ecological and Evolutionary Aspects.

Removal of seeds from frugivore defecations by ants in a Costa Rican rain forest

M. M. Byrne & D. J. Levey
Department of Zoology, University of Florida, Gainesville, FL 32611, USA

Keywords: Ants, *Pheidole*, Seed dispersal, Seed predation, Seed preference, Spatial heterogeneity

Abstract

At our Costa Rican field site, seeds defecated by frugivorous birds usually do not remain where they have been deposited. Many species of ants are attracted to frugivore defecations and remove seeds and/or pulp. *Pheidole* species selectively remove seeds, fungus-growing species (tribe Attini) remove both pulp and seeds. Seeds of many Melastomataceae have an appendage, which we hypothesized is an elaiosome. Indeed, preference trials demonstrated that two species of *Pheidole* selected seeds with the appendage over seeds of the same species in which the appendage had been removed. However, we found that these ants did not take the appendage when it was offered by itself. We conclude that the appendage is not an elaiosome. In further trials, different ant species preferentially selected different seed species. These ants consumed some seeds and deposited others unharmed in refuse piles. We conclude that because the composition of leaf-litter ant communities is highly variable between neighboring square meter plots, and the probability of seed predation depends upon the species of ant, the over-all effect of ants on seed shadows and seed banks is spatially unpredictable.

Introduction

[1]Models and theories of seed dispersal by vertebrates usually assume that dispersed seeds either germinate or die where they are first deposited; although a few studies have examined post-dispersal seed predation (Janzen 1982; Traveset 1990). To a large extent, this assumption is borne out of practicality; it is extremely difficult to map seed shadows generated by seed dispersers and even more difficult to determine the ultimate fate of dispersed seeds. Yet, two observations suggest that we reconsider this assumption for dispersed seeds in neotropical rain forests. First, the number of seeds falling onto a square meter of leaf litter, (e.g. 49 seeds/m^2/month; Denslow & Gomez-Diaz 1990, data from La Selva, Costa Rica), and the density of seeds already in the seed bank (742 seeds/m^2; Putz 1983, data from Barro Colorado Island, Panama), suggest that seeds are a plentiful resource for rainforest granivores. Second, the density of ants is equally impressive. Hölldobler & Wilson (1990) estimate 800 ants/m^2 in the Amazon; ant biomass outweighs vertebrate biomass by approximately four to one. Based on studies of congeners, many of these ants are likely to consume and/or disperse small seeds (e.g. Berg

Addendum: The names of the two *Pheidole* emphasized in this study, *P. nebulosa* and *P. nigricula*, are unpublished names from a generic revision being prepared by E. O. Wilson and W. L. Brown. Their use here is not intended to constitute taxonomic publication but is solely for more precise identification in future ecological research of similar nature.

1975; Bond & Slingsby 1984; Beattie 1985; Horvitz & Schemske 1986).

Given the abundance of dispersed seeds and seed-carrying ants in rain forests, perhaps our emphasis on vertebrate removal of seeds from fruiting plants has been too narrow (see Janzen 1982; Chapman 1989). Ant-seed interactions after vertebrate seed dispersal may have an important influence on the dispersal success of plants. Two types of interactions are most obvious, seed dispersal ('myrmecochory') and predation. Myrmecochory is widespread, occurring in over 60 plant families (Neson 1981; Beattie 1983). Seeds of myrmecochorous species are typified by the presence of an elaiosome, a fleshy lipid-rich appendage that attracts ants. The ants usually carry the seed to their nest, eat the elaiosome, and discard the seed in viable condition (Berg 1966; Beattie 1983). Seed predation by ants is also widespread (Davidson 1977). Although seed predation appears especially prevalent in drier areas, several genera of seed-eating ants are abundant in tropical rain forests (Hölldobler & Wilson 1990).

Here we focus on the interaction between leaf-litter ants and dispersed seeds from several species of Melastomataceae, a speciose family of mostly bird-dispersed understory shrubs and treelets. Ants readily remove melastome seeds from bird defecations (Levey & Byrne 1990). Are these ants playing the role of seed predators, seed dispersers, or both? Pilot experiments and observations led us to believe that they were dispersers. In particular, we noticed that the seeds of many species had an obvious appendage. We suspected this appendage was an elaiosome, since it passed intact through frugivore guts and appeared to attract ants to the seeds. These initial experiments, however, suffered from low sample sizes and our inability to distinguish among ant species and follow the fates of seeds.

We report a series of field and laboratory preference trials designed to determine what component of a frugivore's defecation (pulp vs. seeds) attracts ants, whether the appendage is an elaiosome (which would be evidence of myrmecochory), what seed attributes influence ant preference patterns, and whether species of ants differ in their seed preferences. These trials were conducted both in the field and with captive colonies in the laboratory. They represent a first step towards teasing apart a complex interaction between small-seeded rain forest plants and leaf litter ants.

Methods

Study site and species

The study was conducted in May to July of 1989 and 1990 at Estación Biológica La Selva in Heredia province, Costa Rica. Field experiments were performed in an area of primary forest near the intersection of the 'Camino Central' and the 'Camino Circular Lejano' trails. Laboratory experiments made use of colonies collected from the field and maintained in the lab.

Captive ant colonies were established by removing all ants and brood from a single nest and placing them in a clear plastic petri dish (9 cm diameter) with a layer of Plaster of Paris on the bottom. The Plaster of Paris was kept moist by periodically adding water. Holes in the sides of the dishes allowed ants to enter and exit their 'nest'. Each plate was covered with two layers of red cellophane, and placed in an open plastic box (approximately 35 l × 20 w × 10 h cm). Ants were prevented from escaping by a band of Fluon™ (a slippery, teflon-like substance) painted around the inner sides of the box. They were fed small insects and cookie crumbs, and provided with a cotton-stoppered vial of water.

The ant species used in the laboratory trials, *Pheidole nigricula* and *P. nebulosa*, are abundant in the leaf litter at La Selva (1.32 and 0.37 colonies/m^2, respectively, Byrne 1991). They are small (minor worker size 1–3 mm), and generally live in rotting twigs. Eight colonies of each species were maintained in the laboratory for use in choice experiments. Average size of the colonies was 50 workers for *P. nigricula*, and 90 workers for *P. nebulosa*. All captive colonies contained one queen.

Most of the seed species used were from the

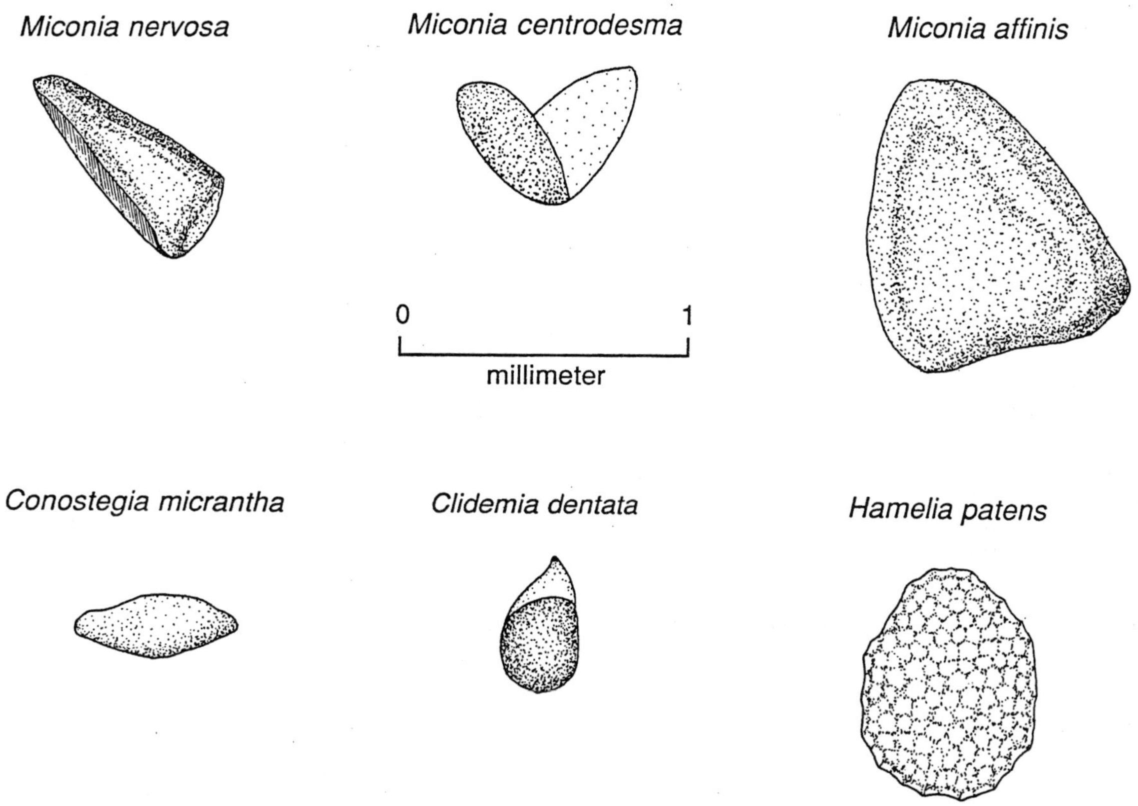

Fig. 1. Seed species used in the preference trials. Except for *Hamelia* (Rubiaceae), all seeds come from Melastomataceae.

family Melastomataceae. The seeds chosen differed in presence and size of the appendage (Fig. 1). However, they also varied in overall size (0.5–1.2 mm). The appendages on most species were firm in texture. The one exception, *Miconia nervosa*, had a fluid-filled appendage. Because this appendage tended to desiccate quickly and we were unsure of whether ants would respond equally to fresh and desiccated appendages, we substituted *Clidemia dentata* and *M. centrodesma* for *M. nervosa* in our 1990 trials.

General methods

All seeds used in the choice experiments were collected daily from fresh bird feces. We fed fruits of the desired species to captive White-collared Manakins (*Manacus candei*), and then removed the seeds from their defecations. For some trials, we removed the seed appendages, being careful not to damage the seed coat. The seeds, appendages, and/or defecated pulp were placed on damp squares (approximately 1.5 × 1.5 cm) of Whatman No. 1 filter paper. Filter paper was chosen as a substrate because it allowed us to see the seeds and ants easily. The ants did not appear to distinguish the filter paper from nearby leaf litter (i.e., ants generally did not pause when they first encountered a piece of filter paper). The seeds were positioned side by side (for two-way choices) or in a triangle (for three-way choices) such that the distance between the seeds was less than the length of an ant's body. We placed the seeds in such close proximity to ensure that ants would be aware of their choices before picking up a seed. Indeed, the ants usually antennated all seeds before selecting one.

The pieces of filter paper were placed on the leaf litter (in the field experiments) or in the bottom of the plastic boxes (in the laboratory experiments). If the filter paper became dry during the experiment, water was added with an eyedropper. We always handled seeds and filter paper with forceps to avoid introduction of human carried scents.

Ants often picked up and then dropped a seed while on the filter paper. However, once the ant had moved off the filter paper with a seed, it rarely dropped it. Thus, in order for an ant to 'select' a given seed, it had to move it completely off the filter paper. After removal of a seed, the paper was replaced by a new one with a complete set of seeds. During the field experiments, all ants that removed seeds were collected and preserved in 70% ethanol. *Pheidole nigricula* and *P. nebulosa* were the most common seed-takers (77% of the collected ants) in the first field experiments, and so were subsequently used in the laboratory trials. *P. nebulosa* is bigger than *P. nigricula* and is thus able to carry slightly larger seeds.

Specific trials

We performed four types of preference tests. The trials were spread equally among the eight captive colonies of each species. Thus, the preferences we report are not artifacts of a single colony's behavior but are likely to represent species-wide preferences (at least at our study site). A minimum of 80 independent choices were obtained in the laboratory trials for each seed and ant species. Sample sizes in the field experiments were lower because we lacked control over which species took seeds, and rates of removal were lower.

Experiment 1
What component of a fruit-eating bird's defecation attracts ants? And, are all ant species attracted to the same component?

To test the assumption that ants were attracted to frugivorous birds' defecations because of the seeds they contained, we first examined an alternative hypothesis that the pulpy matrix adhering to the seeds was the attractant and thus that seeds are removed incidentally by ants. This second hypothesis gains plausibility when one considers that fruit pulp is high in carbohydrates and frugivorous birds have unusually low digestive efficiencies (Levey & Karasov 1989; Karasov & Levey 1990). Thus, defecated pulp may still contain a relatively high density of nutrients.

We ran a three-way choice test with *Miconia nervosa* seeds and pulp. In addition to offering a choice between a seed and a small (approximately 1.5×1.5 mm) clump of defecated pulp, we included a seed whose appendage had been removed. Thus, we tested whether ants preferred seeds or pulp, and if they preferred seeds, whether they preferred them with or without an appendage. These trials were done in the field in 1989.

Experiment 2
Do ants prefer seeds with appendages intact over those with appendages removed?

The field trials described in experiment 1 addressed this question but were limited in scope, lacked control over ant species, and made use of a species with an atypical (fluid-filled) appendage. To avoid these problems, in 1990 we ran two-way choice tests with *P. nigricula* and *P. nebulosa* laboratory colonies using *Miconia centrodesma* seeds with and without appendages.

Experiment 3
Does the appendage itself act as an attractant?

If ants prefer seeds with appendages, it becomes important to determine why. We suggest two non-exclusive explanations: the appendage is nutrient-rich (most elaiosomes are lipid-rich; e.g., Horvitz & Beattie 1980) or it functions as a 'handle' necessary to carry the seed (O'Dowd & Hay 1980). The appendages were too small to collect sufficient quantities for nutrient analyses and our attempts at staining them for lipids yielded inconclusive results. The possible role of a 'handle' is an important one, given that many ant species had difficulty lifting and carrying Melastomataceae seeds (D. Levey & M. Byrne, pers. observation).

To distinguish between these hypotheses, we offered our laboratory colonies two-way choices between either *M. centrodesma* or *Clidemia dentata* seeds with appendage removed and the removed appendage.

Experiment 4
Do ants show significant preferences among different seed species? Does appendage size influence preference? Do preferences differ between *P. nebulosa* and *P. nigricula*?

The above experiments each involved a single species of seed. Here we offered ants either two or three species. We selected species of approximately equal seed size (except for *Miconia affinis* and *Conostegia micrantha*, which were slightly larger and smaller, respectively) but with different appendage size. All species except *Hamelia patens* were in the Melastomataceae. In 1989 we ran two sets of experiments in the field. A two-way preference trial between *Hamelia patens* and *Miconia centrodesma* tested a species with a large appendage (M.c.) against a species without an appendage (H.p.). A three-way trial with *M. centrodesma*, *M. nervosa*, and *C. micrantha*, gave ants a choice among seeds with a large appendage (M.c.), a moderate-sized appendage (M.n.), and no appendage (C.m.).

In 1990 we ran a similar set of trials with laboratory colonies. A two-way test between *M. centrodesma* and *M. affinis* paired species with and without obvious appendages. A three-way test with *M. centrodesma*, *Clidemia dentata*, and *Conostegia micrantha* was similar to the 1989 three-way test among seeds with different appendage sizes (C.d. replaced M.n.).

Table 1. Number of ants of each species chosing *M. nervosa* pulp, seed with appendage, and seed with appendage removed (field trial, 1989).

Ant species	Pulp	Seed with appendage	Seed without appendage
Pheidolini			
Pheidole nigricula	2	32	13
P. nebulosa	0	12	15
P. la selva	0	1	3
P. lancifer	0	0	1
Attini			
Apterostigma	2	0	1
Cyphomyrmex	5	0	0
Trachymyrmex A	1	0	0
Trachymyrmex B	1	0	0

Results

Experiment 1
Both *P. nigricula* and *P. nebulosa* displayed nonrandom choices ($\chi^2 = 29.3$, d.f. = 2, $n = 47$, $P < 0.001$, and $\chi^2 = 14.0$, d.f. = 2, $n = 27$, $P < 0.001$, respectively). Individuals of *P. nigricula* removed intact *M. nervosa* seeds more than twice as often as seeds with the appendage removed (Table 1). They only removed pulp twice. *P. nebulosa* never removed pulp but took the two types of seeds with nearly equal frequency (Ta-

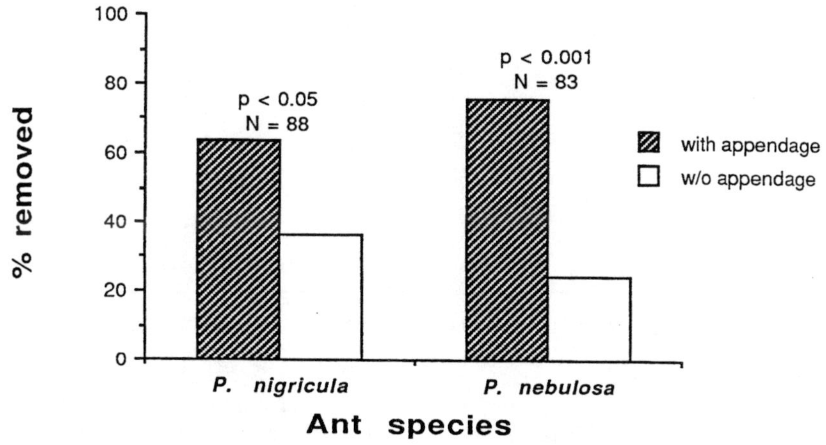

Fig. 2. Laboratory preference trials in which ants chose between *Miconia centrodesma* seeds with and without appendages.

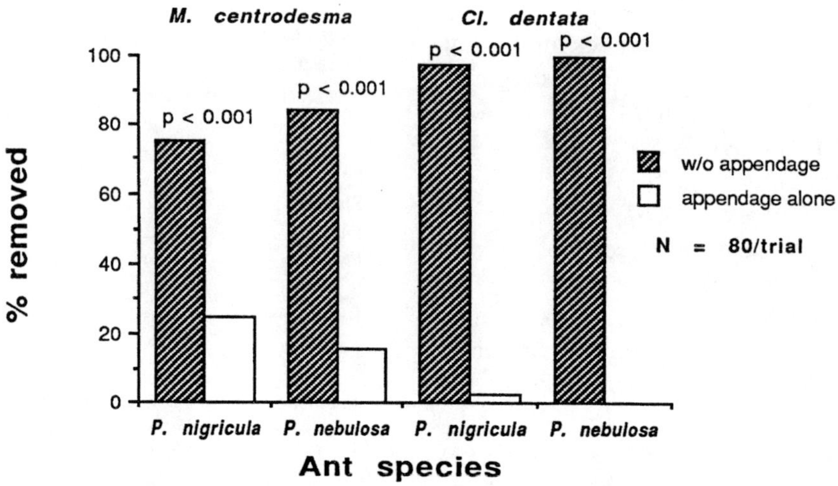

Fig. 3. Laboratory trials in which ants chose between a seed with the appendage removed and the appendage alone. ($n = 80$ for all preference tests)

ble 1). Two other species of *Pheidole*, *P. laselva* and *P. lancifer*, also removed items from the filter paper. Note that they, too, never removed pulp. In contrast, species from the tribe Attini (*Apterostigma* sp., *Cyphomyrmex* sp., and *Trachymyrmex* spp.) almost exclusively removed pulp (Table 1).

Experiment 2

Both species significantly preferred *M. centrodesma* seeds with an appendage over those in which the appendage had been removed (*P. nigricula*: $\chi^2 = 6.54$, d.f. = 1, $n = 88$, $P < 0.001$; *P. nebulosa*: $\chi^2 = 22.28$, d.f. = 1, $n = 83$, $P < 0.001$) (Fig. 2). This behavior contrasts to *P. nebulosa*'s choices

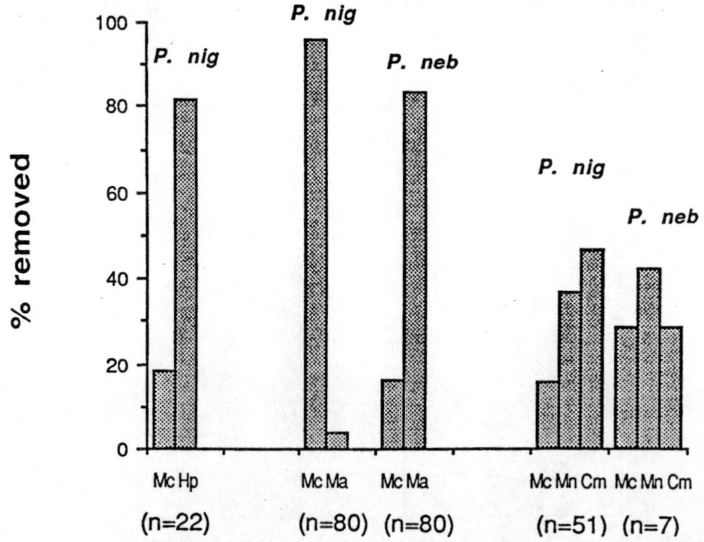

Fig. 4. Preference trials in which ants chose between two or among three seed species. Ant species: *Pheidole nigricula* (P. nig) and *P. nebulosa* (P. neb); and seed species: *Miconia centrodesma* (Mc), *Hamelia patens* (Hp), *M. affinis* (Ma), *M. nervosa* (Mn), and *Conostegia micrantha* (Cm).

in experiment 1, where it showed no preference between seeds with and without their appendage.

Experiment 3

When offered a choice between a *M. centrodesma* seed without the appendage and the appendage itself, both *P. nigricula* and *P. nebulosa* significantly preferred the seed ($\chi^2 = 20.0$, d.f. = 1, $n = 80$, $P < 0.001$; and $\chi^2 = 37.35$, d.f. = 1, $n = 81$, $P < 0.001$, respectively; Fig. 3). They showed the same pattern for *Clidemia dentata* ($\chi^2 = 72.7$, d.f. = 1, $n = 80$, $P < 0.001$; and $\chi^2 = 80.0$, d.f. = 1, $n = 80$, $P < 0.001$, respectively; Fig. 3). Thus, these ants appear to be attracted more to the seed than to the appendage.

Experiment 4. In the field test between *M. centrodesma* and *H. patens*, *P. nigricula* removed *H. patens* significantly more often than *M. centrodesma* ($\chi^2 = 8.91$, d.f. = 1, $n = 22$, $P < 0.01$; Fig. 4). *P. nebulosa* only removed two seeds, both *H. patens*. When offered the three-way preference test in the field, *P. nigricula* displayed nonrandom choices ($\chi^2 = 7.88$, d.f. = 2, $n = 51$, $P < 0.01$; Fig. 4). *C. micrantha* was taken most often, *M. nervosa* slightly less often, and *M. centrodesma* much less often. Removal of seeds by *P. nebulosa* was much less common and more evenly distributed among the three seed species (Fig. 4).

In the laboratory trials, *P. nigricula* significantly preferred *M. centrodesma* over *M. affinis* ($\chi^2 = 68.5$, d.f. = 1, $n = 80$, $P < 0.001$; Fig. 4). *P. nebulosa*, however, showed the opposite pattern, preferring *M. affinis* over *M. centrodesma* ($\chi^2 = 36.5$, d.f. = 1, $n = 80$, $P < 0.001$, Fig. 4).

For the three-way trials in the lab, we analyzed both first and second choices (Fig. 5), which we were unable to do for the field trials due to the slow rate of the ants' return to make a second choice. *P. nigricula* showed nonrandom seed removal choices ($\chi^2 = 73.01$, d.f. = 2, $n = 162$, $P < 0.001$), with *C. micrantha* being chosen first 65% of the time. There was no difference in second choice preferences if *M. centrodesma* or *Cl. dentata* were taken first ($\chi^2 = 0.76$, d.f. = 1, $n = 36$, $P > 0.05$; and $\chi^2 = 0.67$, d.f. = 1, $n = 24$, $P > 0.05$, respectively). When *C. micrantha* was taken first, however, ants significantly chose *M. centrodesma* over *Cl. dentata* ($\chi^2 = 5.04$, d.f. = 1, $n = 105$, $P < 0.05$).

P. nebulosa ($n = 221$) also removed seeds nonrandomly ($\chi^2 = 65.75$, d.f. = 2, $n = 221$, $P < 0.001$), choosing *C. micrantha* first 59% of the time. All second seed selections showed significant differences in seed removal choices. *C. micrantha* was significantly preferred after either *M. centrodesma* and *Cl. dentata* were removed ($\chi^2 = 5.77$, d.f. = 1, $n = 39$, $P < 0.05$; and $\chi^2 = 17.31$, d.f. = 1, $n = 52$, $P < 0.001$, respectively). After *C. micrantha* was taken as a first choice, *Cl. dentata* was most frequently selected second ($\chi^2 = 20.8$, d.f. = 1, $n = 140$, $P < 0.01$).

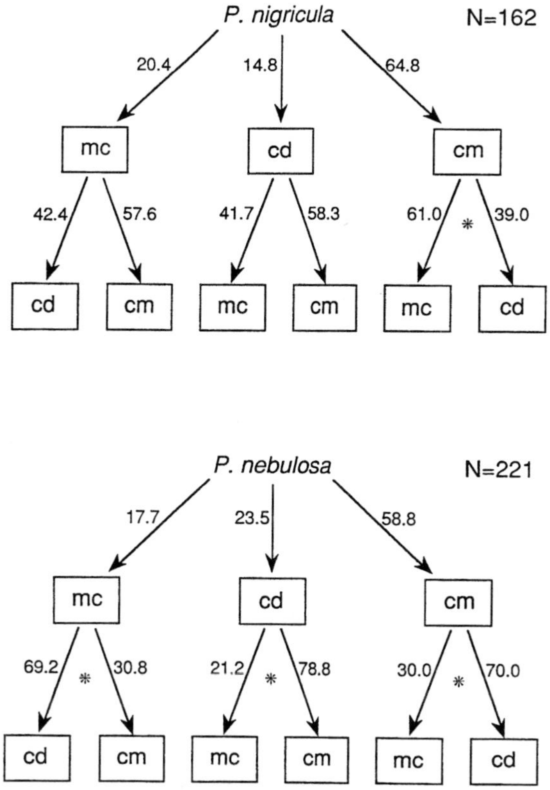

Fig. 5. First and second choice of ants in trials among three seed species. First arrows represent percentage of ants chosing a seed species as a first choice. Second arrows represent percentage of ants chosing each of remaining seed species after first choice had been made. Seed species as in Fig. 4. * Indicates a significant difference in number of ants chosing each seed species.

Discussion

Seed preferences

Influence of appendage

Defecations of fruit-eating birds in the La Selva rain forest were visited by a wide diversity of ants – 13 species of 7 genera. The attine species appeared to have little interest in seeds, they carried away pulp almost exclusively, presumably for their fungus 'gardens'. Although no seeds were associated with pulp in our experiments and hence these ants transported no seeds, seeds and pulp are normally well-mixed and attines usually carry many seeds from defecations (Levey & Byrne, unpubl. data).

In contrast to the attines, all other species (mostly *Pheidole*) removed seeds almost exclusively. These ants generally preferred seeds with appendages over identical seeds whose appendages had been removed. Other studies have documented that many ant species display the same preference for seeds with appendages (i.e., elaiosomes; O'Dowd & Hay 1980; Davidson & Morton 1981; Drake 1981; Lu & Mesler 1981; Bond & Breytenbach 1985). In these studies ants were assumed to be attracted to the appendage. Few experiments have actually tested this assumption, however. Gates (1943) and Kjellsson (1985) did so by offering the ants a choice between an intact seed and an appendage that had been cut from another seed. They found that ants were attracted to and preferred the appendage over the seed. When we conducted the same experiment with *Pheidole nigricula* and *P. nebulosa*, we found they significantly preferred the seed. Thus, despite the preference for seeds with appendages, these ants are not attracted to the appendage itself.

Why do *Pheidole* prefer seeds with appendages if the appendages are not attractive? We offer three explanations. First, like other granivorous ants they may simply prefer larger seeds (Hölldobler 1976; Hansen 1978; M. Kaspari, pers. comm.) – the seeds with the appendage are larger than those without. We reject this explanation because it is inconsistent with our three-way choice trials, in which the smallest seed, *Conostegia micrantha*, was consistently preferred. Second, ants may have taken the seeds with the appendage because they were easier to carry (i.e., the appendage functioned as a 'handle'; O'Dowd & Hay 1980). Although ants had difficulty carrying the seeds of several *Miconia* species, the presence of an appendage did not seem to make the task easier. In fact, the species with the largest appendage, *Miconia centrodesma*, was one of the most difficult to carry and removal of the appendage seemed to make it easier to handle. Third, perhaps the ants detected that our removal of the appendage had damaged the seed and they were simply displaying a preference for undamaged seeds. A fourth explanation is developed and presented below.

Influence of seed and ant species

Many ant species show strong preferences for different types of seeds (Tevis 1958; Mott & McKeon 1977; Culver & Beattie 1978; Whitford 1978; Buckley 1982; Risch & Carroll 1986). The bases of these preferences are complex and not well understood. Seeds may be taken or rejected because of compounds that attract (e.g., Ashton 1979; Marshall *et al.* 1979) or repel (e.g., Buckley 1982), or because of physical attributes such as shape, size, and type of seed coat (O'Dowd & Hay 1980; Hölldobler 1976; Hansen 1978).

P. nigricula and *P. nebulosa* displayed strong preferences among seed species. Our relatively small number of experiments prevents us from generalizing about why they preferred the species they did. We note, however, that both ant species preferred the smallest seed, *Conostegia micrantha*, in three-way choice tests. Also, in a two-way choice test, *P. nebulosa* preferred *M. affinis* over the smaller *M. centrodesma* seed. Because *M. centrodesma* and *C. micrantha* were the only species without an appendage and were preferred over all other species (with the exception of *P. nigricula*'s choice of *M. centrodesma* over *M. affinis* which it could not lift), we speculate that the appendage found on Melastomataceae seeds may contain a compound that deters ants.

Why then did ants prefer seeds with an appendage over seeds that had their appendage removed? We could not remove appendages completely without damaging the seed. So, the manipulated seeds had a small freshly-cut surface of appendage. Any repellent compound may have been more readily volatilized from this surface than from an intact appendage. If so, this could explain why the ants preferred intact seeds; they were not attracted to the appendage on the intact seed but rather repelled by the freshly-cut surface on the seed without an appendage. This explanation is also consistent with the ants' refusal to take appendages when they were offered alone. Presence of a repellent could be tested by determining if ants are repelled or attracted to elaisome extract offered on an inert substrate (e.g., Skidmore & Heithaus 1988).

P. nigricula and *P. nebulosa*'s preferences among seed species were different. In two-way choice tests, *P. nebulosa* preferred *M. affinis* over *M. centrodesma* whereas *P. nigricula* showed the opposite preferencc. In three-way choice tests, although both species preferred *C. micrantha*, *P. nebulosa* typically chose *Cl. dentata* second whereas *P. nigricula* chose *M. centrodesma* second.

Seed dispersal and predation

Although seed dispersal and seed bank composition have been extensively studied (Howe 1986; Murray 1986; Garwood 1989), they have largely been examined independently. Thus, causes of discrepancies in species composition and abundance between seed rain and seed banks have remained for the most part unexplored (Rabinowitz 1981; Denslow & Gomez-Diaz 1990; Garwood 1989). Such discrepancies are almost certainly due to predation or scattering of seeds after they have been dispersed. These processes will alter seed shadows created by dispersers and affect demography and species composition of seed banks (Janzen 1982; Price & Jenkins 1986; Roberts & Heithaus 1986; Clifford & Monteith 1989; Garwood 1989; Louda 1989).

Ants and rodents are responsible for much seed predation and secondary dispersal (Janzen 1982; González-Espinosa & Quintana-Ascencio 1986, Hallwachs 1986). Rodents are generally considered seed predators (Brown *et al.* 1979; Price & Jenkins 1986). Until recently, most seed harvesting of defecated seeds by ants was also assumed to be predation (e.g., Roberts & Heithaus 1986). Several species of ants are now known to remove seeds from frugivore defecations but not damage the seeds of certain plant species (Clifford & Monteith 1989). For both rodents and ants, it is important to note that even seed 'predators' are often seed dispersers as well. Many species, for example, will cache seeds but leave a small number of them intact (Berg 1966; Bullock 1974; Beattie & Lyons 1975; Beattie 1985; Price & Jenkins 1986). These seeds presumably enter the seed bank.

Are *Pheidole nigricula* and *P. nebulosa* predators or dispersers of small seeds in frugivore defecations? They appear to be both. Although they readily consume most seeds that they take, they often deposit viable seeds on refuse piles and leave caches of seeds when they abandon their nests (Levey & Byrne, unpubl. data). These seeds have, in effect, been dispersed by the ants (e.g., Roberts & Heithaus 1986; Rissing 1986). The probability that a seed will be eaten or moved and then abandoned depends upon the species of ant and seed. The ant-seed interaction at La Selva represents a complex continuum between predation and dispersal (see also Andersen 1982; Buckley 1982; Majer 1982).

It is unclear whether secondary dispersal of seeds from frugivore defecations is advantageous to the parent plant. The seeds may benefit because they are placed under ground or in a twig, which could protect them from other seed predators or pathogens (Beattie & Culver 1982). Also, the seeds could benefit from reduction in sibling competition (as they are removed from the defecation clump), which would be intense if the hundreds of seeds in each defecation were not scattered (Howe 1989). On the other hand, seeds might suffer increased levels of interspecific competition if the ants collect many species of seeds and cache them in a common place.

Even if we knew the ultimate effect of secondary seed movement by a given species of ant, predicting the fate of dispersed seeds would still be extremely difficult. Numerous species of ants harvest seeds from bird defecations and the effect of their behavior on seed survivorship varies (Heithaus 1986; Pudlo *et al.* 1980; Levey & Byrne, unpubl. data). Furthermore, the probability that a given ant species will find a defecation is largely unpredictable because ant species composition changes dramatically between nearby areas. For example, at La Selva, the species composition and abundances of twig-dwelling ants changed significantly among three 25 m^2 plots, all within 100 m of each other (Byrne 1991). In addition, primary forest and second growth areas at La Selva overlap very little in composition of the twig-dwelling ant fauna (M. Kaspari & M. Byrne, unpubl. data). Since most of the seed-harvesting species typically forage within 30 cm of their nest (Levey & Byrne, unpubl. data), placement of the defecation relative to ant nests will play a major role in determining which species finds and removes the seeds. Even a 10 cm difference in placement can affect which species discovers the defecation and thus alter the probability of what will happen to the seeds.

One evolutionary implication of this high degree of uncertainty is that any selective forces by frugivores on plant traits may be swamped by the unrelated and highly unpredictable interactions between seeds and ants (see also González-Espinosa & Quintana-Ascencio 1986). This may be one factor contributing to the general lack of tight evolutionary interactions between rain forest frugivores and small-seeded fruiting plants (Herrera 1986).

Another implication of high spatial variation in ant-seed interactions is that interspecific competitive interactions among seedlings may be highly microsite-specific. For example, two of the most common leaf litter ant species at La Selva (*P. nigricula* and *P. nebulosa*) differ in their preferences for two species of abundant seeds (*M. centrodesma* and *M. affinis*). If both seeds are deposited near a *P. nigricula* nest, the treatment of each is likely to be very different than if they had fallen near a *P. nebulosa* nest. These differences will likely affect the species composition and number of seeds that enter the seed bank at each site (González-Espinosa & Quintana-Ascencio 1986). Thus, when conditions for germination are met, the competitive interactions among seedlings will probably vary from one square meter to the next.

Obviously, we are just beginning to uncover the complexity of post-dispersal ant-seed interactions. We suggest that understanding this interaction is necessary for explaining seed bank dynamics and the evolutionary relationship between frugivores and fruiting plants.

Acknowledgements

This project was funded by Earthwatch and its Research Corps. We thank the Organization for Tropical Studies for logistical support, Sylvia Barahona for invaluable assistance and comradarie, and the 31 Earthwatch volunteers who got muddy while trying to see the world from an ant's perspective. We thank T. Fleming, T. C. Moermond, and A. Traveset for helpful comments.

References

Andersen, A. N. 1982. Seed removal by ants in the mallee of northwestern Victoria. In: Buckley, R. C. (ed.), Ant-plant interactions in Australia, pp. 31–43. Junk, The Hague.

Ashton, D. H. 1979. Seed harvesting by ants in forests of *Eucalyptus regnans* F. Muell. in central Victoria. Aust. J. Ecol. 4: 265–277.

Beattie, A. J. 1983. Distribution of ant-dispersed plants. Sonderb. Naturwiss. Ver. Hamb. 7: 249–270.

Beattie, A. J. 1985. The evolutionary ecology of ant-plant mutualisms. Cambridge University Press, Cambridge, UK.

Beattie, A. J. & Culver, D. C. 1982. Inhumation: how ants and other invertebrates help seeds. Nature 297: 627.

Beattie, A. J. & Lyons, N. 1975. Seed dispersal in *Viola* (Violaceae): Adaptations and strategies. Amer. J. Bot. 62: 714–722.

Berg, R. Y. 1966. Seed dispersal of *Dendromecon*: its ecologic, evolutionary, and taxonomic significance. Amer. J. Bot. 53: 61–73.

Berg, R. Y. 1975. Myrmecochorous plants in Australia and their dispersal by ants. Aust. J. Bot. 23: 475–508.

Bond, W. J. & Breytenbach, G. J. 1985. Ants, rodents and seed predation in Proteaceae. S. Afr. J. Zool. 20: 150–154.

Bond, W. J. & Slingsby, P. 1984. Collapse of an ant-plant

mutualism: The argentine ant (*Iridomyrmex humilis*) and myrmecochorous Proteaceae. Ecology 65: 1031–1037.

Brown J. H., Davidson, D. W. & Reichman, O. J. 1979. An experimental study of competition between seed-eating desert rodents and ants. Amer. Zool. 19: 1129–1143.

Buckley, R. C. 1982. Ant-plant interactions: a world review. In: Buckley, R. C. (ed.), Ant-plant interactions in Australia, pp. 111–162. Junk, The Hague.

Bullock, S. H. 1974. Seed dispersal of *Dendromecon* by the seed predator *Pogonomyrmex*. Madrono 21: 378–379.

Byrne, M. M. 1991. Ecology and coexistence mechanisms of tropical twig-dwelling ants. M.S. Thesis, University of Florida.

Chapman, C. A. 1989. Primate seed dispersal: The fate of dispersed seeds. Biotropica 21(22): 148–154.

Clifford, H. T. & Monteith, G. B. 1989. A three phase seed dispersal mechanism in Australian Quinine bush. Biotropica 21: 284–286.

Culver, D. C. & Beattie, A. J. 1978. Myrmecochory in *Viola*: dynamics of seed-ant interactions in some West Virginia species. J. Ecol. 66: 53–72.

Davidson, D. W. 1977. Foraging ecology and community organization in desert seed-eating ants. Ecology 58: 725–737.

Davidson, D. W. & Morton, S. R. 1981. Myrmecochory in some plants (*F. chenopodiaceae*) of the Australian arid zone. Oecologia 50: 357–366.

Denslow, J. S. & Gomez-Diaz, A. E. 1990. Seed rain to treefall gaps in a Neotropical rain forest. J. Can. For. Res. 20: 642–648.

Drake, W. E. 1981. Ant-seed interaction in dry sclerophyl forest on North Stradbroke Island, Queensland. Aust. J. Bot. 29: 293–309.

Garwood, N. C. 1989. Tropical soil seed banks: a review. In: Leck, M. A., Parker, V. T. & Simpson, R. L. (eds), Ecology of soil seed banks, pp. 149–209. Academic Press, Inc. New York.

Gates, B. N. 1943. Caruncúlate seed dissemination by ants. Rhodora 45: 438–445.

González-Espinosa, M. & Quintana-Ascencio, P. F. 1986. Seed predation and dispersal in a dominant desert plant: Opuntia, ants, birds, and mammals. In: Estrada, A. & Fleming, T. H. (eds), Frugivory and seed dispersal, pp. 273–284. Junk, The Hague.

Hallwachs, W. 1986. Agoutis (*Dasyprocta punctata*), the inheritors of guapinol (*Hymenaea courbaril*: Leguminosae). In: Estrada, A. & Fleming, T. H. (eds), Frugivory and seed dispersal, pp. 285–304. Junk, The Hague.

Hansen, S. R. 1978. Resource utilization and coexistence of three species of *Pogonomyrmex* ants in an Upper Sonoran grassland. Oecologia 35: 109–117.

Heithaus, E. R. 1986. Seed dispersal mutualism and the population density of *Asarum canadense*, an ant-dispersed plant. In: Estrada, A. & Fleming, T. H. (eds), Frugivory and seed dispersal, pp. 199–210. Junk, The Hague.

Herrera, C. M. 1986. Vertebrate-dispersed plants: why they don't behave the way they should. In: Estrada, A. & Fleming, T. H. (eds), Frugivory and seed dispersal, pp. 5–20. Junk, The Hague.

Hölldobler, B. 1976. Recruitment behavior, home range creation and territoriality in harvester ants, *Pogonomyrmex*. Behav. Ecol. Sociobiol. 1(4): 405–423.

Hölldobler, B. & Wilson, E. O. 1990. The ants. Belknap Press, Cambridge, MA.

Horvitz, C. C. & Beattie, A. J. 1980. Ant dispersal of *Calathea* (Marantaceae) by carnivorous Ponerines (Formicidae) in a tropical rain forest. Amer. J. Bot. 67: 321–326.

Horvitz, C. C. & Schemske, D. W. 1986. Seed dispersal of a neotropical myrmecochore: variation in removal rates and dispersal distances. Biotropica 18: 319–323.

Howe, H. H. 1986. Seed dispersal by fruit-eating birds and mammals. In: Murray, D. R. (ed.), Seed dispersal, pp. 123–190. Academic Press, Sydney, Australia.

Howe, H. H. 1989. Scatter- and clump-dispersal and seedling demography: hypothesis and implications. Oecologia 79: 417–426.

Janzen, D. H. 1982. Removal of seeds from horse dung by tropical rodents: influence of habitat and amount of dung. Ecology 63: 1887–1900.

Karasov, W. H. & Levey, D. J. 1990. Digestive system tradeoffs and adaptations of frugivorous passerine birds. Physiol. Zool. 63: 1248–1270.

Kjellsson, G. 1985. Seed fate in a population of *Carex pilulifera* L. I. Seed dispersal and ant-seed mutualism. Oecologia 67: 416–423.

Levey, D. J. & Byrne, M. M. 1990. Leaf-litter 'squirrels': Post-dispersal seed predation and scattering by tropical ants. Bull. Ecol. Soc. Am. 71: 229 (abstract).

Levey, D. J. & Karasov, W. H. 1989. Digestive responses of temperate birds switched to fruit or insect diets. Auk 106: 675–686.

Louda, S. M. 1989. Differential predation pressure: a general mechanism for structuring plant communities along complex environmental gradients? Trends Ecol. Evol. 4(6): 158–159.

Lu, K. L. & Mesler, M. R. 1981. Ant dispersal of a neotropical forest floor generiad. Biotropica 13: 159–160.

Majer, J. D. 1982. Ant-plant interactions in the Darling Botanical District of Western Australia. In: Buckley, R. C. (ed.), Ant-plant interactions in Australia, pp. 45–61. Junk, The Hague.

Marshall, D. L., Beattie, A. J. & Bollenbacher, W. E. 1979. Evidence for diglycerides as attractants in an ant-seed interaction. J. Chem. Ecol. 5: 335–344.

Mott, J. J. & McKeon, G. M. 1977. A note on the selection of seed types by harvester ants in northern Australia. Aust. J. Ecol. 2(2): 231–235.

Murray, D. R. 1986. Seed dispersal. Academic Press, Sydney, Australia.

Neson, G. L. 1981. Ant dispersal in *Wedelia hispida* HBK. (Helianthae: Compositae). Southwest. Nat. 26: 5–12.

O'Dowd, D. J. & Hay, M. E. 1980. Mutualism between harvester ants and a desert ephemeral: seed escape from rodents. Ecology 61: 531–540.

Price, M. V. & Jenkins, S. H. 1986. Rodents as seed consumers and dispersers. In: Murray, D. R. (ed.), Seed dispersal, pp. 191–235. Academic Press, Sydney, Australia.

Pudlo, R. J., Beattie, A. J. & Culver, D. C. 1980. Population consequences of changes in an ant-seed mutualism in *Sanguinaria canadensis*. Oecologia 146: 32–37.

Putz, F. E. 1983. Treefall pits and mounds, buried seeds, and the importance of soil disturbance to pioneer tree species in the tropics. Ecology 64: 1069–1074.

Rabinowitz, D. 1981. Buried viable seeds in a North American tall-grass prairie: the resemblance of their abundance and composition to dispersing seeds. Oikos 36: 191–195.

Risch, S. J. & Carroll, C. R. 1986. Effects of seed predation by a tropical ant on competition among weeds. Ecology 67: 1319–1327.

Rissing, S. W. 1986. Indirect effects of granivory by harvester ants: plant species composition and reproductive increase near ant nests. Oecologia 68: 231–234.

Roberts, J. T. & Heithaus, E. R. 1986. Ants rearrange the vertebrate-generated seed shadow of a neotropical fig tree. Ecology 67: 1046–1051.

Skidmore, B. A. & Heithaus, E. R. 1988. Lipid cues for seed-carrying by ants in *Hepatica americana*. J. Chem. Ecol. 14: 2185–2196.

Tevis, L. 1958. Interrelations between the harvester ant, *Vermessor pergandei* (Mayr) and some desert ephemerals. Ecology 39: 695–704.

Traveset, A. 1990. Post-dispersal predation of *Acacia farnesiana* seeds by *Stator vachelliae* (Bruchidae) in Central America. Oecologia 84: 506–512.

Whitford, W. G. 1978. Foraging in seed harvester ants *Pogonomyrmex* spp. Ecology 59: 185–189.

Vegetatio **107/108**: 375–386, 1993.
T. H. Fleming and A. Estrada (eds). Frugivory and Seed Dispersal: Ecological and Evolutionary Aspects.

Predation by vertebrates and invertebrates on the seeds of five canopy tree species of an Amazonian forest

J. Terborgh[1], E. Losos[2], M. P. Riley[1] & M. Bolaños Riley[1]
[1]*Duke University Center for Tropical Conservation, 3705 Erwin Road, Durham, NC 27705, USA;*
[2]*Department of Ecology and Evolution, Princeton University, Princeton, NJ 08544, USA*

Keywords: Seed predation, Distance effect, Exclosures, Mammals, *Astrocaryum*, *Bertholletia*, *Calatola*, *Dipteryx*, *Hymenaea*

Abstract

We studied the pre-germination loss of seeds to invertebrate and vertebrate seed predators of 5 species of Amazonian trees (*Astrocaryum macrocalyx* – Palmae; *Bertholletia excelsa* – Lecithydaceae; *Calatola venezuelana* – Icacinaceae; *Dipteryx micrantha* – Leguminosae (Papilionoidae); *Hymenaea courbaril* – Leguminosae (Caesalpinoidac)). These five species were selected from a large tree flora on several criteria. All possess large (3–10 cm) well-protected seeds that might plausibly be attractive to mammalian seed predators. The reproductive biology of three of the species, or close congeners, had been studied elsewhere in the Neotropics (*Astrocaryum*, *Dipteryx*, *Hymenaea*); one is important to the economy of southeastern Peru (*Bertholletia*); and one, despite large and apparently edible seeds, appeared to suffer no pre-germination loss to predators (*Calatola*). We conducted the research in mature forests in the Manu National Park of southeastern Peru where mammal densities are unperturbed by human activities. Densities of adult trees of the five species in our area range from very high (> 30 per ha: *Astrocaryum*) to very low (≪ 1 per ha: *Hymenaea*).

Loss of seeds to all causes, and to mammalian seed predators in particular, was determined for seeds placed in 2-square meter mammal exclosures and in open controls located at 10 m (near) and 50 m (far) from a large mature individual of the target species (with minor variations in the design for *Astrocaryum* and *Calatola*). The exclosures were of two types: impermeable – designed to exclude all mammals, but not invertebrate seed predators, and semipermeable – designed to admit small (< 500 g), but not large mammals. Experimental and control plots were stocked with apparently viable seeds during the dry-wet transition period (October–November) and scored one year later.

A significant distance effect (higher predation near vs far from a large conspecific adult) was found in only one of the species (*Astrocaryum*), the only one to be attacked with high frequency by invertebrate seed predators. The absence of any detectable distance effect attributable to mammals suggests that mammals, over the course of a year, thoroughly search the forest floor for seeds. Invertebrates may thus be responsible for most pre-germination distance (density) effects. With respect to the treatments, we found three qualitatively distinct results: seeds of three species (*Astrocaryum*, *Bertholletia*, *Dipteryx*) were significantly protected by the impermeable, but not semipermeable exclosures, implicating small mammals in seed loss; the seeds of one species (*Hymenaea*) were significantly protected by exclosures of both types, implicating large mammals; and the seeds of one species (*Calatola*) exhibited 100% survival, whether or not protected by exclosures.

The importance of large mammals as seed predators is generally underestimated in these experiments because semipermeable exclosures may serve as foraging reserves for small mammals. Finally, we noted no relationship between the intensity of mammalian seed predation (as suggested by the survival of unprotected seeds) and the abundance of adults of the five species in the environment. The diversity of results obtained for the five species reveals that large-seeded tropical trees may display a wide range of demographic patterns, and points to the likely importance of post-germination bottlenecks in the population biology of many species, even those that may experience severe pre-germination seed loss.

Introduction

Two decades ago Janzen (1970) and Connell (1971) proposed that recruitment of seedlings near adult trees might be limited by seed predation (and/or post-germination herbivory on seedlings). In these landmark papers it was postulated that seed predators and herbivores would concentrate their activities in zones of high density seed drop around parent trees. Given sufficiently intense predation, seed and seedling survival would be near zero in the vicinity of the parent tree, but would tend to increase farther away. Concurrently, the number of propagules falling to the ground per unit area would decrease due to the limitations of dispersal. Declining seed numbers away from the parent tree, coupled with sharply increasing per capita survivorship, were postulated to generate a zone of optimal recruitment at some unspecified distance from adult trees.

This is an important idea, because it provides an explicit and testable mechanism that could have far-reaching implications for understanding the high tree diversity of tropical forests. A great deal of effort has consequently gone into testing the Janzen-Connell hypothesis. Strong distance or density dependence in seed and seedling survival has been documented in relation to seed predation (e.g., Clark & Clark 1984, Connell *et al.* 1984; Howe *et al.* 1985, Howe 1989; Janzen 1971, 1972; Schupp 1988), fungal damping of seedlings (Augspurger 1984a, 1984b) and herbivory on seedlings (Clark & Clark 1987; Howe 1990; Molofsky & Fisher, in press). Strong negative distance effects have also been found at later stages of the life cycle of some species by Hubbell and Foster (1990) and Hubbell *et al.* (1990) who found reduced growth and survival of saplings and juvenile trees that were nearest neighbors of conspecific adults. Negative influences of conspecifics may thus be expressed at any stage of the life cycle prior to, and possibly including maturity, and perhaps represent a pervasive force in tree recruitment in tropical forests (Terborgh 1988).

In this paper we report on experiments designed to measure pregermination seed survivorship in relation to distance from a conspecific adult in four species of large-seeded, mature phase trees of the Amazonian forest. A fifth large-seeded species was included in experiments not centered around a parent tree. Large seeds are likely to attract mammalian seed predators (Kiltie 1981; Smythe 1986), but often are attacked by invertebrate seed predators as well (Janzen 1971, 1975; Smythe 1989; Wright 1983). We therefore designed the experiments to distinguish seed loss to three classes of seed predators: invertebrates, small mammals (< 500 g., e.g., *Oryzomys*, *Proechimys*, etc.), and large mammals (> 500 g., e.g., *Myoprocta*, *Dasyprocta*, *Agouti*, *Tayassu* spp., etc., see Terborgh *et al.* 1984). We wished to determine, for the seeds of each species, the overall level of pre-germination loss to seed predators, and the extent to which mammals were responsible for seed loss. Finally, given the possibility that the seeds of rare species might experience reduced seed predation by escape in space, the five species were chosen to represent a range of abundances at our site from very common (> 30 adults per hectare) to very uncommon ($\ll$ 1 adult per hectare).

Based on a paper presented at the Second International Symposium on Frugivory and Seed Dispersal, held at Catemaco, Veracruz, Mexico, June 3–6, 1991

Study area and methods

Site

The research was carried out at the Cocha Cashu Biological Station in the Manu National Park, Madre de Dios, Peru. Much information on the flora, fauna, and forest dynamics of this tropical moist forest site is contained in Gentry (1991). Experiments were conducted in mature forest on neutral alluvial soils (inceptisols) in the Manu River floodplain, and on acidic oxisols in the nearby uplands.

Species

We selected five species for study, each for a different specific reason: *Astrocaryum macrocalyx* (*huicungo*: Palmae), because the seeds are an important resource for many animals (Terborgh 1983), and because other species of the genus have been extensively studied in Veracruz, Mexico (Sarrukhán 1978), and on Barro Colorado Island, Panama (Smythe 1989); *Bertholletia excelsa* (Brazil nut or *castaña:* Lecythidaceae) because it is a tree of great economic importance in southeastern Peru; *Calatola venezuelense* (Icacinaceae) because we had noticed that the seeds lie undisturbed for months under the parent trees; *Dipteryx micrantha* (*shihuahuacu:* Leguminosae) because the demography of a closely related species, *D. panamensis* has been investigated in Panama (Steven & Putz 1984; Steven 1988) and Costa Rica (Clark & Clark 1987); and *Hymenaea courbaril* (*azucarhuayo:* Leguminosae) because it is a widespread, economically important species that has been studied in Guanacaste, Costa Rica (Janzen 1975, 1983; Hallwachs 1986).

Dispersal biology

The seeds of all five species are large (3–10 cm) and protected by hard exocarps or endocarps (or both), suggesting that they might be attractive to rodents and other large terrestrial granivores. Observations at Cocha Cashu have provided some information on the reproductive and dispersal biology of each of the five species. We shall review only bare essentials here.

Astrocaryum macrocalyx *(Palmae)*. This is an abundant tree in all mature forest types in the lower Manu basin, often present at densities of = > 30 individuals (> 2 m tall) per hectare (Kiltie & Terborgh 1983). The 4–6 cm long spindle-shaped single-seeded fruits are produced in March and April and are covered with a sweet, apricot-flavored exocarp that is especially favored by capuchin monkeys (Terborgh 1983; Janson 1985). The ripe fruits fall in piles on the ground around the base of the parent tree where they become available to various scatterhoarding rodents (Kiltie 1981). Over 90% of the fruits that are not removed from the vicinity of the parent tree by rodents are attacked by bruchid beetles (Kiltie 1981), a pattern similar to that recorded for a related species of *Astrocaryum* in Panama (Smythe 1989). Preliminary evidence thus suggested that a distance (or density) effect might be expected in the survival of *Astrocaryum macrocalyx* seeds.

Bertholletia excelsa *(Lecythidaceae)*. In our region the Brazil nut tree is exclusively found in upland ('terra firme') forests. The trees attain large size (> 40 m) and are widely scattered at a density of < 1 adult per hectare. The fruits are a hard-walled pyxis, about the size of a coconut. Each fruit contains 20–30 of the familiar 4–5 cm long seeds. The fruits ripen and fall during the latter part of the rainy season (roughly, February to May), whereupon the intact pyxes may lie on the ground for several to many months. Many of them are eventually gnawed opened by agoutis, apparently the only animal in the fauna capable of this feat, whereupon uneaten seeds are scatterhoarded. If not opened by agoutis, the pyxes and the seeds within them eventually rot, suggesting that *Bertholletia* reproductive success is obligately tied to agouti behavior.

Calatola venezuelana *(Icacinaceae).* A medium-sized (up to ca. 25 m) floodplain tree of patchy occurrence, *Calatola venezuelana* predictably ripens 6–8 cm long, yellow, single-seeded fruits in May and June, a time of minimum fruit production by the forest as a whole. The species has thereby been recognized as a 'keystone plant resource' in the local flora (Terborgh 1986). The fleshy exocarp is eaten by titi monkeys (*Callicebus moloch*) and trumpeters (*Psophia leucoptera*: aves) and less enthusiastically by several other vertebrates, but dispersal seems poor, as large numbers of the seeds remain under the parent trees for months after the exocarp has rotted away. The 20–35 g seed is encased in a hard endocarp embellished with a rough grid of longitudinal and transverse ridges that seem designed to deter rodent gnawing. Our attention was attracted to this species because the seeds seem neither to be eaten, nor to be removed from beneath the parent trees by secondary dispersers

Dipteryx micrantha (*Leguminosae: Papilionoidae*). This is perhaps the most prominent and majestic tree of mature forest in the Manu floodplain. Adults occur at densities of 1–>3 per hectare and routinely exceed 50 m in height, with spreads of >30 m. Flowering is in November after the annual rains begin. The 20–40 g single-seeded fruits ripen in May and June at the beginning of the dry season. Bats, particularly *Artibeus jamaicensis*, disperse the fruits, carrying them to scattered feeding roosts, where they chew off the sweet exocarp. Afterwards, the otherwise intact fruits fall to the ground, where they accumulate in small piles that are easy to locate in the forest. Most of the fruits are eventually opened by rodents to extract the 3–4 cm almond-like seed, or scatter-hoarded. Informal observations suggested that the seeds of *Dipteryx*, along with those of *Astrocaryum* and a few other species, are a major resource for vertebrate seed predators in the floodplain forest (Kiltie 1982).

Hymenaea courbaril *(Leguminosae: Caesalpinoidae).* In our area this widespread tree is confined to upland ('terra firme') soils where mature adults are infrequent (density ≪1 per hectare), but may grow to impressive size (height >40 m; dbh to 2 m). The biology of this species in evergreen forest is of interest because Janzen and Martin (1982) listed it as one of a number of species putatively dispersed by gomphotheres. Individual trees do not produce a crop every year. The fruits are typical legumes containing 1 to 5 seeds 2–3 cm across surrounded by a copious, sweet, dry mesocarp. However, the pods are extremely hard, with woody walls 3–4 mm thick. In our area the pods are opened principally by peccaries, which consume the sweet mesocarp, leaving many intact seeds scattered on the ground. Some of these may subsequently undergo secondary dispersal by scatterhoarding rodents. We selected this species because of its large seeds, and because its biology had been studied previously only in the dry forest of Central America (Janzen 1975, 1983).

For the experiments we collected seeds of each of the five species under fruiting trees. Experiments with *Astrocaryum, Calatola* and *Dipteryx* were conducted in mature floodplain forest; those with *Bertholletia* and *Hymenaea* in mature 'terra firme' forest growing on ultisol substrate.

Experimental design

We excluded vertebrate seed predators from 2 × 1 m experimental plots by erecting exclosures constructed of galvanized wire cloth (mesh squares 3/4″ (17 mm) on a side) stapled to sharpened, 1.3-m long stakes made out of the rot- and termite-resistent wood of *Cedrela odorata* (*cedro*: Meliaceae, Spanish cedar). The exclosures were staked at the 4 corners and in the middle of the long sides. The stakes were driven into the ground until the wire was pressed tightly against the earth to preclude passage of small mammals underneath. The exclosures were 85 cm high and open at the top, ensuring normal exposure to rain and sunlight. (Closed tops would have collected twigs and litter, significantly altering the environment of plants inside with respect to controls.)

The exclosures were of 2 kinds, which we

termed impermeable and semipermeable. Impermeable exclosures were made as described above and were intended to exclude all mammals (following Sork 1987; Howe 1990; Molofsky & Fisher, in press). Semipermeable exclosures were identical except that we cut gates 6 cm wide by 7.5 cm tall at 8 cm intervals around the bottom perimeter of the wire cloth. The gates were intended to open the exclosures to small rodents (*Oryzomys*, *Proechimys*, etc.), but not larger mammals (following Brown & Munger 1985). Control plots of two kinds were marked by stakes at the corners, but were open, lacking wire cloth. One set of controls received seeds, the other did not.

The exclosures and control plots were installed in sets of 8, arrayed around a central tree of the target species (Fig. 1). The four types of plots – impermeable, semipermeable, controls with seeds and controls without seeds – were set out along the 4 cardinal directions, one of each type at 10 meters and one at 50 meters from the central tree (except for *Astrocaryum* and *Calatola* see below). The directions used for the treatments were shifted 90 degrees at each successive central tree, so that a complete cycle was obtained with 4 trees. The central trees were large mature individuals with broad crowns, so that the plots at 10 meters from the trunk were well inside the projected crown perimeter in the postulated zone of high seed predator activity. The outer ring of plots was located at 50 meters from the central tree in acknowledgement of the finding of Howe *et al.* (1985) of reduced survival of *Virola surinamensis* seeds at distances out to, but not beyond, 35 m from the parent tree. It was thus hoped that plots located at 50 m from the central tree would be outside the tree's zone of influence on seed predators. However, some of the central trees of *Dipteryx* had conspecific neighbors within 100 m (2 x 50 m), although none of the *Hymenaea* or *Bertholletia* did. To accommodate the smaller size of *Astrocaryum* trees (crown spread of mature

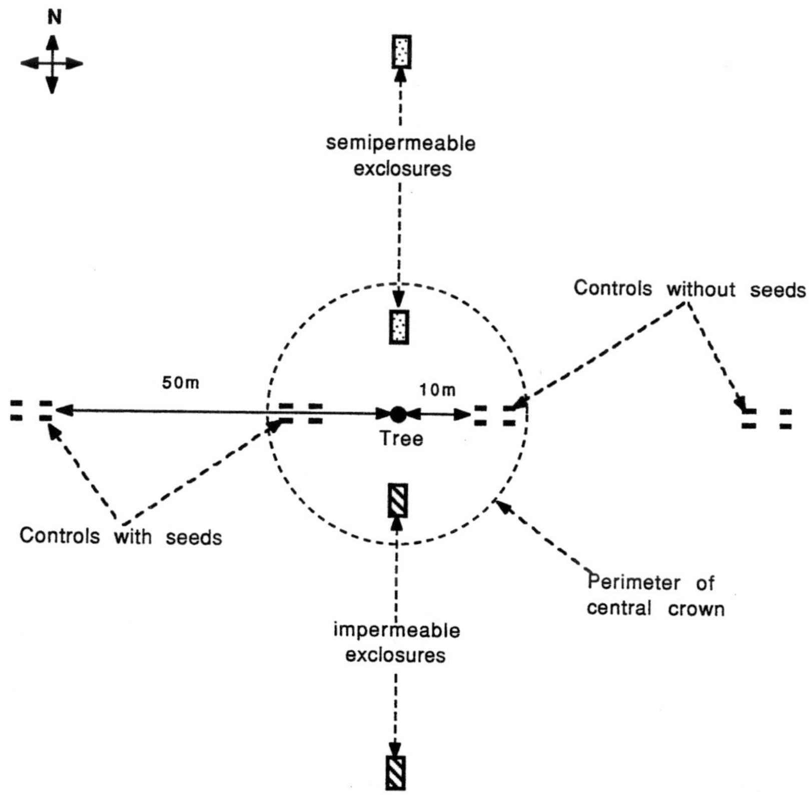

Fig. 1. Schematic showing the array of exclosures and control plots around a central tree (see text for further details).

adults = ca 15 m), we set the plots at 5 and 25 meters from the central trees.

In the cases of *Bertholletia* and *Dipteryx,* 8 experimental and control plots were established around each of 4 central trees for a total of 32 plots per species (4 replicates of each treatment); for *Astrocaryum* and *Hymenaea* we established 8 plots around each of 2 central trees for a total of 16 plots per species (2 replicates per treatment); lacking sufficient wire to install a separate set of plots for *Calatola,* we did not employ central trees but instead added 3 *Calatola* seeds to the arrays of *Dipteryx* seeds in each of the 24 plots around *Dipteryx* trees that received seeds.

All exclosures and half the controls were stocked at one end with seeds of the target species as follows: 75 *Astrocaryum* seeds; 9 *Dipteryx* and 3 *Calatola* seeds; 16 *Bertholletia* seeds; and 16 *Hymenaea* seeds. Fruits of each species were collected in the forest from several to many parent trees and mixed to assure genetic diversity in the seeds used in each trial. (Three of the species have single-seeded fruits – *Astrocaryum*, *Calatola* and *Dipteryx* – in these species entire fruits were set out in the plots. Two of the species have multi-seeded indehiscent fruits with hard protective exocarps – *Bertholletia* and *Hymenaea*. The seeds used in the experiments with these two species were removed from fruits we opened. All seeds were individually examined for any damage, especially insect infestation, and rejected if hollow, lighter than normal, or if possessing noticeable entrance/exit holes. We arrayed seeds in 3 or 4 rows in one end of each experimental plot, spacing them evenly on the surface of the leaf litter. The experiments with *Dipteryx* and *Calatola* were set up in 1988; those with *Astrocaryum*, *Bertholletia*, and *Hymenaea* in 1989. The plots were set up and stocked with seeds at the end of the dry season (October). We scored the results one year later.

Scoring of results

After one year, each plot that had received seeds of the target species was thoroughly searched, including the end that had been reserved as a seed-free control. We carefully sifted through the leaf litter in an effort to uncover all remaining evidence of the stocked seeds. In the impermeable exclosures stocked with *Calatola* and *Dip-*

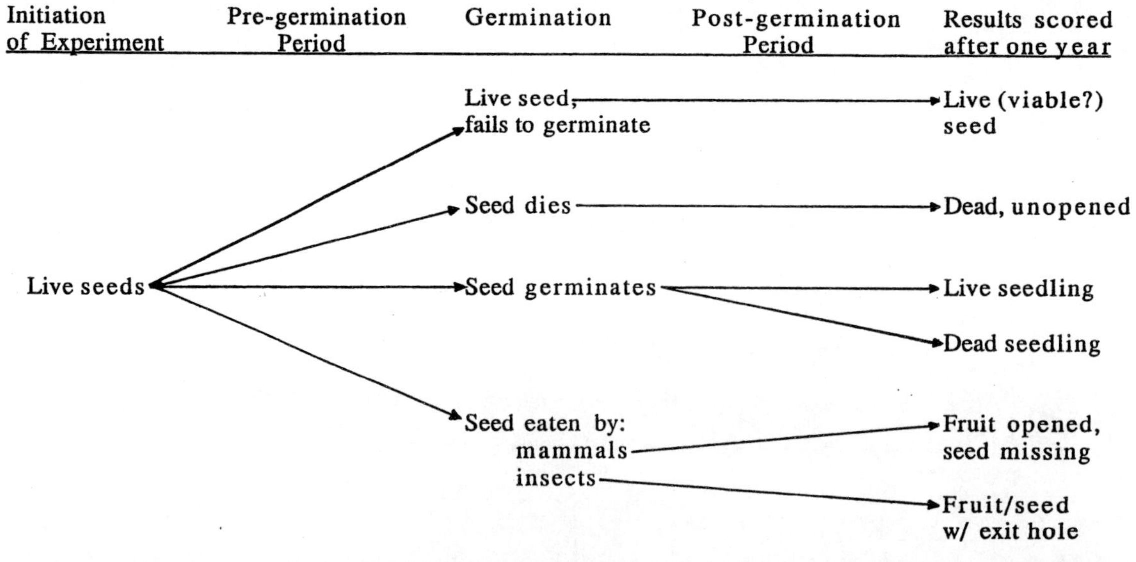

Fig. 2. Possible fates of seeds introduced into experimental and control plots. Criteria for scoring the results after one year are explained in the text.

teryx seeds, it was reassuring to recover >90% of the seeds that had been set out the year before. We sometimes recovered small numbers of seeds from the unseeded ends of the plots, especially in the 'near' plots. These represent the natural seed drop during the year of the experiment. These non-experimental seeds were excluded from the results. Experimental seeds were classified according to the condition in which we found them (Fig. 2): germinated with live seedling present; germinated with remains of dead seedling still evident; seed coat or exocarp present but seed missing; entire fruit/seed, including seed coat and/or exocarp missing; seed/fruit intact but empty with arthropod exit hole apparent; seed/fruit intact but light, indicating seed had died, but without arthropod exit hole; seed/fruit intact and heavy, suggesting persistence of viable seed.

Analysis

For purposes of the analysis, all experimental seeds were classified into three categories: (1) seed survived (germinated, or ungerminated and presumably alive); (2) seed did not survive (fruit/seed missing, or present, but fruit/seed coat opened or with exit hole, or fruit/seed intact but light and presumed dead); (3) seed presumably eaten by a mammal (entire fruit/seed missing; fruit/endocarp opened and seed missing). Category 3 represents a subset of category 2, designated to isolate mammal predation from other causes of seed mortality.

Analyses of variance were performed on the combined data set for all five species using seed survivorship and seed loss to mammalian predators as dependent variables. Replicate trees were a random nested factor within species. Levels of treatment were impermeable exclosures, semipermeable exclosures, controls with seeds, and controls without seeds. When a main effect was significant, post hoc tests based on the full ANOVA model were used to evaluate differences among treatment levels. Variances were stabilized by arcsine square root transformations of proportions.

Results

Distance effects

We first tested the results for each species for evidence of a Janzen/Connell distance effect. A distance effect could result from elevated levels of seed predation near the central tree due to mammals, or to invertebrates, or to both in combination. To examine the results we first asked simply whether seeds survived or did not survive the 1-year exposure to the treatments. Seeds that were intact and still heavy (many *Calatola* and *Hymenaea*), or that had germinated, were classified as having survived; those that were intact but light (e.g., many *Dipteryx*), showed signs of having been eaten, or that were missing were classified as having died.

Of the five species, only *Astrocaryum* showed a distance effect. This resulted from higher levels of invertebrate seed predation in the near (5 m) than in the far (25 m) plots (Table 1). The remaining four species showed negligible levels of invertebrate seed predation at any distance. No cases of seed loss to invertebrates were noted for *Bertholletia*, *Calatola* and *Dipteryx*, and only 5 (out of 96) *Hymenaea* seeds still present in the exclosures after one year showed exit holes that suggested insect attack. No distance effect was detected in any treatment for any of the latter four species in Chi-square tests applied to the data in Table 1. In the absence of any detectable distance effects in these four species, results from the near and far plots were pooled in subsequent analyses.

Treatment effects: survivorship

An ANOVA performed on the survivorship data revealed highly significant effects of treatment, species and the interaction of treatment x species (Table 2). Inspection of the data suggested three qualitatively distinct patterns that were subsequently confirmed by post-hoc Chi-square tests (Fig. 3). The seeds of three species (*Astrocaryum*, *Bertholletia* and *Dipteryx*) survived primarily or exclusively in the impermeable exclosures; seeds

Table 1. Seed survivorship vs. distance from a central tree in five species of large-seeded Amazonian trees.

Species	Treatment					
	Impermeable		Semipermeable		Control w/seeds	
	# Survived	Died	# Survived	Died	# Survived	Died
Astrocaryum						
5 m	38	112	2	148	4	146
25 m	100	50	4	146	14	136
Bertholletia						
10 m	22	26*	0	48*	0	64
50 m	13	19*	0	48*	0	64
Calatola						
10 m	12	0	12	0	12	0
50 m	12	0	12	0	12	0
Dipteryx						
10 m	13	23	3	33	1	35
50 m	15	21	4	32	1	35
Hymenaea						
10 m	24	8	23	9	10	22
50 m	25	7	16	18	6	26

* Several Bertholletia exclosures were overturned and raided by peccaries. These trials were excluded from the data; hence the numbers of seeds do not add to 64 in all treatments.

Table 2. Analysis of variance: survival.

Source	Sum-of-squares	DF	Mean-square	F-ratio	P
Treatment	8.591	2	4.295	32.212	<0.001
Species	42.794	4	10.687	80.145	<0.001
TMT*SP	4.237	8	0.530	3.972	0.001
Error	10.134	76	0.133		

survived in the semipermeable exclosures at low rates indistinguishable from those in the open controls to which seeds had been added. Germination was moderate in all three species, being 46% in *Astrocaryum* seeds that survived intact, 20% in *Bertholletia*, and 45% in *Dipteryx*.

Seed survivorship in the impermeable exclosures was considerably less than 100% for different reasons in each of the three cases. *Astrocaryum* seeds were attacked by invertebrate seed predators, especially in the 'near' plots. Nearly half of the *Dipteryx* seeds died intact without evidence of attack by either vertebrates or invertebrates; a few (14%) had been opened by vertebrates, indicating that some of the impermeable exclosures had been breached by small mammals. *Bertholletia* seeds, which seem to be highly preferred by vertebrate seed predators, were opened or removed from the impermeable exclosures in even greater proportions (56%). The impermeable exclosures nevertheless provided clear protection to these three species compared to the other treatments.

Calatola seeds survived intact under all treat-

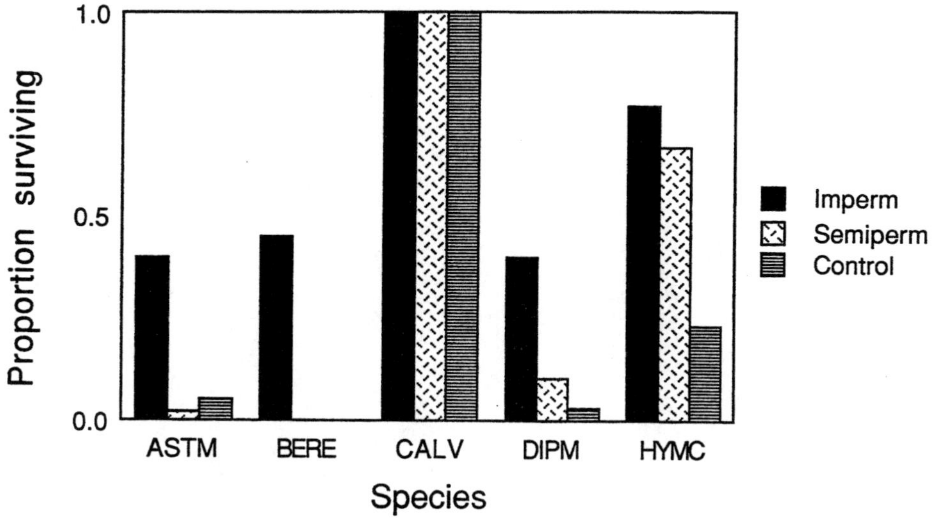

Fig. 3. Proportion surviving of seeds introduced into 2 × 1 m exclosures and control plots. See text for definition of survival. ASTM = *Astrocaryum macrocalyx* (Palmae); BERE = *Bertholletia excelsa* (Lecithydaceae); CALV = *Calatola venezuelana* (Icacinaceae); DIPM = *Dipteryx micrantha* (Leguminosae); HYMC = *Hymenaea courbaril* (Leguminosae).

ments. After one year about a third of the seeds had germinated; the rest were all still heavy and presumably viable. The lack of any evidence of attack, of even unprotected seeds, by either vertebrate or invertebrate seed predators, confirmed informal observations that *Calatola* seeds remain undisturbed for months under the parent tree.

The pattern of seed survivorship in *Hymenaea* differed from that of the other species in that both the impermeable and semipermeable exclosures afforded protection relative to the seeded controls (Chi square = 33; p = <0.001). After one year, only 2 of 128 seeds placed in the two types of exclosures had germinated; 86 (67%) were still intact and had neither germinated nor suffered attack by seed predators. The rest were missing (presumably lost to rodents), or had decomposed. Survival of intact seeds in the unprotected control plots (26%) was much lower than in the exclosures (67%), but high enough to suggest that *Hymenaea* seeds are comparatively unattractive to seed predators in this environment.

Treatment effects: mammal predation

We reanalyzed the data to distinguish loss of seeds to mammalian seed predators (missing or gnawed fruits/seeds) from other fates (survival intact alive or dead, germination, and losses to invertebrates). An ANOVA again indicated very high levels of significance of treatment, species, and the interaction of treatment x species (Table 3). Qualitative examination of these results shows that unprotected seeds of all the species but *Calatola* suffered moderate (*Hymenaea*) to

Table 3. Analysis of variance: mammal predation.

Source	Sum-of-squares	DF	Mean-square	F-ratio	P
Treatment	13.161	2	6.580	50.216	<0.001
Species	43.431	4	10.858	82.857	<0.001
TMT*SP	11.019	8	1.377	10.511	<0.001
Error	9.959	76	0.131		

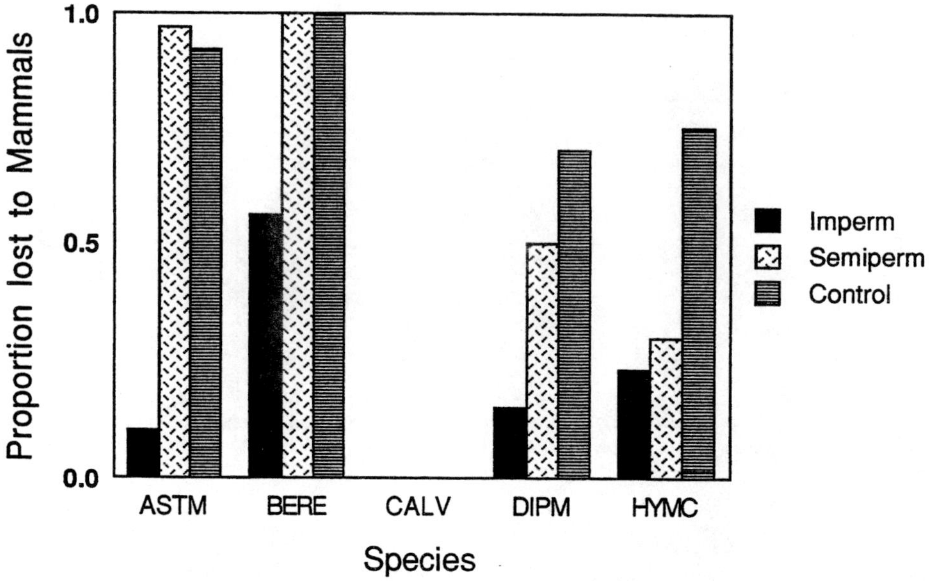

Fig. 4. Proportion lost to mammals of seeds introduced into 2 × 1 m exclosures and control plots. See text for criteria. Abbreviations as in Figure 3.

complete (*Bertholletia*) mortality from mammalian predation (Fig. 4). Again *Hymenaea* was atypical in that seeds in the semipermeable exclosures were as well protected as those in the impermeable exclosures, surviving significantly better than in the open controls (Chi-square = 39.6; p = <0.001).

Discussion

No evidence of a distance effect was noted in any of the three species that were attractive to mammalian but not to invertebrate seed predators (*Bertholletia*, *Dipteryx*, *Hymenaea*). Molofsky and Fisher (in press) similarly found no evidence for distance effects in mammalian predation on the seedlings of three tree species on Barro Colorado Island, Panama.

Of the five species, *Astrocaryum* was the only one that was strongly attacked by invertebrate seed predators. Bruchid beetles killed many more seeds under the central tree than 25 m away, providing the only distance effect noted in the study. Unprotected seeds of *Astrocaryum standleyanum* set out on the soil surface of Barro Colorado Island, Panama were attacked by bruchid beetles and/or mammalian seed predators and suffered 100% mortality (Smythe 1989). Buried seeds, in contrast, enjoyed high survivorship. At Cocha Cashu, Kiltie (1981) found that 98% of *Astrocaryum macrocalyx* seeds collected under adult trees had been attacked by bruchids, while only 22% of scatterhoarded seeds had been attacked. Both species of *Astrocaryum* thus appear to require the services of scatterhoarders for normal regeneration.

Exclusion of mammals from 2-square meter experimental plots resulted in three qualitatively distinct patterns among the five large-seeded species expected to be attractive to mammalian seed predators. One species (*Calatola*) exhibited 100% seed survival in both the exclosures and in open control plots. However, unquantified observations indicate that *Calatola* seedlings suffer heavy post-germination mortality from vertebrate herbivores, suggesting that the bottleneck in recruitment may come in the early post-germination period. Three species suffered heavy seed loss to mammals in semipermeable exclosures and in open control plots, but not in impermeable exclosures (*Astrocaryum*, *Bertholletia*, *Dipteryx*). These results implicate small rodents, such as *Oryzomys* spp. and *Proechimys* spp., as major seed predators of these species. Lastly, seeds of *Hymenaea courbaril* survived well in exclosures of both types, but were mostly removed from the open control plots, suggesting that larger mammals (*Myoprocta*, *Dasyprocta*, *Agouti*, *Tyassu* spp.) may be responsible for most of the vertebrate seed predation on this species.

Although three out of the four species that suffered high rates of mammalian seed predation were taken primarily by small mammals, we must caution that conclusions should not be drawn at face value. By precluding access to mammals larger than *Proechimys* (ca. 300 g), the semipermeable exclosures could act as privileged foraging grounds for small rodents. Seeds that might have been eaten promptly by peccaries or agoutis remain available in these exclosures to small rodents, possibly for months. The true impact of larger mammalian seed predators could thus be considerably greater than suggested by these experiments.

It should be noted that squirrels (*Sciurus* spp., especially the common *S. spadiceus*) might have raided some of the exclosures through their open tops. If so, squirrels could have been responsible for the uncontrolled seed loss from impermeable exclosures containing *Astrocaryum*, *Bertholletia*, *Dipteryx*, and *Hymenaea* seeds. If not, the wire barriers must have been breached occasionally by small terrestrial rodents.

We found no relationship between the apparent appeal of the five species of seeds to mammalian seed predators and the densities of adult trees in the forest. The two most preferred species were *Astrocaryum* and *Bertholletia*. Adults of the former are abundant (> 30 per hectare), and of the latter, quite scarce (< 1 per hectare). *Dipteryx*, which may be slightly less preferred, is moderately common (> 1 per hectare), while least preferred *Hymenaea* is rare ($\ll 1$ per hectare). Curiously, *Calatola*, which appears to suffer zero pregermination seed predation, has a density similar to that of *Dipteryx*, but is of more patchy occurrence. In contrast, Smith (1987) found that the densities of five species of mangroves in Australia were inversely proportional to measured levels of seed predation by grapsid crabs. The lack of such a direct relationship between intensity of seed predation and adult density in the forest at Cocha Cashu could easily be explained if population regulation occurs partly or primarily in the post-germination stages of the life cycle. Unpublished observations on *Calatola* and *Dipteryx* suggest that seedlings of both species are lost in large numbers to both vertebrate and invertebrate herbivores. We conclude that seed predation is clearly important in the demography of tropical trees, but that other processes play significant roles as well.

Acknowledgements

We greatly appreciate the longstanding support of the General Direction of Forestry and Wildlife in the Peruvian Ministry of Agriculture, and of the administration of the Manu National Park, for authorizations to conduct this research at the Cocha Cashu Biological Station. We are also most grateful to S. Joseph Wright for invaluable advice and assistance in data analysis. We thank Juan Quispe and Manuel Sanchez for their stalwart participation in the field work. This research was conducted without benefit of grant support.

References

Augspurger, C. K. 1984a. Seedling survival among tropical tree species: interactions of dispersal distance, light-gaps, and pathogens. Ecology 65: 1705–1712.

Augspurger, C. K. 1984b. Pathogen mortality of tropical tree seedlings: experimental studies of the effects of dispersal distance, seedling density, and light conditions. Oecologia 61: 211–217.

Brown, J. H. & Munger, J. C. 1985. Experimental manipulation of a desert rodent community: food addition and species removal. Ecology 66: 1545–1563.

Clark, D. A. & Clark, D. B. 1984. Spacing dynamics of a tropical rain forest tree; evaluation of the Janzen-Connell model. Amer. Natur. 124: 769–788.

Clark, D. B. & Clark, D. A. 1987. Population ecology and microhabitat distribution of *Dipteryx panamensis*, a neotropical rain forest emergent tree. Biotropica 19: 236–244.

Connell, J. H. 1971. On the role of natural enemies in preventing competitive exclusion in some marine animals and in rain forest trees. In: Dynamics of numbers in populations, den Boer, P. J. & Gradwell, G. R. (eds.), pp. 298–312. Center for Agricultural Publication and Documentation, Wageningen, Netherlands.

Connell, J. H., Tracey, J. G. & Webb, L. J.. 1984. Compensatory recruitment, growth, and mortality as factors maintaining rain forest tree diversity. Ecol. Monogr. 54: 141–164.

Gentry, A. W. (ed.) 1991. Four Neotropical forests. Yale Univ. Press, New Haven.

Hallwachs, W. 1986. Agoutis (*Dasyprocta punctata*): the inheritors of guapinol (*Hymenaea courbaril*: Leguminosae). In: Frugivores and seed dispersal, Estrada A. & Fleming, T. H. (eds.), pp. 285–304. Junk, Dordrecht.

Howe, H. F. 1989. Scatter- and clump-dispersal and seedling demography: hypothesis and implications. Oecologia 79: 417–426.

Howe, H. F. 1990. Survival and growth of juvenile *Virola surinamensis* in Panama: effects of herbivory and canopy closure. J. Trop. Ecol. 6: 259–280.

Howe, H. F., Schupp, E. W. & Westley, L. C. 1985. Early consequences of seed dispersal for a Neotropical tree (*Virola surinamensis*). Ecology 66: 781–791.

Hubbell, S. P., Condit, R. & Foster, R. B. 1990. Presence and absence of density dependence in a neotropical tree community. Phil. Trans. R. Soc. Lond. B 330: 269–281.

Hubbell, S. P & Foster, R. B. 1990. The fate of juvenile trees in a neotropical forest: implications for the natural maintenance of tropical tree diversity. In: reproductive Biology of tropical forest plants, Hadley, M. & Bawa, K. S. (eds.) Man in Biosphere series, Vol. 7, pp. 325–349. Carnforth: UNESCO/IUBS, Paris.

Janson, C. H. 1985. Aggressive competition and individual food comsumption in wild brown capuchin monkeys (*Cebus apella*). Behav. Ecol. Sociobiol. 18: 125–138.

Janzen, D. H. 1970. Herbivores and the number of tree species in tropical forests. Amer. Natur. 104: 501–528.

Janzen, D. H. 1971. The fate of *Scheelea rostrata* fruits beneath the parent tree: predispersal attack by bruchids. Principes 15: 89–101.

Janzen, D. H. 1972. Escape in space by *Sterculia apetala* seeds from the bug *Dysdercus fasciatus* in a Costa Rican deciduous forest. Ecology 53: 350–361.

Janzen, D. H. 1975. Behavior of *Hymenaea courbaril* when its predispersal seed predator is absent. Science 189: 145–147.

Janzen, D. H. 1983. *Hymenaea courbaril* (*Guapinol*, stinking toe). In: Costa Rican natural history, Janzen, D. H. (ed.), pp. 253–256. Univ. Chicago Press, Chicago.

Janzen, D. H. & Martin, P. 1982. Neotropical anachronisms: what the gomphotheres ate. Science 215: 19–27.

Kiltie, R. A. 1981. Distribution of palm fruits on a rain forest floor: why white-lipped peccaries forage near objects. Biotropica 13: 141–145.

Kiltie, R. A. 1982. Bite force as a basis for niche differentiation between rain forest peccaries (*Tayassu tajacu* and *T. pecari*). Biotropica 14: 188–195.

Kiltie, R. A. & Terborgh, J. 1983. Observations on the behavior of rain forest peccaries in Peru: why do white-lipped peccaries form herds? Zeit. Tierpsychol. 62: 241–255.

Molofsky, J. & Fisher, B. L. (in press). The effect of habitat and predation on seedling survival and growth of three shade-tolerant tree species in Panama. Ecology.

Sarrukhan, J. 1978. Studies on the demography of tropical trees. In: Tropical trees as living systems, Tomlinson, P. B. & Zimmermann, M. H. (eds.), pp. 163–184. Cambridge Univ. Press, Cambridge.

Schupp, E. W. 1988. Seed and early seedling predation in the forest understory and in treefall gaps. Oikos 51: 71–78.

Smith, T. J., III. 1987. Seed predation in relation to tree dominance and distribution in mangrove forests. Ecology 68: 266–273.

Smythe, N. 1986. Competition and resource partitioning in the guild of Neotropical, terrestrial, frugivorous mammals. Ann. Rev. Ecol. Syst. 17: 169–188.

Smythe, N. 1989. Seed survival in the palm *Astrocaryum standleyanum*: evidence for dependence upon its seed dispersers. Biotropica 21: 50–56.

Sork, V. L. 1987. Effects of predation and light on seedling establishment in *Gustavia superba*. Ecology 68: 1341–1350.

Steven, D. de 1988. Light gaps and long-term seedling performance of a Neotropical canopy tree (*Dipteryx panamensis*, Leguninosae). Journ. Trop. Ecol. 4: 407–411.

Steven, D. de & Putz, F. E. 1984. Impact of mammals on early recruitment of a tropical canopy tree, *Dipteryx panamensis*, in Panama. Oikos 43: 207–216.

Terborgh, J. 1983. Five Neotropical primates: a study in comparative ecology. Princeton Univ. Press, Princeton.

Terborgh, J. 1986. Keystone plant resources in the tropical forest. In: Conservation Biology: The science of scarcity and diversity, Soule, M. E. (ed), pp. 330–344. Sinauer, Sunderland, Mass.

Terborgh, J. 1988. The big things that run the world – a sequel to E. O. Wilson. Cons. Biol. 2: 402–403.

Terborgh, J., Fitzpatrick, J. & Emmons, L. H. 1984. Annotated checklist of bird and mammal species of Cocha Cashu Biological Station, Manu National Park, Peru. Fieldiana, (Zoology, New Ser.) No. 21, 29 pp.

Wright, S. J. 1983. The dispersion of eggs by a bruchid beetle among *Scheelea* palm seeds, and the effect of distance to the parent palm. Ecology 65: 1016–1021.

Index